Fundamentals of Quantum Computing

Venkateswaran Kasirajan

Fundamentals of Quantum Computing

Theory and Practice

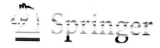

Venkateswaran Kasirajan
Trimble, Inc.
Superior, CO, USA

ISBN 978-3-030-63691-3   ISBN 978-3-030-63689-0   (eBook)
https://doi.org/10.1007/978-3-030-63689-0

Cover image credit: Original cover art by Anuj Kumar.

This Springer imprint is published by the registered company Springer Nature Switzerland AG
The registered company address is: Gewerbestrasse 11. 6330 Cham. Switzerland

This book is dedicated to my parents Meenalosani and Kasirajan.

Acknowledgments

The Ph.D. thesis of Laurens Willems starts with the quote from Sir Isaac Newton, "If I have seen further than others, it is by standing upon the shoulders of giants." How true indeed! I was quite humbled by those noble words. The ninth-century Tamil epic Cīvaka Cintāmaṇi compares the posture of the learned with the bowing stalk of a mature rice plant, "தேர்ந்த நூல் கல்வி சேர் மாந்தரின் இறைஞ்சிக் காய்த்தவே." Education should never make one haughty. So, the second-century BCE Tamil poet Avvaiyar warns, "கற்றது கை மண் அளவு, கல்லாதது உலக அளவு," meaning – what one has learned is equivalent to a handful of sand, and what one is yet to learn is as big as the world. Even if one has gained only a handful of knowledge, as Thiruvalluvar said, "தாமின் புறுவது உலகின் புறக்கண்டு காமுறுவர் கற்றறிந்தார்," the learned will long (for more learning) when they see that while it gives pleasure to themselves, the world also derives pleasure from it.

Whenever I doubted whether I would finish this work or not, the wisdom of the sages kept telling me that I am on the right path. Getting back to the roots, to the things lying at the bottom of my heart, was soul-fulfilling and enchanting. I felt as though I was given a new life.

I have probably read hundreds of books. However, I never knew what it takes to write one. That too on a heavy topic such as quantum computing. What was originally thought of as a 200-page edition in four or five chapters kept expanding, and I had to stop at some point. It came at the expense of family time, and I must thank my family for understanding and support. Without them, this would not be possible. While working at my desk, my daughter Mehala would often stop by to look at which chapter I'm working on. She then would shrug her shoulders, saying, "hmm, still in Chap. 4?" and walk away disappointed as I'm not making significant progress. My wife Vanadurga saw me only at dinnertime. My son Aparajithan reviewed the book despite his busy schedule, identified errors, and pointed out several improvements. His summa cum laude skills readily helped. So, what started as a single man's effort, became a family affair at the end!

There is a famous saying that reminds us not to forget where we came from. I firmly believe that in life, no one can succeed alone. Successful people are mentored and helped all along the way. Forgetting those who contributed to one's success is like kicking the ladder after climbing up, rather than sending the elevator down. I think there will be no better opportunity than this to thank those who helped, influenced, and changed my life. Among those are four personalities I should never forget and be thankful forever – Prof. Parangiri, Mr. Ramprakash, Mr. S. Selvaraj, and Shankar Narayanan. I hope all of them will be delighted to see this publication.

None of this in my life could have been possible without the generous hearts of my parents and the countless sacrifices they made. Words of gratitude or remembrance are insufficient in expression. Mother and father, I am sure you will be proud of this work, and this book is dedicated to your memories.

I am also grateful to all my undergraduate and graduate teachers – Prof. KRB, Prof. Sankaran, Prof. Raman, Prof. Alwan, Prof. Krishnamoorthy, Prof. Sivakumar, Prof. Raju, Prof. TND, Prof. Natarajan, and Prof. Edison – your courses on core physics subjects are still echoing and fresh in my memory as if the lectures given were today.

I am grateful to my management at the office of the Digital Transformation and Horizontal Engineering at Trimble Inc. for giving me space, recognition, and support. My thanks are due to Joy Day and Ulrich Vollath for the time they spent reviewing this book. I also thank Kirk Waiblinger for reviewing the book and pointing out several corrections. I thank my team members – the Trimble Spime team – for their continued journey with me and their confidence. You are my best! I would also like to thank the editorial and production teams at Springer for their excellent work on this book.

My special thanks go to Dr. Laurens Willems for reviewing this book, despite his tight schedule.

In Tim McGraw's words, "When the work put in is realized, one can feel the pride, but must always be humble and kind." I hope this book helps software developers and undergraduate students learn this subject in some small ways. May I remind my readers to study the references cited at the end of each chapter and enhance their knowledge.

Venkateswaran Kasirajan,   வெங்கடேஸ்வரன் காசிராஜன்,
Superior, Colorado.   சுப்பீரியர், கொலராடோ,
September 2020.   செப்டெம்பர் 2020.

1. The author acknowledges the use of the IBM Q for this work. The views expressed are those of the author and do not reflect the official policy or position of IBM or the IBM Q team.
2. 15-qubit backend: IBM Q team, "ibmq_16_melbourne backend specification V2.0.6," (2018). Retrieved from https://quantum-computing.ibm.com
3. 32-qubit simulator: IBM Q team, "ibmq_qasm_simulator simulator specification v0.1.547," (2019). Retrieved from https://quantum-computing.ibm.com
4. 5-qubit backend: IBM Q team, "ibmq_london backend specification V1.1.0," (2019). Retrieved from https://quantum-computing.ibm.com
5. Qiskit: An opensource framework for quantum computing, 2019, H'ector Abraham et al., https://doi.org/10.5281/zenodo.2562110
6. Qiskit/openqasm is released under Apache License 2.0 https://github.com/Qiskit/openqasm/blob/master/LICENSE
7. qiskit-terra/Quantum is released under Apache License 2.0 https://github.com/Qiskit/qiskit-terra/blob/master/LICENSE.txt
8. Microsoft Quantum Development Kit, License Terms. https://marketplace.visualstudio.com/items/quantum.DevKit/license
9. Microsoft Quantum, released under MIT license https://github.com/microsoft/Quantum/blob/master/LICENSE.txt
10. The author of this book acknowledges the research & work of the pioneers in quantum computing, upon their work this book is based. References to the publications from many of these scholars are provided at the end of each chapter. Readers are requested to refer to them for more information on this subject.

Contents

Image Credits

Reference	Description	Credits and Permissions
Figure 1.1	The splendor of the milky way galaxy.	Rich Rudow Printed with permission.
Figure 1.2	The Koh-I-Noor diamond	Wikipedia. G. Younghusband, C. Davenport (circa 1919). Public domain in the source country. See link: https://commons.wikimedia.org/wiki/File:Queen_Mary%27s_Crown.png
Figure 1.3	Simplicity. Boats on blue water.	Image courtesy of Tiket2.com. CC BY 4.0.
Figure 1.4	Water evaporating as lava flows into the ocean.	Vlad Butsky, https://www.flickr.com/photos/butsky/357672637/, CC BY 2.0, CC BY 4.0
Figure 1.5	Lightning creating a plasma channel in the air	Photo by Hallie Larsen, National Park Service. - https://www.facebook.com/USInterior/photos/a.155163054537384.41840.109464015773955/1062000280520319/, Public Domain, https://commons.wikimedia.org/w/index.php?curid=50735060
Figure 1.6	The periodic table	Wikipedia, Author: Offnfopt, Public Domain, Creative Commons CC0 1.0 Universal Public Domain Dedication
Figure 1.7	Picture of the Helium atom	Wikipedia, Author: Yzmo, CC BY-SA 3.0.
Figure 1.9	The standard model of elementary particles	Wikipedia, Author: Cush. This version is released by the author in the public domain. Original author MissMJ released this under CC BY 3.0
Figure 1.11	The blackbody radiation	University Physics Volume 3, CC BY 4.0 https://opentextbc.ca/universityphysicsv3openstax/chapter/blackbody-radiation/
Figure 1.14	IBM System One	IBM Q at CES 2020, Courtesy of International Business Machines Corporation. Unauthorized use not permitted., https://newsroom.ibm.com/image-gallery-research#gallery_gallery_0:21596
Figure 1.16	Propagation of Electromagnetic waves	Image by helder100 from Pixabay https://pixabay.com/service/license/
Figure 1.17	Young's double-slit experiment with a sandblaster gun with Slit 2 closed	
Figure 1.18	Setup with Slit 1 closed	
Figure 1.19	Double-slit experiment with both the slits open	Anuj Kumar
Figure 1.20	Expected behavior	
Figure 1.21	Observed behavior	

(continued)

Reference	Description	Credits and Permissions
Figure 1.23	Electron build up over time	Anuj Kumar
Figure 1.24	Single Electron Beam Experiment	
Figure 1.27	The Copenhagen Interpretation	
Figure 1.31	The vortices of a type-II superconductor	
Figure 3.1	Schrödinger's cat	
Figure 3.2	The cat dies, if the poison potion is broken	
Figure 3.3	Schrödinger's cat in a superposition state	
Figure 3.6	The Stern and Gerlach Experiment	Wikipedia, Current Author: Tatoute, CC BY-SA 4.0
Figure 4.8	Trapped ion qubit	APS/Alan Stonebraker, Copyright 2014 American Physical Society Printed with permission.
Figure 4.29	The quantum dot	Laurens Willems Van Beveren. Printed with permission.
Figure 4.30	Energy band diagram of the heterostructure	
Figure 4.31	Lateral quantum dot device	
Figure 4.32	Schematic view of the quantum dot	
Figure 4.33	Scanning microscope image of the quantum dot	
Figure 4.34	Electron transitions	
Figure 4.35	Schematic of the dilution fridge	Wikipedia, author: Adwaele, CC BY-SA3.0
Figure 4.36	Phase diagram of ^3He–^4He mixture	Wikipedia, Author:Mets501, CC BY-SA3.0
Figure 6.2	Alice and Bob communicating	Alice and Bob illustrated by Anuj Kumar
Figure 6.4	Type-I, Spontaneous Parametric Down Conversion	Anuj Kumar
Figure 6.5	Type-II, Spontaneous Parametric Down Conversion	Anuj Kumar
Figure 6.7	Cipher disk used by Union Army	National Archives

Reading Guide

The following map shall help readers from varied backgrounds and learning needs. Solving the chapter-end exercises is optional; however, it may provide additional learning experiences.

General audience and software developers

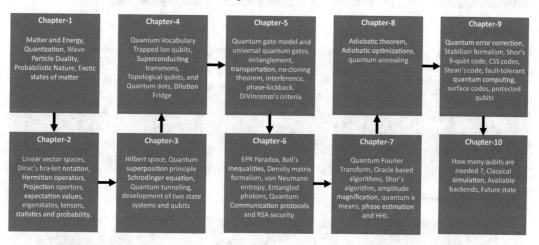

University students with a background of physics and linear algebra

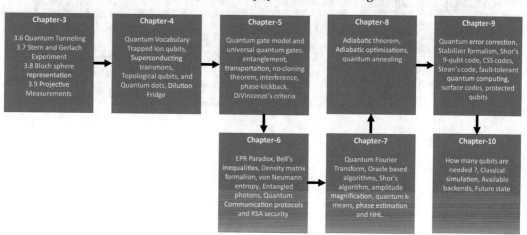

Enthusiasts, self-leaners, and fast trackers

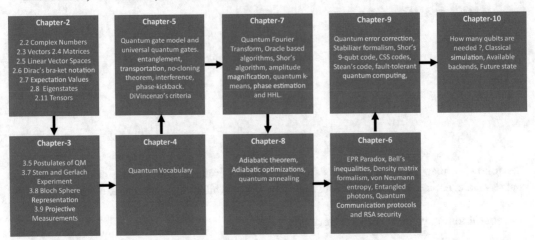

Qubit physics

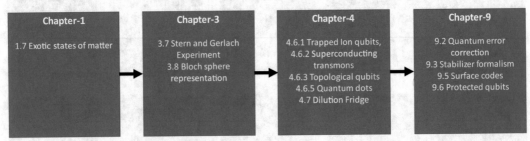

Adiabatic quantum computation

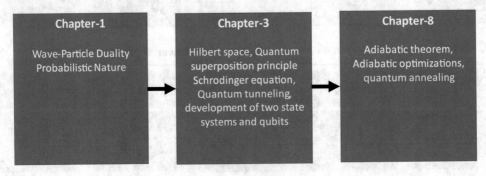

Introduction

Perhaps quantum computing does not need any introduction these days. Whether it is a claim for quantum supremacy/quantum volume or a generous funding a startup received, not a day goes by without a newspaper article on quantum computing. It looks as if the whole world has changed suddenly, and everyone wants to do quantum.

The fact is, the foundational principles of quantum computing were established long ago when personal computers did not exist. The first versions of the no-cloning theorem (it was reinvented in 1982), Holevo's bound, and the first research paper on quantum information theory were all published in the 1970s.

Significant progress was made in the early 1980s when Paul Benioff proposed the quantum Turing machine. Around this time, Richard Feynman and Yuri Manin proposed the basic model of a quantum computer. In 1985, David Deutsch published the concept of the universal quantum computer. In the late 1980s, Bikas K. Chakrabarti established the effectiveness of quantum annealing over classical annealing.

The next decade was buzzing with inventions – Artur Ekert's communication protocol, Deutsch-Jozsa algorithm, Simon's algorithm, Peter Shor's prime number factorization algorithm, Grover's algorithm, and quantum error correction were all invented in the 1990s. In the same period, controlled-NOT gate operations were performed using cold trapped ions. Deutsch and Grover's algorithms were proved using 2 or 3 qubit NMR technology. Experimental proof of quantum annealing was established in the late 1990s. Furthermore, DiVincenzo laid down the minimum requirements of a quantum computer. Alexei Kitaev published his first papers on topological quantum computing, and the first superconducting qubits were reported.

In the last two decades, researchers primarily focused on creating reliable qubit technology. Remarkable progress on quantum communications and experimental proof of Shor's algorithm using NMR were established in research labs. What follows research is commercialization, and quantum computing is no exception!

D-Wave Systems announced the 28-qubit quantum annealer in 2007, which probably was the first usable quantum computer. The landscape changed significantly when IBM announced Quantum Experience for the public in 2016. In the 1980s, IBM brought personal computers to our homes. Nearly 35 years later, IBM took the lead in bringing quantum computing to the public for experimentation! Now, it is a highly competitive space with money pouring in, with many players trying to gain the edge and the claim for quantum supremacy.

The advantage of this competition is that we now have reliable platforms to experiment with. In the last two decades, significant improvements have been made to the qubit designs minimizing the qubit's coupling to unwanted environmental noise sources. The current superconducting circuits have a long coherence time in the order of milliseconds with gate periods close to tens of nanoseconds, providing us the opportunity to experiment quantum algorithms on real quantum devices.

In the last four decades, classical computing has grown exponentially and probably reaching its physical limits. With billions of connected devices, social networking, and digital transformation of the governments and industries, we live in an era where the information overload breaches classical computing limits. So, it is natural for investors and technologists to explore new avenues. Because of the promise of the exponential speedup, quantum supremacy sets in very well in the vacuum building up. It is in this background that this book is published.

The Audience of this Book

Quantum computing is a fascinating scientific advancement, which classical software developers, enthusiasts, and self-learners may want to learn. At the same time, undergraduate students may find this book a useful bridge between classroom recitations, lab work, and advanced textbooks. Quantum computing is one promising area where pulse level programming is available, attracting enthusiasts and self-learners to play with physical devices. Anyone interested to learn this science will find this book easy to start with!

Today, the internet hosts the complete knowledge of humankind. One can learn any subject by simply googling. However, learning quantum computing online may be somewhat unfruitful. A course on quantum computing includes a study of quantum information systems, quantum algorithms, and quantum communications. The courseware is accompanied by lab work on real quantum devices or simulators. Anyone wanting to learn quantum computing will have to go through some prerequisites – courses on linear algebra and quantum mechanics. At an undergraduate level of study, the prerequisite to quantum information systems is a course on quantum mechanics, usually not taught without completing linear algebra, classical mechanics, thermodynamics, and electromagnetic theory. A review of condensed matter physics is required to understand how the qubits work.

Software developers who took computer science as the concentration are generally not introduced to these subjects. Undergoing these intense programs may not be possible for software developers, who have invested a lot in software engineering careers. Learning quantum computing by directly starting from Dirac's bra-ket notation may not make an in-depth study. While most classical developers may draw parallels with the classical computation model (since the qubits are two-state systems) and learn quickly, a foundational knowledge may be helpful to create new quantum algorithms.

Realizing this gap, I set out to write this book about 2 years ago. This book spreads the materials in a way that suits readers with varying backgrounds to get started.

Opportunities for the Readers

Quantum computing will open new job opportunities and job upgrades for software developers and physicists. While physicists research the core of quantum computing, that is, inventing new materials, qubit designs, quantum error correction, and quantum algorithms, system engineers work on control systems, cryogenics, and qubit fabrication. Not to ignore support systems personnel and managerial staff required to keep the systems working. There are many overlapping areas where software developers fit in – programming the control systems, inventing new quantum algorithms, developing transpilers, applying statistical machine learning methods in correcting quantum errors, creating cloud infrastructure, and consuming APIs. They play a critical role at all layers of quantum computing.

Structure of the Book

A fundamental course in computer science is well structured. It begins with the basics – introducing binary digits and digital gates. The course then proceeds into computer system architecture and then into data structures, algorithms, compilers, operating system principles, and application development. Besides, the essence of software development is learned with rigorous lab work by typing code and learning from numerous books of great authors.

This book follows a similar method. The prerequisites are high school level math and physics, with a bit of programming experience. The rest are taught from the beginning.

Throughout the book, theory and practice are intermixed. Source code listings are provided for Qiskit and QDK. The algorithms are explained with illustrations and step by step math. With this level of detailing, readers must be able to port the code on any platform.

Part I

The first part of the book provides the necessary foundations. This part starts with the definition of matter and rapidly takes the reader through the foundational concepts of quantum mechanics and the required math. The concept of quantum mechanical two-state systems is developed in this part and applied to qubits. This part also introduces quantum measurements and Bloch sphere representation of qubits.

Chapter 1 – Foundations of Quantum Mechanics

The first chapter starts by building the necessary foundations of quantum mechanics. This chapter begins by explaining the concept of matter and its composition. It then progresses rapidly into explaining the concept of photons, quantization of energy, electron configuration, wave-particle duality, and probabilistic nature and describes the concept of "wavefunction" to the reader. This chapter then introduces quantum measurement and provides a short description of Heisenberg's Uncertainty Principle. The first chapter ends with a description of some exotic states of matter. Knowledge of the exotic states may be helpful in Chap. 4.

Chapter 2 – Dirac's bra-ket notation and Hermitian Operators

The second chapter refreshes linear algebra and introduces Dirac's bra-ket notation. Short tutorials explaining the concepts of linear operators, Hermitian matrices, expectation values, eigenvalues, eigenfunctions, tensors, and statistics are made with few examples.

Chapter 3 – The Quantum Superposition Principle and Bloch Sphere Representation

The third chapter introduces some advanced topics in quantum mechanics. This chapter introduces the Hilbert space and the Schrödinger equation. The key concepts developed in this chapter are the quantum superposition principle, quantum tunneling, and spin angular momentum. This chapter then builds the concept of qubits as two-state quantum systems. The last section of this chapter introduces the Bloch sphere representation and qubit rotations using Pauli matrices.

Part II

The second part of the book dives into the advanced topics and begins with qubit physics. Once the qubit modalities are explained, this part then briefs quantum gate operations and the quantum circuit model. Elements of quantum computation, quantum algorithms, basic concepts of quantum communication, and quantum error correction are touched upon in the second part.

Chapter 4 – Qubit Modalities

The fourth chapter is devoted to describing the leading qubit modalities. This chapter starts by introducing the quantum computing vocabulary and the Noisy Intermediate Scale Quantum technology available today. Four leading qubit modalities – trapped ion qubits, superconducting transmons, topological qubits, and quantum dots are explained in detail. This chapter ends with a note on the dilution refrigerators.

Chapter 5 – Quantum Circuits and DiVincenzo Criteria

The fifth chapter focuses on the gate model of universal quantum computing. Concepts such as quantum entanglement, no-cloning theorem, quantum teleportation, superdense coding, quantum interference, quantum parallelism, and phase kickback are explained in detail. While studying this chapter, readers shall set up their computers with IBM Q and Microsoft Quantum Development Kit and experiment with some basic circuits. The final section of this chapter talks about DiVincenzo's criteria for quantum computation.

Chapter 6 – Quantum Communications

The sixth chapter focuses on quantum communication and revisits the EPR paradox. The density matrix formalism and von Neumann entropy are two new concepts introduced in this chapter. This chapter also explains how entangled photons can be created in the labs and briefs quantum communication. Two quantum communication protocols – BB84 and Ekert 91 – are introduced to the readers. Before concluding, this chapter examines the premises of RSA security.

Chapter 7 – Quantum Algorithms

The seventh chapter of this book explains some of the well-known quantum algorithms that the reader can try either on IBM Q or Microsoft QDK. The algorithms discussed in this chapter broadly fall under four categories. Deutsch-Jozsa, Bernstein-Vazirani, and Simon's algorithm are oracle-based algorithms. Shor's algorithm, quantum phase estimation, and HHL are algorithms that work on a Fourier basis. Grover's algorithm uses a new method called amplitude magnification. A new category of algorithms that use basic quantum computation principles is evolving. The quantum k-means algorithm uses a swap-test to determine Euclidean distances. This chapter provides a step-by-step derivation of the math and source code for the algorithms discussed.

Chapter 8 – Adiabatic Optimization and Quantum Annealing

The eighth chapter introduces the adiabatic theorem and provides a simple proof. Adiabatic optimization and quantum annealing methods are discussed in this chapter.

Chapter 9 – Quantum Error Correction

The ninth chapter focuses on quantum errors and correction mechanisms. The stabilizer formalism is introduced in this chapter. Shor's 9-qubit error correction code, CSS codes, and Steane's 7-qubit error correction codes are introduced in this chapter. After introducing these stabilizer circuits, this chapter examines the premise of fault-tolerant quantum computation. The concluding sections of this chapter discuss surface codes and protected qubits as means of fault-tolerant quantum computation.

Chapter 10 – Conclusion

The final chapter of this book attempts to look at the promise of quantum supremacy. This chapter discusses how many qubits do we need to perform certain quantum computations, comparison with classical simulations, availability of backends, and the future state.

One important point is there is no source code repository. Readers shall learn quantum computing by typing code from the source code listings in the book, like the old days. Worked-out examples and problems are provided at appropriate places to reinforce the concepts. Each chapter contains a summary, key learning objectives, and a list of sections that can be skipped wherever possible. At the end of each chapter, practice problems and a list of references are provided for further reading. This method shall undoubtedly help the readers retain knowledge, read advanced textbooks, and study research papers.

Reading Guide

An accompanying reading guide helps readers with various backgrounds identify their starting point and fulfill their journey through the book. This mapping shall help readers get quickly onboard and connect with the content important to them.

Wherever a section or the equations can be skipped, a text box highlights the options.

 Note:

Readers can skip the equations and continue to read the text, without compromising the understanding.

If a section has prerequisites, a text box highlights the requirements. This information helps readers who skipped some sections to read the missed topics and get prepared.

Note:

Reading section 1.7 on exotic states of matter may be helpful to read the following sections in this chapter.

Conventions

Indexing Qubits

This book uses a zero-based index for labeling qubits. This indexing scheme is in line with quantum programming languages. In a given quantum circuit, the topmost qubit is labeled q[0]. The indexing proceeds from the top of the circuit to the bottom. The classical bits are also indexed in the same fashion. The topmost classical bit is c[0].

Coloring Scheme for Bloch Spheres

Bloch spheres are used to illustrate the state vector of the qubits. The starting position of the state vector is shown in a lighter shaded green color. The final position of the state vector is shown in a dark shaded green color. The direction in which the state vector is rotated is shown using a dotted line. The arrow of the dotted line shows the direction of rotation. Please see Fig. 5.6 for an example illustration.

Source Code Listings

Source code listings are provided at appropriate places in a highlighted box. Each of the listings is indexed and may be referenced within the body text of the book.

```
q = QuantumRegister(2, 'q')
c = ClassicalRegister(2, 'c')
qc = QuantumCircuit(q, c)

qc.h(q[0])
qc.cx(q[0], q[1])

qc.measure(q[0], c[0])
qc.measure(q[1], c[1])
```

Keeping the Connections

I plan to keep this book updated with new materials, error corrections, and additional volumes. Those who wish to connect with me shall do so through the Good Reads platform.

Venkateswaran Kasirajan,

Superior, Colorado.

September 2020.

வெங்கடேஸ்வரன் காசிராஜன்,

சுப்பீரியர், கொலராடோ,

செப்டெம்பர் 2020.

About the Author

Venkateswaran Kasirajan (Venkat) was a student of physics before studying computer science and taking up a software engineering career. Being a part of two startups, Venkat had the opportunity to work on several core technologies and booted the career of hundreds of engineers. He was also involved in several patents.

Venkat is currently working for Trimble Inc. as an engineering director overseeing a high-profile engineering team.

Venkat's interests are in condensed matter physics. He also researches quantum algorithms and topology and serves as an internal champion for quantum computing at Trimble Inc.

When not working, Venkat is either listening to country music or teaching his daughter. Venkat lives with his family in Colorado.

Part I - Foundations

Foundations of Quantum Mechanics

1

"It proves nothing of the sort!" said Alice. "Why, you don't even know what they're about!"
"Read them," said the King.
The White Rabbit put on his spectacles. "Where shall I begin, please your Majesty?" he asked.
"Begin at the beginning," the King said gravely, "and go on till you come to the end: then stop."

— Lewis Carroll, Alice's Adventures in Wonderland

1.1 Matter

The night sky is fascinating. The beauty of the timeless twinkling stars and the gracious magic of the moon have inspired poets and authors for thousands of years. Yet, we still wonder at its mystery. Who or What created this expanse? Where did it come from, and where is it going? Why was this created? These are some questions we have been asking ourselves since our childhood, and there are no definite answers yet.

Theologists and scientists are attempting to explain creation with their theories—theories that are often are far apart—but still serving as the fundamental point of interaction between religion and science. As we research and learn more, we can someday expect a convergence of these theories.

We can, however, answer some simple questions within our current knowledge of the universe. For example, if one were to ask, what is the universe made of? We know the answer readily, and we learned this in middle school that "**matter is the stuff that makes the universe.**"

In our quest to learn about the fundamentals of quantum computing, this is where we shall start. In the next few sections of this chapter, we shall learn about the stuff that makes the universe. Basic knowledge of this microcosm is essential to understand the foundations of quantum mechanics, which form the basis for quantum computing.

1.1.1 States of Matter

We also learned in middle school that **matter takes up space and has mass. Ordinary matter** exists in four distinct states we can observe today: **solid, liquid, gas**, and **plasma.**

In a **solid**, the constituent particles are densely packed. The particles are held together by strong forces, so they cannot move freely but can only vibrate around their current positions. Thus, solids have a definite shape and volume that can be changed only by breaking or slicing (Fig. 1.2).

V. Kasirajan, *Fundamentals of Quantum Computing*, https://doi.org/10.1007/978-3-030-63689-0_1

Fig. 1.1 The splendor of the Milky Way galaxy, rising in the backdrop of the Nankoweap Trail, Grand Canyon National Park, Arizona. (Photo: Rich Rudow)

Fig. 1.2 Diamonds are the hardest naturally occurring substances on Earth. Pictured here is the famous Koh-I-Noor diamond. (Source: Wikipedia. G. Younghusband; C. Davenport (circa 1919), Author: Cyril Davenport (1848 – 1941), Public Domain)

In a **liquid**, the constituent particles have enough energy to move relative to each other. The ability of the constituent particles to move relative to each other makes the liquids "flow." Hence, the liquids take the shape of the container but retain its volume at a given temperature and pressure (Fig. 1.3).

In a **gas**, the constituent particles are loosely bound. The particles are free to move about and expand to occupy the space of the container. We can compress gases by applying pressure. When

Fig. 1.3 Simplicity. Boats on a blue water. (Image courtesy of Tiket2.com, CC BY)

Fig. 1.4 Water evaporates as lava flows into the ocean. (Source: Vlad Butsky, https://www.flickr.com/photos/butsky/357672637/, CC BY 2.0)

cooled, the gases turn into liquids and then into solids. Some gases, like carbon dioxide, may directly cool into solids. The solid form of carbon dioxide is called dry ice, and the food industry uses it to freeze food. The water cycle is an excellent example of the phase transition between the three states (Fig. 1.4).

Fig. 1.5 Lightning creates a highly conductive plasma channel in the air. (Photo by Hallie Larsen, National Park Service)

We can transform matter from one state to another by changing the temperature and pressure. A simple example we see every day is the melting of ice into water and then into steam. We can reverse this phase transition by cooling the steam. Sublimation is a process where solids convert into gases without melting. Some examples we come across every day are air fresheners and mothballs.

At high temperatures, the electrons rip off from the atoms forming an ionized gas called **plasma**. Like gases, plasma is not defined by any shape. Plasma can conduct electricity and produce electromagnetic fields. We can see plasma every day in lightning and neon lamps (Fig. 1.5).

Matter also exists in some exotic states such as Bose-Einstein condensates, neutron-degenerate matter, and quark-gluon plasma. These occur under extreme conditions of cold, density, and energy.

Nevertheless, what is matter itself made of? The next section answers this question.

1.2   Atoms, Elementary Particles, and Molecules

1.2.1   Atoms

Atoms are the basic building blocks of ordinary matter, and they are, in turn, made up of subatomic particles. These subatomic particles are the **protons, neutrons, and electrons**. Atoms of the same kind make up the given **chemical element**. Atoms are too tiny to be seen with naked eyes. They are so small (around 1×10^{-10} m in size), and classical physics cannot explain their behavior.

We use quantum mechanics to explain the behavior of atoms.

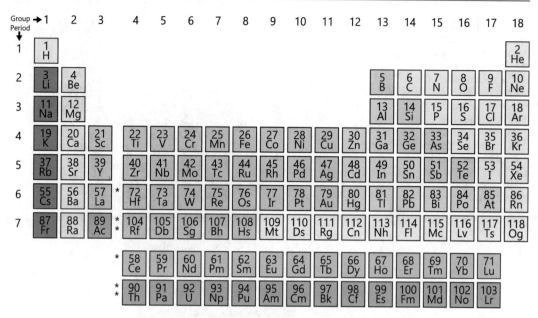

Fig. 1.6 The periodic table. (Image Credits: Offnfopt, Wikipedia, CC BY-SA 4.0)

1.2.2 The Periodic Table

The periodic table lists all chemical elements known to us in a tabular form. The elements are arranged in a particular order (Fig. 1.6).

The Russian Scientist **Dmitri Ivanovich Mendeleev** published the first version of the periodic table in 1869. To commemorate its 150th anniversary, the United Nations declared the year 2019 as the "**International Year of the Periodic Table of the Chemical Elements.**"

The periodic table is an excellent tool to help us understand the properties of the chemical elements. The rows in the periodic table are called "periods." The period number of a chemical element indicates the highest energy level an electron in that chemical element can occupy in the ground state.

The columns are called "groups." The group number of a chemical element indicates the number of valence electrons in it. **Valence electrons** are present in the outermost shell (called **valence shell**) of the atoms. They can be transferred or shared with other atoms. Elements belonging to the same group have similar chemical properties.

With this introduction, let us now explore the structure of the atom in some detail.

1.2.3 Bohr's Model of the Atom

Though this model is somewhat obsolete, we can safely imagine that an atom has a nucleus consisting of protons and neutrons (the nucleons). The electrons are orbiting the nucleus in a finite set of orbits. The number of protons in an atom is its atomic number Z. The atomic number defines which chemical element the atom belongs to.

For example, Hydrogen has one proton, and its atomic number is 1. Helium has two protons, and its atomic number is 2 (Fig. 1.7).

Fig. 1.7 The helium atom at its ground state. The nucleus is at the center, and shaded in pink. The electron cloud distribution is shaded in dark color. The magnified nucleus shows 2 protons (pink) and 2 neutrons (purple). The dark colored bar denotes 1 Angstrom unit which is 10^{-10} m. (Image source: Yzmo, Wikipedia, CC BY-SA 3.0)

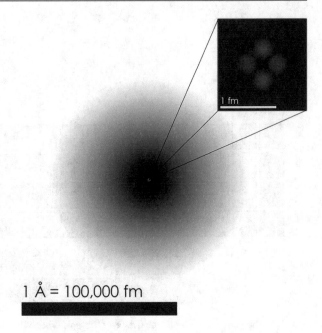

$$1 \text{ Å} = 100,000 \text{ fm}$$

Inside the nucleus, the **protons** (p or p^+) are positively charged particles with a charge of +1e (e here stands for an elementary charge), and **neutrons** have no charge. They both are attracted to each other by nuclear forces. The electrons (e^-) are negatively charged particles with a charge of $-1e$, and they are attracted to the nucleus by electromagnetic forces.

The discovery of the subatomic particles and the development of quantum mechanics is an interesting story [1] in the history of science.

The clues to the presence of electrons were perhaps first noticed by the German physicist **Johann Wilhelm Hittorf** in 1869 when he discovered that the glow emitted from the cathode increased in size when the gas pressure is decreased. Later in 1897, the British physicist **J. J. Thomson** discovered the electrons with his famous cathode ray tube experiments.

In 1886, the German Physicist **Eugen Goldstein** discovered the existence of positively charged rays in the discharge tube by using a perforated cathode. He named them as anode rays or canal rays. In 1899, the British physicist **Ernest Rutherford** discovered **alpha** and **beta rays** from uranium. He later demonstrated that alpha rays are the nuclei of helium atoms.

In 1914, Rutherford discovered that the nucleus of an atom was extremely dense, but only occupied a small fraction of the volume of an atom. He also found that the nucleus was positive in charge. At this time, the scientists concluded that particles opposite in charge to that of electrons must exist to balance the charge of the atoms. Rutherford named this as "positive electron," which later came to be known as "**proton**."

Scientists observed that the atomic mass number A of nuclei was more than twice the atomic number Z for most atoms. This was a puzzle, and there were some proposals to explain the higher value of the atomic mass number. The scientists knew that almost all of the mass of the atom is concentrated in the nucleus (Rutherford's famous gold foil experiment).

By around 1930, it became known that alpha particles that hit beryllium produced radiation, which was not usual. For example, this radiation knocked off high energy photons from paraffin wax. In 1932, a British Scientist **James Chadwick** explained that the radiation from beryllium was a neutral particle with a mass close to a proton.

In 1934, this particle was named **neutron**, which completed the Bohr's model of the atom.

In Bohr's model of the atom, the electrons revolved around the nucleus at certain discrete distances for which the angular momentum is an integral multiple of reduced Planck's constant, that is, $n\hbar$, where $n = 1, 2, 3, \ldots$ is the **principal quantum number**. This provided for the quantization of the angular momentum.

Arnold Johannes Wilhelm Sommerfeld, a German physicist, extended this model by adding two quantum numbers: the **Azimuthal quantum number** l and the **Magnetic quantum number** m_l. The azimuthal quantum number defines the orbital angular momentum, and consequently, this defines the shape of the orbitals. It takes the values $l = 1, 2, 3 \ldots n$. The magnetic quantum number is a projection of the orbital angular momentum along an axis. In other words, it describes the orientation of the orbitals in space. It takes the integer values between $-l$ and l namely, $m_l = -l, -l + 1, -l + 2, \ldots, -2, -1, 1, 2, \ldots, l - 2, l - 1, l$. Hence it can take $2l + 1$ possible values.

1.2.4 Elementary Particles

As new particles were discovered and their behaviors observed, scientists had to create new theories to explain them. Quantum mechanics is a successful theory that explains these behaviors well. However, the advent of quantum mechanics altered the views of classical physics known in the earlier part of the twentieth century. The concept that a single particle can span a field as a wave was quite radical, and it is still a subject of research.

The Standard Model of particle physics is a widely successful theory that tries to explain the fundamental forces of nature: the electromagnetic, strong, and weak interactions. It also classifies the elementary particles known so far. Elementary particles are subatomic particles that are not composed of other particles.

Protons and neutrons are not elementary particles as they are made of quarks. **Quarks** are elementary particles that can combine to form composite particles called hadrons. Protons and neutrons are **hadrons** (Fig. 1.8).

A proton is made of two "*up quarks*" and one "*down quark*". Gluons mediate the force between the quarks. **Gluons** are exchange particles for the strong force between the quarks. **Exchange particles** are virtual particles that interact with particles to create the effects of attraction or repulsion. The **virtual particles** cannot be observed.

Neutrons are made up of two down quarks and one up quark. Gluons also mediate the forces between the quarks.

Fig. 1.8 (**a**) The quark model of the neutron. (**b**) The quark model of the proton

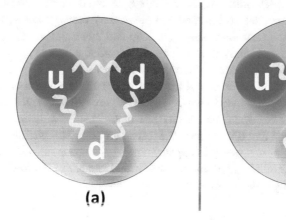

(a)   **(b)**

Standard Model of Elementary Particles

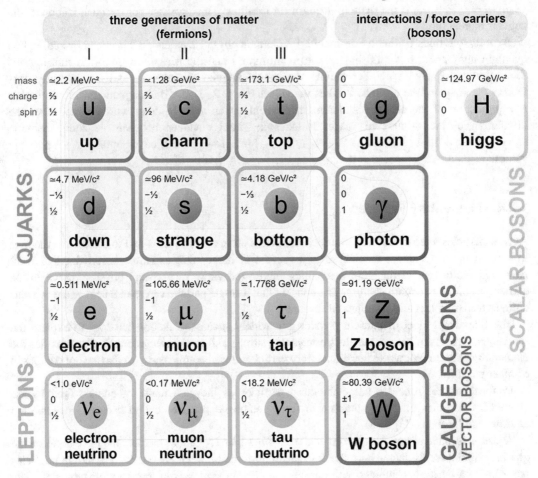

Fig. 1.9 The standard model of elementary particles. (Author: Cush, Source: Wikipedia, Public Domain)

Electrons are considered to be elementary particles. However, under extreme conditions, the electrons in a solid are shown to split into three quasiparticles: **holons** (charge), **spinons** (spin), and **orbitons** (orbital) representing the three degrees of freedom the electron enjoys in a molecule. Figure 1.9 summarizes the standard model of elementary particles known so far.

The first-generation **fermions**, namely, the up and down quarks, the electron, and the electron neutrino, make up the ordinary matter. The higher-order fermions are observed in some extreme environments, and they fall back into the first order fermions.

The **Gauge bosons** do not constitute matter. They are interaction energy or force carriers that act upon matter. For example, the Gluons represent the energy interaction between the up and down quarks in a proton or a neutron and adds to their mass.

The standard model describes bosons, the force-carrying fields well. However, the mass of W and Z bosons seemed to be a mystery to explain. The Higgs boson, also known as the "**god particle**," comes to help here. The **Higgs boson** is a scalar field and adds up mass to the W and Z bosons in a process called the Higgs mechanism.

Fig. 1.10 The Ammonia molecule, sharing 3 of its valence electrons with Hydrogen atoms

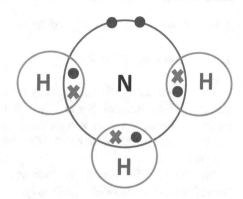

Any particle that interacts with the Higgs field acquires a mass. Particles such as photons that do not interact with the Higgs field do not acquire mass!

According to the classification of elementary particles, electrons are **fermions** with one-half spins (we shall explore the concept of spins in a later section). Fermions are particles that follow **Fermi-Dirac statistics** and obey **Pauli's exclusion principle**. In extreme conditions, the electrons move past each other in a process known as Quantum Tunneling. It has been proved that under these scenarios, the electrons split into its constituent quasiparticles. **Quantum Tunneling** is an everyday phenomenon observed in semiconductors and applied in a class of Quantum Computers known as **Quantum Annealers**.

Quantum Annealers belongs to a class of quantum computing called **Adiabatic Quantum Computation**, which has applications in solving optimization problems where a local minimum is required to be found from a discrete search space. This process uses Quantum Tunneling. We shall explore this concept in detail in a later discussion.

Photons, gluons, W, and Z bosons are force-carrying particles, and they are called **bosons**. Bosons have integer spins and follow **Bose-Einstein statistics**.

1.2.5 Molecules

Molecules are the smallest particle of a substance that retains its physical and chemical properties. Two or more atoms make up the molecules, and they are held together by chemical bonds. The molecules are electrically neutral.

Figure 1.10 illustrates the **Ammonia** molecule, formed by a covalent bond between three Hydrogen and one Nitrogen atom. Each Hydrogen atom shares one electron with the Nitrogen atom. Industrial production of Ammonia involves heating Nitrogen and Hydrogen to a degree of 400 °C (930.2 °F) under high pressure of 200 atmospheres.

It is one of the most energy-consuming and effluent producing processes in the world. Nature produces Ammonia in the root nodules of plants at room temperature without producing any effluent.

An essential application of quantum computers is to simulate atoms and chemical reactions. Someday, quantum computers may help us synthesize enzymes that can create molecules like Ammonia with less energy and less waste.

The preceding sections provided a basic knowledge of matter. An understanding of matter is the first step in learning about energy and electromagnetic radiation. The following sections focus on light and quantization of energy, thereby introducing the foundations of quantum mechanics.

1.3   Light and Quantization of Energy

1.3.1   Blackbody Radiation

In physics, a blackbody is an imaginary object that absorbs all of the radiation incident on it. An ideal blackbody does not reflect any radiation, so it appears dark. The blackbody is also a perfect emitter of radiation. Hence, at a given temperature, the blackbody emits the maximum energy. In other words, the wavelength (or rather the color) of radiation depends upon the temperature of the body and not its composition.

Figure 1.11 shows the relation between the intensity of light emitted as a function of wavelength for various temperatures of a blackbody.

In classical physics, an ideal blackbody at a given temperature is expected to emit radiation in all frequencies, and the energy of the emitted radiation increases with an increase in frequency. In other words, classical physics expects energy to increase or decrease continuously. This does not explain the sharp decrease in the intensity of radiation emitted at shorter wavelengths at lower temperatures. Also, the blackbody is expected to emit an infinite amount of energy, which is against the principle of conservation of energy. This difference between classical theoretical physics and experimental observation is known as the "**ultraviolet catastrophe.**" Therefore, a new theory is essential to explain the behavior of blackbodies. The K here refers to the Kelvin scale in which scientists measure temperature. 0 K (−273.15 °C or −459.67 °F) is known as Absolute Zero, the lowest possible temperature with no heat energy. Water freezes at 273.15 K (0 °C) and boils at 373.15 K (100 °C) under standard atmospheric pressure.

As early as 1872, **Ludwig Boltzmann** theorized that energy states of a physical system could be discrete. He divided the energy of a system into tiny discrete packages [2] in his calculations.

In 1900, the German physicist **Max Planck** explained the blackbody problem with the idea that the discrete energy states are proportional to the frequency of the radiation. He postulated that the atoms and molecules in an object behave like oscillators, and they absorb and emit radiation. The energy of

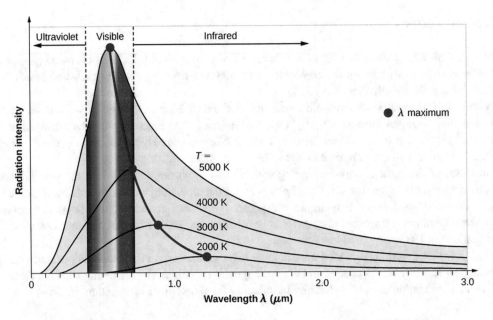

Fig. 1.11 Blackbody radiation. (Source: University Physics Volume 3, Released under CC BY 4.0)

the oscillating atoms and molecules is quantized. It is emitted or absorbed in integral multiples of small steps, known as the "**quantum**." The following equation describes this.

$$E = nh\nu \tag{1.1}$$

where,

E - Energy Element
h - Planck's constant (6.626×10^{-34} Js)
ν - frequency of the radiation.
n – 0, 1, 2, 3. . . . .

This equation means that the energy can increase or decrease only in discrete steps defined by ΔE, the smallest value by which energy can change:

$$\Delta E = h\nu \tag{1.2}$$

1.3.2 Photoelectric Effect and the Concept of Photons

In 1887, the German scientist **Heinrich Hertz** first observed the photoelectric effect. Hertz observed that, when light above a certain frequency (called the threshold frequency) is incident on a metal, electrons are released. This effect is something classical physics cannot easily explain.

In 1905, **Albert Einstein** explained the photoelectric effect using Planck's theory. Einstein received his Nobel Prize in 1921 for explaining the photoelectric effect (Fig. 1.12).

Einstein proposed that the energy in the light is in discrete "packets" called quanta. Each of these packets contained energy defined by this equation:

$$E = h\nu \tag{1.3}$$

where,

E—Energy of light quanta
h—Planck's constant.
ν—frequency of the radiation.

Fig. 1.12 Photoelectric effect

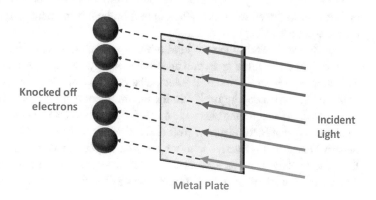

Knocked off electrons

Incident Light

Metal Plate

This is the energy of one quantum of light. This equation also relates energy with the frequency (= color) of light. We now call this quantum of light as a **photon**. Moreover, light is electromagnetic radiation, similar to radio waves.

According to Particle Physics, photons are **bosons,** and they serve as the force carriers for the electromagnetic radiation. Photons are also elementary particles. Hence, photons are not made of other particles. Photons carry no mass. Photons have a spin of 1, and they have no charge. Photons travel at a speed of $c = 299,792,458$ ms^{-1} in a vacuum.

The following equation is called the photoelectric equation:

$$hf = \phi + E_k, \qquad (1.4)$$

where h is the Planck's constant, f is the frequency of the incident light, ϕ is the work function—the minimum energy required to remove an electron from the surface of the metal, and E_k is the maximum kinetic energy of the emitted electrons.

Photons exhibit wave-particle duality, a property we shall explore shortly.

1.3.3   What Happens When Light Hits an Atom?

The Bohr's model of an atom proposes that the electrons move around the nucleus in circular orbits of certain allowed radius, like the planets moving around the sun. The angular momentum (quantity of rotation) of electrons in such orbits meets the following criteria:

$$L_n = m_n v_n r_n \ = \frac{nh}{2\pi} \qquad (1.5)$$

where,

L—angular momentum
m—the mass of the electron
r—radius of the orbit
h—Planck's constant.
n—orbit 1,2,3...

Each of these orbits is assigned a number n. In other words, the angular momentum of an electron is quantized! The closest orbit to the nucleus is $n = 1$, where the energy holding the electron to the nucleus is high.

For example, the hydrogen atom has one electron that occupies the lowest orbit $n = 1$. This lowest orbit is also its "**ground**" state. When light (photons) is incident on an atom, the light can bounce off (reflected), pass through, or absorbed.

The choice depends upon the kind of atom and the frequency (or energy) of the light. If the light has the right energy, the atom can absorb the energy, thereby pushing the electrons to higher orbits. The atom is now in an "**excited**" state. The excited state is not a natural state. Hence, the electron emits the absorbed energy back into the environment and returns to its ground state.

In a hydrogen atom, the energy level E_1 for the ground state ($n = 1$) is determined to be -13.6 eV. The unit eV stands for electron Volt and the value of 1 eV $= 1.6022e - 19$ joules. This energy is negative because it is relative to a free electron at rest. A free-electron at rest is at an infinite distance from the nucleus and has zero energy. The most negative energy of an electron is at the orbitals closest to the nucleus ($n = 1$). This also the most stable orbital and the ground state.

Fig. 1.13 Electron transitions in a Hydrogen atom

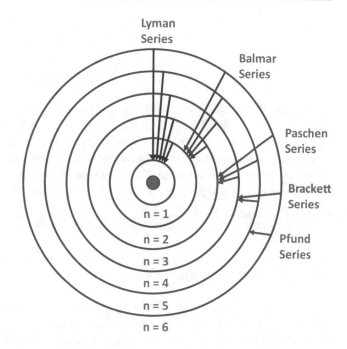

Fig. 1.14 Shown here is a photograph of IBM System One, a 20-qubit circuit-based quantum computer kept at millikelvin temperatures, colder than deep space. (Courtesy: IBM)

The energy level E_2 for the first excited state ($n = 2$) is -3.4 eV. For the second excited state ($n = 3$), the energy E_3 is -1.51 eV, and for the third excited state ($n = 4$), the energy E_4 is -0.85 eV.

Figure 1.13 shows the electron energy levels. It also shows the spectral line series arising due to the electron transition between the levels.

If we must move an electron from the ground state to the first excited state in a Hydrogen atom, we need—

$$\Delta E = E_1 - E_2 = -13.6 \, \text{eV} - -3.4 \, \text{eV}$$
$$\Delta E = -10.2 \, \text{eV} = -1.66626e - 18 \; \text{J} \tag{1.6}$$
$$\text{i.e.,} \, \Delta E = -0.000000000000000000166626 \; \text{J}$$

Compare this number with a 40 W electric bulb, which emits 40 joules of energy per second. The energy to move an electron out of its orbit is so low that we have to isolate the **qubits** (the quantum equivalent of classical bits used in digital computers) from the thermal and electromagnetic background and keep them inside dilution fridges under micro kelvin temperatures.

We use this concept to measure the state of qubits. By shining a measured dose of light on a qubit and measuring the emerging light, the qubit's state can be deterministically measured! Of course, this process destroys the state of the qubit.

We shall learn more of quantum measurements in subsequent sections.

1.4   Electron Configuration

In the previous section, we saw the energy involved when a photon hits an electron and the energy levels occupied by the electrons. The electronic configuration helps describe the distribution of the electrons in the orbitals using the four quantum numbers: the principal quantum number n, the orbital angular momentum quantum number l, the magnetic quantum number m_l, and the spin angular momentum quantum number m_s. The electron configuration is subject to **Pauli's exclusion principle**.

1.4.1   Pauli's Exclusion Principle

According to this principle, no two fermions (particles having half-integer spins) in a quantum mechanical system can occupy the same quantum state. The Austrian physicist **Wolfgang Pauli** formulated this principle in 1925 for electrons. This was later extended to all fermions. In an atom, two electrons residing in the same orbital can have n, l, m_l the same. However, they must have different spin angular momentum (m_s) with values $\frac{1}{2}$ and $-\frac{1}{2}$.

Pauli's exclusion principle does not apply to particles with integer spins (bosons). Bosons can occupy the same quantum state (e.g., photons and Bose-Einstein condensates). This is due to the nature of the **exchange interaction** between bosons, which adds up the probability of two similar particles entering the same state.

1.4.2   Principal Quantum Number *n*

The principal quantum number n denotes the shell or the energy level in which the electron can be found. It takes values $n = 1, 2, 3, \ldots$ The total number of electrons that can be present in a shell is $2n^2$.

1.4.3   Orbital Angular Momentum Quantum Number *l*

The orbital angular momentum quantum number l describes the subshell. The values orbital angular momentum can have are $l = 0, 1, 2, \ldots, n - 1$. The subshells are also described by alphabet labels

Fig. 1.15 Madelung's
rule, electron occupancy
order

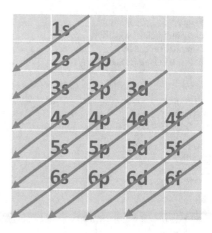

starting from s, that is, $l = s, p, d, f. \ldots$ We can calculate the total number of electrons present in a subshell by using the formula $2(2l + 1)$. Each of these subshells has orbitals with two electrons in each orbital, obeying Pauli's exclusion principle and accounting for the two spin states of the electrons. The s subshell has one orbital, p has 3 orbitals, d has five orbitals, and f has seven orbitals.

Electrons occupy the orbitals in the order of increasing energy as defined by the Aufbau principle:

$$1s < 2s < 2p < 3s < 3p < 4s < 3d < 4p < 5s < 4d < 5p < 6s < 4f < 5d < 6p < 7s < 5f < 6d < 7p$$

Madelung's diagonal rule shown in Fig. 1.15 illustrates this concept.

While describing the electron configuration, a nomenclature, containing the principal quantum number, the subshell, and the number of electrons in the subshell is used.

For example, the electron configuration of neon is $1s^2 2s^2 2p^6$. This means, two electrons are each in the $n = 1, l = 0$ and $n = 2, l = 0$ subshells, and six electrons are in the $n = 2, l = 1$ subshell. Since this list can be quite long, the electron configuration is visualized as the core electrons equivalent to a noble gas in the previous period in the periodic table and the valence electrons. The electron configuration of neon can be written as $[\text{He}]2s^2 2p^6$.

1.5 Wave-Particle Duality and Probabilistic Nature

The photoelectric effect demonstrated that light behaved like a particle. However, some other experiments and theories had a view that light behaves like a wave. Is light a wave or a particle? Well, it depends upon how we measure it!

1.5.1 Maxwell's Electromagnetic Theory

The nineteenth century was bubbling with many inventions around electricity, and magnetism and new theories were developed in explaining them. The invention of the telegraph, the first electric motor, electromagnetic induction, electric bulbs, motion picture, and phonograph were great things that changed the lives of people in that century. However, a unified theory of electricity and magnetism was missing. In this time, the Scottish scientist **James Clerk Maxwell** created his famous

Fig. 1.16 Propagation of electromagnetic waves. (Credits: Image by helder100 from Pixabay)

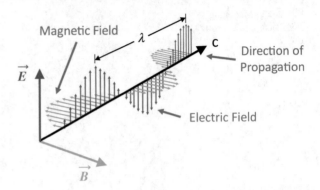

equations of electromagnetism based on experimental observations and theories by other scientists of the century.

Maxwell's equations predicted the presence of electromagnetic radiation - fluctuating electric and magnetic fields, which propagate at a constant speed ($=c$, *the speed of light*) in a vacuum. The spectrum of light at various wavelengths, including infrared, visible light, ultraviolet, x-rays, radio waves, and gamma radiation are propagating electromagnetic waves of alternating electric and magnetic fields—a wave-like behavior (Fig. 1.16).

1.5.2 De Broglie Waves

By the second decade of the twentieth century, it became widely accepted that light had both wave-like and particle-like properties. Matter itself was still thought of particle-like. In 1924, Louis de Broglie challenged this assumption and proposed that electrons too must behave like a wave and must possess a wavelength. He went one step forward and proposed that all matter must have wave-like behavior and their wavelength related to their momentum, as described in the following equation.

$$\lambda = \frac{h}{mv} = \frac{h}{p} \tag{1.7}$$

where,

λ—wavelength, the inverse of frequency
m—the mass of the particle
v—velocity of the particle
h—Planck's constant
p—momentum, a product of m and v

The American scientists, **Clinton Davisson** and **Lester Germer**, proved this later in an experiment where electrons scattered by the surface of a nickel crystal displayed a diffraction pattern—again a wave-like behavior (matter waves!)

To study the probabilistic and wave-like nature of matter, we must perform the classical **Young's double-slit experiment**.

1.5.3   Young's Double-Slit Experiment

Consider the following setup, where a sandblaster gun can randomly fire sand particles at an assembly, which has two slits. We can keep the slits either opened or closed. The sand particles can pass through the slits and hit an observation screen at the back.

In the first run of the experiment, let us close Slit 2 and turn on the sandblaster. The sand particles pass through Slit 1, hit the observation screen, and pile up in a wave-like pattern, as in Fig. 1.17.

If we represent the amplitude of the wave-like pattern as ψ (Greek letter psi) at a point x in the screen, then the probability P_1 of a sand particle hitting the point x in the screen can be written as:

$$P_1(x) = \psi^2 \tag{1.8}$$

Now, let us close Slit 1, open Slit 2, and rerun the experiment.

The sand particles pass through Slit 2, hit the screen, and form a similar pattern. Figure 1.18 shows this pattern. The probability P_2 of a sand particle hitting the screen at a point x in the screen can be written in the same way as:

Fig. 1.17 Young's double-slit experiment with a sandblaster gun with Slit 2 closed. (Credits: Anuj Kumar)

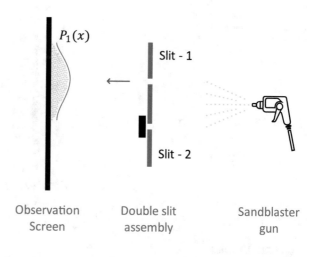

Fig. 1.18 Setup with Slit 1 closed. (Credits: Anuj Kumar)

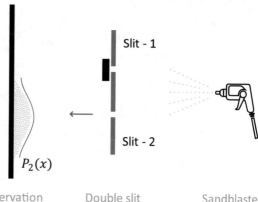

$$P_2(x) = \psi^2 \tag{1.9}$$

Finally, let us open both the slits and rerun the experiment.

The sand particles now have the option of passing through one of the slits and hit the screen. In this final run, the probability P_{12} of a sand particle hitting the screen at a point x in the screen is the sum of the probability of the sand particle passing through Slit 1 and the probability of the sand particle passing through Slit 2 (Fig. 1.19).

The new probability can be written as:

$$P_{12}(x) = P_1(x) + P_2(x) = \psi^2 \tag{1.10}$$

This experimental result concurs with classical physics, and we do not doubt this behavior.

Now, let us replace the sandblaster gun with an electron gun (the ones used in CRTs, if you remember them) and watch the results!

When we replace the sandblaster gun with an electron gun and repeat the experiment, we expect a similar behavior: the electrons pass through the two slits and hit the screen creating two bars like patterns, as drawn in Fig. 1.20.

Fig. 1.19 Double-slit experiment with both the slits open. (Credits: Anuj Kumar)

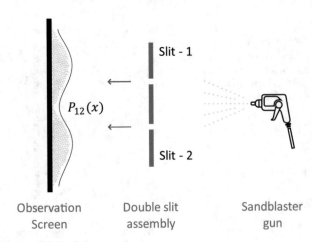

Observation Screen   Double slit assembly   Sandblaster gun

Fig. 1.20 Expected behavior

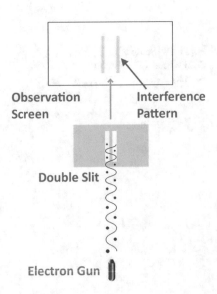

Fig. 1.21 Observed behavior

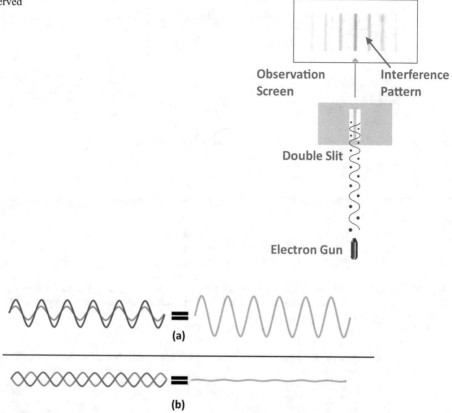

Fig. 1.22 (**a**) Constructive interference when two waves of the same phase meet. (**b**) Destructive interference when two waves with opposite phases meet

However, the observed behavior is entirely different! The electrons pass through the two slits, but form a fringe pattern, as shown in Fig. 1.21. This fringe pattern requires the electron to behave like a wave.

If the electrons are a wave, they shall interfere with each other after passing through the slits. At places where the peaks of the waves meet, we get a bright spot, that is, a constructive interference. At places where a peak meets a low, they cancel each other out, and we get a dark spot, that is, a destructive interference. Intermediates create grey bands. Figure 1.22 illustrates constructive and destructive interferences.

We repeat the experiment with an electron gun, which releases one electron at a time. We can use attenuators to release one electron at a time. Thus, this setup prevents more than one electron from passing through the slits at the same time and causing interference on the other side of the double-slit assembly.

When we run the experiment, the results are quite surprising. The electrons produce a similar pattern of interference fringes (Fig. 1.23)!

The formation of interference fringes cannot be explained unless the electron somehow passes through both slits and interferes with itself on the other side. This is impossible unless the electron behaves like a wave. This is explained in Fig. 1.24a.

Using the analogy before, if $\psi_1(x, t)$ is the probability amplitude of the electron passing through Slit 1 and if $\psi_2(x, t)$ is the probability amplitude of the electron passing through Slit 2, then we can

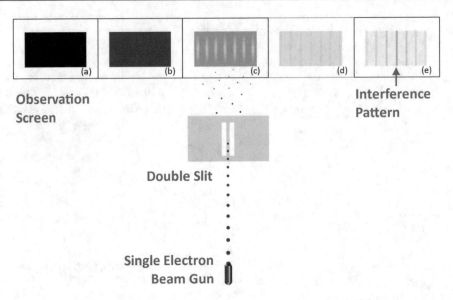

Fig. 1.23 Single-electron beam experiment showing the electron build-up over a period of time—images (**a**) to (**e**) shown in the observation screen. The pattern is not the electron itself. However, it is the probability that the electron shall arrive at a given spot on the screen

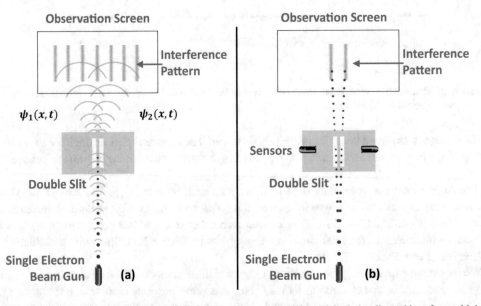

Fig. 1.24 A single electron gun experiment. (**a**) A single electron passes through both the slits and interferes with itself on the other side. (**b**) Experiment repeated with sensors mounted on each path

summarize the probability amplitude of the electron passing through both the slits and hit a point x in time t as:

$$\psi(x,t) = \psi_1(x,t) + \psi_2(x,t), \tag{1.11}$$

where x is the position on the screen, the electron could be found at a given time t.

Now let us mount some sensors at the slits to check whether the electron did pass through both the slits simultaneously!

When we repeat the experiment by mounting two sensors at the slits (Fig. 1.24b), we see that the electrons pass through both slits with a 50% probability in each. Here, the act of **measurement** changes the experiment, and the interference pattern disappears. We shall explore the concept of measurements in a later section.

1.6   Wavefunctions and Probability Amplitudes

We learned in a previous section that an electromagnetic (EM) wave is an alternating wave of Electric field $\vec{E}$ and Magnetic field $\vec{B}$. These two fields are perpendicular to each other and the direction of the propagation of the wave (Fig. 1.16). The energy density (u) of an EM wave is proportional to the square of either of these two fields. The following equation describes this:

$$u = \varepsilon_0 E^2 = \frac{1}{\mu_0} B^2 \qquad (1.12)$$

where,

ε_0—permittivity of free space or the electric constant.
μ_0—permeability of free space or the magnetic constant.

We can safely assume that the greater the energy density, the higher the probability of finding the photon. In other words, the probability of finding the photon is proportional to the square of the fields.

Though the photon and the electron are different kinds of elementary particles, de Broglie has shown that electrons (and other matter as well) behave like waves, and this analogy can be applied to them.

Let us now introduce a term **wavefunction**, or the **probability amplitude** $\psi(x, t)$ where x is the position where the quantum system (electron in our discussion here) can be found at time t.

If we plot the interference pattern we got when we did the double-slit experiment with the electron gun, against the position x on the screen, the graph looks like Fig. 1.25. The peaks and lows in the graph correspond to the bright and dark bands on the screen.

Fig. 1.25 Plot of intensity versus position on the screen (e) in the double-slit experiment with electron gun in Fig. 1.23

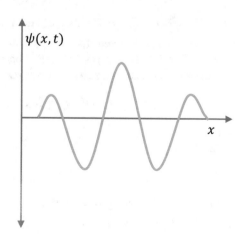

Fig. 1.26 Plot of the absolute squared of the wavefunction

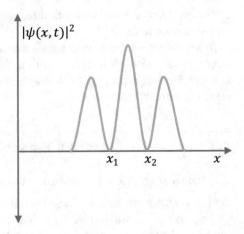

If we take the square of this function and plot it against the x-axis, the graph looks like Fig. 1.26. The height of the graph at any point on the x-axis is proportional to the probability of finding the electron at that point.

Recollect from the lessons on calculus that the probability of finding the electron, somewhere between two points $x1$ and $x2$ on the graph can be written as the integral:

$$P_{12} = \int_{x1}^{x2} |\psi(x,t)|^2 dx \tag{1.13}$$

A more generalized version of this equation in a normalized form, defining the probability of finding an electron somewhere in the universe, is given by:

$$P(x,t) = \int_{-\infty}^{+\infty} |\psi(x,t)|^2 dx = 1 \tag{1.14}$$

Like the EM waves, if we assume that the de Broglie waves contain two alternating parts, we can write the wavefunction as an ordered pair:

$$\psi(x,t) = \{A \cdot \cos(kx - \omega t), A \cdot \sin(kx - \omega t)\}, \tag{1.15}$$

where $k = \frac{2\pi}{\lambda}$, $\lambda = \frac{h}{p}$, h = Planck's constant. p = momentum, ωt - represents the phase factor of the wave, $\omega = 2\pi\nu$, A - represents the amplitude of the wave, and ν—frequency of the wave.

Note: For the sake of brevity [3], this book does not discuss how we arrived at the term $kx - \omega t$. This equation can be written in a complex form as:

$$\psi(x,t) = \{A\cos(kx - \omega t) + iA\sin(kx - \omega t)\} \tag{1.16}$$

where $i = \sqrt{-1}$

We know from the Euler's formula $e^{i\theta} = \cos(\theta) + i\sin(\theta)$, so we can write the above equation as:

$$\psi(x,t) = Ae^{i(kx-\omega t)} \tag{1.17}$$

In this equation, the factor $e^{-i\omega t}$ is a constant for a given time t, and it can be factored out into A. The resulting equation is:

$$\psi(x, t) = Ae^{ikx}, \tag{1.18}$$

where A is a complex number now.

This explains why the wavefunction is a complex quantity. Since $k = \frac{2\pi}{\lambda}$, $\hbar = \frac{h}{2\pi}$, and since $\lambda = \frac{h}{p}$, the above equation can be rewritten as the following momentum eigenfunction:

$$\psi(x, t) = Ae^{\frac{ip_x x}{\hbar}} \tag{1.19}$$

The probability of finding the electron (or in general, a quantum system) can now be written as the absolute square of the probability amplitude:

$$P(x, t) = |\psi(x, t)|^2 \tag{1.20}$$

We shall be using this equality in the forthcoming chapters.

1.6.1 Realism in Physics

In classical physics, realism means the quality of the universe exists independently of ourselves. Even if the result of a possible measurement does not exist before measuring it, it does not mean that it is a creation of the person experimenting.

Quantum mechanics rejects the concept of realism. A quantum system may always exist in all possible states simultaneously until a measurement is made. We do not know the state of the system until we measure it. When we make a measurement, the wavefunction collapses, and the system is localized.

1.6.2 What Happens When We Measure a Quantum System?

In any experiment we do with a quantum computer, we set the qubits to the initial states and perform several operations on the qubits. Finally, we measure the state of the qubits by projecting them into one of the measurement basis.

When we perform a measurement on a quantum system, the measured quantity gets well defined. This is because the wavefunction is a probability density over all possible states. According to the **Copenhagen Interpretation of Quantum Mechanics**, the act of measurement causes the wavefunction to collapse into the corresponding eigenfunction.

If we measured the position of the quantum system to be x_0, then the wavefunction is now the eigenfunction $\delta(x - x_0)$. This function is known as the **Dirac delta function**. It is a linear function which maps the density of a point mass or a point charge to its value at zero (Fig. 1.27).

If we measured the momentum of the quantum system and found it to be p_0, then the wavefunction is now the eigenfunction $e^{\frac{ip_0 x}{\hbar}}$, which we derived earlier.

1.6.3 Heisenberg's Uncertainty Principle

The **Uncertainty Principle** was introduced in 1927 by the German physicist **Werner Heisenberg**. It imposed a limit on how accurately we can measure canonically conjugate observables like position or momentum simultaneously. Time and Energy are another set of examples of canonically conjugate

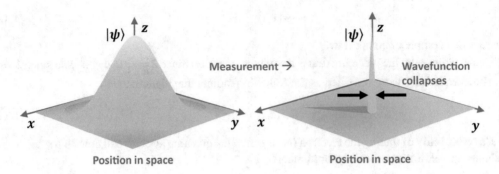

Fig. 1.27 The Copenhagen interpretation and the collapsing of the wavefunction after measurement

observables. According to the uncertainty principle, the more accurately we measure one of them, the less accurate the other becomes and vice versa. The uncertainties with the products of these canonically conjugate observable pairs have a minimum value as defined by the following equations:

$$\Delta p_x \Delta x \geq \frac{\hbar}{2\pi} \text{ and}$$
$$\Delta t \Delta E \geq \frac{\hbar}{2\pi},$$

(1.21)

where Δ -"delta" is a small difference, $\hbar = \frac{h}{2\pi}$, p_x-momentum, x-position, t-time, E-Energy.

The uncertainties are so small that they are significant only with small masses of atomic and subatomic particles. We need not worry about their effects on everyday objects like vehicles and electric bulbs!

Two observables are said to be canonically conjugate if their operators $\hat{A}$ and $\hat{C}$ do not commute. Specifically, they should satisfy the following condition:

$$[\hat{A}, \hat{C}] = \hat{A}\hat{C} - \hat{C}\hat{A} = i\hbar$$

(1.22)

Here, the hat accent is used to denote an operator. $\hbar$ is the reduced Planck's constant and i is an imaginary number. We shall discuss operators in the next chapter.

1.6.4 Weak Measurements

In the last few years, some research papers have been published about the possibility of weak measurements [5] in quantum systems. Weak measurements do not introduce many disturbances in the system, and hence the wavefunction does not collapse. However, we obtain only very little or less accurate information about the observable.

Weak measurements are a challenge to the system engineers designing quantum computers. In Chap. 4, we discuss qubit modalities and learn about the construction and functionality of the qubits. For each modality, we also discuss how to make the measurement. If the system design does a weak measurement of the qubit, the results may be inaccurate. This is a problem the system engineers are required to solve and adds to the complexity of designing quantum computers.

Example 1.1

Prove that the position and momentum operators do not commute.
Solution:

We know that $\hat{x} = x$ and $\hat{p} = \dfrac{\hbar}{i} \dfrac{\partial}{\partial x}$

Note: The momentum operator can be obtained by partial differentiation of the momentum eigenfunction. For brevity, that step is ignored.

$$[\hat{x}, \hat{p}]\psi = \widehat{xp}\psi - \widehat{px}\psi = x \cdot \frac{\hbar}{i} \frac{\partial \psi}{\partial x} - \frac{\hbar}{i} \frac{\partial x \psi}{\partial x}$$

$$= \frac{\hbar}{i} \left[x \frac{\partial \psi}{\partial x} - x \frac{\partial \psi}{\partial x} - \psi \frac{\partial x}{\partial x} \right]$$

$$= -\frac{\hbar}{i} \psi = -\frac{\hbar}{i} = i\hbar$$

Since this value is $i\hbar$, we can say that the position and momentum operators do not commute. Since these two operators do not commute, position and momentum cannot be measured precisely at the same time. Position and momentum are an example pair of **canonically conjugate** variables (Eq. 1.22). The commutative property of operators is discussed in section 2.6.2.5.
◄

1.7   Some Exotic States of Matter

1.7.1   Quasiparticles and Collective Excitations

In quantum mechanics, quasiparticles and collective excitations are two closely related behaviors. They are used to describe the complex dynamics of strongly interacting particles in quantum systems as simpler dynamics of quasi-independent particles in free space. An example is the motion of electrons and holes in a semiconductor. The quasiparticles are not comprised of matter, but they behave like elementary particles with some differences. Like ordinary matter, the quasiparticles can be localized in space, and they can interact with other matter or with other quasiparticles. However, they cannot exist outside a medium, whose elementary excitations like the spin waves are needed for this existence. The term "**quasiparticle**" refers to fermions, and "**collective excitation**" refers to bosons.

1.7.2   Phonons

In quantum mechanics, **phonons** are a type of vibrations in which the crystalline (lattice) structure of atoms or the molecules of an element uniformly oscillate at a single frequency. Phonons are quasiparticles that appear as collective excitations in condensed matter. The phonons play a vital role in the thermal and electrical conductivity of an element. They also exhibit wave-particle duality.

1.7.3   Cooper Pairs (or the BCS Pairs)

The electrons in a metal behave like free particles, and since they are all negatively charged, they repel each other. However, they attract the positively charged ions, which make up the crystalline structure (called lattice) of the metal. This attraction slightly distorts the lattice structure. As electrons move, the lattice structure undergoes continual distortion. The collective motion of the lattice structure is represented as phonons, and the interaction between the electrons and phonons results in a small attraction between the electrons in a metal. The electrons thus pair up and behave like single particles of spin 0 or 1. In other words, they behave like **bosons** and become the **Bose-Einstein Condensate**. This pairing up of electrons is called **Cooper Pairs / BCS Pairs**. American physicist **Leon Cooper** first announced this behavior in 1956. The interaction energy between the paired electrons is quite low $\approx 10^{-3}$ eV, and hence this behavior can be seen only at very low temperatures.

The paired electrons in the condensate are called **superconducting electrons**, and the rest of the electrons are called **normal electrons**. At a temperature lower than the critical temperature (T_C), there are very few normal electrons. As the temperature is increased, there are more normal electrons. At the critical temperature, all electrons are normal electrons.

The paired-up electrons can be quite far apart—up to a few hundred nanometers or so since the interaction is long-ranged. The electron pairs behave coherently. The characteristic distance their number density can change is called **coherence length** (ξ). The wavefunction of the Cooper Pairs is symmetric with integer spins. Therefore, more than one Cooper pair can be in the same state, a property linked with **superconductivity** (a condition where the electrical resistivity vanishes) and **superfluidity** (a condition where the viscosity of a fluid vanishes.) The American scientists **John Bardeen**, **Leon Cooper**, and **John Schrieffer** developed the **BCS theory** of superconductivity for which they won the Nobel prize in 1972.

In 2019, scientists at the Max Planck Institute in Germany announced superconductivity at 250 K using lanthanum hydride under a pressure of 170 gigapascals. This is the highest temperature where a substance exhibits superconductivity. Superconductivity at room temperature is not far away! (Fig. 1.28).

1.7.4   Superconductors

In metals such as copper, the electrical resistance decreases gradually as the temperature is reduced. The thermal coefficient of resistance is 0.00386 per degree C for copper. This means, if the temperature changes by $1\degree$C, the electrical resistance of copper changes by 0.386%.

Fig. 1.28 Diagram illustrating the critical temperature of a superconductor. At the critical temperature, electrical resistance ρ drops rapidly to zero

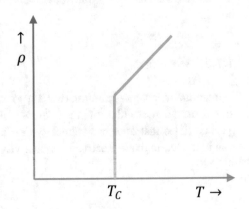

In superconductors, the electrical resistance drops rapidly to zero at a critical temperature T_c. Since there is no resistance, a superconducting loop can retain the electric current almost infinitely! Superconductivity was first observed by the Dutch physicist **Heike Kamerlingh Onnes** in 1911.

Onnes also noticed that the superconductivity could be destroyed by applying an electrical current above a certain critical value I_C, or by applying a magnetic field above the critical field H_C. It is interesting to note that copper, silver, and gold are excellent electrical conductors that we know of, but they do not exhibit superconductivity!

In 1933, **W. Meissner** and **R. Ochsenfeld** discovered another exciting property of superconductors, called the **Meissner effect**. In their experiments, Meissner and Ochsenfeld found that the superconductor expels the magnetic field. This effect is observed not only when the magnetic field is increased at temperatures $T < T_C$, but also when the superconductor is cooled from the normal state with the magnetic field applied (Fig. 1.29). Depending upon how they behave in a magnetic field, superconductors can be classified as type-I or type-II.

A type-I superconductor has a single critical magnetic field H_C, above which superconductivity is lost. If the magnetic field is less than the critical level, the superconductor expels the magnetic field. The critical field is dependent upon the temperature of the superconductor, as described by the following equation.

$$H_C(T) = H_C(0) \left[1 - \left(\frac{T}{T_C} \right)^2 \right] \tag{1.23}$$

The left-hand side of Fig. 1.30 illustrates the phase diagram of the type-I superconductor. Note that the magnetic induction $B = 0$ inside the superconductor. Technically, the magnetic field H does penetrate the superconductor, up to a length called **London's magnetic field penetration depth**, denoted by λ, which is approximately about 100 nm. Screening currents flow in this region between the surface and penetration depth, creating magnetization M inside the superconductor, which balances the applied field H. Thus $B = 0$ inside the superconductor.

Fig. 1.29 The Meissner effect

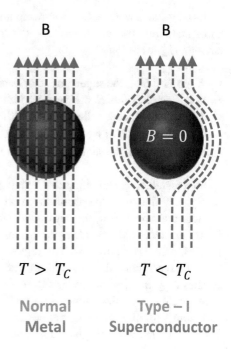

$T > T_C$

Normal
Metal

$T < T_C$

Type – I
Superconductor

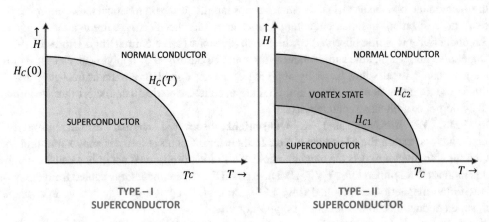

Fig. 1.30 Phase diagram of type-I and type-II superconductors

Fig. 1.31 Vortices of a type-II superconductor

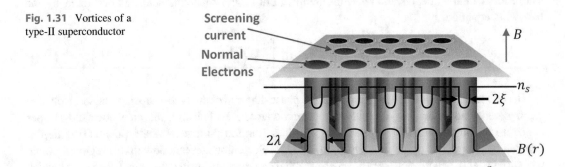

As new alloys were tested for superconductivity, new behaviors were invented. It was found that in some materials, two critical fields—the lower critical field strength H_{C1} and higher critical field strength H_{C2} were found.

- When $H < H_{C1}$, the superconductor behaves like a diamagnet, exhibiting the Meissner effect.
- In the range $H_{C1} < H < H_{C2}$, it was found that the magnetic field penetrates the superconductor forming small tubes called "vortices." The vortices are quantized with the magnetic flux quantum $\Phi_0 = \frac{h}{2e}$. When H is increased, the number and the density of vortices also increase.
- When $H > H_{C2}$, the superconductor returns to the normal state.

Superconductors exhibiting this behavior are called type-II superconductors. The right-hand side of Fig. 1.30 illustrates the phase diagram of the type-II superconductors.

Note that above H_{C1}, the superconductor does not make a sudden switch into the normal state. However, a thin cylindrical region of normal material appears to be passing through the material, parallel to the axis of the field. This region is called the **normal core**. As the applied field is increased, more of these normal cores appear, and they get more closely packed. When the applied field reaches H_{C2}, the material returns to its normal state. The state between H_{C1} and H_{C2} is called the vortex state or the mixed state. Note that both H_{C1} and H_{C2} depend upon the temperature.

Figure 1.31 illustrates the vortices formed in a type-II superconductor in a magnetic field. Each vortex contains one flux quantum. The core of the vortices contains normal electrons, and the superconducting order parameter is zero at the center of the vortices. The coherence length ξ is a function of the number of superconducting electrons n_s. For type-II superconductors, the coherence length ξ is generally lower than the penetration depth λ. For type-I superconductors, $\xi > \lambda$.

1.7.5   s-Wave, d-Wave, and p-Wave Superconductors

The superconducting wavefunction for a Cooper pair is asymmetric. It has components for the motion around the center of mass (k), spin (S), and orbital (L) angular momentum, in the order as shown in the following equation.

$$\psi(r_1, \sigma_1, r_2, \sigma_2) = e^{ikR} f(r_1 - r_2)\chi(\sigma_1 - \sigma_2)$$
$$= -\psi(r_2, \sigma_2, r_1, \sigma_1) \tag{1.24}$$

The spin states can be a "0" or "1." The orbital angular momentum can have $L = 0$ (s), $L = 1$ (p), $L = 2$ (d), and $L = 3$ (f) states. The **order parameter of superconductivity** $\approx \Delta(k)e^{i\phi(k)}$, where $\Delta(k)$ is the superconducting gap, k is the momentum of the center of mass of the Cooper pair, and $\phi(k)$ is the phase of the superconducting order. The **superconducting gap** is the energy required to break a Cooper pair.

The **spin-singlet** is a state in which all the available electrons are paired. In an **s-wave** superconductor, the spin-singlet ($S = 0$) states are caused when two electrons in a BCS pair have opposite momenta ($\frac{1}{2}$ and $\frac{-1}{2}$). Due to this, the center of mass momentum (k) is zero. However, being fermions, the electrons do not commute and can support only asymmetric exchange in Eq. (1.24). Hence if the spin is asymmetric, the orbital angular momentum of the wavefunction must be symmetric and vice versa.

When the spin wavefunction is asymmetric, the orbital wavefunction is symmetric and has even parity ($L = 0, 2, 4 \ldots$) denoting $s, d, \ldots$ states.

$$\chi(\sigma_1, \sigma_2) = \frac{1}{\sqrt{2}}(|\uparrow\downarrow\rangle - |\downarrow\uparrow\rangle) \tag{1.25}$$

When the spin wavefunction is symmetric, the orbital wavefunction is asymmetric and has odd parity ($L = 1, 3, 5, \ldots$) denoting $p, f, \ldots$ states.

$$\chi(\sigma_1, \sigma_2) = \begin{cases} |\uparrow\uparrow\rangle \\ \frac{1}{\sqrt{2}}(|\uparrow\downarrow\rangle + |\downarrow\uparrow\rangle) \\ |\downarrow\downarrow\rangle \end{cases} \tag{1.26}$$

Here the arrows denote the direction of the spin, and the vertical bar together with the right-facing angular bracket represents the ket vector in Dirac's bra-ket notation. In the upcoming chapters, this convention and the concept of spin shall be explained.

The bound state caused by the isotropic electron-electron interaction in an s-wave superconductor is symmetric when the electrons change positions. This interaction only depends upon $|k|$ and has angular momentum $L = 0$. The spin wavefunction is asymmetric with the spin-singlet state $S = 0$, $L = 0$. The resulting spin-singlet Cooper pair is ($k\uparrow, -k\downarrow$). In this mode, the pairing wavefunction is spherically symmetric. $\Delta(k)$ in an s-wave superconductor is a constant irrespective of the direction of k and has the same phase in all directions, that is, isotropic. Spin-singlet states are characterized by a single spectral line arising due to the spin-singlet state. Elements such as Al, Nb, Pb exhibit s-wave superconductivity.

We get a **d-wave** spin-singlet when $S = 0, L = 2$. The d-wave superconductor breaks rotational symmetry and the phase changes by 180 degrees at the boundaries. In a d-wave superconductor, the superconducting gap becomes zero in certain directions. Cuprate is an example of a high-temperature d-wave superconductor.

A **spin-triplet** is a state with $S = 1$, and hence the spin angular momentum can take three values $m_s = -1, 0, 1$. Spin-triplet states are characterized by the threefold splitting of spectral lines. Spin-triplet Cooper pairs are realized in ^3He ($S = 1, L = 1$). Strontium Ruthanate (Sr_2RuO_4) is another example of a p-wave superconductor. In a **p-wave** superconductor, the electron-electron interaction is mediated by magnetic spin fluctuations.

There is some evidence for f-wave ($S = 1, L = 3$) superconductivity in the exotic high-fermion superconductor UPt3.

1.7.6 Anyons

Anyons are particles that exist in a two-dimensional space [4]. They are neither bosons nor fermions, but something in between and can follow either of the two statistics. However, like fermions, they cannot occupy the same space. Between two layers of Aluminum Gallium Arsenide, a thin layer of two-dimensional electron gas (2DEG) forms at extremely cold temperatures. By cooling the 2DEG further, in the presence of a strong magnetic field, anyons form out of the excitations (which we shall discuss in Chap. 4) of the 2D lattice. Anyons have peculiar properties under particle exchange: they are classified as abelian and non-abelian depending upon their behavior during particle exchange. When we exchange two anyons, if they do not leave any measurable difference, except for a phase factor, they belong to the **abelian** group. In systems with at least two degenerate ground states, the exchange of two particles can rotate the state to another wavefunction in the same degenerate space. Anyons of this type are **non-abelian** anyons. It is this property of non-abelian anyons that interests quantum computing, and we shall explore this in detail in Chap. 4. Anyons have been observed in the **Fractional Quantum Hall Effect (FQHE)**, a quantum version of the **Hall Effect**. Recall that the Hall effect produces the electrical voltage difference across an electrical conductor in a direction transverse to the flow of the electric current and perpendicular to the direction of the magnetic field. Under extremely cold temperatures in high-quality materials, the 2DEG exhibits collective states, in which the electrons bind magnetic flux lines to form quasiparticles and excitations with fractional elementary charges of magnitude $e^* = \frac{e}{q}, q = 3, 5, 7\cdots$ and the Hall conductance forms a series of plateaus at fractional values of $\frac{e^2}{h}$. This brings anisotropic features in the 2D space and a new topological order exploited in the topological quantum computing.

1.7.7 Majorana Fermions

Majorana Fermions are quasiparticle excitations, and they follow non-abelian statistics. The Majorana fermions are antiparticle for themselves. These particles were first proposed by the Italian physicist **Ettore Majorana** in 1937. The fermions defined in the Standard Model (Fig. 1.9) are all Dirac's fermions except in the case of neutrinos, which are unsettled. In the theory of **Supersymmetry**, every boson has a fermion as the **superpartner** and vice-versa. The superpartners have the same electric charge and a few other common properties, but they have different mass and spin. The hypothetical **neutralinos** are the superpartners for the gauge bosons and Higgs bosons, and they are thought to be Majorana particles.

Majorana fermions are expected to be found as quasiparticles in superconductors, as a quasiparticle is its antiparticle in a superconductor. Majorana bound states have been predicted to emerge as localized zero-energy modes in superconductors, and there is some experimental evidence to this in iron-based superconductors. The ground level degeneracy in the topological order of Majorana fermions is a property used for realizing topological qubits. In Chap. 4, we shall discuss the topological qubits in detail.

1.8 Summary

In this chapter, we learned about some of the building blocks of quantum mechanics. We started with the definition of the matter and went on learning about what matter is made of. Quantization of energy, wave-particle duality, and the probabilistic behavior of nature were the next few topics we read about. We concluded this chapter by learning about wavefunctions, probability amplitudes, and some exotic states of matter.

The following chapter is a math refresher that focuses on linear algebra and introduces Dirac's bra-ket notation.

Practice Problems

1. A source emits gamma rays of wavelength $\lambda = 3 \times 10^{-12}$ m. If a burst of gamma rays from the source contained 2 J of energy, calculate the number of photons in the burst. (use: Planck's constant $h = 6.626 \times 10^{-34}$ Js and speed of light $c = 3 \times 10^8$ ms^{-1}.)
2. What is the minimum uncertainty in an electron's speed, if the uncertainty of its position is 100 nm? (use: mass of electron $= 9.10938356 \times 10^{-31}$ kilograms.)
3. A quantum measurement experiment measures the position and wavelength of a photon simultaneously. The wavelength of the photon is measured to be 500 nm, with an accuracy of 10^{-6} (that is, $\frac{\Delta\lambda}{\lambda} = 10^{-6}$). What is the minimum uncertainty in the position of the photon.
4. Photons with energy 6.0 eV are incident on a photoelectric plate. This produces photoelectrons with a kinetic energy of 3.2 eV. What is the work function of the photoelectric plate? If we double the energy of the incident photons, does the kinetic energy of the emerging photoelectrons double?
5. Describe the quantum numbers of all orbitals in the 4f subshell.
6. Write the ground state electron configuration of Silicon.
7. Calculate the energy and wavelength of the photon emitted, when an electron undergoes a transition from its first excited state ($n = 2$) to its ground state ($n = 1$). (use: Energy of a particle in an infinite potential well of width L and state n is given by: $E_n = \frac{n^2 h^2}{8mL^2}$.)
8. A stationary electron is accelerated through a potential of 10 MeV. Find its de Broglie wavelength. (use: Velocity of a charged particle is $\frac{\sqrt{2eV}}{m}$ where $e = 1.6 \times 10^{-19}$ Coulombs.)
9. Light of wavelength 500 nm is incident on a double-slit with a distance 0.01 mm between the slits. Calculate the angle made by the first order bright (constructive interference) fringe ($n = 1$). (use: $\lambda = \frac{d \sin\theta}{n}$.)

10. In a double-slit experiment, the distance between the slits was $d = 0.2$mm. The screen was kept at a distance of $L = 1$m from the slits. The second bright fringe was found at a distance of 5 cm from the center. Find the wavelength of light used. (use: $\lambda = \frac{xd}{nL}$.)

References

1. The Second Creation: Makers of the Revolution in Twentieth-Century Physics by Robert P. Crease, Charles C. Mann (1986)
2. http://philsci-archive.pitt.edu/1717/2/Ludwig_Boltzmann.pdf
3. Wave Mechanics. For further reading, please refer the following text book: Quantum Wave Mechanics, 2/e, Larry Reed, 2018.
4. Toric Codes and the theory of anyons – For further reading, Introduction to abelian and non-abelian anyons, Sumathi Rao, 2016, https://arxiv.org/pdf/1610.09260.pdf
5. Weak Measurements: https://arxiv.org/ftp/arxiv/papers/1702/1702.04021.pdf, http://www-f1.ijs.si/~ramsak/seminarji/KnaflicSibka.pdf, https://www.scientificamerican.com/article/particle-measurement-sidesteps-the-uncertainty-principle/

Dirac's Bra-ket Notation and Hermitian Operators

2

"Contrariwise," continued Tweedledee, "if it was so, it might be; and if it were so, it would be; but as it isn't, it ain't. That's logic."
— Lewis Carroll, Alice's Adventures in Wonderland.

This chapter is a refresher course on linear algebra and probability theory. Those who have recently studied quantum mechanics or linear algebra may skip this chapter and proceed to the following chapters. For those who continue to read this chapter, the learning objectives are as follows:

▶ **Learning Objectives**
- Scalars, vectors, Hermitian, unitary, and Pauli matrices
- Linear vector spaces
- Dirac's bra-ket notation
- Linear, Hermitian, and unitary operators
- Inner and outer products, projection operators
- Expectation values
- Eigenvalues, eigenvectors, and eigenfunctions
- Tensors
- Statistics and probability

2.1 Scalars

By definition, a scalar is a physical quantity described by a single field called magnitude, which is mostly a real number. Scalars may have a unit of measure. Beyond the magnitude, scalars do not have any other property.

Examples of scalars are mass of an object, speed of a vehicle, money in our wallet, marks scored in the mid-term, and monthly salary.

A scalar is often represented by a lower-case letter.

2.2 Complex Numbers

We used the concept of complex numbers in Chap. 1 when we worked out the wavefunction of an electron passing through two slits and hitting a screen. In this section, we shall briefly study the complex numbers.

© The Author(s), under exclusive license to Springer Nature Switzerland AG 2021
V. Kasirajan, *Fundamentals of Quantum Computing*, https://doi.org/10.1007/978-3-030-63689-0_2

Fig. 2.1 Complex number
on a graph

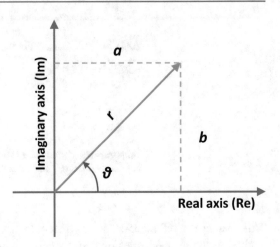

A complex number has two parts: a real and an imaginary part. The complex numbers are plotted on a graph by representing the x-axis with the real part and the y-axis with the imaginary part. Figure 2.1 shows this representation. In the **rectangular form**, the complex numbers are written as $a + ib$ where a and b are real numbers. a is the real part and ib is the imaginary part. i is an imaginary number with a value $i = \sqrt{-1}$. Complex numbers are used to solve polynomial equations that real numbers cannot solve.

In the polar form, the complex number is represented as $r \angle \theta$, which is to be read as r at angle θ. Here r is the length (or norm) of the complex number, and θ is the angle the norm makes with the horizontal axis. (Fig. 2.2).

The coordinates a and b can be written in terms of r and θ as:

$$a = r \cos \theta$$
$$b = r \sin \theta \qquad (2.1)$$

Hence, the complex number can be described in **polar form**:

$$z = r(\cos \theta + i \sin \theta) \qquad (2.2)$$

Complex numbers can also be expressed in **exponential form**. We shall request our readers to work out the proof:

$$z = re^{i\theta}, \qquad (2.3)$$

where, the component $e^{i\theta}$ is the phase factor.

De Moivre's Theorem:

De Moivre's theorem helps us to find the powers of complex numbers. For a complex number z and an integer n, de Moivre's theorem states that:

$$z^n = (r(\cos \theta + i \sin \theta))^n = r^n (\cos n\theta + i \sin n\theta) \qquad (2.4)$$

Fig. 2.2 Graph showing a vector $\vec{A}$ about the origin making an angle θ and extending to the point (5,4)

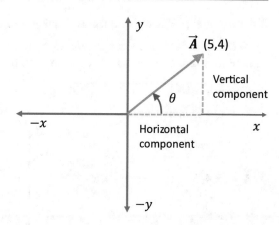

2.2.1 Complex Conjugate

The complex conjugate of a number has the same real part; the imaginary part has the same magnitude but opposite in sign.

For example, the complex conjugate of $z = a + ib$ is $z^* = a - ib$.

2.2.2 Norm of a Complex Number

The following equation gives the norm or the length of a complex number:

$$|z| = \sqrt{a^2 + b^2} \tag{2.5}$$

The norm of a non-zero complex number is always a positive number. We can easily verify that the square of the norm is equal to the product of the complex number and its conjugate:

$$|z|^2 = zz^* \tag{2.6}$$

Calculating this in the exponential form, we get:

$$|z|^2 = \left|re^{i\theta}\right|^2 = \left(re^{i\theta}\right)\left(r^*e^{-i\theta}\right) = r^2 \tag{2.7}$$

We shall use these properties in subsequent sessions.

2.3 Vectors

Vectors are quantities having both magnitude and direction. Examples are force, weight (the force gravity exerts on a body), and velocity. Vectors are often represented using an upper-case letter or a bold typeface. In a two-dimensional space, vectors are drawn as arrows; the arrows have a starting point, and their length represents the magnitude.

In a 2D plane, the angle the vector makes from the x-axis in the anticlockwise direction is its direction. A vector can also be written using its cartesian coordinates by measuring the horizontal and vertical components. The vector in Fig. 2.2 can be written as $\vec{A} = [5, 4]$. The arrow above the letter

A denotes that A is a vector quantity, having a directional component. Similarly, a vector in a 3D space can be written by its coordinates along the x, y, z axes, for example, $\vec{A} = [5, 4, 3]$.

In other words, a vector is a list of numbers. It can be either a row vector or a column vector.

Example 2.1

Row and Column Vectors

$\vec{A} = [5, 4, 3, 2, 1]$ is a row vector. $\vec{A} = \begin{bmatrix} 5 \\ 4 \\ 3 \\ 2 \\ 1 \end{bmatrix}$ is a column vector.

◀

2.3.1   Magnitude of Vectors

The magnitude of a vector is its length. For a vector $\vec{V} = [v_1, v_2, v_3]$ in three dimensions, the following equation describes its magnitude:

$$\left| \vec{V} \right| = \sqrt{v_1{}^2 + v_2{}^2 + v_3{}^2} \tag{2.8}$$

2.3.2   Unit Vector

A unit vector has a magnitude (length) of 1. We use $\vec{i}, \vec{j}, \vec{k}$ to denote the unit vectors along the Cartesian coordinate axes x, y, z. Often, the arrow accent may be omitted, and we may denote these unit vectors as mere i, j, k. Some texts may also use the hat accent to denote the unit vectors: $\hat{i}, \hat{j}, \hat{k}$.

2.3.3   Addition and Subtraction of Vectors

Two vectors can be added or subtracted by adding or subtracting the components of the two vectors. Let,

$$\vec{V} = [v_1, v_2, v_3], \text{and} \tag{2.9}$$

$$\vec{U} = [u_1, u_2, u_3] \text{ then} \tag{2.10}$$

$$\vec{V} \pm \vec{U} = [v_1 \pm u_1, v_2 \pm u_2, v_3 \pm u_3] \tag{2.11}$$

Vector addition and subtraction can also be done graphically by using the head-to-tail method. To add, we draw the first vector from the origin. The second vector is drawn by placing its tail on the head of the first vector. If there are more vectors to be added, we follow the same process. Finally, we draw the vector sum from the origin to the head of the last vector. The vector sum is shown in Fig. 2.3.

Fig. 2.3 Addition of two
vectors

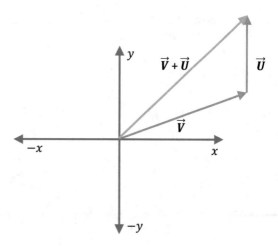

Fig. 2.4 Subtraction of
two vectors

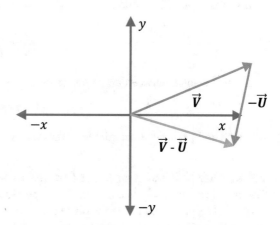

Subtraction is done similarly, but by flipping the direction of the second vector. Vector subtraction is shown in Fig. 2.4.

Scalar Multiplication of Vectors:
When we multiply a vector by a scalar, we multiply every component of the vector by the scalar. Consequently, the magnitude gets multiplied by the scalar. However, the direction remains the same.

Figure 2.5 illustrates the multiplication of the vector $\vec{V}$ by a scalar 0.5.

2.3.4 Multiplication of Vectors

Vectors exhibit two different properties of multiplication. Two vectors can be multiplied to produce a scalar result, or they can be multiplied to produce a vector result. We shall study both these two methods in the following sections.

2.3.4.1 Dot Product or Scalar Product of Vectors

The dot product or the scalar product of two vectors is the sum of the product of the components. It is a single scalar value.

Fig. 2.5 Scalar
multiplication of a vector

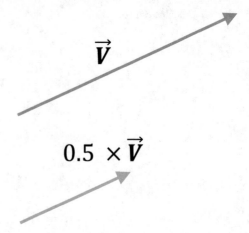

If $v = [v_1, v_2, \ldots, v_n]$ and $w = [w_1, w_2, \ldots, w]$, then the dot product of v and w is defined as:

$$v \cdot w = \sum_{i=1}^{n} v_i w_i \qquad (2.12)$$

If θ is the angle between the two vectors, then the dot product can be written as:

$$v \cdot w = |v||w| \cos \theta, \qquad (2.13)$$

where $|v|$ and $|w|$ are the magnitude of the two vectors v and w.

2.3.4.2 Vector Product or Cross Product of Vectors

The cross product of two vectors v and w, represented as $v \times w$ (to be pronounced as v cross w), is a vector that is, perpendicular to the plane containing v and w. Hence, the cross product of two vectors is defined in a 3D space. The cross product is zero, if the vectors are collinear or if one of them is of length zero. If θ be the angle between two vectors v and w, then the cross product between them is defined by the following equation:

$$v \times w = |v||w| \sin (\theta) \widehat{n}, \qquad (2.14)$$

where $|v|$ and $|w|$ are the magnitude of the vectors v and w. $\widehat{n}$ is a unit vector perpendicular to the plane containing v and w. The direction of $\widehat{n}$ is defined by the right-hand rule. (Fig. 2.6).

The vector product is anticommutative and distributive over addition, that is,

$$\begin{aligned} v \times w &= -w \times v \\ u \times (v + w) &= u \times v + u \times w \end{aligned} \qquad (2.15)$$

We can show that the unit vectors we saw earlier satisfy the following cross product relations:

$$\begin{aligned} i \times j &= k, & j \times i &= -k \\ j \times k &= i, & k \times j &= -i \\ k \times i &= j, & i \times k &= -j \\ i \times i &= j \times j &= k \times k = 0 \end{aligned} \qquad (2.16)$$

Fig. 2.6 Cross product of two vectors v and w

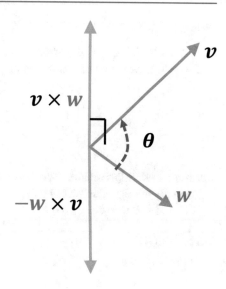

2.4 Matrices

A matrix is a rectangular array of numbers. The following is matrix A.

$$A = \begin{bmatrix} a_{11} & a_{12} & \cdots & a_{1n} \\ a_{21} & a_{22} & \cdots & a_{2n} \\ \vdots & \vdots & \vdots & \vdots \\ a_{m1} & a_{m2} & \cdots & a_{mn} \end{bmatrix} \tag{2.17}$$

In the above matrix, the numbers a_{ij} are its elements (row i, column j). The dimension of the matrix is $m \times n$. The matrix is called a square matrix if $m = n$. We can easily figure out that a matrix of dimensions $m \times 1$ is a column vector, and a matrix of dimensions $1 \times n$ is a row vector. With this definition, let us revise some of the key properties of matrices.

2.4.1 Properties of Matrices

Equality: Two matrices are equal if they both have the same dimensions, and the corresponding elements are the same.

That is, given two matrices A and B, $A = B$ if and only if $a_{ij} = b_{ij}$; $\forall\, i$ and j.

Note: The symbol $\forall$ is to be read as "for all."

Scalar Multiplication: In order to multiply a matrix by a scalar, we have to multiply all the elements of the matrix by the scalar.

Example 2.2

Scalar Multiplication of a Matrix

If $B = \alpha A$, where α is a scalar, then $b_{ij} = \alpha a_{ij}$; $\forall\, i$ and j.

$$5 \begin{bmatrix} 3 & 2 \\ 0 & -1 \end{bmatrix} = \begin{bmatrix} 5 \times 3 & 5 \times 2 \\ 5 \times 0 & 5 \times (-1) \end{bmatrix} = \begin{bmatrix} 15 & 10 \\ 0 & -5 \end{bmatrix} \tag{2.18}$$

◀

Addition and Subtraction: We can add or subtract two matrices of the same dimensions by individually adding or subtracting the elements.

If $C = A + B$ then $c_{ij} = a_{ij} + b_{ij}$.

Example 2.3

Matrix Addition

$$\begin{bmatrix} 4 & 3 \\ -1 & 5 \end{bmatrix} + \begin{bmatrix} 2 & 6 \\ 7 & -3 \end{bmatrix} = \begin{bmatrix} 4+2 & 3+6 \\ -1+7 & 5+(-3) \end{bmatrix} = \begin{bmatrix} 6 & 9 \\ 6 & 2 \end{bmatrix} \tag{2.19}$$

◀

Subtraction can be thought of as adding the negative. The following properties are applicable for matrix additions:

$$A + B = B + A \;\text{(Commutative Property of addition)}$$
$$A + (B + C) = (A + B) + C \;\text{(Associative Property)} \tag{2.20}$$

Multiplication of Two Matrices: The product C of two matrices A with dimension $m \times p$ and B with dimensions $p \times n$ yields an $m \times n$ matrix. This is given by the formula:

$$c_{ij} = \sum_{k=1}^{p} a_{ik} b_{kj} \tag{2.21}$$

Example 2.4

Matrix Multiplication

$$\text{Consider } \begin{bmatrix} 2 & 3 \\ 4 & 5 \end{bmatrix} \times \begin{bmatrix} 5 & 3 & 4 \\ 2 & 1 & 6 \end{bmatrix} \tag{2.22}$$

$$= \begin{bmatrix} 2 \times 5 + 3 \times 2 & 2 \times 3 + 3 \times 1 & 2 \times 4 + 3 \times 6 \\ 4 \times 5 + 5 \times 2 & 4 \times 3 + 5 \times 1 & 4 \times 4 + 5 \times 6 \end{bmatrix} \tag{2.23}$$

$$= \begin{bmatrix} 16 & 9 & 26 \\ 30 & 17 & 46 \end{bmatrix} \tag{2.24}$$

◀

It must be noted that the multiplication of two matrices is generally not commutative. This means $AB \neq BA$.

The following properties are applicable for matrix multiplication:

$$A(BC) = (AB)C \text{ (Associative Property)}$$
$$A(B+C) = AB + AC \text{ (Distributive Property)} \tag{2.25}$$

The Identity Matrix: The identity matrix I is a square matrix with 1's along the diagonal and 0's at other places.

$$I_3 = \begin{bmatrix} 1 & 0 & 0 \\ 0 & 1 & 0 \\ 0 & 0 & 1 \end{bmatrix} \tag{2.26}$$

The identity matrix satisfies the following equality:

$$AI = IA = A \tag{2.27}$$

The Transpose of a Matrix: A^T denotes the transpose of a matrix A and it is obtained by swapping the rows with columns, that is, $a_{ij}^T = a_{ji}$.

Example 2.5

Transpose of a Matrix

The following example is a matrix composed of complex numbers:

$$\text{Let } A = \begin{bmatrix} 1+3i & 5-2i \\ 6-3i & -3+5i \\ 5+7i & 4+2i \end{bmatrix} \tag{2.28}$$

$$A^T = \begin{bmatrix} 1+3i & 6-3i & 5+7i \\ 5-2i & -3+5i & 4+2i \end{bmatrix}$$

◀

The transpose operation preserves the property:

$$(AB)^T = B^T A^T \tag{2.29}$$

The Trace (Tr) of a Square Matrix is defined as the sum of its diagonal elements. This is defined by the following equation:

$$Tr(A) = \sum_i a_{ii} \tag{2.30}$$

Example 2.6

The trace of the following matrix is:

$$Tr\left(\begin{bmatrix} 1 & 3 \\ 4 & 5 \end{bmatrix}\right) = 1 + 5 = 6 \tag{2.31}$$

◀

Exercise 2.1

Prove that **Tr(I) = n**

◀

Exercise 2.2

Prove that **Tr(AB) = Tr(BA)**

◀

The Determinant of a Matrix: The determinant of a matrix is a scalar value derived from the elements. The determinant of a matrix is considered the volume enclosed by the row vectors of the matrix. This property is applicable only for square matrices, and two encapsulating vertical bars are used to denote a determinant. The determinant of a 2×2 matrix is defined by the following equation.

$$\det(A) = |A| = \begin{vmatrix} a & b \\ c & d \end{vmatrix} = ad - bc \tag{2.32}$$

Calculating the determinant of a 3×3 matrix follows a similar method outlined by the following equation:

$$|A| = \begin{vmatrix} a & b & c \\ d & e & f \\ g & h & i \end{vmatrix} = a\begin{vmatrix} e & f \\ h & i \end{vmatrix} - b\begin{vmatrix} d & f \\ g & i \end{vmatrix} + c\begin{vmatrix} d & e \\ g & h \end{vmatrix} \tag{2.33}$$

Upon expanding this equation, we get:

$$|A| = aei - afh - bdi + bfg + cdh - ceg \tag{2.34}$$

A similar method can be used for matrices of a higher order. Determinants have the following properties, and these can be easily verified.

1. $\det(I) = 1$
2. $\det(A^T) = \det(A)$
3. $\det(AB) = \det(A) \times \det(B)$
4. $\det(A^{-1}) = \frac{1}{\det(A)}$, $\det(A) \neq 0$
5. $\det(cA) = c^n \det(A)$ where n is the dimension of the matrix.

The Adjugate of a Matrix: The adjugate of a matrix is the transpose of its cofactor. The cofactors of a matrix are obtained by taking out the intersecting row and the column of a given element, as shown below, and forming a new matrix. Note the alternating signs. The cofactor of a 3×3 matrix is calculated as follows:

$$\text{Let } A = \begin{bmatrix} a & b & c \\ d & e & f \\ g & h & i \end{bmatrix} \tag{2.35}$$

The first step is to calculate the cofactor of A:

$$C = \begin{bmatrix} +\begin{vmatrix} e & f \\ h & i \end{vmatrix} & -\begin{vmatrix} d & f \\ g & i \end{vmatrix} & +\begin{vmatrix} d & e \\ g & h \end{vmatrix} \\ -\begin{vmatrix} b & c \\ h & i \end{vmatrix} & +\begin{vmatrix} a & c \\ g & i \end{vmatrix} & -\begin{vmatrix} a & b \\ g & h \end{vmatrix} \\ +\begin{vmatrix} b & c \\ e & f \end{vmatrix} & -\begin{vmatrix} a & c \\ d & f \end{vmatrix} & +\begin{vmatrix} a & b \\ d & e \end{vmatrix} \end{bmatrix} \tag{2.36}$$

The next step is to calculate the transpose of the cofactor matrix.

$$C^T = \begin{bmatrix} +\begin{vmatrix} e & f \\ h & i \end{vmatrix} & -\begin{vmatrix} b & c \\ h & i \end{vmatrix} & +\begin{vmatrix} b & c \\ e & f \end{vmatrix} \\ -\begin{vmatrix} d & f \\ g & i \end{vmatrix} & +\begin{vmatrix} a & c \\ g & i \end{vmatrix} & -\begin{vmatrix} a & c \\ d & f \end{vmatrix} \\ +\begin{vmatrix} d & e \\ g & h \end{vmatrix} & -\begin{vmatrix} a & b \\ g & h \end{vmatrix} & +\begin{vmatrix} a & b \\ d & e \end{vmatrix} \end{bmatrix} \tag{2.37}$$

Exercise 2.3

Prove that **adj(I) = I**

The Inverse of a Matrix: The inverse of a square matrix A is denoted by A^{-1}, and it satisfies the following relation:

$$AA^{-1} = A^{-1}A = I \tag{2.38}$$

One way to find the inverse of a matrix is through its adjugate matrix. This is shown in the following equation:

$$A^{-1} = \frac{1}{|A|} \, \text{adj}(A) \tag{2.39}$$

Not all matrices can be inverted, however!

The Symmetric Matrix: A matrix A is symmetric if $A^T = A$, that is, $a_{ij} = a_{ji} \; \forall \, i$ and j.

The Self-Adjoint Matrix: A matrix A is self-adjoint if $A^* = A$, that is, $a_{ij} = a_{ji}^* \; \forall \, i$ and j.

Note that A^* is the **conjugate transpose** of A. We obtain this by taking the complex conjugate of each element in A and then transposing.

2.4.2 Permutation Matrices

A permutation matrix is a square matrix. It is obtained by repeated exchange of rows of an identity matrix of the same size. Each row and column of a permutation matrix contains one non-zero value, which is "1." A square matrix of size n can have $n!$ number of permutation matrices. For example, there are two 2×2 permutation matrices, six 3×3 matrices, and so on. The following is an example of a permutation matrix. It represents the matrix form of a CNOT quantum gate.

$$\text{CNOT} = \begin{bmatrix} 1 & 0 & 0 & 0 \\ 0 & 1 & 0 & 0 \\ 0 & 0 & 0 & 1 \\ 0 & 0 & 1 & 0 \end{bmatrix} \tag{2.40}$$

2.4.3 Hermitian Matrix

Now, we can learn about Hermitian matrices.

A Hermitian matrix is a square matrix composed of complex numbers, and it is equal to its conjugate transpose. A Hermitian matrix is written with a superscript of letter H or the dagger $\dagger$ symbol, and it is self-adjoint.

If A is a Hermitian matrix, then $A = A^\dagger$. The following equation denotes this equivalence:

$$a_{ij} = \bar{a}_{ji}; \forall i \text{ and } j. \tag{2.41}$$

The bar $^-$ on a denotes the complex conjugation, a superscript of $*$ is also used to denote the complex conjugation.

The following is a Hermitian matrix.

$$\begin{bmatrix} 2 & 3 - i2 & 6 + i5 \\ 3 + i2 & -1 & 4 + i \\ 6 - i5 & 4 - i & 5 \end{bmatrix} \tag{2.42}$$

Note that the diagonal elements are real, as they have to be self-conjugate.

In quantum mechanics, most operators are Hermitian. We shall return to this subject and examine the properties of Hermitian operators in later sections.

2.4.4 Unitary Matrices

Unitary matrices are complex square matrices whose Hermitian conjugate is also their inverse. Hence, they satisfy the following condition:

$$AA^\dagger = A^\dagger A = I \tag{2.43}$$

2.4.5 Pauli Matrices

The following set of three matrices with dimensions 2×2 are called Pauli matrices. They are used to represent spin angular momentum, which we shall look in detail in Chap. 3. In quantum computing, we use Pauli matrices to set the rotational parameters for qubits and measure quantum errors.

$$\sigma_1 = \sigma_x = X = \begin{bmatrix} 0 & 1 \\ 1 & 0 \end{bmatrix} \tag{2.44}$$

$$\sigma_2 = \sigma_y = Y = \begin{bmatrix} 0 & -i \\ i & 0 \end{bmatrix} \tag{2.45}$$

$$\sigma_3 = \sigma_z = Z = \begin{bmatrix} 1 & 0 \\ 0 & -1 \end{bmatrix} \tag{2.46}$$

Pauli matrices are traceless, Hermitian, and unitary.

Exercise 2.4

Prove that the Pauli matrices are traceless, Hermitian, and unitary.

◀

Exercise 2.5

Prove that the Pauli matrices satisfy the following equation.

$$X^2 + Y^2 + Z^2 = I \tag{2.47}$$

◀

We shall revisit Pauli matrices when we discuss spin angular momentum, and the gate model of quantum computing.

We now have sufficient background to learn about vector spaces, and start using Dirac's bra-ket notation.

2.5 Linear Vector Spaces

A linear vector space $\mathbb{V}$ is a collection of vectors that meets the following conditions:

1. There exists a vector addition operation $+$ such that, $\forall |X\rangle, |Y\rangle \in \mathbb{V}$
   (a) $|X\rangle + |Y\rangle = |Y\rangle + |X\rangle \in \mathbb{V}$ (commutative and closure property)
   (b) $|X\rangle + (|Y\rangle + |Z\rangle) = (|X\rangle + |Y\rangle) + |Z\rangle$ (associative property)
   (c) There exists a *null vector* $|0\rangle$ such that $|X\rangle + |0\rangle = |X\rangle$
   (d) For every vector $|X\rangle$, there exists an additive inverse $|-X\rangle$ such that $|X\rangle + |-X\rangle = |0\rangle$. This can be written as $|X\rangle - |X\rangle = |0\rangle$.
2. There exists a multiplication operation with scalars such as $a, b, c, \ldots$ such that
   (a) $a(|X\rangle + |Y\rangle) = a|X\rangle + a|Y\rangle$ (distributive property over vectors)

(b) $a(b\,|X\rangle) = ab\,|X\rangle$ (associative property)
(c) $(a+b)\,|X\rangle = a\,|X\rangle + b\,|X\rangle$ (distributive property over scalars)
(d)   There exists a multiplicative identity $|1\rangle$ such that $|1\rangle\,|X\rangle = |X\rangle$
(e) $0\,|X\rangle = |0\rangle$

In these definitions, we have used the Dirac's ket notation $|\quad\rangle$ denoted by the vertical bar, and the right facing angular bracket. The ket represents a column vector, and we shall learn this a bit more in subsequent sections.

The scalars $a, b, c, \ldots$ belong to the field $\mathcal{F}$ over which the vector is defined.

If the field is composed of all real numbers, then the vector space is a Real Vector Space $\mathbb{R}$. If it is composed of complex numbers, then it is a Complex Vector Space $\mathbb{C}$.

2.5.1   Linear Independence

A finite set of vectors $\{\,|v_1\rangle,\ |v_2\rangle,\ |v_3\rangle,\ \ldots,\ |v_n\rangle\,\}$ are said to be **linearly dependent**, if there exists a set of corresponding scalars $\{a_1 a_2, a_3, \ldots, a_n\}$ with at least one of them having a non-zero value, such that:

$$\sum_{i=1}^{n} a_i |v_i\rangle = |0\rangle \tag{2.48}$$

Otherwise, the vectors are said to be **linearly independent**.

If, for example, a_1 is non-zero, and if we can write,

$$|v_1\rangle = \frac{-a_2}{a_1}|v_2\rangle + \frac{-a_3}{a_1}|v_3\rangle + \frac{-a_4}{a_1}|v_4\rangle + \ldots + \frac{-a_n}{a_1}|v_n\rangle \tag{2.49}$$

The vector $|v_1\rangle$ is said to be in a linear combination of other vectors. This is an important property we shall use when we describe the Quantum Superposition Principle.

2.5.2   The Dimensionality of the Vector Space

A vector space of n linearly independent vectors is said to have a dimension of n. It is written as $\mathbb{V}^n$. More specifically, $\mathbb{R}^n$ for real vector spaces and $\mathbb{C}^n$ for complex vector spaces.

It must be noted that a vector $|V\rangle$ in an n-dimensional vector space can be written as a linear combination of n linearly independent vectors $\{\,|v_1\rangle,\ |v_2\rangle,\ |v_3\rangle,\ \ldots,\ |v_n\rangle\,\}$.

2.5.2.1 Basis
A set of n linearly independent vectors in an n-dimensional vector space is called a basis.

$$|V\rangle = \sum_{i=1}^{n} a_i |v_i\rangle \tag{2.50}$$

Here, the vectors $|v_i\rangle$ form a **basis**, labeled by the integers $i = 1, 2, 3, \ldots$. The coefficients a_i are called the **components of the state vector** $|V\rangle$ on the basis $|v_i\rangle$. The components are usually complex numbers.

2.5.2.2 Standard Basis

Unit vectors $\hat{i}, \hat{j}, \hat{k}$ along the cartesian coordinate axes x, y, z form a mutually orthogonal (meaning perpendicular) basis vectors in $\mathbb{R}^3$. These vectors are often represented as i, j, k dropping the hat above them and called **Standard Basis**. In a matrix form, they can be written as:

$$i = \begin{bmatrix} 1 \\ 0 \\ 0 \end{bmatrix}, j = \begin{bmatrix} 0 \\ 1 \\ 0 \end{bmatrix}, k = \begin{bmatrix} 0 \\ 0 \\ 1 \end{bmatrix} \tag{2.51}$$

2.5.2.3 Addition of Two Vectors

In a given vector space, if we can write—

$$|V_1\rangle = \sum_{i=1}^{n} a_i |v_i\rangle \text{ and} \tag{2.52}$$

$$|V_2\rangle = \sum_{i=1}^{n} b_i |v_i\rangle \text{ then} \tag{2.53}$$

$$|V_1\rangle + |V_2\rangle = \sum_{i=1}^{n} (a_i + b_i) |v_i\rangle \tag{2.54}$$

2.5.2.4 Multiplication by a Scalar

To multiply a vector by a scalar, we multiply its components by the scalar.

$$\alpha |V\rangle = \alpha \sum_{i=1}^{n} a_i |v_i\rangle = \sum_{i=1}^{n} \alpha\, a_i |v_i\rangle \tag{2.55}$$

2.5.3 Inner Product or the Scalar Product

The inner product or the scalar product of two vectors $|X\rangle$ and $|Y\rangle$ is denoted by the notation $\langle X| \, Y\rangle$. The inner product has the following properties.

1. $\langle X|Y\rangle = \langle Y|X\rangle^*$
2. Positive Definite Property: $\langle X|X\rangle \geq 0$, where the equality holds iff $|X\rangle = |0\rangle$.
3. $\langle \alpha X|Y\rangle = \alpha \langle X|Y\rangle$
4. Linearity: $\langle X + Y|Z\rangle = \langle X|Z\rangle + \langle Y|Z\rangle$
5. $\langle X|Y + Z\rangle = \langle X|Y\rangle + \langle X|Z\rangle$
6. $\langle X + Y|X + Y\rangle = \langle X|X\rangle + \langle X|Y\rangle + \langle Y|X\rangle + \langle Y|Y\rangle$

The inner product produces a scalar value, in most cases, a complex number, and it is equivalent to the dot product in vectors.

2.5.3.1 Orthogonal Vectors

Two vectors are said to be orthogonal or perpendicular if their inner product is zero.

2.5.3.2 Norm of the Vector

The following equation describes the length or the norm of the vector:

$$\sqrt{\langle V|V \rangle} = |V| \tag{2.56}$$

If this value is 1, then the vector is said to be a normalized vector.

2.5.3.3 Orthonormal Basis

If a set of basis vectors are normalized and if they are orthogonal to each other, they can be termed as an orthonormal basis.

Let us assume:

$$|V_a\rangle = \sum_{i=1}^{n} a_i |v_i\rangle \tag{2.57}$$

$$|V_b\rangle = \sum_{j=1}^{n} b_j |v_j\rangle \tag{2.58}$$

Then the inner product of the two vectors can be written as:

$$\langle V_a|V_b\rangle = \sum_i \sum_j a_i^* b_j \langle v_i|v_j\rangle \tag{2.59}$$

If the basis vectors $|v_i\rangle$ and $|v_j\rangle$ are orthonormal, then the following relation holds good:

$$\langle v_i|v_j\rangle = \begin{cases} 1, i = j \\ 0, i \neq j \end{cases} = \delta_{ij}. \tag{2.60}$$

The function δ_{ij} is known as the "**Kronecker**" delta function. This function gives a value of 1 if the variables i and j are the same. Otherwise, the function returns a value of 0.

By applying this in the previous Eq. (2.59), we get:

$$\langle V_a|V_b\rangle = \sum_i a_i^* b_i \tag{2.61}$$

Since both the vectors $|V_a\rangle$ and $|V_b\rangle$ are uniquely specified by their respective components in the respective basis, we can write $|V_a\rangle$ and $|V_b\rangle$ as column vectors.

$$|V_a\rangle = \begin{bmatrix} a_1 \\ a_2 \\ \vdots \\ a_n \end{bmatrix} \text{ and } |V_b\rangle = \begin{bmatrix} b_1 \\ b_2 \\ \vdots \\ b_n \end{bmatrix} \tag{2.62}$$

The inner product of the two vectors $|V_a\rangle$ and $|V_b\rangle$ can be written as the product of the transpose conjugate of $|V_a\rangle$ and $|V_b\rangle$.

$$\langle V_a | V_b \rangle = \begin{bmatrix} a_1^* & a_2^* & \cdots & a_n^* \end{bmatrix} \begin{bmatrix} b_1 \\ b_2 \\ \vdots \\ b_n \end{bmatrix} \tag{2.63}$$

This readily gives a scalar result.

If we assume two continuous functions, $a_i = \psi_a(x)$ and $b_i = \psi_b(x)$, we can replace the sum by an integral:

$$\langle V_a | V_b \rangle = \int \psi_a^*(x)\, \psi_b(x) dx \tag{2.64}$$

This means the basis vectors are all possible values of x, which is an important property in quantum mechanics.

Example 2.7

Assume $|V\rangle$ in a subspace $\mathbb{R}^2$ with the basis $\left\{ \begin{bmatrix} 3 \\ 2 \end{bmatrix}, \begin{bmatrix} 4 \\ -6 \end{bmatrix} \right\}$. Find an orthonormal basis of $|V\rangle$.

Solution:

$$\text{Let } v_1 = \begin{bmatrix} 3 \\ 2 \end{bmatrix} \text{ and } v_2 = \begin{bmatrix} 4 \\ -6 \end{bmatrix} \tag{2.65}$$

Let us first check if they are orthogonal by verifying the inner product.

$$\langle v_1 | v_2 \rangle = \begin{bmatrix} 3 & 2 \end{bmatrix} \begin{bmatrix} 4 \\ -6 \end{bmatrix} = 3 \times 4 + 2 \times -6 = 0 \tag{2.66}$$

The inner product is zero, hence they are orthogonal. The set $\{v_1, v_2\}$ is an orthogonal basis for $|V\rangle$. We can now normalize $\{v_1, v_2\}$ to get the orthonormal basis.

$$\begin{aligned} |v_1| &= \sqrt{3^2 + 2^2} = \sqrt{13} \\ |v_2| &= \sqrt{4 + (-6)^2} = \sqrt{52} = \sqrt{4 \times 13} = 2\sqrt{13} \end{aligned} \tag{2.67}$$

The orthonormal basis is:

$$\left\{ \frac{1}{\sqrt{13}} \begin{bmatrix} 3 \\ 2 \end{bmatrix}, \frac{1}{\sqrt{13}} \begin{bmatrix} 2 \\ -3 \end{bmatrix} \right\} \tag{2.68}$$

◀

Exercise 2.6

Prove that the vectors of the standard basis are orthogonal.

◀

2.6 Using Dirac's Bra-ket Notation

In the preceding sections, we have seen the process of writing an abstract vector $|V\rangle$ as a column vector in a basis $|v_i\rangle$. This we can call as ket $|V\rangle$, and we can write it as:

$$|V\rangle = \begin{bmatrix} a_1 \\ a_2 \\ \vdots \\ a_n \end{bmatrix} \tag{2.69}$$

We have also seen how to take the adjoint (transpose conjugate) of $|V\rangle$ and write it as a row vector. We can call the conjugate transpose vector as bra $\langle V|$ on the basis $\langle v_i|$. We can write the bra $\langle V|$ as:

$$\langle V| = [a_1^* \quad a_2^* \quad \cdots \quad a_n^*] \tag{2.70}$$

Thus, for every ket vector, there is a corresponding bra vector and vice-versa. Together, the bra and ket vectors form dual vector spaces.

In a given basis, we can write this linear conversion from the ket to the bra and vice-versa as:

$$|V\rangle = \begin{bmatrix} a_1 \\ a_2 \\ \vdots \\ a_n \end{bmatrix} \leftrightarrow [a_1^* \quad a_2^* \quad \cdots \quad a_n^*] = \langle V| \tag{2.71}$$

The basis ket $|v_i\rangle$ can be written as a column matrix with the ith component being "1" and the rest "0." Similarly, the basis bra $\langle v_i|$ can be written as a row matrix with the ith component being "1" and the rest "0."

Hence, the ket $|V\rangle$ can be written as:

$$|V\rangle = \sum_{i=1}^{n} a_i |v_i\rangle \tag{2.72}$$

$$|V\rangle = a_1 \begin{bmatrix} 1 \\ 0 \\ \vdots \\ 0 \end{bmatrix} + a_2 \begin{bmatrix} 0 \\ 1 \\ \vdots \\ 0 \end{bmatrix} + \cdots + a_n \begin{bmatrix} 0 \\ 0 \\ \vdots \\ 1 \end{bmatrix} \tag{2.73}$$

With this introduction about bra and ket vectors, we can now move on to study operators and their properties.

2.6.1   Operators

An operator is a rule that transforms a vector $|V\rangle$ into another vector $|V'\rangle$. For simplicity, in this book, we shall assume that the transformation happens in the same vector space. We can write this transformation as follows:

$$\widehat{O}|V\rangle = |V'\rangle \tag{2.74}$$

$$\langle V|\widehat{O} = \langle V'| \tag{2.75}$$

We use the cap ($\hat{}$) accent to denote an operator.

Examples of operators are the momentum operator $-i\hbar\frac{d}{dx}$, the differential operator $\frac{d}{dx}$, square root, and log.

2.6.2 Properties of Operators

2.6.2.1 Addition and Subtraction

Two operators $\hat{O}$ and $\hat{P}$ can be added or subtracted. This is defined by the following set of equations.

$$\left(\hat{O}+\hat{P}\right)|V\rangle = \hat{O}|V\rangle + \hat{P}|V\rangle \tag{2.76}$$

$$\left(\hat{O}-\hat{P}\right)|V\rangle = \hat{O}|V\rangle - \hat{P}|V\rangle \tag{2.77}$$

2.6.2.2 Product of Two Operators

The product of two operators $\hat{O}$ and $\hat{P}$ is defined by:

$$\hat{O}\hat{P}|V\rangle = \hat{O}\left(\hat{P}|V\rangle\right) \tag{2.78}$$

2.6.2.3 Identity Operator

The identity operator $\hat{I}$ does not perform any transformation.

$$\hat{I}|V\rangle = |V\rangle \tag{2.79}$$

2.6.2.4 Associative Law

Operators obey the associative law.

$$\hat{O}\hat{P}\hat{Q}|V\rangle = \hat{O}\left(\hat{P}\hat{Q}|V\rangle\right) \tag{2.80}$$

2.6.2.5 Commutative Property

In general, $\hat{O}\hat{P} \neq \hat{P}\hat{O}$.

The commutator of $\hat{O}$ and $\hat{P}$ is defined as:

$$\left[\hat{O},\hat{P}\right] \equiv \hat{O}\hat{P} - \hat{P}\hat{O} \tag{2.81}$$

If $\hat{O}$ and $\hat{P}$ commute, then $\left[\hat{O},\hat{P}\right] = 0$.

2.6.2.6 Anticommutator

The anticommutator of two operators is:

$$\left\{\hat{O},\hat{P}\right\} = \hat{O}\hat{P} + \hat{P}\hat{O} \tag{2.82}$$

If $\hat{O}$ and $\hat{P}$ anticommute then $\hat{O}\hat{P} = -\hat{P}\hat{O}$.

Table 2.1 Commutators and anticommutators of the Pauli matrices

Commutators	Anticommutators
$[\sigma_x, \sigma_y] = 2i\sigma_z$	$\{\sigma_x, \sigma_y\} = 0$
$[\sigma_y, \sigma_z] = 2i\sigma_x$	$\{\sigma_y, \sigma_z\} = 0$
$[\sigma_z, \sigma_x] = 2i\sigma_y$	$\{\sigma_z, \sigma_x\} = 0$
$[\sigma_x, \sigma_x] = [\sigma_y, \sigma_y] = [\sigma_z, \sigma_z] = 0$	$\{\sigma_x, \sigma_x\} = \{\sigma_y, \sigma_y\} = \{\sigma_z, \sigma_z\} = 2I$

Then $\left\{\widehat{O}, \widehat{P}\right\} = 0$

Exercise 2.7

Prove the Table 2.1 of commutators and anticommutators of the Pauli matrices.

◀

Example 2.8

Prove that any matrix that commutes with all the three Pauli matrices is a multiple of the Identity matrix.

Solution:

Any 2×2 matrix can be written as a linear combination of the Pauli matrices.

$$M = \begin{bmatrix} a & b \\ c & d \end{bmatrix}$$
$$= \frac{1}{2}\left[(a+d)I + (b+c)\sigma_x + (b-c)\sigma_y + (a-d)\sigma_z\right] \tag{2.83}$$

Recall the following facts.

- The Identity matrix I commutes with all matrices, including the three Pauli matrices.
- From Table 2.1, Pauli matrices do not commute among themselves.

Since the Pauli matrices do not commute among themselves, M cannot have any component from the Pauli matrices. Hence, $(b+c) = (b-c) = (a-d) = 0$. This means $b = c = 0$ and $a = d$.

Hence from Eq. (2.83), we can say that any matrix that commutes with all three Pauli matrices must be a multiple of the Identity matrix.

◀

2.6.2.7 Power

The nth power of an operator is obtained by successively applying the operator n times.

$$\widehat{O}^3|V\rangle = \widehat{O}\widehat{O}\widehat{O}|V\rangle \tag{2.84}$$

2.6.3 Linear Operators

The operators that we use in quantum mechanics are all linear operators, they can be written in a matrix form, and they meet the following criteria:

$$\widehat{O}a|V\rangle = a\widehat{O}|V\rangle \tag{2.85}$$

$$\widehat{O}(|V\rangle + |W\rangle) = \widehat{O}|V\rangle + \widehat{O}|W\rangle \tag{2.86}$$

Example 2.9

Prove that the momentum operator given in the following equation is linear.

$$\widehat{p} = -i\hbar \frac{d}{dx} = -i\hbar \nabla \tag{2.87}$$

Solution:

From Eq. (2.85), we can write the action of the momentum operator on $a|V\rangle$ as:

$$-i\hbar \frac{d}{dx}\, a|V\rangle = -i\hbar a \frac{d}{dx}|V\rangle \tag{2.88}$$

Using the sum law of differentiation, we can derive,

$$-i\hbar \frac{d}{dx}(|V\rangle + |W\rangle) = -i\hbar \frac{d}{dx}|V\rangle - i\hbar \frac{d}{dx}|W\rangle \tag{2.89}$$

This proves the Eq. (2.86). Hence, we can say that the momentum operator is linear.

◀

2.6.3.1 The Inverse of an Operator

In a vector space $\mathbb{V}^n$, we have seen that an operator $\widehat{O}$ transforms a unique vector $|V\rangle$ into $|V'\rangle$.

$$\widehat{O}|V\rangle = |V'\rangle \tag{2.90}$$

If there exists an operator $\widehat{P}$, such that

$$\widehat{P}|V'\rangle = |V\rangle \tag{2.91}$$

Then, we can say that $\widehat{P}$ is the inverse of $\widehat{O}$. Not all operators have an inverse, however!

$$\widehat{P} = \widehat{O}^{-1} \tag{2.92}$$

$$\widehat{O}\widehat{P} = \widehat{O}\widehat{O}^{-1} = I \tag{2.93}$$

Also, note that,

$$\left(\widehat{O}\widehat{P}\right)^{-1} = \widehat{O}^{-1}\widehat{P}^{-1} \tag{2.94}$$

2.6.4   Hermitian Operators

Let us now learn about the concept of Hermitian operators. Assume two arbitrary vectors $|V_a\rangle$ and $|V_b\rangle$ on a given basis.

It is a practice in quantum mechanics to write the bra and ket vectors using the subscript (or the label) only and moving forward, we shall adopt that convention.

Adopting the subscript (label) convention and following the expansion of arbitrary vectors from Eq. (2.50), we can write these two vectors:

$$|a\rangle = \sum_{j=1}^{n} a_j |j\rangle, |b\rangle = \sum_{j=1}^{n} b_j |j\rangle \tag{2.95}$$

If we assume $|b\rangle$ is obtained after the action of the operator $\widehat{O}$ on $|a\rangle$, we can write $|b\rangle$ as:

$$\widehat{O}|a\rangle = |b\rangle = \widehat{O} \sum_{j=1}^{n} a_j |j\rangle = \sum_{i=1}^{n} b_j |j\rangle \tag{2.96}$$

Multiplying both sides by $\langle i|$, we get

$$\sum_{j=1}^{n} \langle i|\widehat{O}|j\rangle a_j = \sum_{j=1}^{n} \langle i|j\rangle b_j = b_i \tag{2.97}$$

Due to the orthonormality of the basis vectors and j, $\langle i|\, j\rangle = \delta_{ij}$. Hence, we can write the matrix elements:

$$O_{ij} = \langle i|\widehat{O}|j\rangle \tag{2.98}$$

where O_{ij} are the elements of a matrix describing the linear operator $\widehat{O}$ in the basis $\{i, j\}$. Plugging this in Eq. (2.97), we get:

$$b_j = \sum_{j=1}^{n} O_{ij} a_j \tag{2.99}$$

This equation can be written in a matrix form as follows.

$$\begin{bmatrix} b_1 \\ b_2 \\ \vdots \\ b_n \end{bmatrix} = \begin{bmatrix} O_{11} & O_{12} & \cdots & O_{1n} \\ O_{21} & O_{22} & \cdots & O_{2n} \\ \vdots & \vdots & \cdots & \vdots \\ O_{n1} & O_{n2} & \cdots & O_{nn} \end{bmatrix} \begin{bmatrix} a_1 \\ a_2 \\ \vdots \\ a_n \end{bmatrix} \tag{2.100}$$

The next step is to apply the linear operator $\widehat{O}$ on the inner product of two vectors $|V_a\rangle$ and $|V_b\rangle$.

$$\langle V_a|\widehat{O}V_b\rangle = \sum_{i=1}^{n} \sum_{j=1}^{n} a_i^* O_{ij} b_j \tag{2.101}$$

Note that this summation can be converted into an integral:

$$\langle a|\widehat{O}|b\rangle = \int \psi_a^*(x)\, \widehat{O}\, \psi_b(x) dx \tag{2.102}$$

We shall return to the integral form a little later in the derivation. Note that we have now switched to using the subscript (label) convention. We can also write this equation in the matrix form as:

$$= \begin{bmatrix} a_1^* & a_2^* & \cdots & a_n^* \end{bmatrix} \begin{bmatrix} O_{11} & O_{12} & \cdots & O_{1n} \\ O_{21} & O_{22} & \cdots & O_{2n} \\ \vdots & \vdots & \cdots & \vdots \\ O_{n1} & O_{n2} & \cdots & O_{nn} \end{bmatrix} \begin{bmatrix} b_1 \\ b_2 \\ \vdots \\ b_n \end{bmatrix} \tag{2.103}$$

From the properties of the inner product of vectors:

$$\langle a|\hat{O}|b\rangle = \langle \hat{O}b|a\rangle^* = \langle b|\hat{O}^\dagger|a\rangle^* = \langle \hat{O}^\dagger a|b\rangle \tag{2.104}$$

Let us now apply this logic to basis vectors $|i\rangle$ and $|j\rangle$.

$$\langle \hat{O}i|j\rangle = \langle i|\hat{O}^\dagger j\rangle \tag{2.105}$$

$$\langle j|\hat{O}i\rangle^* = \langle i|\hat{O}^\dagger j\rangle \tag{2.106}$$

From (2.98), the elements of the basis vectors as:

$$O_{ji}^* = O_{ij}^\dagger \tag{2.107}$$

This representation is nothing but the conjugate transpose of the same matrix O. From our earlier discussion on Hermitian matrices, we can say that matrix O is a Hermitian matrix.

$$O = O^\dagger \tag{2.108}$$

In other words, the operator $\hat{O}$ is a **Hermitian operator**. In quantum mechanics, all observables are represented by Hermitian operators.

Returning to the integral form, it follows that integration over an infinite space is a requirement of a Hermitian operator.

$$\int \psi_j^*(x)\hat{O}\,\psi_i(x)dx = \int \psi_i(x)\left(\hat{O}\,\psi_j(x)\right)^* dx \tag{2.109}$$

Using Dirac's bra-ket notation:

$$\langle i|\hat{O}|j\rangle = \langle j|\hat{O}|i\rangle^*; \forall i \text{ and } j \tag{2.110}$$

In quantum mechanics, the position operator $\hat{x}$, the momentum operator $\hat{p} = -i\hbar\frac{d}{dx}$, and the Hamiltonian operator $\hat{H} = \frac{\hat{p}^2}{2m} + \hat{V}(x)$ are basic properties, and we shall prove that these operators are Hermitian in the next sections.

The Hamiltonian operator $\hat{H}$ represents the total energy of the system. The term $\frac{\hat{p}^2}{2m}$ describes the Potential Energy (P.E.) Operator $\hat{T}$ of the system and $\hat{V}(x)$ represents the Kinetic Energy (K.E.) Operator $\hat{V}$ of the system. In its simple form, the Hamiltonian represents the total energy (T.E.) of a single particle system.

An understanding of the Hamiltonian operator is essential for developing quantum algorithms. The quantum annealing algorithm and the variational quantum eigensolver describe the system state as a set of Hamiltonians.

Example 2.10

Prove that the position operator $\widehat{x}$ is Hermitian.

Solution:

Let us start with the following equation.

$$\int \psi_j^*(x)\widehat{x}\,\psi_i(x)dx \tag{2.111}$$

By rearranging the functions, we get

$$= \int \psi_i(x)\widehat{x}\,\psi_j^*(x)dx \tag{2.112}$$

Since $\widehat{x}^* = \widehat{x}$,

$$= \int \psi_i(x)\left(\widehat{x}\,\psi_j(x)\right)^* dx \tag{2.113}$$

QED.

◀

Example 2.11

Prove that the momentum operator $\widehat{p}$ is Hermitian.

Solution:

$$\int \psi_j^*(x)\widehat{p}\,\psi_i(x)dx = \int \psi_j^*(x) - i\hbar\frac{d}{dx}\psi_i(x)\,dx \tag{2.114}$$

$$= -i\hbar \int \psi_j^*(x)\,d\psi_i(x) \tag{2.115}$$

By integrating by parts (that is, using $\int u\,dv = uv - \int v\,du$)

$$= -i\hbar \left(\left[\psi_j^*(x)\psi_i(x)\right]_{x=-\infty}^{x=\infty} - \int \psi_i(x)\,d\psi_j^*(x) \right) \tag{2.116}$$

Since the wavefunction vanishes at infinity, the first term is zero. We can now write this as:

$$= i\hbar \int \psi_i(x)\,d\psi_j^*(x)) \tag{2.117}$$

Since $\left(-i\hbar\frac{d}{dx}\right)^* = i\hbar\frac{d}{dx}$, we can write as:

$$= \int \psi_i(x)\left(\widehat{p}\psi_j(x)\right)^* dx \tag{2.118}$$

QED.

◀

Example 2.12

Prove that the Hamiltonian operator $\widehat{H}$ is Hermitian.

Solution:

For a single particle system, the Hamiltonian operator can be written as:

$$\widehat{H} = \frac{\widehat{p}^2}{2m} + \widehat{V}(x) = \widehat{T} + \widehat{V} \tag{2.119}$$

Since $\widehat{V}$ is a function of x, it becomes Hermitian by default. So, we must prove that the Potential Energy Operator $\widehat{T}$ is Hermitian.

$$\int \psi_j^*(x)\widehat{T}\,\psi_i(x)dx = \frac{1}{2m}\int \psi_j^*(x)p^2\,\psi_i(x)dx \tag{2.120}$$

By rearranging the terms,

$$= \int \psi_i(x)\left(\widehat{T}\psi_j(x)\right)^* dx \tag{2.121}$$

QED.

◄

2.6.4.1 Anti-Hermitian Operators

An operator $\widehat{O}$ is said to be antihermitian, if $\widehat{O}^\dagger = -\,\widehat{O}$.

2.6.4.2 Unitary Operators

An operator $\widehat{U}$ is said to be unitary, if it satisfies the following relation.

$$\widehat{U}\,\widehat{U}^\dagger = \widehat{U}^\dagger\widehat{U} = I \tag{2.122}$$

Exercise 2.8

Prove that if H *is a Hermitian matrix, then there exists a unitary matrix* U, *such that the matrix* $U^\dagger HU$ *is diagonal.*

◄

Exercise 2.9

If A and B are two operators, prove that $(AB)^\dagger = B^\dagger A^\dagger$.

◄

2.6.5   Outer Product or the Projection Operators

Consider an operator $(|i\rangle\,\langle i|)$, which is made up of the outer product of a ket vector and a bra vector in an orthonormal basis. We call this outer product the **projection operator** P_i of the ket $|i\rangle$. Applying the projection operator to an arbitrary vector $|V\rangle$, we get:

$$P_i|V\rangle = |i\rangle\langle i|V\rangle = |i\rangle a_i \tag{2.123}$$

From this equation, we can understand that P_i acting on $|V\rangle$ multiplies the components a_i of $|V\rangle$ by a factor of $|i\rangle$. This is a sort of projection of the components of $|V\rangle$ along $|i\rangle$. Due to this reason, P_i is called a projection operator along $|i\rangle$.

Projection operators can also work on the bra vectors in a similar fashion:

$$\langle V|P_i = \langle V|i\rangle\langle i| = a_i^*\,\langle i| \tag{2.124}$$

Property 1: *Sum of all projection vectors is the identity operator.*

We can write $|V\rangle$ as an expansion of the projection operator in a given orthonormal basis in a given vector space.

$$|V\rangle = \sum_{i=1}^{n} |i\rangle\langle i| \, |V\rangle \tag{2.125}$$

This is not possible unless the sum of the ket-bra operator behaves like an identity operator.

$$\sum_{i=1}^{n} |i\rangle\langle i| = I = \sum_{i=1}^{n} P_i \tag{2.126}$$

This equation illustrates an important property that the sum of all projection operators is the identity operator itself.

This equation is called "**completeness relation**" for the chosen orthonormal basis. This equation is also called "**resolution of the identity.**"

Property 2: *Multiple application of the projection operator yields the same.*

This property is a condition for a Hermitian operator to be a projection operator. Let us examine this property:

$$P_i P_i = |i\rangle\langle i|i\rangle\langle i| = |i\rangle\langle i| = P_i \tag{2.127}$$

This means that once we project $|V\rangle$ along $|i\rangle$, further projections along $|i\rangle$ makes no difference.

Property 3: *Once projected along $|i\rangle$, further projections on another perpendicular direction $|j\rangle$ results in zero.*

To test this property, let us apply the projection P_i first and then P_j.

$$P_i P_j = |i\rangle\langle i|j\rangle\langle j| = \delta_{ij} P_j \tag{2.128}$$

Once we project out along $|i\rangle$, further projection along a perpendicular axis $|j\rangle$ results in a zero as there are no more components in that direction. Since $i \neq j$, this equation results in a zero.

We use this property in quantum communications to measure light polarized in one direction.

Exercise 2.10

If P *is a projection operator, prove that* $(1 - P)$ *is also a projection operator.*

◄

2.7   Expectation Values and Variances

In Chap. 1, we learned that the probability of finding a quantum system could be written as the square of the probability amplitude, as shown in the following equation.

$$P(x,t) = |\psi(x,t)|^2 \tag{2.129}$$

If we make a large number of independent measurements of the position (or displacement) of identical quantum systems, the results from each of these measurements may be quite different. The mean of these measurements can be described using the Probability theory as:

$$\langle x \rangle = \int\limits_{-\infty}^{+\infty} \hat{x}\, |\psi|^2 \, dx, \tag{2.130}$$

where $\langle x \rangle$ denotes the **expectation value** of the position x, and $\hat{x}$ denotes the position operator. The expectation value is the average measurement of x for the quantum state ψ. Like the average value of a fair 6-faced dice, 3.5, representing an impossible outcome, the expectation value $\langle x \rangle$ does not need to make physical sense.

Equation (2.130) can be rewritten:

$$\langle x \rangle = \int\limits_{-\infty}^{+\infty} \hat{x}\, \psi^*(x,t)\, \psi(x,t)\, dx \tag{2.131}$$

We assume that $\psi(x, t)$ is normalized. It means $\int_{-\infty}^{+\infty} \psi^*(x,t)\, \psi(x,t) dx = 1$.
Expecting the linearity of the operators, we can write this again as:

$$\langle x \rangle = \int\limits_{-\infty}^{+\infty} \psi^*(x,t)\, \hat{x}\, \psi(x,t)\, dx \tag{2.132}$$

This can also be written conveniently using the Dirac's bra-ket notation as:

$$\langle x \rangle = \langle \psi | \hat{x} | \psi \rangle \tag{2.133}$$

The various measurements of the position x are scatted around $\langle x \rangle$, as this is described by the same wavefunction. The degree of the scatter is termed as the **variance** of x, and it is given by the following relation:

$$\sigma_x^2 = \langle x^2 \rangle - \langle x \rangle^2 \tag{2.134}$$

The square root of this quantity σ_x is known as the **standard deviation** of x. The concept of expectation value applies to any observable in the quantum system.

Interested readers can explore quantum errors in qubits as variances in measurement and develop new logic for quantum error correction.

In a quantum system where we are deterministically able to measure the qubits, the standard deviation of the measurement $\sigma = 0$. Note that a quantum measurement collapses the wavefunction, and the qubit is localized.

2.8   Eigenstates, Eigenvalues, and Eigenfunctions

We have learned that when a linear operator $\hat{O}$ acts on a ket $|V\rangle$, it transforms that ket into a new ket $|V'\rangle$. It is needless to say that, unless the operator is an identity operator or a multiple of identity, the transformed ket will be in a different shape.

There may be certain kets, such that the action of an operator results in a linear multiple of the ket. Mathematically, this is written as the following equation:

$$\widehat{O}|\psi\rangle = a|\psi\rangle \tag{2.135}$$

Such kets are known as **eigenkets**. In the above example, $|\psi\rangle$ is an eigenket of the operator $\widehat{O}$. The wavefunction ψ is called **eigenstate**. This is because the ket in the Eq. (2.135) is independent of any observable. When we project this ket into an observable basis (say position or momentum), ψ becomes a function of that observable. In such cases, ψ is called an **eigenfunction**. The value a is the **eigenvalue**. Such equations, which produce scalar times a function, are known as **eigenvalue equations**. The set of eigenvalues of an operator are called its **spectrum**. Note that the eigenvalues of an operator need not be distinct. The eigenvectors corresponding to the distinct eigenvalues form a sub-space called the **eigenspace**.

Example 2.13

Time Independent Schrödinger Equation (TISE)
The general form of the time-independent Schrödinger equation is given by the following equation:

$$\widehat{H}\psi = E\psi, \tag{2.136}$$

where, $\widehat{H}$ is the Hamiltonian operator of the **eigenfunction** ψ. E is the **eigenvalue**. In this equation, the Hamiltonian operates upon the wavefunction to produce energy (which is a scalar) times the wavefunction. This is an example of an eigenvalue equation.

Supposing that $\widehat{O}$ be a Hermitian operator corresponding to a physical observable O, ψ_a be the eigenstate and a be the eigenvalue, we can then write the expectation value of O from Eq. (2.132):

$$\langle O \rangle = \int_{-\infty}^{+\infty} \psi_a^* \widehat{O}\, \psi_a \, dx = a \int_{-\infty}^{+\infty} \psi_a^* \psi_a \, dx = a \tag{2.137}$$

Similarly,

$$\langle O^2 \rangle = \int_{-\infty}^{+\infty} \psi_a^* \widehat{O}^2 \, \psi_a \, dx = a^2 \int_{-\infty}^{+\infty} \psi_a^* \psi_a \, dx = a^2 \tag{2.138}$$

Now, the variance is

$$\sigma = \langle O^2 \rangle - \langle O \rangle^2 = a^2 - a^2 = 0 \tag{2.139}$$

Since the variance is zero, we can deterministically measure the quantity O to the same value a. It also implies that the eigenstate ψ_a is uniquely attached to the eigenvalue a for the Hermitian operator $\widehat{O}$.

When two eigenstates ψ_a and ϕ_a of the same Hermitian operator $\widehat{O}$ correspond to the same eigenvalue a, then the eigenstates are said to be **degenerate**. Besides, the eigenstates ψ_a and ϕ_a of a Hermitian operator, corresponding to different eigenvalues a_1 and a_2 are orthogonal.

◀

Example 2.14

Prove that the eigenvalues of a Hermitian operator are all real.
Solution:
From (2.109) for Hermitian operators,

$$\int \psi_j^*(x)\widehat{O}\,\psi_i(x)dx = \int \psi_i(x)\left(\widehat{O}\,\psi_j(x)\right)^* dx \qquad (2.140)$$

If we assume $\psi_i(x) = \psi_j(x) = \psi$ and if we assume ψ is normalized, then by referring to (2.137) this equation reduces to

$$a = a^* \qquad (2.141)$$

which is not possible, unless a is real.
  QED.

◀

Let us consider two eigenfunctions $\psi_i(x)$ and $\psi_j(x)$ of Hermitian operator $\widehat{O}$ with corresponding eigenvalues a_i and a_j.

Since $\widehat{O}$ is Hermitian, the eigenvalues are all real. Besides, the eigenfunctions are orthogonal, if $a_i \neq a_j$.

$$\int \psi_i^*(x)\psi_j(x) = 0 \qquad (2.142)$$

Therefore, even if they are degenerate eigenfunctions, they are orthogonal.

The eigenfunctions form a complete set. Any function, even if it is not an eigenfunction, can be written as a linear combination of eigenfunctions.

$$\psi(x) = a_1\psi_1(x) + a_2\psi_2(x) + \ldots \qquad (2.143)$$

where the weights a_1, a_2 are complex numbers and normalized constants. $\psi_i(x)$ are properly normalized (that is, mutually orthogonal) eigenfunctions.

This property is generalized as the "**Quantum Superposition**" principle, which we shall explain in the next chapter. $\psi(x)$ can now be written as:

$$\psi(x) = \sum_{i=1}^{n} a_i\psi_i(x), \qquad (2.144)$$

where the sum of the squares of the constants is 1.

$$\sum_{i=1}^{n} |a_i|^2 = 1 \qquad (2.145)$$

Condition Number:
The condition number of a symmetric matrix is the ratio between its maximum and minimum of the eigenvalues (that is, $\frac{\max(\lambda_i)}{\min(\lambda_9)}$, where λ_i are the eigenvalues.)

2.9   Characteristic Polynomial

The characteristic polynomial of a square matrix is a polynomial, whose roots are the eigenvalues of the square matrix.

By definition, for a given square matrix A, we can say that λ is its eigenvalue, if there exists a non-zero vector v, called eigenvector such that:

$$Av = \lambda v \tag{2.146}$$

or, in another form:

$$(A - \lambda I)v = 0 \tag{2.147}$$

The eigenvalues of A are the roots of $det(A - \lambda I)$, which is a polynomial in λ.

Example 2.15

Find the eigenvalues and eigenvectors of the Pauli matrices.
Solution:
The Pauli matrices are given by the Eqs. (2.44), (2.45), and (2.46).
The first Pauli matrix is given by:

$$X = \begin{bmatrix} 0 & 1 \\ 1 & 0 \end{bmatrix} \tag{2.148}$$

From our discussions in the previous section, the determinant of the characteristic equation of X is:

$$|X - \lambda I| = \left| \begin{bmatrix} 0 & 1 \\ 1 & 0 \end{bmatrix} - \begin{bmatrix} \lambda & 0 \\ 0 & \lambda \end{bmatrix} \right| = \begin{vmatrix} -\lambda & 1 \\ 1 & -\lambda \end{vmatrix} \tag{2.149}$$

This determinant can be expanded to:

$$= \lambda^2 - 1 = (\lambda + 1)\,(\lambda - 1) \tag{2.150}$$

This implies that the eigenvalues of X are $+1$ and -1. To find the eigenvector, we substitute the values of λ, into the characteristic equation. For $\lambda = +1$, this equation is:

$$\begin{bmatrix} -1 & 1 \\ 1 & -1 \end{bmatrix} \begin{bmatrix} v_1 \\ v_2 \end{bmatrix} = \begin{bmatrix} 0 \\ 0 \end{bmatrix} \tag{2.151}$$

Expanding this equation, we get the following system of equations.

$$\begin{aligned} -v_1 + v_2 &= 0 \\ v_1 - v_2 &= 0 \end{aligned} \tag{2.152}$$

Solving this, we get $v_1 = 1$ and $v_2 = 1$. Hence, the eigenvector is $\begin{bmatrix} 1 \\ 1 \end{bmatrix}$. This eigenvector can be normalized. The norm of this eigenvector is:

$$\left\| \begin{matrix} 1 \\ 1 \end{matrix} \right\| = \sqrt{1^2 + 1^2} = \sqrt{2} \tag{2.153}$$

Table 2.2 Summary of eigenvalues and eigenvectors of Pauli matrices

#	Pauli matrices	Eigenvalue	Eigenvectors	Eigenvalue decomposition
1	$\sigma_1 = \sigma_x = X = \begin{bmatrix} 0 & 1 \\ 1 & 0 \end{bmatrix}$	$+1, -1$	$\frac{1}{\sqrt{2}}\begin{bmatrix} 1 \\ 1 \end{bmatrix}, \frac{1}{\sqrt{2}}\begin{bmatrix} 1 \\ -1 \end{bmatrix}$	$\lvert 0 \rangle \langle 1 \rvert + \lvert 1 \rangle \langle 0 \rvert$
2	$\sigma_2 = \sigma_y = Y = \begin{bmatrix} 0 & -i \\ i & 0 \end{bmatrix}$	$+1, -1$	$\frac{1}{\sqrt{2}}\begin{bmatrix} 1 \\ i \end{bmatrix}, \frac{1}{\sqrt{2}}\begin{bmatrix} 1 \\ -i \end{bmatrix}$	$i\lvert 1 \rangle \langle 0 \rvert - \lvert 0 \rangle \langle 1 \rvert$
3	$\sigma_3 = \sigma_z = Z = \begin{bmatrix} 1 & 0 \\ 0 & -1 \end{bmatrix}$	$+1, -1$	$\begin{bmatrix} 1 \\ 0 \end{bmatrix}, \begin{bmatrix} 0 \\ 1 \end{bmatrix}$	$\lvert 0 \rangle \langle 0 \rvert - \lvert 1 \rangle \langle 1 \rvert$

The normalization constant is $\frac{1}{\sqrt{2}}$. Hence, the eigenvector for $\lambda = +1$ is $\frac{1}{\sqrt{2}}\begin{bmatrix} 1 \\ 1 \end{bmatrix}$. Now let us try this for $\lambda = -1$.

$$\begin{bmatrix} 1 & 1 \\ 1 & 1 \end{bmatrix}\begin{bmatrix} v_1 \\ v_2 \end{bmatrix} = \begin{bmatrix} 0 \\ 0 \end{bmatrix} \tag{2.154}$$

Expanding this equation, we get the following system of equations.

$$\begin{aligned} v_1 + v_2 &= 0 \\ v_1 + v_2 &= 0 \end{aligned} \tag{2.155}$$

Solving this, we get $v_1 = 1$ and $v_2 = -1$. Hence, the eigenvector is $\begin{bmatrix} 1 \\ -1 \end{bmatrix}$. Like before, we can normalize this eigenvector. The norm of this eigenvector is:

$$\left\| \begin{bmatrix} 1 \\ -1 \end{bmatrix} \right\| = \sqrt{1^2 + (-1^2)} = \sqrt{2} \tag{2.156}$$

Here again, the normalization constant is $\frac{1}{\sqrt{2}}$. Hence, the eigenvector for $\lambda = -1$ is $\frac{1}{\sqrt{2}}\begin{bmatrix} 1 \\ -1 \end{bmatrix}$.

Likewise, we can calculate the eigenvalue and eigenvectors for the rest of the Pauli matrices. Table 2.2 summarizes these values.

2.10 Definite Symmetric Matrices

2.10.1 Positive Semidefinite

A square symmetric matrix $H \in \mathbb{R}^{n \times n}$ is **positive semidefinite** if $v^T H v \geq 0;\ \forall\ v \in \mathbb{R}^n$. It is **positive definite**, if $v^T H v > 0;\ \forall\ v \in \mathbb{R}^n$.

The eigenvalues of a positive semidefinite matrix are non-negative.

2.10.2 Negative Semidefinite

A square symmetric matrix $H \in \mathbb{R}^{n \times n}$ is **negative semidefinite** if $v^T H v \leq 0;\ \forall\ v \in \mathbb{R}^n$. It is **negative definite**, $v^T H v < 0;\ \forall\ v \in \mathbb{R}^n$.

2.11 Tensors

Tensors are mathematical objects that are used to describe physical properties, such as elasticity and electrical conductivity in physics. Tensors obey certain transformation rules about the physical properties they represent and are considered generalizations of scalars, vectors, and matrices. A scalar is a tensor of rank (or order) zero. A vector is a tensor of rank one. The rank of a tensor is the number of directions (and hence the dimensionality of the matrix) required to describe the object it represents. A tensor of rank n in an m dimensional space has n indices and m^n components. A scalar has zero indices, and a vector has one index.

2.11.1 Direct Sum

Recall our earlier discussions that a vector can be described in terms of basis (refer Eq. (2.50)) vectors. Let us consider two vectors $|V\rangle$ and $|W\rangle$ that are expanded in some orthonormal basis $\{|i\rangle\}$ and $\{|j\rangle\}$ respectively, as shown in the following equation.

$$|V\rangle = \sum_{i=1}^{n} v_i |i\rangle$$
$$|W\rangle = \sum_{j=1}^{m} w_j |j\rangle \tag{2.157}$$

If we define $n + m$ basis vectors, that is, $\{i = 1, 2, 3, \ldots, n, j = 1, 2, 3, \ldots, m\}$, then we can say that the vectors space spanned by the basis vectors is $V \oplus W$, which is the direct sum of $|V\rangle$ and $|W\rangle$. From Eq. (2.62), we can write these vectors as:

$$|V\rangle = \begin{bmatrix} v_1 \\ v_2 \\ \vdots \\ v_n \end{bmatrix} \text{ and } |W\rangle = \begin{bmatrix} w_1 \\ w_2 \\ \vdots \\ w_m \end{bmatrix} \tag{2.158}$$

These two vectors can be expanded as elements of the direct sum by filling the unused entries as zeros making a total of $n + m$ elements in each.

$$|V\rangle = \begin{bmatrix} v_1 \\ v_2 \\ \vdots \\ v_n \\ 0_1 \\ 0_2 \\ \vdots \\ 0_m \end{bmatrix} \text{ and } |W\rangle = \begin{bmatrix} 0_1 \\ 0_2 \\ \vdots \\ 0_n \\ w_1 \\ w_2 \\ \vdots \\ w_m \end{bmatrix} \tag{2.159}$$

Then, the direct sum of these two vectors is:

$$|V\rangle \oplus |W\rangle = \begin{bmatrix} v_1 \\ v_2 \\ \vdots \\ v_n \\ w_1 \\ w_2 \\ \vdots \\ w_m \end{bmatrix} \qquad (2.160)$$

Example 2.16

If $|V\rangle = \begin{bmatrix} 1 \\ 2 \end{bmatrix}$ *and* $|W\rangle = \begin{bmatrix} 3 \\ 4 \\ 5 \end{bmatrix}$, *find the direct sum* $|V\rangle \oplus |W\rangle$.

Solution:

$$|V\rangle \oplus |W\rangle = \begin{bmatrix} 1 \\ 2 \end{bmatrix} \oplus \begin{bmatrix} 3 \\ 4 \\ 5 \end{bmatrix} = \begin{bmatrix} 1 \\ 2 \\ 3 \\ 4 \\ 5 \end{bmatrix} \qquad (2.161)$$

◀

2.11.2 Tensor Product

The tensor product is also known as Kronecker product or the direct product. We use the tensor to describe the system state of two or more qubits (Chap. 5). For the two vectors we used in the previous section, let us define a new set of basis vectors nm, which is defined by the tensor product $i \otimes j$ of the basis vectors. This space is bilinear, meaning it is linear in both $|V\rangle$ and $|W\rangle$. We can write the tensor product of $|V\rangle$ and $|W\rangle$ as:

$$\begin{aligned} |V\rangle \otimes |W\rangle &= \sum_{i=1}^{n} v_i |i\rangle \otimes \sum_{j=1}^{m} w_j |j\rangle \\ &= \sum_{i=1}^{n} \sum_{j=1}^{m} v_i w_j (|i\rangle \otimes |j\rangle) \end{aligned} \qquad (2.162)$$

This equation can be interpreted as a vector $|V\rangle \otimes |W\rangle$ that spans a basis vector $|i\rangle \otimes |j\rangle$ whose coefficients are $v_i w_j$. The following equation describes this.

$$|V\rangle \otimes |W\rangle = \begin{bmatrix} v_1 \\ v_2 \\ \vdots \\ v_n \end{bmatrix} \otimes \begin{bmatrix} w_1 \\ w_2 \\ \vdots \\ w_m \end{bmatrix} = \begin{bmatrix} v_1 w_1 \\ v_1 w_2 \\ \vdots \\ v_1 w_m \\ v_2 w_1 \\ v_2 w_2 \\ \vdots \\ v_2 w_m \\ \vdots \\ v_n w_m \end{bmatrix} \qquad (2.163)$$

Example 2.17

If $|V\rangle = \begin{bmatrix} 1 \\ 2 \end{bmatrix}$ *and* $|W\rangle = \begin{bmatrix} 3 \\ 4 \\ 5 \end{bmatrix}$ *, find the tensor product* $|V\rangle \otimes |W\rangle$ *.*

Solution:

$$|V\rangle \otimes |W\rangle = \begin{bmatrix} 1 \\ 2 \end{bmatrix} \otimes \begin{bmatrix} 3 \\ 4 \\ 5 \end{bmatrix} = \begin{bmatrix} 1 \times 3 \\ 1 \times 4 \\ 1 \times 5 \\ 2 \times 3 \\ 2 \times 4 \\ 2 \times 5 \end{bmatrix} = \begin{bmatrix} 3 \\ 4 \\ 5 \\ 6 \\ 8 \\ 10 \end{bmatrix} \qquad (2.164)$$

◀

2.11.2.1 Tensor Product of 2 × 2 Matrices

The tensor product of 2×2 matrices is calculated using the following method. The same method can be extended to matrices of higher dimensions. We use this method when we apply two or more unitary operations in a sequence on a qubit in Chap. 5.

$$\begin{bmatrix} a & b \\ c & d \end{bmatrix} \otimes \begin{bmatrix} e & f \\ g & h \end{bmatrix} = \begin{bmatrix} ae & af & be & bf \\ ag & ah & bg & bh \\ ce & cf & de & dg \\ cg & ch & dg & dh \end{bmatrix} \qquad (2.165)$$

Exercise 2.11

Prove that $(\langle a| \otimes \langle b|)(|c\rangle \otimes |d\rangle) = \langle a|c\rangle \langle b|d\rangle$ *.*

◀

2.12 Statistics and Probability

In the following chapters, we shall learn that probability is built into nature. For example, when we make a quantum measurement—it involves several parameters—the number of degrees of freedom the system possesses, interaction with the environment and other qubits, decoherence of the qubits, the initial status of the measurement equipment, and so on. With the uncertainties associated with the variables involved, it is good to start with some probabilistic assumptions. In Eq. (2.133), the expectation value of the observable $\hat{x}$ can be interpreted as the probability that a measurement shall give a certain value. How exactly probability enters quantum mechanics is an intriguing study. In this introductory section, we shall refresh some of the basics of statistics and probability, which may be helpful with the following chapters.

2.12.1 Definitions

The following **frequency table** illustrates the number of COVID-19 cases in the first ten counties in Colorado, as on 8/1/2020 (Table 2.3).

This table is one way of collecting and analyzing data. We do not know the probability of survival of a certain case. However, we can calculate the percentage of people who survived based on data.

In statistics, the quantity that is studied is called a **random variable**. A possible outcome of the study is called **observable**. All such observables constitute the **population**. **Data** is a collection of observables. Typically, we study a sub-set of the population, called the **sample**. To ensure that the distribution is uniform, we perform a **random sample**. It is a good idea to illustrate data graphically. In the following chapters, we shall be using histograms to plot the outcome of quantum measurements we perform after each experiment. Figure 2.7 is a plot of Table 2.3.

Exercise 2.12

For the data contained in Table *2.3, identify the random variable and an observable. Also, form some random samples of the data. What can be inferred from this data?*

Table 2.3 Number of COVID-19 cases in some counties in Colorado

County	Adams	Alamosa	Arapahoe	Archuleta	Baca	Bent	Boulder	Broomfield	Chaffee	Cheyanne
Cases	6018	224	6932	33	14	5	1890	406	286	8

Fig. 2.7 Histogram of number of COVID-19 cases in some counties in Colorado

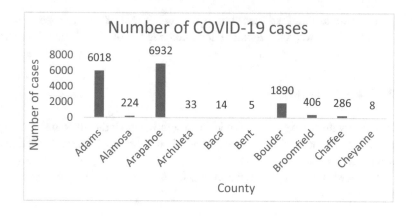

2.12.2 Measures of Location

The sample **mean** of the data is described as follows:

$$\bar{x} = \frac{x_1 + x_2 + \ldots + x_n}{n} = \frac{1}{n} \sum_{i=1}^{n} x_i, \tag{2.166}$$

where n is the total number of data elements x_i. As it appears, the mean is the same as the average of the data. If the data is tabulated as a frequency table, if the outcomes y_1, y_2, ... correspond to frequencies $f_1, f_2, \ldots$, the formula for calculating the mean is:

$$\bar{x} = \frac{y_1 f_1 + y_2 f_2 + \ldots + y_n f_n}{n} = \frac{1}{n} \sum_{i=1}^{n} y_i f_i \tag{2.167}$$

In a data set, the sample **median** is the middle of the data element, when the data is sorted in ascending order. The median is defined as follows:

$$\text{median} = \begin{cases} x_{\frac{n+1}{2}} & \text{if } n \text{ is odd} \\ \frac{1}{2}\left(x_{\frac{n}{2}} + x_{\frac{n}{2}+1}\right) & \text{if } n \text{ is even} \end{cases} \tag{2.168}$$

The **mode** is the value that occurs with the highest frequency.

$$\text{mode} = \{y_i | f_i = \max\{f_i\}\} \tag{2.169}$$

2.12.3 Measures of Dispersion

The **range** is the difference between the largest and smallest data elements. If the data is sorted in ascending order, we can write the range as:

$$\text{range} = x_n - x_1 \tag{2.170}$$

The sample **variance** is the average squared distance of the observation from the mean.

$$s^2 = \frac{1}{n-1} \sum_{i=1}^{n} (x_i - \bar{x})^2 \tag{2.171}$$

The sample **standard deviation** is the square root of the variance.

The ratio between the standard deviation and the mean is called the **coefficient of variation** $\frac{s}{\bar{x}}$.

2.12.4 Probability

Probability theory is useful to model situations in which the outcome of an experiment is random. The possible outcomes of the experiment are called the **sample space**, defined as the set S. A subset of the sample space is called an **event**.

Consider the sample space $S = \{a, b, c, d, e, f\}$ and two events $A = \{a, b, c, e\}$ and $B = \{b, d, e, f\}$.

The **union** of A and B is defined as $A \cup B = \{a, b, c, e\} \cup \{b, d, e, f\} = \{a, b, c, d, e, f\}$. In our example, $A \cup B = S$.

The i**ntersection** of A and B is defined as $A \cap B = \{a, b, c, e\} \cap \{b, d, e, f\} = \{b, e\}$.

The **complement** of A is defined as the events in the space S that did not occur. In our example, the complement of A is $\overline{A} = \{d, f\}$.

Two events A and B, are **disjoint** if they do not intersect, that is, $A \cap B = \{\ \}$.

Exercise 2.13

Prove that $A \cap \overline{A} = \{\ \}$ *and* $A \cup \overline{A} = S$.

◀

Let us define a function P called **probability measure**, which acts on the subsets of S. In an experiment, the following limit determines the probability that an event A can occur:

$$P(A) = \lim_{n \to \infty} \frac{r}{n}, \tag{2.172}$$

where r is the number of times the event A occurred in n repeats of the experiment. If the sample space is constructed in such a way that each of the outcomes has an equal probability of occurring, then we can rewrite the above equation:

$$P(A) = \frac{\text{Number of ways } A \text{ can occur}}{\text{Total number of outcomes}} \tag{2.173}$$

Examples of this type of problem are the dice, deck of cards. We can define the following properties.

- In an experiment, one of the outcomes occur, that is, $P(S) = 1$. Note that $P(\{\ \}) = 0$.
- The probability is positive definite. If $A \subseteq S$, then $P(A) \geq 0$.
- The probability of two or more experiments adds up, that is, $P(A \cup B) = P(A) \cup P(B)$.
- $P(A \cup B) = P(A) + P(B) - P(A \cap B)$. This property is also known as the addition rule.
- $P(\overline{A}) = 1 - P(A)$.
- It follows that if $A \subseteq B$, then $P(A) \leq P(B)$.
- If there are n experiments, and if $p_1, p_2, \ldots p_n$ are the likely outcomes in each of the experiments, then the likely possible outcomes of the n experiments are $\prod_i p_i$. This property is also known as the multiplication principle.
- The conditional probability of A if the probability of B is given is: $P(A|B) = \frac{P(A \cap B)}{P(B)}$, iff $P(B) > 0$.
- If $A \subseteq B$, then $P(A|B) = \frac{P(A)}{P(B)}$
- $P(A \cap B) = P(B) \times P(A|B) = P(A) \times P(B|A)$
- From the above statement, we can write $P(A|B) = \frac{P(A) \times P(B|A)}{P(B)}$, which is also called the **Bayes Theorem**.
- Two events A and B, are said to be **independent events** if $P(A \cap B) = P(A) \times P(B)$.

Example 2.18

A dice has six sides. Each time the dice is thrown, there are six possible outcomes. If we throw a dice two times, the possible outcomes are $6 \times 6 = 36$.

The probability of getting a one in both the throws is $\frac{1}{6} \times \frac{1}{6} = \frac{1}{36}$.

The probability of getting a one in one of the throws is $\frac{1}{6} + \frac{1}{6} - \frac{1}{36} = \frac{11}{36}$, using the addition rule.

◀

2.12.5 Permutation

Assume we have a set of n natural numbers $N = \{1, 2, \ldots, n\}$, if we have to select r unique numbers from this set, we have to repeat the experiment r times. The first attempt has n outcomes. The second experiment has $n - 1$ outcomes. The rth experiment has $n - r + 1$ outcomes. The number of ordered ways in which the r numbers can be selected from the set is called the **permutation**, and it can be derived from the multiplication principle.

$$
\begin{aligned}
P(n, r) = P_r^n &= (n) \times (n - 1) \times (n - 2) \times \ldots \times (n - r + 1) \\
&= \frac{(n) \times (n - 1) \times (n - 2) \times \ldots \times 3 \times 2 \times 1}{(n - r) \times (n - r - 1) \times \ldots \times 3 \times 2 \times 1} \\
&= \frac{n!}{(n - r)!}
\end{aligned}
\tag{2.174}
$$

Note that if the numbers need not be unique, then we can do this in n^r times.

Example 2.19

A password scheme requires four unique digits. Calculate how many ways the ten digits $0 \ldots 9$ can be arranged to create a password so that each digit is used only once.

$$
P_r^n = \frac{10!}{(10 - 4)!} = \frac{10 \times 9 \times 8 \times 7 \times 6 \times 5 \times 4 \times 3 \times 2 \times 1}{6 \times 5 \times 4 \times 3 \times 21 \times} = 5040
$$

◄

2.12.6 Combinations

Assume we have a set of n natural numbers $N = \{1, 2, \ldots, n\}$, and we must select r numbers from this set. If the order of the selection does not matter, the number of possible ways of selecting r numbers, called **combinations**, is defined as:

$$
C(n, r) = C_r^n = \binom{n}{r} = \frac{P_r^n}{P_r^r} = \frac{n!}{r!(n - r)!}
\tag{2.175}
$$

This number is also called the **binomial coefficient** or the **combinatorial number**.

Exercise 2.14

Prove that $C(n, r) = C(n, n - r)$

◄

2.13 Summary

This chapter was intended to be a math refresher. It started with the simple definition of scalars and then proceeded with complex numbers, vectors, and matrices. The concept of linear vector spaces and Dirac's bra-ket notation was introduced next. This was followed by an introduction to Hermitian operators and outer products. After this stage, this chapter briefed expectation values, eigenstates, eigenvalues, and tensor math. This chapter ended with a short none on statistics and probability.

Practice Problems

1. Find the real part of $(\cos 0.5 + i \sin 0.5)^{12}$. (Hints: Write this in the exponential form.)
2. Solve $x^2 - 32 = 0$. (Hints: write this in polar form and use de Moivre's theorem.)
3. Find a vector that points to the same direction of $\vec{x} = i + 3j + 9k$, but with a magnitude of 5. (Hints: Normalize the vector and scale it to a magnitude of 5.)
4. On a 2D surface, an ant starts from the origin and walks for 10 cm at a $45°$ angle, stops for a while, and then walks for another 10 cm at a $45°$ angle. What is the final coordinate of the ant?
5. If A and B are symmetric matrices of the same order, prove that AB is symmetric, iff A and B commute.
6. Given a square matrix A, if A^2 is invertible, prove that A is also invertible.
7. If A and B are square matrices of the same order, prove that $tr(AB) = tr(BA)$.
8. If $\{u, v, w\}$ are linearly independent vectors in a vector space $\mathbb{V}$, prove that $\{u + v, u + w, v + w\}$ is also linearly independent.
9. Evaluate whether the vectors $(1, 0, -1)$, $(2, 1, -1)$, and $(-2, 1, 4)$ form a basis in $\mathbb{R}^3$.
10. Find the eigenvalues of the matrix $\begin{bmatrix} 3 & 4 & 2 \\ 0 & 1 & 2 \\ 0 & 0 & 0 \end{bmatrix}$ and its eigenvectors.
11. Evaluate $\lim\limits_{x \to 4} \frac{2 - \sqrt{x}}{4 - x}$. (Hints: Multiply numerator and denominator by the conjugate of the numerator.)
12. Calculate $\frac{dy}{dx}$ if $y = (x + 1)^5 e^{-5x}$.
13. Calculate the integral $\int (x^2 + x + 1)^3 (4x + 2) \, dx$. (Hints: Assume $u = x^2 + x + 1$.)
14. Prove that $(|u\rangle \langle v|)^\dagger = |v\rangle \langle u|$.
15. Prove that $|u\rangle \langle v|$ is a linear operator.
16. Two systems A and B have angular momentum $L_A = 1$ and $L_B = 2$. Find the eigenstates of the total angular momentum.
17. Hermitian operators A, B, C meet the following conditions $[A, B] = 0$, $[B, C] = 0$, and $[A, B] \neq 0$. Prove that one of the eigenvalues of C must degenerate.
18. If $\langle u| = \sum_i b_i \langle i|$ and $|v\rangle = \sum_i a_i |i\rangle$, then prove that $\langle u|v\rangle = \sum_i a_i b_i$.
19. A computer program generates two-digit random numbers (that is, between 10 and 99.) What is the probability that a generated number is a multiple of five.
20. Assume that there are three coins. Two of them are regular coins, and one of them is a toy coin with both sides as heads. A coin is randomly picked and tossed to get a head. What is the probability that it is a regular coin? If all three coins are tossed together. What is the probability of getting all three heads?
21. In how many ways can we arrange n plates of r chocolate cookies and s butter cookies.

References

1. Mathematical Methods for Physicists - A Comprehensive Guide, 7/e, Arfken, Weber and Harris.
2. Algebra 2/e, Michael Artin.
3. A modern introduction to Probability and Statistics, Understanding Why and How, F.M. Dekking, C. Kraaikamp, H.P. Lopuhaa, L.E. Meester.

The Quantum Superposition Principle and Bloch Sphere Representation

3

"Curiouser and curiouser!" cried Alice (she was so much surprised, that for the moment she quite forgot how to speak good English);

— *Lewis Carroll, Alice in wonderland*

In Chap. 2, we took a short break to refresh our knowledge on linear algebra and start using Dirac's notation. In this chapter, we return to building up further on the principles of quantum mechanics. Those who have studied quantum mechanics in recent times can directly go to section 3.8 on Bloch sphere representation and read from there. Others will indeed find the discussions quite interesting. Griffith [1], Shankar [2], and Neilsen & Chuang [3] shall serve as references for further reading.

▶ **Learning Objectives**

- Hilbert space
- Schrödinger equation and the cat state
- Quantum superposition principle
- Postulates of quantum mechanics
- Quantum tunneling
- Stern and Gerlach experiment and the development of spin angular momentum
- Concept of qubits
- Bloch sphere representation
- Quantum measurements
- Qudits and qutrits

This chapter begins with the definition of Euclidian space, condition for completeness, and applies them to Hilbert space.

3.1 Euclidian Space

We are familiar with the two- or three-dimensional Euclidian space defined by the principles of Euclidian geometry. The **Euclidian space** is a vector space of real numbers $\mathbb{R}^n$. The Cartesian coordinate system in the two-dimensional x, y plane, the three-dimensional x, y, z space, the formula to determine the Euclidian distance between two points in these two coordinate systems, are the basic principles of this system. The square of the Euclidian distance is an essential parameter in statistical methods, and it is used in the notion of least squares in regression analysis.

© The Author(s), under exclusive license to Springer Nature Switzerland AG 2021
V. Kasirajan, *Fundamentals of Quantum Computing*, https://doi.org/10.1007/978-3-030-63689-0_3

3.2   Metric Space

A **metric space** is a set of objects with a metric, usually a function that measures the distance between two objects (also called points) in the set. The distance function d satisfies the following criteria:

- The distance between a point and itself is zero, $d(x_m, x_m) = 0$.
- The distance between two different points m and n is positive, $d(x_m, x_n) > 0$.
- The distance between x_m to x_n is the same as the distance between x_n and $x_{m.}$, that is, $d(x_m, x_n) = d(x_n, x_m)$.
- The distance between x_m to x_n is less than the distance through another point x_o.

The Euclidian space is a metric space. We denote a metric space by (X, d).

3.2.1   Completeness and Cauchy Sequences

A sequence can be defined as a list of elements in a particular order. A list of positive even integers is an infinite sequence of elements. The Fibonacci numbers (0, 1, 1, 2, 3, 5, 8, 13, 21, 34, ...) form a sequence. Each number in the Fibonacci sequence is the sum of two preceding numbers.

By definition, a sequence of real numbers $a_1, a_2, a_2, \ldots$ converges to a real number a, if, for every $\varepsilon > 0$, there exists $N \in \mathbb{N}$, such that for all $n \geq N$, $|a_n - a| < \varepsilon$. This can be written as a limit $\lim_{n\to\infty} a_n = a$.

For example, the following sequence:

$$\frac{1}{2}, \frac{2}{3}, \frac{3}{4}, \ldots, \frac{n}{n+1}, \ldots \tag{3.1}$$

converges to 1, as $\lim_{n\to\infty} \left(\frac{n}{n+1}\right) = 1$.

In a **Cauchy sequence**, the elements become arbitrarily close to each other as the sequence progresses. For every $\varepsilon > 0$, there exists $N \in \mathbb{N}$, such that for all $m, n \geq N$, $|a_m - a_n| < \varepsilon$. A sequence converges iff it is Cauchy.

By definition, in a metric space (X, d), if every Cauchy sequence converges to some element in X, then that metric space is said to be **complete**. Here, d the metric is a distance function.

Example 3.1

Prove that the sequence $\left\{\frac{1}{n}\right\}$ is a Cauchy sequence.

Solution:

Let us assume that there exists a positive real number $\varepsilon > 0$. Assume that there exists a positive integer $N \in \mathbb{N}$, such that $N > \frac{2}{\varepsilon}$. Consider two integers m and n such that $m, n \geq N$.

Using the triangle inequality, we can write:

$$|x_n - x_m| = \left|\frac{1}{n} - \frac{1}{m}\right| \leq \left|\frac{1}{n}\right| + \left|\frac{1}{m}\right| \tag{3.2}$$

Since $m, n \geq N > \frac{2}{\varepsilon}$, we have:

$$\frac{1}{n} \leq \frac{1}{N} < \frac{\varepsilon}{2} \text{ and } \frac{1}{m} \leq \frac{1}{N} < \frac{\varepsilon}{2} \tag{3.3}$$

This means,

$$\frac{1}{n} + \frac{1}{m} < \frac{\varepsilon}{2} + \frac{\varepsilon}{2} < \varepsilon \tag{3.4}$$

Hence the sequence converges, and we can say $\{\frac{1}{n}\}$ is a Cauchy sequence.

◀

3.3 Hilbert Space

The concept of **Hilbert space** is an extension of the mathematical methods of the two- or three-dimensional Euclidian space into finite or infinite dimensions. In quantum mechanics, we represent the state of a quantum system as a vector (called state vector) in the Hilbert space (called state space). The Hilbert space $\mathbb{H}$ is a complex vector space $\mathbb{C}^n$, with an inner product. The inner product provides for the orthogonality of the linear state space. It helps us visualize the geometry of the finite or infinite number of coordinate axes that are orthogonal to each other. The inner product gives rise to an associated norm defined by:

$$\text{for } x \in \mathbb{H}, |x| = \sqrt{\langle x|x \rangle} \tag{3.5}$$

The Hilbert space is complete for this norm. Here, the term complete means that every Cauchy sequence of elements in the Hilbert space $\mathbb{H}$ converges to an element in the same space $\mathbb{H}$, while the norm of differences between successive terms in the sequence approaches zero.

In the previous chapter, we found that the measurement of a physical observable scatters around an expectation value, and it is probabilistic. We also learned that the probability of finding a quantum system is the absolute squared of the wavefunction. Therefore, the wavefunction must be square-integrable (that is, a L^2 function) such that the integral of the absolute value squared of the wavefunction over an infinite space converges, meeting Cauchy's convergence criteria. The Hilbert space meets this condition. The Hilbert space is also linear, which lays the foundation for the quantum superposition principle. These are some reasons why we chose to represent quantum systems as vectors in the Hilbert space.

3.4 Schrödinger Equation

The Schrödinger equation is a fundamental equation of quantum mechanics. **Erwin Schrödinger**, an Austrian physicist, published this equation in 1925. He was awarded a Nobel Prize for his work on wave mechanics in 1933. Schrödinger studied the de Broglie waves and thought that if the electrons behaved like waves, it should be possible to write a wave equation for them, like Maxwell's equation for electromagnetic waves.

In classical mechanics, the total energy of a particle is the sum of its kinetic and potential energies. It is defined by the following equation.

$$E = \frac{p^2}{2m} + V(x), \tag{3.6}$$

where,

E—Energy of the particle
p—the momentum of the particle
m—the mass of the particle
V—Potential energy of the particle (assuming one dimension)

If we assume that the particle is confined to a certain space by a potential around it, the particle needs minimum energy to escape the potential. If the potential changes from one place to another, assuming that the particle's energy is a constant, only the momentum changes. If the momentum changes, so must the de Broglie wavelength. Hence, the momentum and the de Broglie wavelength are related:

$$\lambda = \frac{h}{p},$$
(3.7)

where,

λ—de Broglie wavelength
p—the momentum of the particle
h—Planck's constant

To represent this phenomenon, Schrödinger suggested a wavefunction $\psi(x, t)$, which varied with position and time. Replacing the momentum with a differential operator, Schrödinger created a partial differential equation which solves for the wavefunction $\psi(x, t)$.

$$i\hbar \frac{\partial \psi}{\partial t}(x,t) = (-\frac{\hbar^2}{2m}\frac{\partial^2}{\partial x^2} + V(x,t))\psi(x,t)$$
(3.8)

This is the time-dependent form of the Schrödinger equation for a particle of mass m, acted upon by a potential $V(x, t)$ while moving along the $x-$axis at non-relativistic speeds at time t. We have chosen one dimension (the $x-$axis) for convenience. This equation can be defined for multiple coordinate axes or basis vectors in the Hilbert space.

The time-dependent form of the Schrödinger equation is a first-order differential equation in time. It has the following property that applies to all spaces where the particle can be found: If we can define the wavefunction at an arbitrary starting time $t = 0$, then we can determine the wavefunction for any future time t for the particle.

The square of the wavefunction, that is, ψ^2 defines the probability of finding the particle at a given place and time.

The transient term (that is, the left-hand side) of this equation has an imaginary component; hence this is, a complex equation. It is convenient if we can find an interpretation of the wavefunction in real values. By seeking an analogy with the norm-squared of the complex numbers we saw in Chap. 2, Eq. (2.6), we can write:

$$|\psi(x,t)|^2 = \psi^*(x,t)\psi(x,t)$$
(3.9)

We call this the **Probability Density** $P(x, t)$, which we came across in Chap. 1, Eq. (1.20). This definition is one of the axioms of **Born rule**, a key postulate of quantum mechanics.

$$P(x,t) = |\psi(x,t)|^2 = \psi^*(x,t)\psi(x,t)$$
(3.10)

The probability density is always a positive number, and it defines the probability of finding the particle between a tiny space $[x, x + \Delta x]$ at time t. Δx is a small incremental value to x.

This property requires the wavefunction to be normalized, and the integral of the probability density overall space must be 1 since the particle should be somewhere in the space. This integral is expressed in the following equation:

$$\int_{-\infty}^{+\infty} |\psi(x,t)|^2 \, dx = \int_{-\infty}^{+\infty} \psi^*(x,t)\psi(x,t) \, dx = 1 \tag{3.11}$$

The Schrödinger equation is a linear equation for ψ. If we assume $\psi_1, \psi_2, \ldots$ are its solutions, then from our discussions on linear vector spaces in Chap. 2, a linear combination of these solutions is a solution as well:

$$\psi = a_1\psi_1 + a_2\psi_2 + \ldots, \tag{3.12}$$

where the coefficients $a_1, a_2, \ldots$ are complex numbers having a magnitude and a phase.

This property is known as the **Linear Superposition Principle**. We can write this using Dirac's bra-ket notation as:

$$|\psi\rangle = a_1|\psi_1\rangle + a_2|\psi_2\rangle + \ldots, \tag{3.13}$$

where $\{|\psi_1\rangle, |\psi_1\rangle \ldots\}$ are the basis vectors. From Eq. (2.144), we can rewrite this as a summation as follows:

$$|\psi\rangle = \sum_{i=1}^{n} a_i|\psi_i\rangle \tag{3.14}$$

The probability of projecting (or measuring) the state $|\psi\rangle$ into a basis state $|\psi_i\rangle$ is equal to the square of the absolute value of the corresponding probability amplitude, that is, $|a_i|^2$. The sum of the square of the absolute value of the probability amplitudes is 1.

$$\sum_{i=1}^{n} |a_i|^2 = 1 \tag{3.15}$$

The quantum state $|\psi\rangle$ is characterized by two n-dimensional vectors: the probability amplitudes $(a_1, a_2, a_3, \ldots, a_n)$, which are complex numbers and the probabilities, which is a vector of real numbers $(|a_1|^2, |a_2|^2, |a_3|^2 \ldots, |a_n|^2)$.

This means a quantum state can be described as a sum of two or more quantum states. In other words, we can say that a quantum system can always be in all possible states until a measurement is done. When a measurement is made, the wavefunction collapses into one of the probable states, and the quantum system is localized. This concept can be explained better using Schrödinger's cat, a thought experiment.

3.4.1 Schrödinger's Cat Thought Experiment (Do Not Try This at Home!)

The Schrödinger's cat thought experiment was created by Ervin Schrödinger in 1935 as part of the discussions around the EPR paradox (which we shall see in a later chapter). The experiment intends to illustrate the superposition principle in a larger system that is dependent on a quantum system in a superposition state.

In this thought experiment, a cat is put in a steel chamber. Inside the steel chamber, there is an experimental setup. The setup has a small amount of radioactive material with a 50% probability of decaying in the next 1 hour (Fig. 3.1).

Fig. 3.1 Schrödinger's
cat, thought experiment

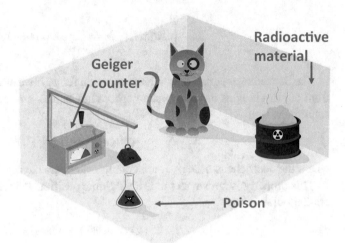

Fig. 3.2 The cat dies, if
the poison potion is broken.
(Credits: Anuj Kumar)

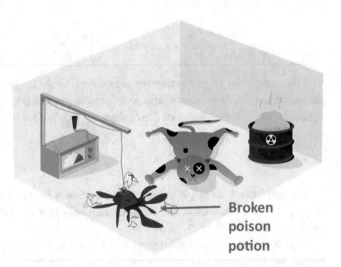

The Geiger counter inside the chamber can detect the radiation. If there is radiation in the chamber, the Geiger counter releases a weight block that breaks a poison potion. If the potion is broken, it releases a deadly hydrogen cyanide gas, killing the cat instantaneously (Fig. 3.2).

If the radioactive material does not decay, there is no radiation, and the poison potion is not broken. In this case, the cat lives. Now, we put the cat inside this experimental chamber and lock it.

One hour passes... Is the cat alive or dead?

If we do not open the chamber, according to Copenhagen Interpretation of Quantum Mechanics, the cat is both alive and dead at the same time—meaning the cat is in a superposition state (Fig. 3.3).

When we open the chamber, we make an observation; at this point, the wavefunction of the cat collapses into one of the states—dead or alive.

There are many interpretations of this thought experiment, and we leave it to our readers for further reading and imagination!

The quantum superposition principle is an important concept we apply in quantum computing. Before jumping into setting the qubits in superposition states, we should learn two important concepts in quantum mechanics—quantum tunneling and the experiment that established spin angular

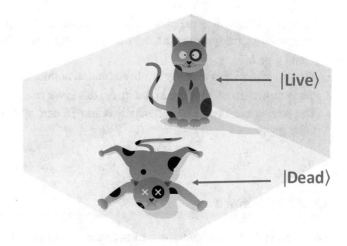

Fig. 3.3 Schrödinger's cat in a superposition state of being dead and alive at the same time

momentum and the randomness of quantum systems. These two concepts shall help us understand the quantum systems better, as we start representing the qubit states as Bloch spheres.

3.4.1.1 Cat State

The definition of the cat state draws from Schrödinger's thought experiment of the cat being "dead" and "alive" at the same time. Formally the cat state is described as follows:

$$|\text{cat}\rangle = \frac{1}{\sqrt{2}}(|\text{dead}\rangle + |\text{alive}\rangle) \tag{3.16}$$

The cat state is a state of a quantum system that is composed of two diametrically opposite states at the same time. For example, in Schrödinger's thought experiment, the state of the cat is dead and alive at the same time. These are two diametrically opposite states! This equation will become quite familiar to us in the forthcoming chapters, of course, we will not experiment with cats!

3.5   Postulates of Quantum Mechanics

Quantum mechanics is a well-established theory. Its mathematical background is based on certain fundamental principles. Understanding these principles gives us a fair idea of how quantum mechanics works. Since we have come across Hilbert space and Schrödinger equation, it is good to recall those postulates and re-emphasize our understanding of quantum mechanics at this juncture. There is no consensus on the list of basic principles needed to describe the functioning of quantum mechanics in the literature. However, the following postulates are widely accepted by various authors as the guiding principles of quantum mechanics.

3.5.1   Postulate 1

A quantum mechanical system is described by a complex wavefunction $\psi\left(\vec{r}, t\right)$, which contains all the information that can be known about that system. The wavefunction is a continuous, square-integrable, and single-valued function of position and time of the system.

- The probabilistic interpretation of the wavefunction is $\psi^*(\vec{r},t)\psi(\vec{r},t)dV$. According to this interpretation, the quantum mechanical system must be within the volume element dV at position $\vec{r}$ and time t.
- The wavefunction is spatially localized (that is, normalized). Since the wavefunction must find the quantum system somewhere in the space, this gives rise to its norm: $\int_{-\infty}^{\infty}\psi^*(\vec{r},t)\psi(\vec{r},t)dV = 1$.
- The state-space of the wavefunction is the Hilbert space. Hence the superposition principle holds good.

3.5.2 Postulate 2

Every observable in quantum mechanics corresponds to a linear Hermitian operator.

- The expectation value of an operator corresponding to an observable is a real value. Hence the operator must be Hermitian (Example 2.14) In other words, the measured quantities are real values.
- Every operator has got some states which do not change when the operator operates on them, that is, $\widehat{A}\,|\psi\rangle = a\,|\psi\rangle$. The states are just multiplied by a constant. Such states are called eigenstates. The constants a are the eigenvalues of the operator.

3.5.3 Postulate 3

In a quantum measurement of an observable, the only possible outcome is one of the eigenvalues of the corresponding operator.

We can write the system state as a linear combination of probability amplitudes, as shown in Eq. (2.50).

$$|\psi\rangle = \sum_{i=1}^{n} c_i|\psi_i\rangle \tag{3.17}$$

Here, the value of n can scale to infinity. Hence, we can write the eigenequation as a complete set of eigenvectors $\widehat{A}\,|\psi_i\rangle = a_i|\psi_i\rangle$. Even though we may have an infinite number of eigenvalues a_i, the measurement operation produces only one eigenvalue with a probability 1. The probability of which eigenvalue we measure is given by the absolute square of the corresponding coefficient, $|c_i|^2$.

When we measure $|\psi\rangle$, we get an eigenvalue a_i for a certain value of i. This causes the wavefunction to collapse to $|\psi_i\rangle$, and $|c_i|^2 = 1$. If a_i is degenerate (refer to earlier Sect. 2.8), the wavefunction collapses to one of the degenerate subspaces.

The probability of measuring a certain eigenvalue a_i can be defined as:

$$P(a_i) = |\langle\psi_i|\psi\rangle|^2, \tag{3.18}$$

where $|\psi_i\rangle$ is an eigenket of A, corresponding to the eigenvalue a_i to which the system collapses after the measurement.

3.5.4 Postulate 4

The average value or the expectation value of an observable is defined as $\langle A \rangle = \int_{i=0}^{n} \psi^*(\vec{r}, t) A \psi(\vec{r}, t) dV$.

The expectation value is the average of all measurements made on quantum systems prepared with the same state. Sometimes, we also use the standard deviation or the uncertainty of an observable. We discussed these terminologies in Sect. 2.7.

3.5.5 Postulate 5

The wavefunction of an isolated quantum system evolves in time following the time-dependent Schrödinger equation: $\hat{H}\psi(\vec{r}, t) = i\hbar \frac{\partial}{\partial t} \psi(\vec{r}, t)$, where $\hat{H}$ is the Hamiltonian operator of the system.

Once we are given an initial state of the system $\psi(\vec{r}, t_0)$, it is possible to derive the state of the system at time t, if no observations are made. Note that performing a measurement alters the state of the system, which cannot be described by the time-dependent Schrödinger equation. When we measure a physical quantity of the system, the state vector undergoes a probabilistic change, which can be observed in the measurement outcome.

Example 3.2

Consider the system evolution $|\psi'\rangle \to e^{i\theta}|\psi\rangle$, where the system picks up a global phase $e^{i\theta}$. The probability of measuring a certain a_i at the initial and final state of the system can be calculated as follows:

$$\text{Initial state } |\psi\rangle \quad P(a_i) = |\langle \psi_i | \psi \rangle|^2$$

$$\text{Final state } |\psi'\rangle \quad P(a_i) = |\langle \psi_i | e^{i\theta} \psi \rangle|^2 \tag{3.19}$$
$$= |e^{i\theta}|^2 |\langle \psi_i | \psi \rangle|^2$$
$$= |\langle \psi_i | \psi \rangle|^2$$

Thus, the states $|\psi\rangle$ and $|\psi'\rangle$ are equivalent. The global phase is not observable and insignificant. We shall discuss global and relative phases in detail in later sections and provide additional proof.

3.5.6 Symmetric and Antisymmetric Wavefunctions

Consider a system of two identical particles. Since the particles are indistinguishable, under the exchange of coordinates (which includes spin), the probability density should not be affected. For such systems, the probability density of the wavefunctions describing the two particles must be identical:

$$|\psi(r_1, r_2)|^2 = |\psi(r_2, r_1)|^2 \tag{3.20}$$

There are two scenarios where this is possible.

1. The wavefunction is symmetric, that is, $\psi(r_1, r_2) = \psi(r_2, r_1)$.
2. The wavefunction is antisymmetric, that is, $\psi(r_1, r_2) = -\psi(r_2, r_1)$.

Bosons have zero or integral intrinsic spin. The wavefunctions representing bosons are symmetric. Fermions (such as electrons) have half-integral spins, and their wavefunctions are antisymmetric under particle exchange. Hence, any number of bosons can occupy the same state, whereas two fermions cannot occupy the same state. Pauli's exclusion principle derives from the antisymmetric wavefunctions (Sect. 1.4.1).

3.6   Quantum Tunneling

 Note:
Readers who cannot follow the math can skip the derivation and continue to read the text.

Quantum tunneling is an exciting quantum mechanical phenomenon in which particles can penetrate through a potential barrier, even if the height of the potential barrier is higher than the total energy of the particle. This behavior is in total violation of classical physics. However, we observe this phenomenon every day in semiconductor devices and the sun. Quantum tunneling is used in a class of quantum computers known as "Quantum Annealers." Quantum annealing is a method used to solve combinatorial optimization problems, employing the principles of quantum tunneling. In this section, we shall explore quantum tunneling in detail as it is a useful application of the Schrödinger's wavefunction we learned in the previous section. We shall first look at a classical pedagogical example, much recited in classrooms and textbooks.

Assume that we kick a football with about 100 J of energy. As the ball rolls, it encounters a hill. Assume that it takes about 200 J of energy to reach the summit of the hill. Since the ball has energy lower than the energy needed to climb up to the summit, it has no way of climbing up the hill and appearing on the other side. No matter how many times we try this experiment, the ball will not make it to the other side of the hill. At some point (It is an excellent exercise to calculate the maximum height the ball can climb up the hill.), while climbing up the hill, the ball will lose its energy and be pulled back by gravity. Classically, the probability of the ball making it to the other side of the hill is 0%, while the probability of the ball being reflected by the hill is 100%.

In quantum mechanics, there is a possibility for the ball to appear on the other side of the hill. The reason is, the ball also has wave-like properties. Even though the ball can be localized, its wavefunction is defined all over the space. If the wave packet of the ball is larger than the width of the hill, the ball has a probability of appearing on the other side of the hill!

The equivalent of a hill in quantum mechanics is the potential barrier shown in Fig. 3.4. The function describing the potential energy of the barrier is described in the following equation.

$$V(x) = \begin{cases} 0, & \text{when } x < 0 \\ V_0, & \text{when } 0 \le x \le L \\ 0, & \text{when } x > L \end{cases} \qquad (3.21)$$

Fig. 3.4 The potential
energy barrier

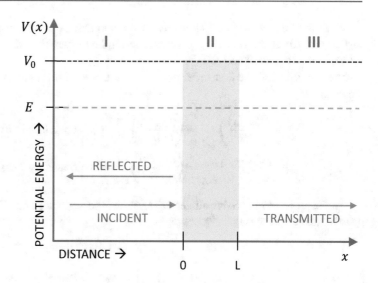

Figure 3.4 has three distinct regions. In the first region, the distance $x < 0$. This region is potential free, and the incoming particles (and their wave packets) move freely. There may be some particles that were reflected by the barrier in this region.

In the second region, some of the particles that were not reflected at the barrier boundary $x = 0$ propagate as a transmitted wave with a constant potential $V(x) = V_0$. If the width L and height V_0 of the barrier are finite, part of the wave packet penetrates the barrier boundary and propagates through the barrier. Inside the barrier, the wave packet is gradually attenuated. The wave packet *tunnels* through the second region and emerges as a transmitted wave packet at $x = L$.

The third region is a potential-free zone. In this region, the particle has tunneled through the potential barrier at $x > L$ and moves as a free particle.

The probability of a particle to tunnel through the barrier depends on three factors: the barrier width L, the height of the barrier V_0, and the energy E of the particle. We can calculate this by solving the time-independent form of the Schrödinger equation for the potential barrier described above.

$$-\frac{\hbar^2}{2m} \frac{d^2\psi(x)}{dx^2} + V(x)\psi(x) = E\psi(x); \text{ where } -\infty < x < +\infty \qquad (3.22)$$

We assume that the energy E of the incoming particle is less than the barrier height V_0. We solve the Schrödinger equation for $\psi(x)$ in such a way that the wavefunction is continuous, and the first derivative of the solution is continuous at the boundary between the regions. This solution can provide a probabilistic interpretation with $|\psi(x)|^2 = \psi^*(x)\psi(x)$ as the probability density.

For the three regions, we can write the Schrödinger equation as follows:

$$
\begin{array}{ll}
\text{REGION I} & -\frac{\hbar^2}{2m} \frac{d^2\psi_1(x)}{dx^2} = E\psi_1(x) \\
-\infty < x < 0 &
\end{array}
$$

$$
\begin{array}{ll}
\text{REGION II} & -\frac{\hbar^2}{2m} \frac{d^2\psi_2(x)}{dx^2} + V_0\psi_2(x) = E\psi_2(x) \\
0 \leq x \leq L &
\end{array} \qquad (3.23)
$$

$$
\begin{array}{ll}
\text{REGION III} & -\frac{\hbar^2}{2m} \frac{d^2\psi_3(x)}{dx^2} = E\psi_3(x) \\
L < x \leq +\infty &
\end{array}
$$

Note that the wavefunction needs to be continuous at the region boundaries. This means, $\psi_1(0) = \psi_2(0)$ at the boundary between the regions I and II, and $\psi_2(L) = \psi_3(L)$ at the boundary between the regions II and III.

We also said that the first-order derivative of the solution is continuous at the region between the boundaries:

$$
\begin{aligned}
\left(\frac{d\psi_1(x)}{dx}\right)_{x=0} &= \left(\frac{d\psi_2(x)}{dx}\right)_{x=0} \quad \text{at the boundary between I and II} \\
\left(\frac{d\psi_2(x)}{dx}\right)_{x=L} &= \left(\frac{d\psi_2(x)}{dx}\right)_{x=L} \quad \text{at the boundary between II and III}
\end{aligned}
\tag{3.24}
$$

The solutions to the Schrödinger equation for the regions I and III can easily be found, and they take the following form:

$$
\psi_1(x) = Ae^{ikx} + Be^{-ikx}
\tag{3.25}
$$

$$
\psi_3(x) = Fe^{ikx} + Ge^{-ikx},
$$

where the wavenumber $k = \frac{\sqrt{2mE}}{\hbar}$, $e^{\pm ikx} = \cos(kx) \pm \sin(kx)$ are the oscillations, and the constants $A, B, F,$ and G may be complex numbers. We can notice two sets of oscillations for each of the regions, if we examine the above equation. In region I, we see the incident and reflected waves. Therefore, the coefficients A and B must be representing these two waves. Hence, we can write the wavefunction explicitly in terms of these two waves. $\psi_{in}(x) = Ae^{ikx}$ and $\psi_{ref}(x) = Be^{-ikx}$. In region III, there is only one wave—the transmitted wave. Therefore, G must be 0. We can write the transmitted wave as $\psi_{trans}(x) = Fe^{ikx}$.

The amplitude of the incident wave can be calculated as follows:

$$
\begin{aligned}
|\psi_{in}(x)|^2 &= \psi_{in}^*(x)\psi_{in}(x) = \left(Ae^{ikx}\right)^* Ae^{ikx} = A^* e^{-ikx} Ae^{ikx} \\
&= A^*A = |A|^2
\end{aligned}
\tag{3.26}
$$

We can derive this for the reflected and transmitted wave as $|B|^2$ and $|F|^2$, respectively. The square of the amplitude is proportional to the intensity of the wave. So, if we want to know how much of the wave penetrates the barrier, we must find the ratio between the intensity of the transmitted wave and the intensity of the incident wave. This ratio is known as the **transmission probability** or the **tunneling probability**.

$$
T(L, E) = \frac{|\psi_{trans}(x)|^2}{|\psi_{in}(x)|^2} = \frac{|F|^2}{|A|^2} = \left|\frac{F}{A}\right|^2
\tag{3.27}
$$

Now, let us return to region II. We can rewrite the equation:

$$
\begin{aligned}
-\frac{\hbar^2}{2m}\frac{d^2\psi_2(x)}{dx^2} &= E\psi_2(x) - +V_0\psi_2(x) \\
\frac{d^2\psi_2(x)}{dx^2} &= \frac{2m}{\hbar^2}(V_0 - E)\,\psi_2(x) \\
\frac{d^2\psi_2(x)}{dx^2} &= \beta^2\psi_2(x); \quad \text{where } \beta^2 = \frac{2m}{\hbar^2}(V_0 - E)
\end{aligned}
\tag{3.28}
$$

Fig. 3.5 Diagram showing the quantum tunneling effect

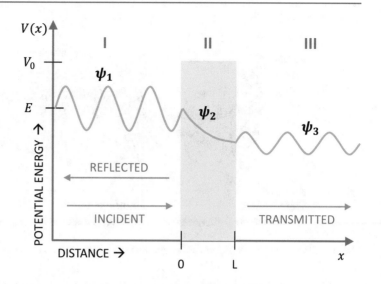

Since $V_0 > E$, β^2 is a positive real number. The solutions to this equation do not contain oscillatory terms. They contain exponentials that denote the gradual attenuation of $\psi_2(x)$:

$$\psi_2(x) = Ce^{-\beta x} + De^{\beta x} \tag{3.29}$$

This is illustrated in Fig. 3.5. With a little bit of math work, we can derive the approximate value of the tunneling probability.

$$T(L,E) = 16\frac{E}{V_0}\left(1 - \frac{E}{V_0}\right)e^{\beta L} \tag{3.30}$$

Note that the tunneling probability is dependent upon the barrier width L and the potential barrier V_0.

Figure 3.5 illustrates the three solutions to the Schrödinger equation for a particle encountering a potential barrier. In regions I and III, the particle moves freely. In region II, the particle moves with potential V_0, and it is attenuated exponentially.

Quantum tunneling was first examined by the German physicist **Friedrich Hund** in 1927. Later, the Russian–American theoretical physicist **George Gamow** explained the radioactive α-decay of atomic nuclei by quantum tunneling. In 1957, tunnel diode, which worked on the principles of quantum tunneling, was invented by **Leo Esaki**, **Yuriko Kurose**, and **Takashi Suzuki** at Tokyo Tsushin Kogyo (now Sony.)

3.7 Stern and Gerlach Experiment

In Chap. 1, we saw that the orbital angular momentum quantum number l takes only certain discrete values. This restriction means, the angular momentum vector can only have certain orientations called "**space quantization.**" In 1922, the German physicists **Otto Stern** and **Walther Gerlach** set out to prove whether the spatial orientation of the angular momentum is quantized. An orbiting electron must be producing a magnetic moment proportional to the orbital angular momentum. They

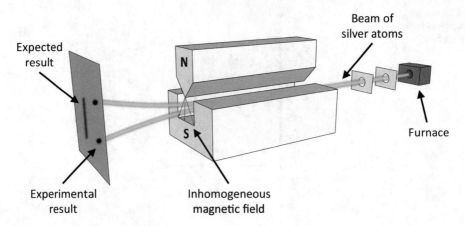

Fig. 3.6 Stern and Gerlach Experiment setup. (Image source: Wikipedia, Author: Tatoute.CC BY-SA 4.0)

thought that by measuring the magnetic moment of the atoms, they could prove whether spatial quantization existed.

In their experimental setup, a collimated beam of neutral silver atoms emanating from a furnace passes through an inhomogeneous magnetic field and hits a target screen (Fig. 3.6). Let the direction of the beam be along the y-axis and let the direction of the magnetic field be along the z-axis. The electrons in a circular orbit have an angular momentum $\vec{L} = m\omega r^2$. Since the electrons in orbit carry a charge, they produce a small current loop around them, which creates a dipolar magnetic field and a magnetic moment $\mu = -\frac{2}{2m}\vec{L}$.

The magnetic field $\vec{B}$ created by the experiment creates a torque $\mu \times \vec{B}$ around this dipole in such a way that the angular momentum $\vec{L}$ starts to precess along the direction of the magnetic field. Since the magnetic field is inhomogeneous along the direction z, the atoms experience a force defined by the following equation:

$$F_z = \nabla_z\left(\mu \cdot \vec{B}\right) = \mu_z\frac{\partial \vec{B}}{\partial z} \tag{3.31}$$

Due to the random thermal effects inside the oven, we can expect the magnetic moment vectors of the silver atoms to be randomly oriented in different angles in the space as they emerge from the oven. Once they pass through the inhomogeneous magnetic field, they are deflected along the $+z$ and $-z$ directions. On the screen, we expect a maximum distribution at an angle of zero deflection with a decreasing distribution on both sides. What was seen was different. The incoming beam was deflected into two distinct beams (Fig. 3.7). Hence the force that deflects the beam must have certain discrete values, and Stern and Gerlach thought that it meant "**Spatial Quantization**" is true.

As quantum mechanics advanced with the theories of Heisenberg and Schrödinger, it was found that the experiment was right, but the inference was wrong. Stern and Gerlach assumed that the angular momentum L was 1 for silver atoms. They thought that the magnetic moment of atoms with $L = 1$ had two components along the direction of the magnetic field $\mu_z = \pm\frac{eh}{4m}$, which caused the two deflected beams. However, the silver atoms are in $L = 0$ state. If the silver atoms are in $L = 1$ state, then the magnetic quantum number m must have $2L + 1 = 3$ states, as we have seen in Chap. 1. This must have caused the beam to split into three parts. However, we got two.

Fig. 3.7 Results of the Stern and Gerlach experiment

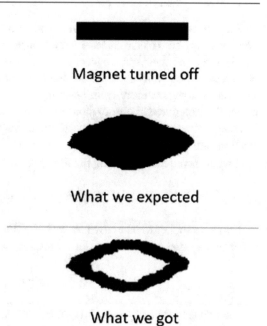

Fig. 3.8 The spin of fermions

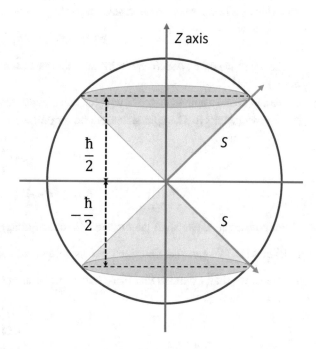

In 1925, the Dutch–American physicists **George Eugene Uhlenbeck** and **Samuel Abraham Goudsmit** proposed an alternate theory that solved this problem. They proposed that in addition to the orbital angular momentum L, the electrons have an intrinsic angular momentum or the **spin angular momentum** S with the value of $\frac{\hbar}{2\pi}$ (Fig. 3.8).

The total angular momentum of electrons is the sum of the orbital and spin angular momentum.

$$J = L + S \tag{3.32}$$

The electron configuration of silver is 4d10 5s1. So, it has one valence electron in the $n = 5$ shell. The remaining core electrons have a total angular momentum of zero. Their orbital angular momentum is zero, and their spins cancel each other out. Hence, they do not contribute to experimental observation. The valence electron does not have orbital angular momentum. Its total angular momentum comes entirely from its spin. The magnetic moment must be from this electron, which means the experimental observation is due to the spin. Hence, $L = 0$ for silver atoms.

Based on the experimental observation, the spin can have two different values, and we say that the electrons are spin-$\frac{1}{2}$ particles. The same is true for other fermions.

If we measure the spin along the direction of the z-axis, the two spin states are as follows:

$$S_z = \pm \frac{1}{2}\hbar \tag{3.33}$$

The same applies to other axes as well. We could have experimented with the magnetic field along either of the other axes and got similar results.

$$S_x = \pm \frac{1}{2}\hbar, \qquad S_y = \pm \frac{1}{2}\hbar, \tag{3.34}$$

The two spin states $S_z = +\frac{1}{2}\hbar$ (spin upward $\uparrow$) and $S_z = -\frac{1}{2}\hbar$ (spin downward $\downarrow$) of the electron can be described in terms of two basis vectors, $\{|\uparrow_z\rangle, |\downarrow_z\rangle\}$ in a two dimensional Hilbert space. We can write this as a linear superposition state using Dirac's bra-ket notation:

$$|\psi\rangle = a|\uparrow_z\rangle + b|\downarrow_z\rangle \tag{3.35}$$

The coefficients a and b are arbitrary complex numbers meeting the normalization criteria $|a|^2 + |b|^2 = 1$.

We can define a spin operator $\widehat{S}_z$ with the two eigenstates (spin upward and spin downward) having the eigenvalues $\pm \frac{1}{2}\hbar$. The spin states can be written as:

$$\widehat{S}_z|\uparrow_z\rangle = +\frac{1}{2}\hbar|\uparrow_z\rangle \tag{3.36}$$

$$\widehat{S}_z|\downarrow_z\rangle = -\frac{1}{2}\hbar|\downarrow_z\rangle \tag{3.37}$$

For convenience (we shall later relate this to the qubits), we can refer the basis states $|\uparrow_z\rangle$ and $|\downarrow_z\rangle$ as $|0\rangle$ and $|1\rangle$. We can also describe them as column vectors $\begin{bmatrix} 1 \\ 0 \end{bmatrix}$ and $\begin{bmatrix} 0 \\ 1 \end{bmatrix}$ respectively. In the next few steps, we shall establish that these kets form an orthonormal basis.

$$|\uparrow_z\rangle = |0\rangle = \begin{bmatrix} 1 \\ 0 \end{bmatrix} \tag{3.38}$$

$$|\downarrow_z\rangle = |1\rangle = \begin{bmatrix} 0 \\ 1 \end{bmatrix} \tag{3.39}$$

The basis bra vectors can be obtained by taking the conjugate transpose.

$$\langle\uparrow_z| = \langle 0| = [1 \quad 0] \text{ and } \langle\downarrow_z| = \langle 1| = [0 \quad 1] \tag{3.40}$$

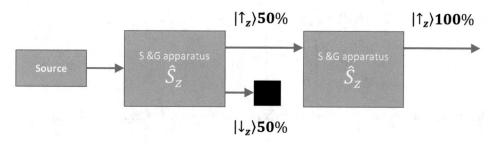

Fig. 3.9 Cascading of two S & G apparatuses

Applying this to Eq. (3.35), we get

$$|\psi\rangle = a|\uparrow_z\rangle + b|\downarrow_z\rangle = a|0\rangle + b|1\rangle \tag{3.41}$$

$$|\psi\rangle = a\begin{bmatrix} 1 \\ 0 \end{bmatrix} + b\begin{bmatrix} 0 \\ 1 \end{bmatrix} = \begin{bmatrix} a \\ b \end{bmatrix} \tag{3.42}$$

Let us consider an experimental setup, where we have cascaded two Stern & Gerlach experimental apparatuses, as shown in Fig. 3.9. Both the apparatuses used in this setup have the inhomogeneous magnetic field along the z-axis. The first apparatus produces $|\uparrow_z\rangle$ and $|\downarrow_z\rangle$ beams, each with a 50% probability. This is because the atoms are randomly oriented in the oven. We feed the $|\uparrow_z\rangle$ output to the second apparatus, leaving out the $|\downarrow_z\rangle$ beam.

We notice that the output of the second apparatus has only the $|\uparrow_z\rangle$ components. The $|\downarrow_z\rangle$ component is entirely absent. This observation is what we expect because we blocked out the $|\downarrow_z\rangle$ component from the first apparatus. This suggests that we have an orthonormal basis.

The output of the second apparatus consisted only of $|\uparrow_z\rangle$ components. This means the incoming beam did not have any $|\downarrow_z\rangle$ components. Besides, the second apparatus did not produce any other component other than $|\uparrow_z\rangle$.

From this observation, we can derive the condition for an orthonormal basis by relating to Chap. 2, Eq. (2.60).

$$\begin{array}{ll} \langle\uparrow_z|\downarrow_z\rangle = \langle 0|1\rangle = 0 & \langle\uparrow_z|\uparrow_z\rangle = \langle 0|0\rangle = 1 \\ \langle\downarrow_z|\uparrow_z\rangle = \langle 1|0\rangle = 0 & \langle\downarrow_z|\downarrow_z\rangle = \langle 1|1\rangle = 1 \end{array} \tag{3.43}$$

From Eq. (2.100), we can write the matrix form of the spin operator as:

$$\widehat{S}_z = \begin{bmatrix} \langle\uparrow_z|\widehat{S}_z|\uparrow_z\rangle & \langle\uparrow_z|\widehat{S}_z|\downarrow_z\rangle \\ \langle\downarrow_z|\widehat{S}_z|\uparrow_z\rangle & \langle\downarrow_z|\widehat{S}_z|\downarrow_z\rangle \end{bmatrix} \tag{3.44}$$

Since there are only two spin states, this is a 2×2 matrix. From Eqs. (3.36), (3.37), and (3.43), we can write the individual components of this matrix as:

$$\begin{aligned} \left\langle\uparrow_z|\widehat{S}_z|\uparrow_z\right\rangle &= \frac{\hbar}{2}\langle\uparrow_z|\uparrow_z\rangle = \frac{\hbar}{2} \\ \left\langle\uparrow_z|\widehat{S}_z|\downarrow_z\right\rangle &= -\frac{\hbar}{2}\langle\uparrow_z|\downarrow_z\rangle = 0 \\ \left\langle\downarrow_z|\widehat{S}_z|\uparrow_z\right\rangle &= \frac{\hbar}{2}\langle\downarrow_z|\uparrow_z\rangle = 0 \\ \left\langle\downarrow_z|\widehat{S}_z|\downarrow_z\right\rangle &= -\frac{\hbar}{2}\langle\downarrow_z|\downarrow_z\rangle = -\frac{\hbar}{2} \end{aligned} \tag{3.45}$$

Substituting this in Eq. (3.44), we get

$$\widehat{S}_z = \begin{bmatrix} \dfrac{\hbar}{2} & 0 \\ 0 & -\dfrac{\hbar}{2} \end{bmatrix} = \frac{\hbar}{2}\begin{bmatrix} 1 & 0 \\ 0 & -1 \end{bmatrix} = \frac{\hbar}{2}\sigma_z \tag{3.46}$$

This result is the same as the Pauli's matrix σ_z, which we saw in Chap. 2, Eq. (2.46).

3.7.1 Ladder Operators (or Raising/Lowering Operators)

Raising/lowering operators increase or decrease the eigenvalue of another operator. They are also called **creation** and **annihilation operators** in quantum mechanics.

The raising and lowering operators for spin angular momentum are given by:

$$\begin{aligned} \widehat{S}_+ &= \widehat{S}_x + i\widehat{S}_y \\ \widehat{S}_- &= \widehat{S}_x - i\widehat{S}_y \end{aligned} \tag{3.47}$$

Note: For brevity, we are not detailing how these equations were derived.

With a little math work, we can rewrite these two equations as:

$$\begin{aligned} \widehat{S}_x &= \frac{1}{2}\left(\widehat{S}_+ + \widehat{S}_-\right) \\ \widehat{S}_y &= \frac{1}{2i}\left(\widehat{S}_+ - \widehat{S}_-\right) \end{aligned} \tag{3.48}$$

The general form of the ladder operators for the spin angular momentum is:

$$\begin{aligned} \widehat{S}_+|s,m\rangle &= \hbar\sqrt{s(s+1) - m(m+1)}\,|s,m+1\rangle \\ \widehat{S}_-|s,m\rangle &= \hbar\sqrt{s(s+1) - m(m-1)}\,|s,m-1\rangle \end{aligned} \tag{3.49}$$

Note: For brevity, we are not detailing how these two equations were derived.

Considering $s = \frac{1}{2}$ and $m = \pm\frac{1}{2}$, we get,

$$\begin{aligned} \widehat{S}_+\left|\frac{1}{2}, -\frac{1}{2}\right\rangle &= \hbar\left|\frac{1}{2}, \frac{1}{2}\right\rangle \\ \widehat{S}_-\left|\frac{1}{2}, \frac{1}{2}\right\rangle &= \hbar\left|\frac{1}{2}, -\frac{1}{2}\right\rangle \end{aligned} \tag{3.50}$$

Other combinations cancel out. We can write these operators in a matrix form using Chap. 2, Eq. (2.100), and applying orthonormality. With a little bit of math work, we get:

$$\begin{aligned} \widehat{S}_+ &= \hbar\begin{bmatrix} 0 & 1 \\ 0 & 0 \end{bmatrix} \\[1em] \widehat{S}_- &= \hbar\begin{bmatrix} 0 & 0 \\ 1 & 0 \end{bmatrix} \end{aligned} \tag{3.51}$$

By applying these two matrices in Eq. (3.48), we get:

$$\hat{S}_x = \frac{\hbar}{2}\begin{bmatrix} 0 & 1 \\ 1 & 0 \end{bmatrix} = \frac{\hbar}{2}\sigma_x$$

$$\hat{S}_y = \frac{\hbar}{2}\begin{bmatrix} 0 & -i \\ i & 0 \end{bmatrix} = \frac{\hbar}{2}\sigma_y$$

(3.52)

σ_x and σ_y are the remaining two Pauli matrices we didn't derive so far. We shall be using the Pauli matrices in the next few chapters.

Exercise 3.1

Verify the following commutation relationships for the spin operators.

$$\left[\hat{S}_y, \hat{S}_z\right] = i\hbar\hat{S}_x$$

$$\left[\hat{S}_z, \hat{S}_x\right] = i\hbar\hat{S}_y$$

$$\left[\hat{S}_x, \hat{S}_y\right] = i\hbar\hat{S}_z$$

Hints: Apply Eq. (2.81), Chap. 2.

◄

3.8 Bloch Sphere Representation

In the previous section, we have seen that the spin-$\frac{1}{2}$ system is a two-state quantum system, and it is represented in a two-dimensional Hilbert space. The two-states of this system are the spin-up (↑)and spin-down (↓) states. We have also seen that irrespective of the measurement axis, the spin is always $\frac{\hbar}{2}$. In the theory of Quantum Information Systems, this kind of a two-state system is referred to as the **"qubit"** meaning **quantum bits** by drawing an analogy with the digital bits of classical computers. In the following chapters, we shall learn more about qubits. However, for now, we have to introduce some additional math work in this chapter, which shall help further discussions.

Let us recollect from our previous discussions that the state vector of the qubit (we shall use this term moving forward for systems expressed in the two-dimensional Hilbert space) can be written using the linear superposition principle in an orthonormal basis of $\{|\uparrow\rangle, |\downarrow\rangle\}$ as:

$$|\psi\rangle = a|\uparrow\rangle + b|\downarrow\rangle = \begin{bmatrix} a \\ b \end{bmatrix}$$

(3.53)

Alternatively, if we use an orthonormal basis of $\{|0\rangle, |1\rangle\}$, called **computational basis**, we can write the wavefunction as:

$$|\psi\rangle = a|0\rangle + b|1\rangle = \begin{bmatrix} a \\ b \end{bmatrix},$$

(3.54)

where the weighing factors, a and b, are probability amplitudes and normalized complex numbers, meeting the normalization criteria: $|a|^2 + |b|^2 = 1$. The relative phase between a and b is responsible for **Quantum Interference,** and we shall explore that in detail in Chap. 5, Sect. 5.12.

The matrix form of the kets $|0\rangle$ and $|1\rangle$ can be derived from Eq. (3.56), and are the same as Eqs. (3.38) and (3.39):

$$|0\rangle = \begin{bmatrix} 1 \\ 0 \end{bmatrix} \text{ and } |1\rangle = \begin{bmatrix} 0 \\ 1 \end{bmatrix} \tag{3.55}$$

For convenience, this state vector can be drawn as a vector pointing to the surface of a unit sphere called the **Bloch sphere**, as shown in Fig. 3.10. The spin-up state $|0\rangle$ is at the north pole, and the spin-down state $|1\rangle$ is at the south pole, as shown in Fig. 3.11. The various equal superposition states are

Fig. 3.10 Bloch sphere representation of the qubits

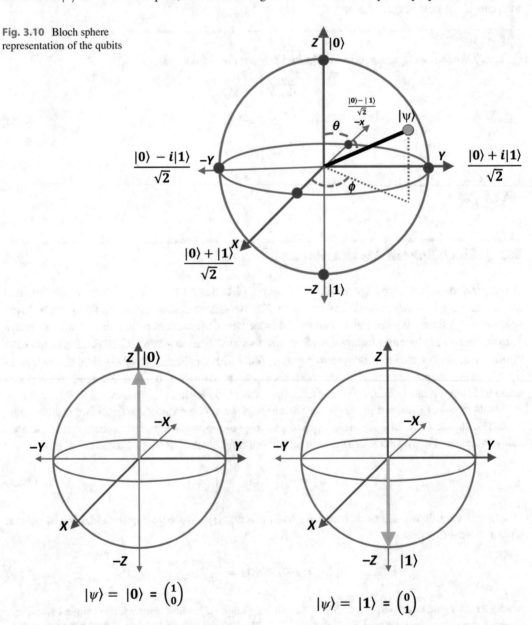

Fig. 3.11 Bloch sphere showing states "0" and "1". In state "0" the state vector points to the north pole and in state "1" the state vector points to the south pole

along the equator, as shown in Fig. 3.10. Assume for now that the radius of the Bloch sphere to be 1. We shall explain this in the forthcoming Sect. 6.3.2.

Using the spherical coordinate system, an arbitrary position of the state vector of a qubit can be written in terms of the angles θ (elevation, the state vector makes from z-axis) and ϕ (azimuth, the angle of projection of the state vector in the xy plane from the x-axis) it makes in the Bloch sphere as:

$$|\psi\rangle = \cos\left(\frac{\theta}{2}\right)|0\rangle + e^{i\phi}\sin\left(\frac{\theta}{2}\right)|1\rangle = \begin{bmatrix} \cos\left(\dfrac{\theta}{2}\right) \\ e^{i\phi}\sin\left(\dfrac{\theta}{2}\right) \end{bmatrix} \quad (3.56)$$

We can measure the direction of the qubit's state vector by directing a magnetic field $\vec{B}$ along the z-axis and measuring the energy. This is the same direction as the expectation value vector of $\widehat{S}_z$.

The qubit is in a superposition state described by Eqs. (3.53) and (3.54). When we measure the qubit, it can measure either "0" or "1" equivalent to the digital bits. The probability of the outcome $|0\rangle$ which measures "0" is equal to $|a|^2$, and the probability of the outcome $|1\rangle$ which measures "1" is equal to $|b|^2$.

Example 3.3

If the state vector of a qubit is $|\psi\rangle = \begin{bmatrix} a \\ b \end{bmatrix}$, *calculate the expectation value of* $\widehat{S}_z$.

Solution:

From Eq. (2.133), using Eq. (3.46), we can write the expectation value of $\widehat{S}_z$ as:

$$\langle \hat{S}_z \rangle = \langle \psi \mid \hat{S}_z \mid \psi \rangle = [a^* \ b^*] \cdot \frac{\hbar}{2}\begin{bmatrix} 1 & 0 \\ 0 & -1 \end{bmatrix} \cdot \begin{bmatrix} a \\ b \end{bmatrix} \quad (3.57)$$

This simplifies to:

$$\left\langle \widehat{S}_z \right\rangle = \frac{\hbar}{2}(a^*a - b^*b) = \frac{\hbar}{2}\left(|a|^2 - |b|^2\right) \quad (3.58)$$

◄

3.8.1 Qubit Rotations

The Pauli matrices can be used to rotate the state vector of the qubit by 180° along the respective axes. Pauli rotations are performed by applying a measured dose of a π rotational pulse along the respective axis. These operations are called "**gates**". We shall learn about the qubit gate operations in the next chapter. For now, since we discussed the Bloch sphere representation and spin matrices, we should look at the Pauli rotations on the qubits.

3.8.1.1 Pauli *X*-Gate

The application of this gate rotates the qubit by 180° along the x-axis. This gate is also called NOT gate, as it flips the qubit state from $|1\rangle$ to $|0\rangle$ and vice versa. Figure 5.6 illustrates the operation of the X-gate.

$$X|0\rangle = \begin{bmatrix} 0 & 1 \\ 1 & 0 \end{bmatrix} \begin{bmatrix} 1 \\ 0 \end{bmatrix} = \begin{bmatrix} 0 \\ 1 \end{bmatrix} = |1\rangle$$

$$X|1\rangle = \begin{bmatrix} 0 & 1 \\ 1 & 0 \end{bmatrix} \begin{bmatrix} 0 \\ 1 \end{bmatrix} = \begin{bmatrix} 1 \\ 0 \end{bmatrix} = |0\rangle$$

(3.59)

3.8.1.2 Pauli Y-Gate

The application of this gate rotates the qubit by 180° along the y-axis. This gate is also called phase-shift gate as it shifts the phase of the qubit.

$$Y|0\rangle = \begin{bmatrix} 0 & -i \\ i & 0 \end{bmatrix} \begin{bmatrix} 1 \\ 0 \end{bmatrix} = \begin{bmatrix} 0 \\ i \end{bmatrix} = i \begin{bmatrix} 0 \\ 1 \end{bmatrix} = i|1\rangle$$

$$Y|1\rangle = \begin{bmatrix} 0 & -i \\ i & 0 \end{bmatrix} \begin{bmatrix} 0 \\ 1 \end{bmatrix} = \begin{bmatrix} -i \\ 0 \end{bmatrix} = -i \begin{bmatrix} 1 \\ 0 \end{bmatrix} = -i|0\rangle$$

(3.60)

3.8.1.3 Pauli Z-Gate

The application of this gate rotates the qubit by 180° along the z-axis. This gate is also called a phase flip gate. This gate leaves state $|0\rangle$ as such but flips the state $|1\rangle$ to $-|1\rangle$.

$$Z|0\rangle = \begin{bmatrix} 1 & 0 \\ 0 & -1 \end{bmatrix} \begin{bmatrix} 1 \\ 0 \end{bmatrix} = \begin{bmatrix} 1 \\ 0 \end{bmatrix} = |0\rangle$$

$$Z|1\rangle = \begin{bmatrix} 1 & 0 \\ 0 & -1 \end{bmatrix} \begin{bmatrix} 0 \\ 1 \end{bmatrix} = \begin{bmatrix} 0 \\ -1 \end{bmatrix} = -\begin{bmatrix} 0 \\ 1 \end{bmatrix} = -|1\rangle$$

(3.61)

Figure 5.12 illustrates the operation of Pauli-Y and Pauli-Z gates.

3.8.1.4 Rotation Operator

The general form of the rotation operator $\widehat{R}$ is given by the following equation:

$$\widehat{R} = e^{-i\frac{\theta}{2}(\widehat{n}\cdot\sigma)} = \cos\left(\frac{\theta}{2}\right) - i\,(\widehat{n}\cdot\sigma)\sin\left(\frac{\theta}{2}\right),$$

(3.62)

where, $\widehat{n}$ is the unit vector along the axis of rotation, θ is the angle to be rotated, and σ is the corresponding Pauli matrix. To perform a rotation along, say, y-axis, we replace the term $(\widehat{n}\cdot\sigma)$ with Y.

Calculating this for the three rotational axes, we get:

$$\widehat{R}_x(\theta) = \begin{bmatrix} \cos\left(\frac{\theta}{2}\right) & -i\sin\left(\frac{\theta}{2}\right) \\ -i\sin\left(\frac{\theta}{2}\right) & \cos\left(\frac{\theta}{2}\right) \end{bmatrix}$$

(3.63)

$$\widehat{R}_y(\theta) = \begin{bmatrix} \cos\left(\dfrac{\theta}{2}\right) & -\sin\left(\dfrac{\theta}{2}\right) \\ \sin\left(\dfrac{\theta}{2}\right) & \cos\left(\dfrac{\theta}{2}\right) \end{bmatrix} \tag{3.64}$$

$$\widehat{R}_z(\theta) = \begin{bmatrix} e^{-i\frac{\theta}{2}} & 0 \\ 0 & e^{i\frac{\theta}{2}} \end{bmatrix} \tag{3.65}$$

Some authors consider rotations to be calculated by taking the negative of the angles. Applying that convention, note that the Pauli matrices are equal to the phase factor i times the general rotation matrices for $-180°$.

$$\sigma_x = -i\widehat{R}_x(180°) = i\widehat{R}_x(-180°) \tag{3.66}$$

$$\sigma_y = -i\widehat{R}_y(180°) = i\widehat{R}_y(-180°) \tag{3.67}$$

$$\sigma_z = -i\widehat{R}_z(180°) = i\widehat{R}_z(-180°) \tag{3.68}$$

Example 3.4

Let us perform a $-90°$ (actually $90°$, but by using the negative angle convention) rotation along the x-axis on a qubit at state $|0\rangle$.

By substituting the angle directly in Eq. (3.63) and applying the rotation over the state $|0\rangle$ we get,

$$\widehat{R}_x(-90)|0\rangle = \begin{bmatrix} \cos\left(\dfrac{-90}{2}\right) & -i\sin\left(\dfrac{-90}{2}\right) \\ -i\sin\left(\dfrac{-90}{2}\right) & \cos\left(\dfrac{-90}{2}\right) \end{bmatrix} \begin{bmatrix} 1 \\ 0 \end{bmatrix} \tag{3.69}$$

$$= \begin{bmatrix} \dfrac{1}{\sqrt{2}} & i\dfrac{1}{\sqrt{2}} \\ i\dfrac{1}{\sqrt{2}} & \dfrac{1}{\sqrt{2}} \end{bmatrix} \begin{bmatrix} 1 \\ 0 \end{bmatrix} = \begin{bmatrix} \dfrac{1}{\sqrt{2}} \\ i\dfrac{1}{\sqrt{2}} \end{bmatrix} = \dfrac{1}{\sqrt{2}}\left(\begin{bmatrix} 1 \\ 0 \end{bmatrix} + \begin{bmatrix} 0 \\ i \end{bmatrix}\right) \tag{3.70}$$

$$= \dfrac{1}{\sqrt{2}}\left(\begin{bmatrix} 1 \\ 0 \end{bmatrix} + i\begin{bmatrix} 0 \\ 1 \end{bmatrix}\right) \tag{3.71}$$

$$= \dfrac{1}{\sqrt{2}}\left(|0\rangle + i|1\rangle\right) \tag{3.72}$$

The qubit is rotated to a position intersecting the y-axis on the surface of the Bloch sphere, which is readily seen in Fig. 3.12.

◀

Qubit rotations are important gate operations in quantum computing. In Chap. 5, we learn about a few rotational operations and apply them in Chap. 7. It is interesting to know that arithmetic can be done by rotations. However, that is in Chap. 7!

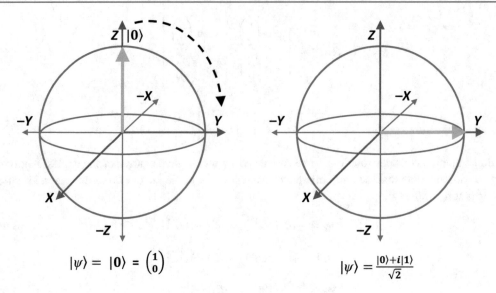

$$|\psi\rangle = |0\rangle = \begin{pmatrix} 1 \\ 0 \end{pmatrix} \qquad\qquad |\psi\rangle = \frac{|0\rangle + i|1\rangle}{\sqrt{2}}$$

Fig. 3.12 Bloch sphere showing a qubit's initial state vector at $|0\rangle$ and after a $\widehat{R}_x(-90)$ to the final state of $\frac{1}{\sqrt{2}}(|0\rangle + i|1\rangle)$

3.9 Projective Measurements

We now have an idea about the qubits, which are two-state systems. We also learned about some basic rotational operations on the qubits. We shall learn about the construction of the qubits and gate operations in the forthcoming chapters. In this introductory section, we learn about the measurement of the qubits.

We know that a quantum system can be described using the quantum superposition principle as $|\psi\rangle = a|0\rangle + b|1\rangle$ in the computational basis. Under this definition, the probability of measuring the system in state $|0\rangle$ is $|a|^2$, and the probability of measuring the system in state $|1\rangle$ is $|b|^2$. Measurements are an essential activity in quantum computation and performed most often at the end of the computation. Sometimes, we perform measurements during the computation to correct errors or to implement "if" statements. In the computational basis, measurements are done along the z-axis. However, we may use other basis or measure several qubits at the same time. Therefore, it is advantageous to define a mathematical formalism for making quantum measurements.

In this section, we discuss projective (or von Neumann) measurements and von Neumann's description of the measurement process.

3.9.1 Projection Operators

In quantum mechanics, projective measurement is described by an observable M. The operator M is defined in the same Hilbert space in which the system is observed. If the eigenvalues m are the possible outcomes of the measurement, then we can write the operator M as:

$$M = \sum_m m P_m, \tag{3.73}$$

where P_m are the complete set of orthogonal projectors onto the eigenspace of M. If we assume the state of the system before the measurement is $|\psi\rangle$, then the following axioms hold good.

Axiom—1:

The total probability of all projective measurements sums to 1, that is, $\sum_m P_m = I$. Here, I represents the identity matrix.

Axiom—2:

The probability of observing a certain measurement m is $p(m) = \langle \psi | P_m | \psi \rangle$. Let us derive this state. Consider the following equation for a system in state $|\psi\rangle = a|0\rangle + b|1\rangle$.

$$\langle 0|\psi\rangle = \langle 0|(a|0\rangle + b|1\rangle) = a\langle 0|0\rangle + b\langle 0|1\rangle = a \tag{3.74}$$

We can similarly prove $\langle 1|\psi\rangle = b$. From this equation, we can write the probability of finding the quantum system in state $|0\rangle$ as follows:

$$
\begin{aligned}
p(0) &= |\langle 0|\psi\rangle|^2 \\
&= \langle 0|\psi\rangle^* \langle 0|\psi\rangle = \langle \psi|0\rangle\langle 0|\psi\rangle \\
&= \langle \psi|P_0|\psi\rangle
\end{aligned} \tag{3.75}
$$

We can derive the same for state $|1\rangle$.

Axiom—3:

The state of the system immediately after the measurement is given by:

$$|\psi'\rangle \rightarrow \frac{P_m\psi}{\sqrt{p(m)}} = |m\rangle \tag{3.76}$$

Measurement causes the system to collapse into a state that is consistent with the measurement. This abrupt change is often called the **projection postulate.** It states that the outcome of a measurement is a new state, which is a normalized projection of the original system ket into the ket corresponding to the measurement. Note that the projective measurements can be repeated. Hence, the outcome of a projective measurement is deterministic. However, repeated measurements do not have much significance in quantum computation and often violated. For example, in a quantum communication channel, the encoded photons arriving at the end of the channel get absorbed by the photodetectors. They are no longer available for repeat measurements. In quantum computation, we are usually not interested in the post-measurement state of the system. Quantum algorithms are mostly concerned with the probabilities of the measurement outcomes, and we shall see them in the algorithms we explore in Chap. 7.

Exercise 3.2

Prove that projective measurements can be repeated.

◀

The operators of the projective measurements are called **projectors.** We learned about them in the previous chapter. Projective measurements are thought to be the ideal measurements of a quantum system. Projective operators satisfy the equality $P^2 = P$, and they can be written as outer products.

For example, in the computational basis of $\{|0\rangle, |1\rangle\}$, we can define two measurement operators P_0 and P_1 as follows:

$$P_0 = |0\rangle\langle 0|$$
$$(|0\rangle\langle 0|)^2 = |0\rangle\langle 0||0\rangle\langle 0| = |0\rangle\langle 0|0\rangle\langle 0| = |0\rangle\langle 0| \tag{3.77}$$

Similarly,

$$P_1 = |1\rangle\langle 1|$$
$$(|1\rangle\langle 1|)^2 = |1\rangle\langle 1||1\rangle\langle 1| = |1\rangle\langle 1|1\rangle\langle 1| = |1\rangle\langle 1| \tag{3.78}$$

Besides, these two operators form a complete set, and we infer this from Eq. (2.126):

$$P_0^\dagger P_0 + P_1^\dagger P_1 = |0\rangle\langle 0|0\rangle\langle 0| + |1\rangle\langle 1|1\rangle\langle 1| = |0\rangle\langle 0| + |1\rangle\langle 1| = I \tag{3.79}$$

We have successfully defined two operators, P_0 and P_1, in the computational basis. Let us now return to the system state $|\psi\rangle = a|0\rangle + b|1\rangle$, and verify that these operators work.

The probability of measuring the system in state $|0\rangle$ is given by:

$$\begin{aligned}
\langle\psi|P_0|\psi\rangle &= \langle\psi|(|0\rangle\langle 0|)|\psi\rangle = \langle\psi|0\rangle\langle 0|\psi\rangle \\
&= (a^*\langle 0| + b^*\langle 1|)|0\rangle\langle 0|\,(a|0\rangle + b|1\rangle) \\
&= |a|^2
\end{aligned} \tag{3.80}$$

This derivation is according to our earlier statement of the probability of measuring the quantum system in state $|0\rangle$. The state of the system after the measurement is:

$$|\psi'\rangle \rightarrow \frac{P_0\psi}{\sqrt{\langle\psi|P_0|\psi\rangle}} = \frac{|0\rangle\langle 0|(a|0\rangle + b|1\rangle)}{|a|} = \frac{a}{|a|}|0\rangle \tag{3.81}$$

Note that $\frac{a}{|a|} = e^{i\phi}$ for some phase angle ϕ. This global phase cannot be measured and has no significance (Refer to Example **3.2**). Hence, we can say that the final state of the system is $|0\rangle$. In a similar way, we can derive the system state after a measurement of $|1\rangle$ as $\frac{b}{|b|}|1\rangle$.

Using the same procedure, we can create measurement operators for other basis that may be required. For example, the measurement operators for the polar basis $\{|+\rangle, |-\rangle\}$ is given in the following equation. We shall learn about the polar basis in forthcoming chapters.

$$\begin{aligned}
P_+ &= |+\rangle\langle +| = \frac{1}{\sqrt{2}}\,(|0\rangle + |1\rangle)\,(\langle 0| + \langle 1|) \\
P_- &= |-\rangle\langle -| = \frac{1}{\sqrt{2}}\,(|0\rangle - |1\rangle)(\langle 0| - \langle 1|)
\end{aligned} \tag{3.82}$$

Axiom—4:

The average value of the measurement is:

$$\begin{aligned}
\langle M\rangle &\equiv \sum_m m\, p(m) \\
&= \sum_m m\,\langle\psi|P_m|\psi\rangle = \left\langle\psi\left|\sum_m m P_m\right|\psi\right\rangle \\
&= \langle\psi|M|\psi\rangle
\end{aligned} \tag{3.83}$$

With this definition, we can define the standard deviation of the observed values.

$$\sigma_M{}^2 = \Delta M^2 = \langle M^2 \rangle + \langle M \rangle^2 \tag{3.84}$$

3.9.2 Measuring Multi-Qubit Systems

Quantum circuits are constructed with multiple qubits. In Chaps. 5 and 7, we learn about the quantum circuits in detail. Section 5.4.5 explains the method of writing multi-qubit systems using tensor products. In this section, we focus on measuring the multi-qubit systems. Our readers may want to jump to that section, read ahead, and come back here.

To keep our discussions simple, we use a two-qubit system to explain the concept. The method outlined here can be extended to larger circuits. Consider the following two-qubit state explained in Sect. 5.4.5. The combined state is written as a tensor product of the two qubits. Assume $|\psi_1\rangle = a_1|0\rangle + b_1|1\rangle$ and $|\psi_2\rangle = a_2|0\rangle + b_2|1\rangle$.

$$
\begin{aligned}
|\psi\rangle &= |\psi_1\rangle \otimes |\psi_1\rangle \\
&= a_1 a_2 |00\rangle + a_1 b_2 |01\rangle + b_1 a_2 |10\rangle + b_1 b_2 |11\rangle \\
&= \alpha |00\rangle + \beta |01\rangle + \gamma |10\rangle + \delta |11\rangle
\end{aligned} \tag{3.85}
$$

By drawing an analogy with the single-qubit measurement, we can define a complete set of four measurement operators for each of the states contained in the ket vectors of this equation.

$$
\begin{aligned}
P_{00} &= |00\rangle\langle 00| & P_{01} &= |01\rangle\langle 01| \\
P_{10} &= |10\rangle\langle 10| & P_{11} &= |11\rangle\langle 11|
\end{aligned} \tag{3.86}
$$

In most scenarios, we may have to measure only one of the qubits. For example, if we measure the first qubit alone, we must distinguish between state "00" and "01." Hence, the measurement operators must be able to project to all states consistent with the measurement. Therefore, the operator that measures the first qubit in the state $|0\rangle$, must have projectors for the states "00" and "01." These operators are defined as follows:

$$
\begin{aligned}
P_0^{(0)} &= P_{00} + P_{01} = |0\rangle\langle 0| \otimes |0\rangle\langle 0| + |0\rangle\langle 0| \otimes |1\rangle\langle 1| \\
&= |0\rangle\langle 0| \otimes (|0\rangle\langle 0| + |1\rangle\langle 1|) \\
&= |0\rangle\langle 0| \otimes I \\
P_1^{(0)} &= P_{10} + P_{11} = |1\rangle\langle 1| \otimes |0\rangle\langle 0| + |1\rangle\langle 1| \otimes |1\rangle\langle 1| \\
&= |1\rangle\langle 1| \otimes (|0\rangle\langle 0| + |1\rangle\langle 1|) \\
&= |1\rangle\langle 1| \otimes I
\end{aligned} \tag{3.87}
$$

We have now defined the projectors that can measure a system of multiple qubits. We have used a notation $\{P_m^{(j)}\}$, where m denotes the measurement outcome, and j denotes the qubit the operator operates upon. Note that, in this book, qubits are numbered relative to 0, that is, the first qubit is named $q[0]$. This numbering scheme is in line with the programing languages we use in later chapters. We can now calculate the probability of the first qubit being measured in state $|0\rangle$.

$$p(0)^{(0)} = \langle \psi \mid P_0^{(0)} \mid \psi \rangle$$
$$= (\alpha^* \langle 00| + \beta^* \langle 01| + \gamma^* \langle 10| + \delta^* \langle 11|) \, (|0\rangle \langle 0| \otimes I) \tag{3.88}$$
$$(\alpha|00\rangle + \beta|01\rangle + \gamma|10\rangle + \delta|11\rangle)$$

The γ and δ terms cancel out. We get

$$= (\alpha^* \langle 00| + \beta^* \langle 01|) \, (\alpha|00\rangle + \beta|01\rangle)$$
$$= |\alpha|^2 + |\beta|^2 \tag{3.89}$$

The probability of the first qubit being measured in state $|0\rangle$ is the sum of the probabilities of measuring $|00\rangle$ and $|01\rangle$. The system state after this measurement is as follows:

$$|\psi'\rangle \rightarrow \frac{P_0^{(0)} \psi}{\sqrt{\langle \psi \mid P_0^{(0)} \mid \psi \rangle}}$$
$$= \frac{(|0\rangle \langle 0| \otimes I) \, (\alpha|00\rangle + \beta|01\rangle + \gamma|10\rangle + \delta|11\rangle)}{\sqrt{|\alpha|^2 + |\beta|^2}} \tag{3.90}$$
$$= \frac{\alpha|00\rangle + \beta|01\rangle}{\sqrt{|\alpha|^2 + |\beta|^2}}$$

Therefore, when we measure the first qubit in state $|0\rangle$, the probability of the system being in states $|10\rangle$ and $|11\rangle$, no longer exists. We can similarly derive the first qubit in state $|1\rangle$. We leave that as an exercise for our audience.

3.9.3   Measurements

When we measure the voltage in an electrical circuit, we bring a voltmeter across the circuit and perform the measurement. Von Neumann proposed a similar procedure for measuring quantum systems. To measure an observable Q of a quantum system S, we bring it in proximity with a meter M and allow them to interact linearly. At the time of measurement, assume that the system is in a superposition state $|\psi\rangle = \sum_i a_i \, |x_i\rangle$, and the meter is in state $|m\rangle$. The composite state of the system $S + M$ is given by the tensor product $\sum_i a_i \, |x_i\rangle \otimes |m\rangle$. $|x_i\rangle$ are the eigenvectors of Q, and the possible outcomes of the measurement are q_i. When the meter registers the measurements q_i, the state of the system is $\sum_i \gamma_i \, |x_i\rangle \otimes |m_i\rangle$, where the m_i are orthonormal vectors, that is, $a|0\rangle|m_0\rangle + b|1\rangle|m_1\rangle$. In this state, the meter is entangled (Sect. 5.6) with the quantum system.

Due to the measurement process, the system "collapses" from the state $\sum_i \gamma_i \, |x_i\rangle \otimes |m_i\rangle$ into the state $|x_i\rangle \otimes |m_i\rangle$, for some i.

A density matrix function can describe the state of the meter. From Eq. (6.16), we can write this as:

$$\rho_M = |a|^2 |m_0\rangle \langle m_0| + |b|^2 |m_1\rangle \langle m_1|, \tag{3.91}$$

where ρ_M is the density matrix of the meter, m_0 is the state of the meter if the quantum system measures $|0\rangle$, and m_1 is the state of the meter if the quantum system measures $|1\rangle$. This equation represents a probability distribution over all possible values the meter can measure. The probability is

$|a|^2$ for the meter to be in state $|m_0\rangle$ and $|b|^2$ for the meter to be in state $|m_1\rangle$. This is the physics of the measurement.

3.9.4 Positive Operator Valued Measure (POVM)

Positive Operator Valued Measures are concerned about the statistics of the measurement, and not about the post-measurement state of the system. There are many use cases of this measurement, for example, quantum algorithms in which the measurement is performed at the end of the circuit.

Assume that we measure a quantum system $|\psi\rangle$ with a measurement operator M_m, and the outcome of the measurement be m. The probability of the outcome is given by $p(m) = \langle \psi | M_m^\dagger M_m | \psi \rangle$. If we can define $A_m = M_m^\dagger M_m$, it follows that $\sum_i A_m = I$ and $p(m) = \langle \psi | A_m | \psi \rangle$.

The operators A_m are positively well defined.

Note that since $p(m) = \langle \psi | A_m | \psi \rangle$, we can use the set of operators $\{A_m\}$ to estimate probabilities of the outcomes, instead of $\{M_m\}$. The complete set of $\{A_m\}$ is known as POVM. If we consider a projective measurement done by a complete set of projectors P_m, then $A_m \equiv P_m^\dagger P_m = P_m$. This means projective measurements are an example of the POVM formalism.

3.10 Qudits

Qudits [4] are basic units of quantum computation, like the qubits. While the qubits store quantum information in a superposition of two-states, the qudits work on a superposition of d states ($d > 2$). There are also **qutrits** that have a dimension of three. Extending the dimensionality of the computational units brings a lot of computing power. A system of two qudits, each with 32 dimensions, is equal to 20 qubits.

Qudits are early in the research, and researchers have reported few applications. Qudits with $d = 10$ have been experimentally realized in labs. Qudits are considered as substitutes for the ancilla qubits. This reduces the number of interactions between information carriers in multi-qubit gates. Qudits can create alternate hardware topologies reducing circuit depth (Chap. 5.) Qudits can also improve quantum communication. If we can encode more information in one photon, the channel efficiency can increase tremendously.

3.11 Summary

In this chapter, we continued our learning on quantum mechanics. We started with the definition of Hilbert space and learned about the Schrödinger equation. After refreshing our knowledge on the postulates of quantum mechanics, we learned about quantum tunneling and the concept of two-state systems using the Stern and Gerlach experiment. We applied this learning into building the concept of qubits, the basic unit of information in quantum computing. We learned how to use Pauli operators to rotate the state vector of the qubits and performing projective measurements.

The next chapter is an essential milestone in our learning, and it focuses on the physical systems—how qubits are constructed.

Practice Problems

1. If x is the position operator and p is the momentum operator, calculate the commutators $[x, p]$, $[x^2, p]$, and $[x, p^2]$.

2. Characterize the operator $\sigma_\pm = \sigma_x \pm i\sigma_y$.

3. Consider a system of 4 qubits (spin-$\frac{1}{2}$ states.) How many states are exhibited by the system?

4. Find the eigenvalues of the spin operator s of an electron, along a direction, perpendicular to the x-z plane.

5. The Hamiltonian of a quantum state $|\psi\rangle = \begin{bmatrix} 1 \\ 2 \\ 1 \end{bmatrix}$ is given by $H = \begin{bmatrix} 2 & 0 & 3 \\ 0 & 1 & 0 \\ 3 & 0 & 2 \end{bmatrix}$. Supposing, if we measure the quantum state in an arbitrary basis, which eigenvalue is the most probable? Also, calculate the expectation value of that state.

6. In a region, a particle of mass m, and zero kinetic energy has a wavefunction $\psi(x) = Axe^{-x^2/k^2}$, where A and k are constants. Find its potential energy.

7. A particle of energy 2 eV is incident on a potential barrier of width 0.1 nm and height 10 eV. Find the probability of the particle to tunnel through the barrier.

8. The basis vectors of a qutrit can be defined as $\{|0\rangle, |1\rangle, |2\rangle\}$. In matrix form, the basis states can be defined as $|0\rangle = \begin{bmatrix} 1 \\ 0 \\ 0 \end{bmatrix}$, $|1\rangle = \begin{bmatrix} 0 \\ 1 \\ 0 \end{bmatrix}$, and $|2\rangle = \begin{bmatrix} 0 \\ 0 \\ 1 \end{bmatrix}$. If the state of the system is defined as $|\psi\rangle = \frac{1}{\sqrt{38}}(2|0\rangle + 3|1\rangle + 5|2\rangle)$. If a measurement operator $\begin{bmatrix} 2 & 0 & 0 \\ 0 & 3 & 0 \\ 0 & 0 & -6 \end{bmatrix}$ operates on the system, what are the possible measurements? What are their probabilities?

9. Given an orthogonal projector P, and a normalized state $|\psi\rangle$, prove that the expectation value $\langle P \rangle$ is in the range $0 \le \langle P \rangle \le 1$.

10. The wavefunction of a linear harmonic oscillator at a certain state is given by $|\psi\rangle = a|\psi_0\rangle + b|\psi_1\rangle$, where $\psi_0(x)$ and $\psi_1(x)$ are its ground and excited states, a and b are constants. Show that $\langle x \rangle$ is not zero.

References

1. Introduction to Quantum Mechanics, 3/e, David J. Griffiths (Author), Darre; F. Schroeter (Author.)
2. Principles of Quantum Mechanics, 2/e, R. Shankar (Author.)
3. Quantum Computation and Quantum Information (Cambridge Series on Information and the Natural Sciences), 1.e, Michael A. Nielsen (Author), Isaac L. Chuang (Author.)
4. For more information of Qudits: (a) Scalable quantum computing with qudits on a graph, Kiktenko et al., 2020, https://arxiv.org/pdf/1909.08973.pdf; (b) The geometry of quantum computation, Dowling and Neilsen, 2008, https://arxiv.org/pdf/quant-ph/0701004.pdf; (c) High-dimensional optical quantum logic in large operational spaces, Imany, P., Jaramillo-Villegas, J.A., Alshaykh, M.S. et al. npj Quantum Inf 5, 59 (2019). https://doi.org/10.1038/s41534-019-0173-8

Part II - Essentials

Qubit Modalities

4

> *"Well, in our country,"* said Alice, still panting a little, *"you'd generally get to somewhere else—if you run very fast for a long time, as we've been doing."*
> *"A slow sort of country!"* said the Queen. *"Now, here, you see, it takes all the running you can do, to keep in the same place. If you want to get somewhere else, you must run at least twice as fast as that!"*
>
> — Lewis Carroll, *Alice's Adventures in Wonderland.*

The previous chapters gave an introduction to the principles of quantum mechanics. We learned about Hilbert space, wavefunctions, quantum superposition principle, two-state systems (qubits), Bloch sphere representation, and Pauli matrices. We used Dirac's bra-ket notation in explaining these concepts. This chapter takes us through the next steps in our journey and focuses on qubit modalities. We begin this chapter by building up a vocabulary of quantum computing.

▶ **Learning Objectives**
- The vocabulary of quantum computing
- Difference between classical bits and quantum bits
- Noisy intermediate-scale quantum technology and qubit metrics
- Qubit physics
  - Trapped ion qubits
  - Superconducting transmons
  - Topological qubits
  - Quantum dots
- Dilution refrigerators

4.1   The Vocabulary of Quantum Computing

Qubits
The **qubits** (meaning quantum bits) are the quantum equivalent of classical digital bits. The qubits work on the principles of quantum mechanics and are in the superposition state. We must use the principles of quantum mechanics to change the state of the qubits. At the end of the computation, we can measure the qubit's state by projecting them into classical digital bits.

V. Kasirajan, *Fundamentals of Quantum Computing*, https://doi.org/10.1007/978-3-030-63689-0_4

Universal Quantum Computer

By definition [1], a **Quantum Turing Machine,** or a **Universal Quantum Computer,** *is an abstract machine used to model the effect of a quantum computer* [2]. Analogous to the classical Turing Machine, any quantum algorithm can be expressed formally as a specific quantum Turing Machine. The internal states are represented by quantum states described in the Hilbert space. The transition function is a set of unitary matrices in the Hilbert space.

Quantum Annealing

It is a procedure for finding a heuristic algorithm that identifies a global minimum from a finite set of candidate solutions using **Quantum Fluctuations**. Combinatorial optimization problems with a discrete search space with many local minima, such as the traveling salesman problem, can be solved using **Quantum Annealing**. The system starts with the superposition of all candidate states using quantum parallelism and evolved using the time-dependent Schrödinger equation. By adjusting the transverse field (a magnetic field perpendicular to the axis of the qubit), the amplitudes of all states can be changed, causing **Quantum Tunneling** between the states. The goal is to keep the system close to the ground state of the Hamiltonian. When the transverse field is finally turned off, the system reaches its ground state, which corresponds to the solution of the optimization problem. In 2011, D-Wave systems demonstrated the first-ever Quantum Annealer.

Quantum Speedup

An optimal scenario, where there is a proof that no classical algorithm can outperform a given quantum algorithm. Besides factorization and discrete logarithms, there are a handful of quantum algorithms with a polynomial speedup. Grover's algorithm is one such algorithm that we shall see in an upcoming chapter. Simulation algorithms for physical processes in quantum chemistry and solid-state physics have been reported. An approximation algorithm [3] for Jones polynomial with a polynomial speedup and a solution to Pells's equation, and the principal ideal problem in polynomial time [4] have been published [5]. This space is evolving.

Quantum Advantage

The computational advantage of quantum computers. A notion that quantum computers can perform some computations faster than classical computers.

Quantum Supremacy

The potential ability [6] of quantum computers to solve problems that classical computers practically cannot.

Decoherence

The process in which the qubit loses the quantum information in time, due to interactions with the environment.

Quantum Volume

A pragmatic [7] way to measure and compare progress toward improved system-wide gate error rates for near-term quantum computation and error-correction experiments. It is a single-number metric that a concrete protocol can measure using a near-term quantum computer of modest size $n \lesssim 50$.

4.2   Classical Computers—A Recap

The computers we use today store information in a binary format—as 0s and 1s using transistor switches, called bits. The bits are the basic computational units in a digital computer. There are billions of such switches integrated into minuscule circuits in a computer today. Each time we do a task on a computer system, these switches are turned on/off a billion times!

By forward biasing the transistor, the transistor can be turned ON, equivalent to the switch being in the CLOSED state. When the switch is closed, the current drain in the circuit denoting a binary state of "0."

By reverse biasing the transistor, the transistor can be turned OFF. This is equivalent to the switch being turned OPEN. When the switch is open, it denotes a binary state of "1" (Figs. 4.1 and 4.2).

By creating arrays of these switches in 8-bit, 16-bit, 32-bit, or 64-bit sizes, large numbers can be stored and processed. Figure 4.3 shows an 8-bit register storing the alphabet "A" in the standard ASCII encoding.

The digital gate operations help us manipulate this binary information. A sequence of these gates is created to perform basic mathematical operations such as addition, subtraction, multiplication, or division. In Chap. 5, we shall construct a quantum circuit to perform a basic addition operation.

Any operation that we do at the application level—indexing, sorting, statistical calculations, and user interface—they all boil down to a sequence of gate operations at the processor level.

Figure 4.4 shows the set of digital gates, called universal digital gates and their truth table. The truth table summarizes the various combinations of inputs and outputs. These gates are called

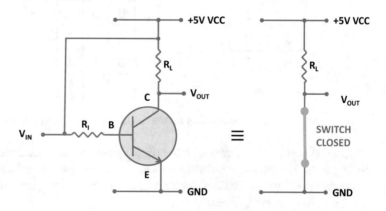

Fig. 4.1 A transistor switch closed—binary state of "0."

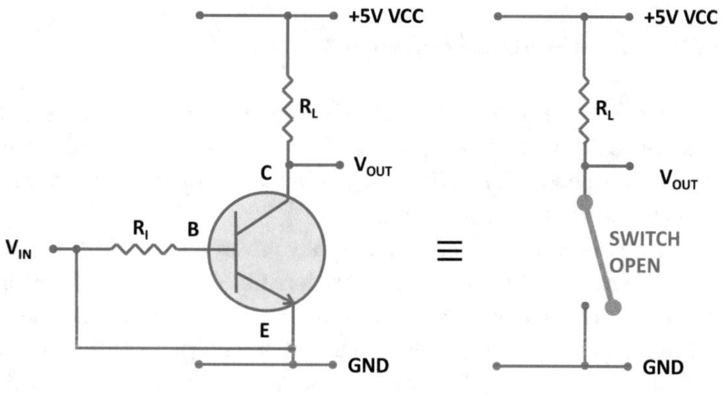

Fig. 4.2 A transistor switch in an open state—binary state of "1."

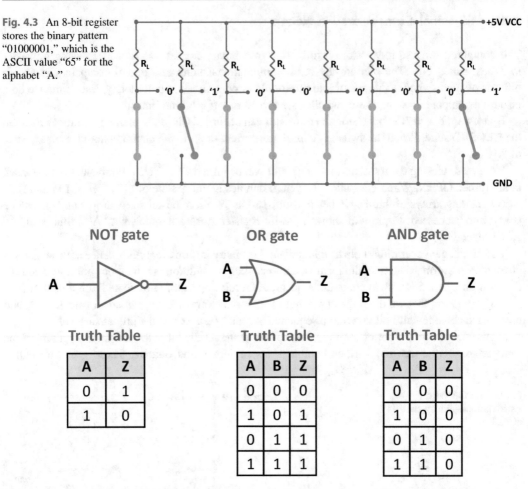

Fig. 4.3 An 8-bit register stores the binary pattern "01000001," which is the ASCII value "65" for the alphabet "A."

Fig. 4.4 Universal digital gates and their truth tables. All operations can be derived from these gates

universal digital gates because all other digital gates can be constructed from them (or NAND and NOR gates). The transistor switches we saw earlier are used to construct these gates. To keep our discussions minimal on this topic, we do not show how these gates are constructed [8]. Chapter 5 describes the quantum universal gates.

4.2.1 Fundamental Difference

A fundamental difference between the digital bits and the qubits is the way they operate. The digital bits can be deterministically set in a "0" or a "1" state, and read (or measured) any number of times as long as it is powered. Reading a digital bit does not destroy its state. Moreover, the digital bit retains its state as long as it is powered.

The qubits are probabilistic, and they are in a superposition state of "0" and "1" with different probabilities. They possess the characteristics of both the states simultaneously, at all times, until measured. The qubits lose their internal state when they are measured. Besides, due to the laws of nature, the qubits interact with the environment, pick-up Rabi oscillations, and "**decohere**" over a period of time. We shall learn more about the characteristics of the qubits in subsequent sections (Fig. 4.5).

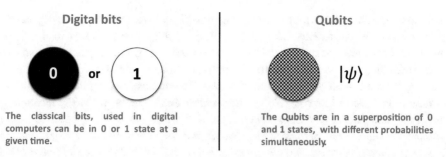

Fig. 4.5 Digital bits and qubits

4.3   Qubits and Usability

The digital bits we saw earlier are designed as transistor circuits in silicon substrates. It is a mature semiconductor technology now, and it took several decades to get us to this stage. For the past 30 years or so, research labs and the industry are researching the right modality for implementing commercial-grade qubits. Several physical systems have been researched so far: Nuclear Magnetic Resonance devices, valency centers in diamonds, photonics, trapped ion devices, superconducting transmons, quantum dots, and topological qubits. The research in this space is expanding, and new technologies are reported by research labs quite frequently.

The digital bits are fabricated on a single technology, namely semiconductors. The physical systems for qubits are based on technologies that differ in the way how they store, manipulate, and measure the quantum state. Irrespective of the underlying technology, we expect the qubits to have a longer coherence time and high gate fidelity. Scalability and quantum error correction are other essential factors. Another important factor is the coupling between the qubits. Coupling is needed, at least at 2-qubit levels. Quantum algorithms require coupling between any combinations of qubits. If that is not possible, SWAP operations may be required to exchange states between qubits for running the algorithm (like data exchange between two variables in classical programming). This constraint limits the length of the quantum computation we can perform before the system decoheres. Several methods are under research for qubit coupling.

The physical system must not be just confined to research labs for research work. It should be possible to implement commercial-grade real-time systems. For this, the physical system must be scalable to thousands of qubits. Elaborate error correction mechanisms and coupling schemes should be built so that developers can focus on creating quantum algorithms rather than creating code to fix quantum errors and manage coupling schemes.

This task is quite challenging for system engineers to characterize and improvise qubits. It may take a few years before large-scale fault tolerance systems are available for commercialization. In the next few sections, we shall learn the most researched and industry adopted qubit modalities, all of which have a high potential for scalability in the future.

4.4   Noisy Intermediate Scale Quantum Technology

Noisy Intermediate Scale Quantum Technology is a term [9] coined by Prof. John Preskill, Caltech, to denote the current quantum computing era. Though some of the algorithms we can run on the currently available 50 or so qubit quantum processors cannot be simulated efficiently on classical computers, the algorithms known to us today are sensitive to quantum errors. They cannot be run reliably in real-time environments. The susceptibility of the current devices to decoherence and quantum noises limits the systems' usability to demonstrations or prototypes in current times. This situation is frustrating, we know the quantum systems are more powerful than the classical computers in some ways, but we cannot put the quantum systems to good use beyond the research labs.

According to Prof. John Preskill, *"Noisy Intermediate-Scale Quantum (NISQ) technology will be available in the near future. Quantum computers with 50–100 qubits may be able to perform tasks which surpass the capabilities of today's classical digital computers, but noise in quantum gates will limit the size of quantum circuits that can be executed reliably. NISQ devices will be useful tools for exploring many-body quantum physics, and may have other useful applications, but the 100-qubit quantum computer will not change the world right away—we should regard it as a significant step toward the more powerful quantum technologies of the future. Quantum technologists should continue to strive for more accurate quantum gates and, eventually, fully fault-tolerant quantum computing."*

So, we are there, but not quite yet. In our discussions so far, we touched upon decoherence and quantum errors a few times. We know they can impact quantum computation. If we are given a quantum processor with a certain number of qubits, how do we measure the qubits' quality? The next section describes the metrics adopted by the industry and facilitated by research. Once we know the device metrics, we can determine what algorithms can be potentially run on them.

4.5 Qubit Metrics

4.5.1 Spin or Energy Relaxation Time (T_1)

This measure is the characteristic time taken by a qubit to return to its ground state from an excited state due to spin-environment coupling. We measure this by inversion recovery. The qubit is first prepared in the state $|0\rangle$. We then apply an X transform and wait for a time t to measure the probability of the qubit to be in state $|1\rangle$, that is, relaxation did not occur (Fig. 4.6).

Fig. 4.6 Characteristic T_1 time

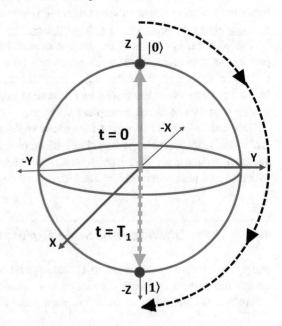

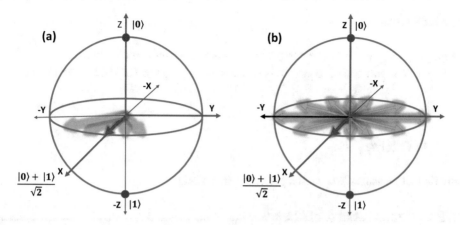

Fig. 4.7 (a) The Bloch vector starts to decohere. (b) A fully decohered Bloch Vector (T_2)

4.5.2 Dephasing Time ($T_\varnothing$)

It is the time taken by the Bloch vector to diffuse to the equator.

4.5.3 Decoherence Time (T_2) [10, 11]

Assume a qubit in an equal superposition state $\frac{|0\rangle+|1\rangle}{\sqrt{2}}$. When the qubit is not interacting with the environment, it undergoes a constant precession at a frequency ω, known as the qubit frequency at an azimuthal angle $\varnothing$ (Fig. 3.10 in Chap. 3). Due to environmental influence, the Bloch vector starts to decohere (Fig. 4.7a). The time it takes to decohere completely (Fig. 4.7b) is known as the Decoherence Time, and it is a measure of the qubit's usable lifetime. The decoherence time can be measured using Hann Echo or Ramsay Experiment.

4.5.4 Hann Echo (T_2)

In this method, we prepare the qubit in the equal superposition state $\frac{|0\rangle+|1\rangle}{\sqrt{2}}$ (by applying a Hadamard gate), and wait for half the time $\frac{t}{2}$. After this, we apply an X transform and wait for the remaining time $\approx \frac{t}{2}$. Finally, we apply a Hadamard gate again to measure its probability of being in state $|0\rangle$. The wait time gives the qubit the time to start to decohere.

4.5.5 Ramsay Experiment $\left(T_2^*\right)$

In this method, we prepare the qubit in the equal superposition state $\frac{|0\rangle+|1\rangle}{\sqrt{2}}$ (by applying a Hadamard gate) and wait for the time t. After the wait time, we apply a Hadamard gate to measure its probability of being in state $|0\rangle$.

4.5.6   Gate Time

The number of gate operations one can perform before the qubit completely decoheres and loses its usable lifetime. The gate time is equal to $\frac{T_2}{T_\varnothing}$, and it is in the range of 100–10,000 for the leading qubit modalities.

4.5.7   Gate Fidelity

It is a metric that measures how well a gate operation works.

4.6   Leading Qubit Modalities

Qubits are two-state systems in which information equivalent to one classical bit can be stored. Any quantum mechanical system that can exhibit two distinct states and coupling can be used to construct a qubit. Since the 1990s, many physical systems have been researched and prototyped as qubits. Some of the notable implementations are NMR, nitrogen valence centers in diamonds, trapped ion qubits, superconducting devices, photonics, semiconductor-based quantum dots, etc. Each of these implements is benchmarked against standard metrics such as the ability to store and measure quantum states, the fidelity of operations, and continual improvement. Scalability, speed, and the probability of errors and recovery are paramount conditions of usability.

This section studies four leading qubit modalities: trapped ion, superconducting transmons, topological, and quantum bits.

> 📖 **Note:**   Reading Sect. 1.7, on exotic states of matter may be helpful to read the following sections in this chapter.

4.6.1   Trapped Ion Qubit

In Chap. 1, we learned that we could create ions by stripping off electrons from atoms. This qubit modality follows a similar method. A small amount of calcium is molten in a vacuum, and the evaporating vapor is hit with a high energy electron beam. This process strips off one electron from each of the calcium atoms and creates $^{43}Ca^+$ ions. A Paul trap attracts some of these ions. The Paul trap is powered by DC electrodes and RF fields. The RF field is set up in such a way that the confinement is alternating between the x and y axes. The oscillating force is minimal at the center of the trap and grows significantly toward the edges. This oscillating force confines the ions to the center of the trap, where the total energy is minimal. The DC electrodes provide the needed potential so that the ion does not escape along the z-axis. A blue laser cools the ion trap into millikelvin degrees and stabilizes the qubit.

Any additional thermal energy acquired or lost by the ions from the environment must be very small compared to the kinetic energies involved in the controlled operation of the qubit. If this is not possible, the ion is not stable, and the Bloch vector diffuses rapidly. The energy gained by a trapped ion by absorbing or emitting a photon by interacting with the environment is about

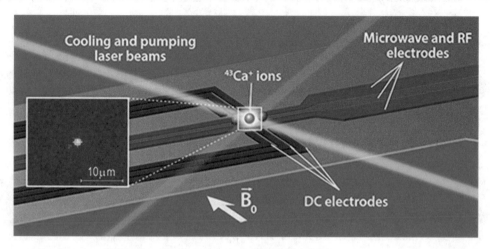

Fig. 4.8 Schematic of a trapped ion qubit. (Image: APS/Alan Stonebraker, Copyright 2014 American Physical Society. Printed with permission)

0.00000000000000000000000000000002 joules. This thermal energy is so low that we must cool the qubit to millikelvin temperatures (Fig. 4.8).

In atoms, the hyperfine structure in the spectral lines happens due to the interaction between the magnetic field caused by the electron's movement and the nuclear spin. $^{43}\text{Ca}^+$ ions have two hyperfine ground states at $S_{\frac{1}{2}}(F = 3, M_F = 0)$ and $S_{\frac{1}{2}}(F = 4, M_F = 0)$. Here F is the total angular momentum of the atom and $M_F = -F, -F + 1, \ldots, F - 1, F$. The energy-level separation between the two hyperfine ground states is 2.3 GHz, and a minimum magnetic field of about 146 G can create this. The hyperfine states are large enough, and they are not susceptible to magnetic field fluctuations. Since the hyperfine structure is at the ground state, its lifetime is higher. These are some reasons for the choice of $^{43}\text{Ca}^+$ ions for this modality. Some organizations have reported the use of ytterbium (Yb) atom, however. The two hyperfine structures represent the $|0\rangle$ and $|1\rangle$ states.

We can prepare the qubit to the desired $|0\rangle$ or $|1\rangle$ states by applying measured RF pulses through the RF electrodes. By applying measured doses of RF pulses along the required axes, we can perform gate operations. The trapped-ion qubits are coupled through mutual Coulomb repulsion or by using lasers. The state of the qubit is read by monitoring the fluorescence. Only the state $|1\rangle$ results in fluorescence, and hence it is easy to implement the measurement part. The coherence time of the trapped-ion qubits is ≈50 s and have a 99.97% error rate in 150,000 operations. These are good numbers for trying out quantum algorithms.

The trapped-ion qubits are constructed as micro-circuits using the same processes used to make silicon-based integrated circuits. The diode lasers and the electrodes used in the circuit are fabricated in the same ways the integrated circuits are made. The setup must be kept in a high vacuum. The challenge with this modality is that it employs too many lasers, and qubit coupling is difficult. The sources of noise are the heating of the qubits, environmental noises, and instability of the laser fields. We do not yet have much proof that the trapped ion qubits are the way forward, but yes, they are an excellent alternative to superconducting transmons. The Spanish physicist **Ignacio Cirac** and the Austrian physicist **Peter Zoller** first reported the trapped-ion qubits in a research paper in 1995. Several research labs and companies are using this modality. The following table summarizes them [26] (Table 4.1).

Table 4.1 Organizations using trapped ion qubits

Alpine Quantum Technologies	Duke University	University of Maryland	University of Sussex
Honeywell	IonQ	University of Waterloo	NextGenQ
MIT	NQIT	Oxford Ionics	Oxford University
Sandia National Laboratories	Weizmann Institute	Quantum Factory	Universal Quantum

4.6.2   Superconducting Transmons

▦ **Note:** Readers can skip the equations and continue to read the text, without compromising the understanding.

4.6.2.1 Josephson Junction (JJ)

A Josephson Junction is a junction formed between two superconductors, which are separated by a thin insulator. The insulator is thin enough ≈ 10 nm, such that the Cooper pairs can tunnel through. This tunneling gives rise to the Josephson current I_J.

In Chap. 1, we learned about Cooper pairs and superconductivity.

Typical implementations of Josephson Junction are fabricated using aluminum wiring and aluminum–amorphous aluminum oxide–aluminum (Al–AlOx–Al) on silicon or sapphire based substrates. The superconducting critical temperature T_c of aluminum is 1.2 K, and it can be achieved using commercially available dilution fridges. The qubits based on Al–AlOx–Al Josephson Junctions can be fabricated using the same technology that makes integrated circuits and has the highest coherence rates of around 100 μS.

Implementations based on niobium–amorphous aluminum oxide–niobium (Nb–AlOx–Nb) are used in Quantum Annealers. The superconducting critical temperature T_c of niobium is 7.2 K. The Josephson Junctions made of Nb–AlOx–Nb layers are fabricated on a niobium layer with interconnections to other silicon-based circuits. This interconnection causes multilayer losses, and the coherence times are in the range of 10 to 100 nS.

Figure 4.9 illustrates a Josephson Junction formed between two superconductors labeled 1 and 2. A thin insulating junction (the shaded region) separates them. The phase difference $\theta_2 - \theta_1$ between the superconductors causes the Josephson current to flow in the circuit. The Josephson Junction behaves like a nonlinear inductor. We can use this property to construct a qubit.

If we assume ψ_1 and ψ_2 are the wavefunctions defining the superconducting states of the superconducting islands 1 and 2, then using Schrödinger equations, we can write the dynamics of these two wavefunctions [12] as:

Fig. 4.9 Schematic of a Josephson Junction

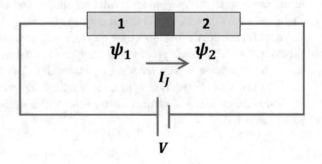

$$i\hbar \frac{\partial \psi_1}{\partial t} = E_1 \psi_1 + K \psi_2$$

$$i\hbar \frac{\partial \psi_2}{\partial t} = E_2 \psi_2 + K \psi_1, \tag{4.1}$$

where E_1 and E_1 are the lowest energies the Cooper pairs on the superconducting islands can occupy. K is a coupling constant.

We assume the wavefunctions as:

$$\psi_1 = \sqrt{n_1} \, e^{i\theta_1} \quad \text{and}$$

$$\psi_2 = \sqrt{n_2} \, e^{i\theta_2}, \tag{4.2}$$

where n_1 and n_2 are the density of Cooper pairs in the islands. θ_1 and θ_2 are the phases.

With this, Eq. (4.1) can be written as:

$$\hbar \frac{\partial n_1}{\partial t} = -\hbar \frac{\partial n_2}{\partial t} = 2K \sqrt{n_1 n_2} \sin(\theta_2 - \theta_1)$$

$$-\hbar \frac{\partial (\theta_2 - \theta_1)}{\partial t} = E_2 - E_1 \tag{4.3}$$

The application of voltage V shifts the energy level by $E_2 - E_1 = 2$ eV. The time derivative of the charge density n_1 gives the Josephson current or the supercurrent I_J in the circuit. The constant $I_c = \frac{2K}{\hbar} \sqrt{n_1 n_2}$ defines the critical current of the junction, and the superconducting phase difference is defined by $\gamma = \theta_1 - \theta_2$. With this, we can rewrite Eq. (4.3) as:

$$I_J = I_c \sin \gamma$$

$$V = \frac{\hbar}{2e} \frac{\partial \gamma}{\partial t} = \frac{\Phi_0}{2\pi} \frac{\partial \gamma}{\partial t}, \tag{4.4}$$

where $\Phi_0 = \frac{h}{2e} = 2.07 \times 10^{-15}$ T m^2 is the "**superconducting magnetic flux quantum**." The magnetic flux in a superconducting loop is quantized with value $n \cdot \Phi_0$, where $n = 0, 1, 2, \ldots$.

By differentiating the first equation and substituting $\frac{\partial \gamma}{\partial t}$ with V, we get

$$\frac{dI_J}{dt} = I_c \cos \gamma \frac{2\pi}{\Phi_0} V. \tag{4.5}$$

This equation describes an inductor. Since $V = L \frac{dI}{dt}$, we can write the Josephson Inductance as:

$$L_J = \frac{\Phi_0}{2\pi I_c \cos \gamma}. \tag{4.6}$$

The term $\frac{1}{\cos \gamma}$ means that we have a nonlinear behavior with the inductance, which we shall use in the next steps. Now that the nonlinear behavior is established, we can focus our attention on the creation of nondegeneracy required to create a two-state system. For that, we need to use the inductance in an LC circuit to make it behave like a **Quantum Harmonic Oscillator**.

Consider the *LC* circuit, shown in Fig. 4.10b. This circuit is made by connecting an inductor of inductance L in parallel with a capacitor of capacitance C. Let V be the voltage across the circuit and let I be the current flowing in the circuit. Let ϕ be the magnetic flux. This circuit is an example of a simple harmonic oscillator, with a characteristic frequency $= \frac{1}{\sqrt{LC}}$. The voltage and current in this

Fig. 4.10 The Quantum
Harmonic Oscillator. (**a**)
Energy states; (**b**)
equivalent LC circuit

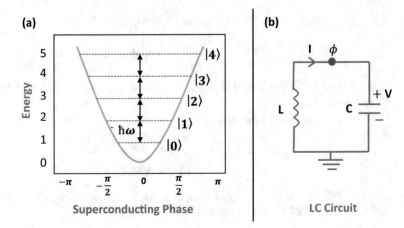

circuit oscillate at the same frequency, but the voltage leads the current by a phase difference of $\frac{\pi}{2}$. It can be shown that the classical energy of this circuit is given by the following equation:

$$E = \frac{1}{2}CV^2 + \frac{1}{2}LI^2. \tag{4.7}$$

The first term in this equation represents the electrical energy stored in the capacitor when it is charging. The second term represents the energy stored in the magnetic field when the current I flows through the inductance.

It can be shown that the Hamiltonian for an *LC* circuit, equivalent to the classical energy in Eq. (4.7), can be written by the following equation [13]:

$$H = 4E_c n^2 + \frac{1}{2}E_L\phi^2, \tag{4.8}$$

where $E_c = e^2/(2C)$ is the charging energy required to add each electron in the Cooper pair into the island and $E_L = \frac{(\Phi_0/2\pi)^2}{L_J}$ is the inductive energy. The reduced flux $\phi \equiv \frac{2\pi\Phi}{\Phi_0}$, is known as the "**gauge-invariant phase**" across the inductor. The reduced charge $n = \frac{Q}{2C}$ is the excess number of Cooper pairs on the island.

This equation is identical to the Hamiltonian of a single particle Quantum Harmonic Oscillator in one dimension. The first term represents the kinetic energy stored in the inductor, and the quadratic second term represents the potential energy stored in the capacitor. We can solve this equation as an eigenvalue problem on the coordinate basis, with ϕ as the position coordinate, for the wavefunction $\langle\phi|\psi\rangle = \psi(\phi)$. The solution leads to an orthogonal polynomial called Hermite functions with an infinite set of eigenstates $|n\rangle$, $n = 0, 1, 2, \ldots$.

The corresponding eigenenergies are:

$$E_n = \hbar\omega\left(n + \frac{1}{2}\right), \quad \text{where} \quad n = 0, 1, 2, \ldots, \tag{4.9}$$

where $\frac{\sqrt{8E_cE_L}}{\hbar} = \frac{1}{\sqrt{LC}}$.

This has few inferences—the energies are quantized (which is good!), and equally spaced by $\hbar\omega$. The lowest energy in the ground state (or the **vacuum state**) represented by $n = 0$ is called "**zero-point energy**." Quantum mechanically, this does not represent a zero particle. Instead, it must be treated as a superposition of $n = 0, 1, 2, \ldots$ states. Such wavefunctions are called **quantum fluctuations,** and the wavefunction corresponding to the lowest energy is called **zero-point fluctuation.** These energy states are shown in Fig. 4.10.

Since there are many equally spaced energy states, the transitions between these states are degenerate. Hence this is not suitable for a two-state system (qubit). We can have an effective two-state system only if the frequency ω_{01} that drives transitions between the qubit states $|0\rangle \leftrightarrow |1\rangle$ is different from the frequency ω_{12} for transitions $|1\rangle \leftrightarrow |2\rangle$. This requires us to introduce nonlinearity into the circuit, and we have already seen that the Josephson Junction is a nonlinear component.

Out of this nonlinearity, we can form three types of Josephson qubits—phase, flux, and charge. Phase qubits are based on the two different energy states in one well of the Josephson Junction potential. The flux qubits are based on the two flux states of the superconducting quantum interference device (SQUID). The charge qubits are based on the two different charge states of the Cooper pairs in a Cooper pair box.

In this book, we focus on the charge qubits as they are the basis for the commercially implemented superconducting transmons.

If we replace the linear inductor L in Fig. 4.10 with the nonlinear Josephson iunction and adopting Eq. (4.4) into Eq. (4.8), we get the following:

$$H = 4E_c n^2 - E_J \cos(\phi), \tag{4.10}$$

where $E_c = \frac{e^2}{2C_\Sigma}$. $C_\Sigma = C_s + C_j$. C_s is the shunt capacitance. The capacitance of the junction is C_J. $E_j = \frac{I_c \Phi_0}{2\pi}$. With this change, the energy spectrum is no longer harmonic, and it is nondegenerate. Figure 4.11 illustrates the nonlinearity.

The energy level separation between the states $|0\rangle \leftrightarrow |1\rangle$ is different from other interlevel energy differences in the system. Therefore, this circuit forms a two-state system.

Thus, the energy level spectra are entirely configurable using circuit elements. In other words, the qubit can be designed to produce atom-like spectral lines. For this reason, the superconducting qubits are also called **"artificial atoms."**

4.6.2.2 Charge Qubit/Cooper Pair Box

In this qubit arrangement, a superconducting island (the box in Fig. 4.12 with dashed boundaries) is coupled with a capacitor with capacitance C_g and a Josephson Junction with Josephson energy E_J to a grounded superconducting reservoir (the bulk). The charge states can be adjusted through a bias

Fig. 4.11 Quantum harmonic oscillator with nonlinear inductance: **(a)** the energy states and **(b)** the Josephson junction resonator

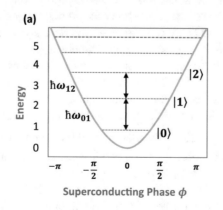

(a)

Superconducting Phase ϕ

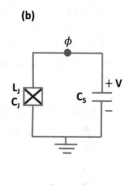

(b)

Josephson Junction Resonator

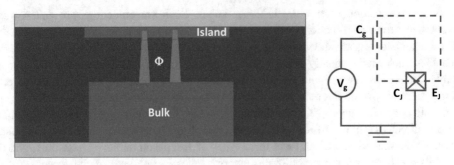

Fig. 4.12 The charge qubit or the Cooper pair box, schematic

voltage V_g. The quantum state is determined by the number of Cooper pairs that have tunneled through the junction (N). The below equation outlines the Hamiltonian of the charge qubit:

$$H = E_c \left(N - N_g\right)^2 - E_j \cos\left(\phi\right),\tag{4.11}$$

where the charging energy $E_c = \frac{(2e)^2}{2C_\Sigma}$, $C_\Sigma = C_g + C_J$, the gate charge $N_g = \frac{C_g V_g}{2e}$, Josephson energy $E_j = \frac{I_c \Phi_0}{2\pi}$, and ϕ is the guage invariant phase difference between the two junctions.

The bias voltage V_g is adjusted until the charge degeneracy point $N_g = \frac{1}{2}$ is reached. This results in a broader separation between states $|0\rangle \leftrightarrow |1\rangle$ than other states. Hence, this is a candidate for a two-state system.

The coherence times are in the order of 1–2 μs. Readouts are performed by coupling the island to a single electron transistor (SET).

The sources of noise in superconducting qubits can arise from fluctuating charges (quasiparticles, electron hopping, electric dipole flipping), fluctuating magnetic spins, and fluctuating magnetic vortices.

At low temperatures, fluctuating magnetic spins, and fluctuating magnetic vortices are the sources of flux noises. The charge noises arise out of the fluctuating charges.

4.6.2.3 Transmon Qubits

The superconducting transmon qubits are a variant of the charge qubits. The transmon was first developed by scientists [14] at the Yale University & Universit'e de Sherbrooke in 2007. The term "**transmon**" is an abbreviation for "transmission line shunted plasma qubit."

If we take a closer look at Eq. (4.11), we can notice that the Hamiltonian H is determined by one of the dominating energies E_c or E_J. In the circuit designs where $E_c \gg E_J$, the qubit is susceptible to charge noise. The charge noise is difficult to contain than the flux noise. Hence the focus was on creating circuit designs where $E_J \gg E_c$. The transmon qubit's design approaches this problem by creating a shunt capacitor across the junction with large capacitance $C_B \gg C_J$, which reduces the charge noise significantly. With the addition of the shunt capacitance C_B, the charging energy $E_c = \frac{(2e)^2}{2C_\Sigma}$, $C_\Sigma = C_g + C_J + C_B$ is significantly reduced, and the regime moves toward $E_J \gg E_c$.

This design improves the charge qubit by having two Josephson Junctions (these two junctions can be symmetric or asymmetric) in a loop, forming a dc-superconducting quantum interface device (dc-SQUID). The effective current between the two arms of the dc-SQUID can be controlled using an external magnetic flux Φ_{ext}. This control helps in the tuning of the qubit. In addition to the shunt

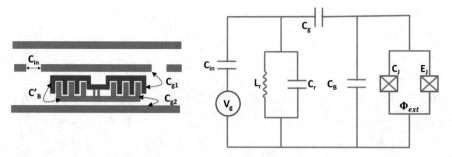

Fig. 4.13 Superconducting split symmetric transmon. (Ref: Koch et al., 2007)

capacitance C_B across the superconductors, a proportional gate capacitance C_g is added to the circuit (Fig. 4.13).

We treat the split Josephson Junctions as a single junction with energy E'_J. The Hamiltonian of this qubit is given by the following equation, ignoring the effects of the transmission line introduced in the circuit:

$$H = 4E_c (N - N_g)^2 - 2E_J |\cos (\varphi_e)| \cos (\phi),$$ (4.12)

where $\varphi_e = \frac{\pi \Phi_{ext}}{\Phi_0}$ and $E'_J(\varphi_e) = 2E_J |\cos (\varphi_e)|$.

The ratio E_J/E_C determines the anharmonicity of the energy levels and the charge dispersion. The charge dispersion relates to the susceptibility to charge noise, and the anharmonicity in the energy levels correspond to the device's ability to perform as a two-state system.

From the second term of the above equation, we can infer that the Josephson energy E'_J ranges from 0 to $2E_J$, and it can be tuned using the external flux Φ_{ext}. While this helps to reduce charge dispersion, it increases the qubit's sensitivity to the flux. A solution to this problem is by using asymmetric junctions, which suppresses the flux sensitivity across the range. The Hamiltonian of an asymmetric split transmon is given by the equation (4.13):

$$H = 4E_c (N - N_g)^2 - E_{J\Sigma} \sqrt{\cos^2(\varphi_e) + d^2 \sin^2(\varphi_e)} \cos (\phi),$$ (4.13)

where $E_{J\Sigma} = E_{J1} + E_{J2}$, the respective junction energies. The junction asymmetry parameter $d = (\gamma - 1)(\gamma + 1)$ with $\gamma = \frac{E_{J2}}{E_{J1}}$.

Here again, we can treat the two junctions as a single junction with energy $E'_J(\varphi_e) = E_{J\Sigma} \sqrt{\cos^2(\varphi_e) + d^2 \sin^2(\varphi_e)}$.

When $d = 0$, this equation reduces to the symmetric case as in Eq. (4.12).

When $|d| \rightarrow 1$, the asymmetric energy $E'_J(\varphi_e) \rightarrow E_{J\Sigma}$, and the equation reduces to the single junction Eq. (4.11).

This setup suppresses the flux sensitivity across the range and also reduces the range itself. So the qubit can be calibrated using a small tuning frequency against fabrication variances. This small tuning range decreases the susceptibility to flux noise and hence improves coherence.

In recent times, new qubit designs that protect against noise at the hardware level—charge-parity qubits [31], fluxon-pair qubits [28], and $0 - \pi$ qubits [30] have been reported. Section 9.6 provides an introduction to these qubits with fault-tolerance at the hardware level.

Because of the scalable potential of this technology, several companies and research labs are working on this technology. Some of them are listed in the table below [26] (Table 4.2).

Table 4.2 List of research institutes and companies working on superconducting qubits

CAS-Alibaba Quantum Computing Laboratory	Bleximo	Chalmers University of Technology
Google	IBM	Intel
IQM Finland	MDR	MIT
Origin Quantum Computing	Oxford Quantum Circuits	Oxford University
Quantic	Quantum Circuits Inc	Qutech
Raytheon BBN	Rigetti	SeeQC.EU
University of California, Santa Barbara	University of Science & Technology of China	University of Waterloo—IQC
University of Wisconsin	Yale Quantum Institute	

4.6.3 Topological Qubits

In the quest for building fault-tolerant quantum computers, several techniques are researched. In the last section, we came across several improvements to a standard Josephson Junction—the shunt capacitance, the transmon, and the usage of asymmetric Josephson Junctions. Since quantum computing is very niche at this time of writing, it receives much funding for research, and researchers experiment with many avenues to achieve quantum supremacy. One such technology currently being researched is based on the application of Majorana fermions into a modality called **Topological qubits**, which is novel and promising. Topological qubits use Majorana zero modes to create highly degenerate ground states in which quantum information can be encoded nonlocally, protecting it from local noise. At this time of writing, there are no functional topological qubits to experiment with. However, the underlying physics is quite fascinating, and this section is devoted to exploring this modality from the theoretical perspective.

4.6.3.1 Majorana Fermions

Majorana's original theory proposed that neutral spin-$\frac{1}{2}$ particles are identical to their antiparticles. Particles of this kind are known as **Majorana fermions** and can be described by a real wave equation, called the **Majorana equation**. This is a real wave equation because the wavefunctions of antiparticles can be obtained from complex conjugation of the wavefunctions of particles.

We can describe the difference between Dirac fermions and Majorana fermions from the corresponding creation and annihilation operators. The creation operator $a_j^\dagger$ creates a fermion in state j. The annihilation operator a_j annihilates the fermion or, in other words, creates an antiparticle. The creation and annihilation operators obey the following conditions:

$$\left\{a_i, a_j^\dagger\right\} = \delta_{ij}$$
$$\left\{a_i, a_j\right\} = \left\{a_i^\dagger, a_j^\dagger\right\} = 0 \tag{4.14}$$

The Majorana fermions are antiparticle to themselves. Hence, we can say that the Majorana fermions are created by the operator given below:

$$\gamma_i = a_i + a_i^\dagger, \tag{4.15}$$

which means, $\gamma_i = \gamma_i^\dagger$. Thus, the creation and annihilation operators are the same for Majorana fermions, while for a Dirac fermion, they are different. This behavior is somewhat difficult to comprehend. We cannot easily imagine the physical systems where this scenario is possible. For example, if we consider the case of electrons, the creation operator $a_j^\dagger$ creates an electron of charge $-e$. The annihilation operator a_i creates a hole of charge +e. This means, γ_i creates a superposition of two different charges, which is not ordinarily possible. However, this is possible in superconductors, where electrons pair up to form Cooper pairs (which are bosons), which condensates. We shall study this possibility in detail later in this chapter.

4.6.3.2 Topological Quantum Computing

Topology is a mathematical study of the properties of manifolds (that is, geometrical objects—set of points that forms a topologically closed surface), which are preserved under local smooth deformations such as stretching, twisting, crumpling, and bending, but not abrupt global deformations such as tearing or gluing. Such smooth deformations can change the object locally, but the topology remains the same.

In the past chapters, we saw that coherency is an essential criterion for quantum computing. The ability to store quantum states for extended periods and unitarily evolve them to the desired state is critically essential. To do that, we have to ward off local perturbations and sources of noise. If we can store the quantum state using the topological properties of matter, and evolve them using topology, such systems are protected from local perturbations. They are resilient to local sources of errors such as electron-phonon interaction and hyperfine electron–nuclear interaction, which are common error sources in superconducting devices.

The Russian-American Physicist **Alexei Kitaev** proposed that surface codes for quantum error correction can be viewed as spin-lattice models, where the elementary excitations are anyons. The quantum states of these excitations can be changed by moving the anyons on topologically equivalent paths. These topological paths could be used for constructing quantum gates, and hence universal quantum computing can be done.

In Chap. 1, we learned about anyons and their behavior when two particles are exchanged (or technically, braided). Let us examine the anyons in some detail.

Assume two indistinguishable particles A and B, whose wavefunctions are ψ_1 and ψ_2. The two wavefunctions explore all possible paths the particles can take on their journey between the initial and final positions. Figures 4.14 and 4.15 show the trajectories[1] of particles A and B in a 2 + 1 dimensional space-time as worldlines or strands. In Fig. 4.14, we show two different paths, the particles A and B can take in time, called the **worldline**. In figure (a), the first particle starts at A and reaches C. The second particle starts at B and reaches D. Figure (b) illustrates an elementary braid operation. In this braid, the first particle starts at A and reaches D. The second particle starts at B and reaches C.

Since the particles are indistinguishable, the end states are similar. However, their worldlines have different topologies. If we assume that the particles are *interchanged* as shown in Fig. 4.14b, we can write the state transition:

$$|\psi_1\psi_2\rangle = e^{i\theta}|\psi_2\psi_1\rangle. \tag{4.16}$$

The phase factor $e^{i\theta}$ is 1 for bosons (which obey Bose–Einstein statistics) and -1 for fermions (which obey Fermi–Dirac statistics). We can interpret this equation as follows: when two particles rotate anticlockwise about each other for half a turn, they return to the same quantum state, except by getting multiplied by a complex phase factor $e^{i\theta}$. For clockwise rotations, it would be $e^{-i\theta}$.

---

[1] See Appendix for braid word labeling convention.

Fig. 4.14 Worldline of
two indistinguishable
particles. (**a**) Point A
travels to C. Point B travels
to D. (**b**) Elementary braid
operation

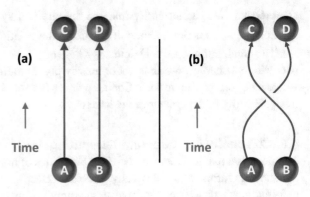

Fig. 4.15 Anyons
interchanged twice, in the
same way, gives rise to the
identity

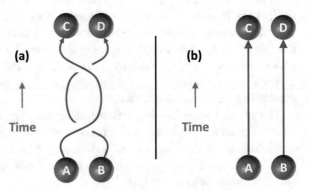

This process works very well in a two-dimensional space, where the clockwise and anticlockwise directions make sense. When θ is 0, it represents bosons, and when it is π, it represents fermions. However, between these two states ($\theta = 0$ for bosons and $\theta = \pi$ for fermions), there could be other particles which we call **anyons**, and we refer them as anyons with statistics θ.

The anyons have a peculiar property that, if they revolve about each other for half turn in the anticlockwise direction to swap places, and if they revolve again for half turn in the clockwise direction to go back to their original places, their wavefunction is not the same, but multiplied by a phase factor $e^{2i\theta}$. This is a sort of memory that helps us to track the trajectory of the particles. By controlling the path, the anyons can take, we can implement a two-state system. This system is one way to implement a topological qubit.

4.6.3.3 Abelian and Non-Abelian Anyons

In a system of N indistinguishable anyons, as time evolves from $t = t_i$ to $t = t_f$, the particles move from the initial positions $R_1, R_2, \ldots, R_N$ to the final positions $R_1, R_2, \ldots, R_N$, slowly in such a way that they do not excite the system out of the ground state. Their worldlines appear like a spaghetti. The history of each of the N particles' trajectory forms a **braid**, and the system forms a braid group B_N. The elements of the braid group $\{b_1, b_2, \ldots, b_{N-1}\}$ satisfy the following conditions:

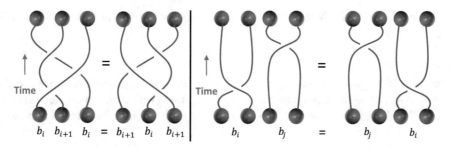

Fig. 4.16 Two braid operations of Eq. (4.17)

$$b_i b_j = b_j b_i \text{ for } |i - j| \geq 2 \leftarrow \text{Commutativity}$$
$$b_i b_{i+1} b_i = b_{i+1} b_i b_{i+1} \text{ for } 1 \leq i < N \leftarrow \text{Braid relation} \qquad (4.17)$$

These two braid operations are shown in Fig. 4.16.

The braids can be deformed to one another, without cutting the strands. An arbitrary element of the braid group is given by $e^{im\theta}$, where m is the sum of the total number of times the particle winds around another particle in clockwise and anticlockwise directions. The order of the braiding is not essential here. Particles of this type are **abelian anyons**.

Now, let us focus on the non-abelian anyons. Consider a set of g degenerate states of N particles at fixed positions $R_1, R_2, \ldots, R_N$. Consider an orthonormal basis ψ_i, $i = 1, 2, \ldots, g$ of these degenerate states. If we exchange two particles, say 1 and 2, the state can be transformed into another wavefunction $\psi_i' = U\psi_i$ where U is a $g \times g$ transformation matrix. Instead, if we exchange another set of particles, say 3 and 4, this can give rise to another wavefunction $\psi_i' = V\psi_i$. The kind of particles for which these two matrices U and V do not commute are known as **non-abelian anyons**. The transformation matrix is diagonal and ± 1 for fermions and bosons. For abelian anyons, the diagonal elements are $e^{i\theta}$. For non-abelian anyons, the transformation matrix is nondiagonal.

A minimum of two degenerate states is required for a particle to be a non-abelian anyon. Hence, we see that the ground states of non-abelian anyons are degenerate. If the ground states can be separated from the other states, the topological order of the ground state can be used for quantum computing. If the total number of degenerate states is less than two or if the matrices U and V commute, then the particles belong to the abelian group.

4.6.3.4 Fusion Properties

In Chap. 1, we saw that two electrons (fermions) combine to form bosons (Cooper pairs). Similarly, the anyons can combine to form a composite anyon. We saw that an anyon exchange leads to a phase factor of $e^{i\theta}$. In a system of N particles, we can generalize this as $e^{iN^2\theta}$. If there are two abelian anyons with statistics θ, we can make a bound state of these two particles with statistics 4θ. If we can bring two such abelian anyons close together, they can be approximated as a single particle by combining the quantum numbers. This process is called **fusion**. When we bring together an anyon and an anti-anyon, it results in a state of zero particle called **vacuum or the trivial particle** denoted in boldface as **1**. This process applies to abelian anyons, and they form larger composites of $\frac{\pi}{m}$ particles.

When two non-abelian anyons are brought together, the outcomes are probabilistic, as there is no unique way of combining the topological quantum numbers. For example, two spin-$\frac{1}{2}$ particles can combine to form a spin-0 or a spin-1 particle, and the outcome of this combination is probabilistic. The different possibilities in which the non-abelian anyons combine to form new particles are through different **fusion channels**. Each of these fusion channels is responsible for the degenerate multiparticle states.

Fusion Space

In an anyon model spanned by some particles $M = \{1, a, b, c, \ldots\}$, where a, b, c are topological charges carried by the anyons when a particle of species a fuses with another particle of species b, the fusion channel is defined as a sum of all possible anyons in the system, including the trivial particle $c = \mathbf{1}$:

$$a \times b = \sum_c N_{ab}^c \quad c \tag{4.18}$$

The fusion coefficient[2] N_{ab}^c is the total number of distinct quantum states through which a and b can merge. The result of this fusion can be a particle c if the fusion coefficient $N_{ab}^c \neq 0$.

In the case of non-abelian anyons, there is at least one a,b for which there are multiple fusion channels c with $N_{ab}^c \neq 0$. For abelian anyons, for each a and b, there exists a c such that $N_{ab}^c = 1$ and $N_{ab}^c = 0$ for everything else. The fusion process corresponds to a Hilbert space consisting of all distinct orthonormal quantum states of a and b defined as $\langle ab; c | ab; d \rangle = \delta_{cd}$.

Hence, a system of N distinct fusion channels in the presence of a pair of particles exhibits N fold degeneracy, called **fusion space**. This fusion space is a protected low-energy, decoherence-free subspace, in which local perturbations do not affect the degeneracy. This subspace is a collective nonlocal property of the anyons, and it can be used to encode quantum information.

The fusion states $|ab; c\rangle$ and $|ab; d\rangle$ belong to two different topological charge sectors defined by c and d. Hence, they cannot be used to create a superposition. Therefore, we cannot use these two fusion states to create a qubit. We need more than two anyons which can fuse in different ways but produce the same result.

Consider three anyons a, b, and c, which may fuse in some order producing intermediate outcomes, but always produce d at the end. There are two distinct basis for these three anyons, as shown in Fig. 4.17. In the first scenario, a and b fuses to produce i. In this case, the basis states are labeled by i and denoted by $|(ab)c; ic; d\rangle$. In the second scenario, b and c fuse to produce j. In this case, the basis states are labeled by j and denoted by $|a(bc); aj; d\rangle$. These two are related by a fusion matrix F_{abc}^d, which is determined by the **Pentagon identities**. This relation is shown in the following equation:

$$|(ab)c; ic; d\rangle = \sum_j \left(F_{abc}^d\right)_i^j |a(bc); aj; d\rangle \tag{4.19}$$

This is summed over all the anyons b and c can fuse to subject to the condition $N_{bc}^j \neq 0$, which we saw earlier. Thus, the F matrix describes the fusion space.

Figure 4.18 shows the fusion outcome 5 of four anyons 1, 2, 3, 4. The first diagram has intermediate outcomes a and b. The two F moves in the upper path lead to the final fusion diagram with intermediate outcomes c and d. The alternative path on the lower part of the diagram has three F moves and produces the same outcome. We require that these two paths are equivalent and produce the same outcome. These two paths are stated in the following pentagon identity equation:

---

[2] Note the potential notation difference: in some literature, the subscript and superscript can be reversed.

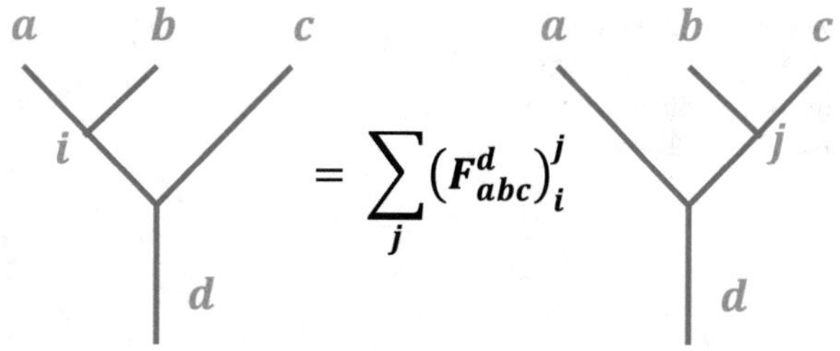

Fig. 4.17 Fusion diagram or the **F-move** of three anyons

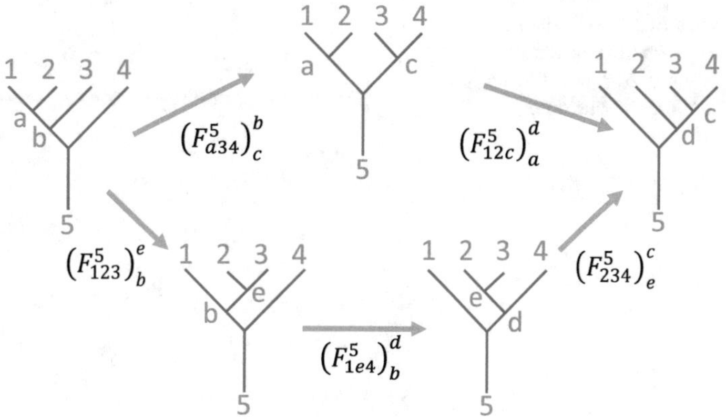

Fig. 4.18 The pentagon identity

$$\left(F_{12c}^5\right)_a^d \left(F_{a34}^5\right)_c^b = \sum_e \left(F_{234}^d\right)_e^c \left(F_{1e4}^5\right)_b^d \left(F_{123}^5\right)_b^e. \tag{4.20}$$

The summation e is over all types of possible particles in the system.

Braiding Operations

When anyons are exchanged (braided), the fusion space undergoes a unitary evolution. We saw earlier that in the case of abelian anyons, the exchange operation results in a complex factor. We also saw that the phase factor depends upon the type of the anyon and the direction of the exchange, but not on the order of exchange. We can now define an exchange operator $R_{ab} = e^{i\theta_{ab}}; 0 \geq \theta_{ab} \geq 2\pi$, which applies to abelian anyons.

In the case of non-abelian anyons, the resultant phase depends upon the type of the anyons and the fusion channel. We can define a corresponding exchange operator as $R_{ab}^c = e^{i\theta_{ab}^c}$. Here, we see that braiding assigns different phases to different fusion channels that depend on the orientation of the exchanges and the order of the exchange. These R matrices (derived from a set of identities known as **Hexagon identities**) describe all unitary evolutions happening in a fusion space. Considering the basis $|(ab)c; ic; d\rangle$ from Eq. (4.19), the clockwise exchange of a and b can be written as a unitary evolution:

$$|(ba)c; ic; d\rangle = \sum_j R_{ab}^j \, \delta_{ij} \, |(ab)c; ic; d\rangle. \tag{4.21}$$

Fig. 4.19 Clockwise exchange of non-abelian anyons producing a phase factor—the **R-move**

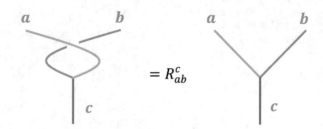

Fig. 4.20 The hexagon identities. Two sequences of alternating fusion and braiding operations leading to the same outcome

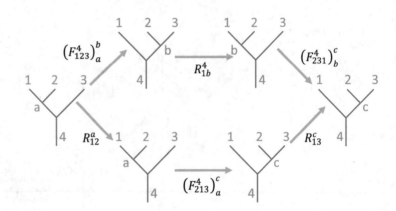

Figure 4.19 shows the clockwise exchange of non-abelian anyons a and b, producing the fusion outcome c with a phase factor R_{ab}^c.

Consider the anyons 1, 2, 3, which fuse to produce 4. The fusion channel is through a. By applying an alternate sequence of F and R moves, we can interchange the fusion order of the anyons and create two distinct paths producing the same fusion outcome. These two paths are shown in Fig. 4.20.

The two paths are required to yield the same outcome, and we can describe this in the hexagon equation as follows:

$$\sum_b \left(F_{231}^4\right)_b^c R_{1b}^4 \left(F_{123}^4\right)_a^b = R_{13}^c \left(F_{213}^4\right)_a^c R_{12}^a. \tag{4.22}$$

Summary

Thus, an anyon model can be described by three primary factors—the fusion coefficients N_{ab}^c, which defines the number of distinct anyons and how they fuse, the F matrices that define the structure of the fusion space, and the R matrices, which defines the mutual statistics of the anyons. The pentagon and hexagon identities are self-sufficient in describing the non-abelian models, and we do not need further conditions. However, these two identities are trivial for abelian anyons, which can have arbitrary statistical phases. For a given set of fusion rules, the solution to these two equations is a discrete set of F and R matrices known as **Ocneanu rigidity** [15], a state without deformations. The discrete nature of topological models is thus resilient to local perturbations.

We have now learned about the theoretical model of anyons. We can now apply the model to two types of non-abelian anyons considered for quantum computing.

Exercise 4.1

Derive the braiding unitary for two anyons that do not have a fusion channel. (Hint: Use F- moves to rearrange the order of fusion.)

◄

Exercise 4.2

Derive the equivalent of the pentagon equation by using anticlockwise exchange operator R^{-1}.

◄

4.6.4 Anyon Model

4.6.4.1 Fibonacci Anyons

The Fibonacci anyons [16] are a family of non-abelian anyons with labels 1 and τ. The following fusion rules apply to them:

$$
\begin{aligned}
1 \times 1 &= 1 \\
1 \times \tau &= \tau \\
\tau \times \tau &= 1 + \tau \ ,
\end{aligned}
\tag{4.23}
$$

where the $\times$ symbol denotes the fusion operation, and the $+$ symbol denotes the possible fusion channels.

The fusion rules of the Fibonacci anyons are shown in Fig. 4.21. The τ anyons are antiparticles to themselves, and two τ anyons can behave like a single τ anyon. The repeated application of the fusion rule grows the dimension of the fusion space, which looks like the Fibonacci series, and hence the name. This is shown in Eq. (4.24):

$$
\begin{aligned}
\tau \times \tau \times \tau &= 1 + 2\tau \\
\tau \times \tau \times \tau \times \tau &= 2 \cdot 1 + 3\tau \\
\tau \times \tau \times \tau \times \tau \times \tau &= 3 \cdot 1 + 5\tau.
\end{aligned}
\tag{4.24}
$$

As in Eq. (4.19), we need three τ anyons to encode a qubit. The basis for this fusion space is given by the states $\{|(\tau\,\tau\,)\tau;\ 1\tau;\ \tau\,\rangle$ and $|(\tau\,\tau\,)\tau;\ \tau\,\tau;\ \tau\,\rangle\}$.

The F matrix and R matrices are given by:

$$
\begin{aligned}
F = F^\tau_{\tau\tau\tau} &= \begin{bmatrix} \Phi^{-1} & \Phi^{\frac{-1}{2}} \\ \Phi^{\frac{-1}{2}} & \Phi^{-1} \end{bmatrix} \\
R = \begin{bmatrix} R^1_{\tau\tau} & 0 \\ 0 & R^\tau_{\tau\tau} \end{bmatrix} &= \begin{bmatrix} e^{\frac{4\pi i}{5}} & 0 \\ 0 & e^{\frac{-3\pi i}{5}} \end{bmatrix},
\end{aligned}
\tag{4.25}
$$

where $\Phi = (1 + \sqrt{5})\,/\,2$ is the **Golden ratio of the Fibonacci series**.

Fig. 4.21 Fusion trees of the Fibonacci anyons

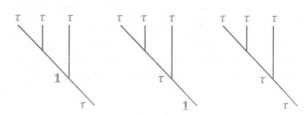

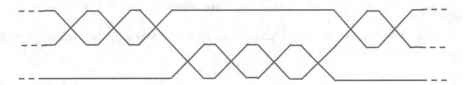

Fig. 4.22 The quantum H gate operation by the braiding of non-abelian anyons

When two τ anyons are fused, from Eq. (4.23), we can say that the fusion outcome **1** and τ occurs each with some probability. We know that the non-abelian anyons have degenerate ground states. In a system of M anyons, the dimensionality of the ground state is d_a^{M-2}. The term d_a is the **asymptotic degeneracy** or the **quantum dimension** of the particle of charge a. Abelian anyons have one-dimensional ground state degeneracy irrespective of the number of particles. Hence, $d_a = 1$ for abelian anyons. The quantum dimension of non-abelian anyons need not be an integer. The quantum dimension of particle **1** is $d_1 = 1$. The quantum dimension of the particle τ is $d_\tau = \Phi$, the golden ratio. The probability of the fusion outcome **1** is $P_1 = \frac{1}{\Phi^2}$. The probability of the fusion outcome τ is $P_\tau = \frac{\Phi}{\Phi^2} = \frac{1}{\Phi}$.

By braiding these anyons, we can evolve the system unitarily, making these anyons suitable for universal quantum computing. There are some limitations, however. Since there is no tensor product in the structure, if we have to implement two qubits, we need six anyons. Besides, simple gate operations require several braiding operations (Fig. 4.22). Furthermore, realizing these anyons in the laboratories is quite challenging, and some methods are under research.

4.6.4.2 Ising Anyons

The term Ising is named after the German physicist **Ernst Ising**, who proposed a mathematical model for ferromagnetism in statistical mechanics. We can assume a system in equilibrium as an ensemble of many states s, each with their weighted probability P_s. If we want to measure an observable $X(s)$ in the ensemble, its sum runs over all probabilities $\langle X \rangle = \sum_s X(s)P_s$. In a quantum system, the total number of probabilities grows exponentially with the number of particles. So, this is not an ideal model for computation. One way we can solve this problem is by deriving behavior by structuring the ensemble. The Ising model solves the many-body problem by bringing in a structure.

The Ising model is a lattice of sites. Each site has index i, and can have a value ± 1. Let us assume that each site has an unpaired electron. If the electron at a given site is spinning up, then $\sigma_i = +1$. If the electron is spinning down, then $\sigma_i = -1$. These two assumptions correspond to the eigenvalues of Pauli matrices. Each of these electron spins acts as a microscopic magnet with a magnetic moment. If all the individual spins are aligned in one direction, then the lattice has a net magnetic moment, and it behaves like a large magnet. If the individual spins are randomly aligned, they cancel each other out. There is no net magnetic moment, and the lattice is nonmagnetic. We can say that the lattice is a **toy model** of the magnet.

In the Ising model, if we have to calculate the Hamiltonian, we must consider the Hamiltonian arising due to external fields and interactions. In the presence of an external magnetic field, each lattice site feels the external field, and its spin energy is modified. We must calculate the sum over all sites to find the net effect of the external magnetic field. The interaction energy arises due to the interaction between adjacent sites in the lattice. Since each spin behaves like a tiny magnetic dipole, it exerts its magnetic field on its neighbors. The spins either align in one direction or align in random. This spin–spin interaction happens between two sites, and we must sum over pairs of lattice sites to calculate this. Ernst Ising solved the one-dimensional Ising model in 1925.

Let us now return to the Ising anyons and apply our learning of the Ising model.

The family of Ising anyons has three particles $\mathbf{1}$, σ (anyon), ψ (fermion), with the following fusion rules:

$$
\begin{aligned}
&\mathbf{1} \times \mathbf{1} = \mathbf{1} \quad \psi \times \psi = \mathbf{1} \quad \sigma \times \sigma = \mathbf{1} + \psi \\
&\mathbf{1} \times \sigma = \sigma \quad \psi \times \sigma = \sigma \\
&\mathbf{1} \times \psi = \psi
\end{aligned}
\tag{4.26}
$$

These fusion rules can be explained better in the context of a p-wave superconductor (Sect. 1.7.5) In a p-wave superconductor, the trivial particle $\mathbf{1}$ behaves like a Cooper pair condensate. The fermions ψ are **Bogoliubov quasiparticles**, which pair up to become Cooper pairs and become trivial particles. The σ anyons are Majorana zero modes bound to vortices. The Majorana zero modes correspond to $\frac{1}{2}$ of a complex fermion mode. Hence, we need a pair of vortices to host a single nonlocal fermion mode (ψ). The vortices are either unoccupied if the fusion role $\sigma \times \sigma = \mathbf{1}$ is used, or occupied if the fusion rule $\sigma \times \sigma = \psi$ is used. Hence this is a two-dimensional fusion space associated with two fusion channels $\{|\sigma\sigma, \mathbf{1}\rangle, |\sigma\sigma, \psi\rangle\}$, namely, the **vacuum channel** and the **Majorana channel** respectively. These two fusion channels represent two different ψ-parity sectors and hence cannot be used for creating quantum superposition. Hence, we need three or more anyons, which can fuse to a single σ or $\mathbf{1}$ in two different ways using repeated application of the fusion rules:

$$
\begin{aligned}
&\sigma \times \sigma \times \sigma = 2\sigma \\
&\sigma \times \sigma \times \sigma \times \sigma = 2 \cdot \mathbf{1} + 2\sigma \\
&\sigma \times \sigma \times \sigma \times \sigma \times \sigma = 4\sigma.
\end{aligned}
\tag{4.27}
$$

From these fusion rules, we can infer that each time a pair of σ anyons are added to the system, the dimensionality of the fusion space doubles. The basis for the fusion of two left most anyons into $\mathbf{1}$ or ψ, is $\{|(\sigma\,\sigma)\sigma;\, \mathbf{1}\sigma;\, \sigma\rangle$ and $|(\sigma\,\sigma)\sigma;\, \psi\sigma;\, \sigma\rangle\}$. The fusion matrix for changing the basis from right to left is given by:

$$
F = F^{\sigma}_{\sigma\sigma\sigma} = \frac{1}{\sqrt{2}} \begin{bmatrix} 1 & 1 \\ 1 & -1 \end{bmatrix}.
\tag{4.28}
$$

The fusion matrix resembles the Hadamard gate, which we shall see in the next chapter and creates an equal superposition. We see that the tensor product is inbuilt into the Ising anyons, and the fusion orders correspond to the basis we need to create a qubit.

The R matrix corresponding to the exchange of two left most anyons in the clockwise direction is given by:

$$
R = \begin{bmatrix} R^{\mathbf{1}}_{\sigma\sigma} & 0 \\ 0 & R^{\psi}_{\sigma\sigma} \end{bmatrix} = e^{-i\frac{\pi}{8}} \begin{bmatrix} 1 & 0 \\ 0 & i \end{bmatrix}.
\tag{4.29}
$$

To create the qubits, we pull out pairs of σ anyons from the vacuum. We need four such σ anyons to form one qubit and six to form two qubits. Since the pairs are initialized in the $\sigma \times \sigma = \mathbf{1}$ fusion channel, the system generally belongs to the vacuum sector, and the qubits are initialized to the state $|0\rangle$.

Gate operations can be performed by braiding operations. For example, the X gate can be created using a $F^{-1}R^2F \otimes \mathbf{1}$, and the Z gate can be implemented by a $R^2 \otimes \mathbf{1}$ transaction (working this out is a good exercise.) It has been proved that the braiding operations of Ising anyons can only create the quantum gates belonging to the **Clifford group**. Hence, it cannot implement a universal quantum computer [16], without additional nontopological elements. One way is to bring the anyons closer nontopologically and then precisely separate them, in a process where we can create a non-Clifford $\frac{\pi}{8}$ phase gate. However, any nontopological interaction between the anyons cancels the degeneracy of the fusion space, and the system dephases quickly. Nevertheless, since the Ising anyons can be realized as Majorana zero modes in nanowires, it is a candidate for experimenting topological quantum computing.

To perform a measurement, we bring the anyon pairs physically closer and measure the fusion outcome. Since the σ anyons can fuse to $\mathbf{1}$ or ψ, the corresponding energy states are detectable, and hence we can measure the qubit on the computational basis.

Errors can happen due to stray anyons that can get introduced into the system. These stray anyons can cause unexpected braiding operations and tweak the results of the experiments. Unexpected braiding can be avoided by containing the stray anyons to the vacuum channels. Besides, the anyons can tunnel into neighboring topological structures, which requires isolation.

4.6.4.3 Sources of Anyons

In Chap. 1, we saw that anyons are found in two-dimensional spaces, where the quasiparticle excitations are above the ground state. Such systems are in the topological phase when all observable properties are invariant under smooth deformations of the space-time manifold. The **topological phase of matter** that supports anyons is categorized into two classes: symmetry-protected topological order and intrinsic topological order depending upon whether they preserve symmetry.

Topological states which preserve some of the fundamental symmetries are known as **symmetry protected topological order (SPTs)**. Quantum spin Hall states, superconductors and topological insulators are some examples which protect symmetry. When these systems have defects such as vortices (the natural defects in superconductors) in superconductors or lattice dislocations, they may bind local zero-energy states which behave like anyons. In topological superconductors, the zero-energy modes are half of the fermion modes called Majorana fermions. The Majorana zero modes behave like non-abelian anyons, which can occur in SPT states of free fermions.

Intrinsic topological order does not preserve symmetry and does not require defects. This property occurs in fractional Hall effect states and spin liquids, which support intrinsic quasiparticle excitations as anyons. The intrinsic topological order has two properties: topological ground state degeneracy and topological entanglement entropy, which we see next.

Topological ground state degeneracy applies to any intrinsic topological state, including abelian and non-abelian anyons. The level of degeneracy depends upon the anyon model and the spatial manifold. For example, the system state on a sphere would have a unique ground state, as the spatial manifold has no holes. On a torus, the ground state would degenerate, as the spatial manifold has a hole. By designing the spatial manifold in such a way to create a certain level of degeneracy, **quantum memories** can be implemented. We explore this in detail in Chap. 9.

In general, a quantum state can be split into two disjoint regions A and B. The degree of entanglement between these two regions can be measured by the **topological entanglement entropy**, which is proportional to the length of the boundary $|\partial A|$ between them. The topological entanglement entropy can be written as an equation:

$$S = \alpha|\partial A| - \gamma, \tag{4.30}$$

where α is a constant, and γ is nonzero in intrinsic topological order.

In the first property—topological ground state degeneracy, the degeneracy depends upon the manifold's topology. In the second property—topological entanglement entropy, γ depends upon the anyon model. Unfortunately, these two are not unique properties of anyon models. Hence the ground state degeneracy must be derived using some advanced methods, and currently, we do not have any viable methods for quantum computation that uses intrinsic topological order.

N. Read and **Dimitry Green** from the Yale University established [18] in 1999 that in a spinless p-wave superconductor, the weak and strong pairing of the fermions in two dimensions with broken charge (C) and time-reversal (T) symmetry is topological. The weak pairing phase exhibits non-abelian statistics, and the strong pairing phase exhibits abelian modes. The transition between the two phases involves a Majorana fermion. We do not yet have much experimental proof for this method at this time of writing, but what is promising is a Majorana wire.

When nanowires made of **half metals** (half metals conduct electrons of one type of spin and insulate the other) such as strontium ruthenate are placed on top of a p-wave superconductor with unidirectional spin, they host Majorana modes at the end of the wires or at the domain walls (that is, between topological and trivial phases) under certain orientation of the magnetic field. This behavior is an example of an SPT state, and the Majorana modes are Ising anyons. These nanowires are topologically free from noises and are candidates for quantum computing (Fig. 4.23).

Kitaev's one-dimensional toy model [19] explains the Majorana modes at the ends of the nanowires. Assume a one-dimensional lattice, occupied by composite fermions at L sites. Let us define a pair of annihilation and creation operators $a_j, a_j^\dagger$ for each of these fermions. We have seen that the Majorana mode is half of a composite fermion mode. So, we need two Majorana modes to make up a composite fermion mode. The two Majorana operators $\gamma_{2j-1}, \gamma_{2j}$ pertaining to the site j can be written as:

$$\gamma_{2j-1} = a_j + a_j^\dagger, \qquad \gamma_{2j} = \frac{a_j - a_j^\dagger}{i} \quad , \quad \text{where } j = 1, 2, 3, \ldots, N. \tag{4.31}$$

If the fermionic mode is not occupied (that is, $a_j^\dagger a_j = 0$), then the corresponding Majorana mode is the vacuum. If the fermionic mode is occupied (that is, $a_j^\dagger a_j = 1$), then the corresponding Majorana mode is the particle ψ.

If the Majorana modes γ_1 and γ_2 exist as quasiparticle excitations localized far apart, then the composite fermion shared by them is nonlocal.

Figure 4.24 illustrates this one-dimensional lattice structure.

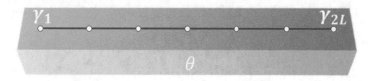

Fig. 4.23 A Majorana wire placed on top of a three-dimensional p-wave superconductor with spin-triplet pairing. The ends of the wire display Majorana mode

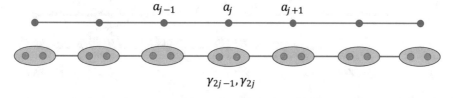

Fig. 4.24 Kitaev's one-dimensional toy model showing the occupation of composite fermions at sites along a one-dimensional lattice and their corresponding Majorana modes

The Hamiltonian for the superconducting fermions along this one-dimensional wire of length L is given by the following equation:

$$H = \sum_{j=1}^{L} \left[-w \left(a_j^\dagger a_{j+1} + a_{j+1}^\dagger a_j \right) - \mu \left(a_j^\dagger a_j + \frac{1}{2} \right) + \Delta a_j^\dagger a_j + \Delta^* a_{j+1}^\dagger a_j \right], \qquad (4.32)$$

where w is the tunneling amplitude, μ is the chemical potential, and $\Delta = |\Delta| e^{i\theta}$ is the superconducting pairing potential. For simplicity, the phase factor θ can be encoded into the Majorana operators, which can be rewritten as:

$$\gamma_{2j-1} = e^{\frac{i\theta}{2}} a_j + e^{-\frac{i\theta}{2}} a_j^\dagger, \qquad \gamma_{2j} = -i e^{\frac{i\theta}{2}} a_j + i e^{-\frac{i\theta}{2}} a_j^\dagger, \qquad (4.33)$$

where, $j = 1, 2, 3, \ldots, L$.

With this, the Hamiltonian can be rewritten in terms of the Majorana operators as:

$$H = \frac{i}{2} \sum_{j=1}^{L} \left[-\mu \gamma_{2j-1} \gamma_{2j} + (w + |\Delta|) \gamma_{2j} \gamma_{2j+1} + (-w + |\Delta|) \gamma_{2j-1} \gamma_{2j+2} \right]. \qquad (4.34)$$

Two possible cases can be deduced from this equation.

Case 1

The trivial case in which $|\Delta| = w = 0$, $\mu < 0$. The Hamiltonian, in this case, simplifies to the following:

$$H = \frac{i}{2} (-\mu) \sum_{j=1}^{L} \gamma_{2j-1} \gamma_{2j}. \qquad (4.35)$$

In this case, the Majorana operators $\gamma_{2j-1} \gamma_{2j}$ from the site j are paired together to form a ground state $(a_j^\dagger a_j = 1)$ at every site. This state arises due to the localization of the fermion modes at physical sites. Hence this state is topologically trivial. This state is illustrated in Fig. 4.25.

Case 2

In this case, $|\Delta| = w > 0$, $\mu = 0$. The Hamiltonian for this case can be written as:

$$H = iw \sum_{j=1}^{L} \gamma_{2j} \gamma_{2j+1}. \qquad (4.36)$$

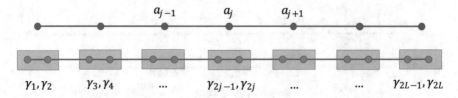

Fig. 4.25 Case #1, Trivial phase arising from the fermion modes binding at physical sites

In this case, we find that the Majorana operators from adjacent sites $\gamma_{2j}\gamma_{2j+1}$ pair together. We can define a set of new annihilation and creation operators $\tilde{a}_j = \frac{1}{2}\left(\gamma_{2j} + i\gamma_{2j+1}\right)$ and $\tilde{a}_j^\dagger = \frac{1}{2}\left(\gamma_{2j} + i\gamma_{2j+1}\right)$ that work on adjacent sites j and $j + 1$. The Hamiltonian can now be written as:

$$H = 2w\sum_{j=1}^{L-1}\left(\tilde{a}_j\tilde{a}_j^\dagger - \frac{1}{2}\right) \tag{4.37}$$

In this case, the Majorana operators γ_1 and γ_{2L} are no longer represented in the Hamiltonian. The Hamiltonian now describes the interactions between $L - 1$ composite fermions only. We can define a new operator that describes the missing fermion as $d = e^{-i\frac{\theta}{2}}\frac{1}{2}(\gamma_1 + i\gamma_{2L})$. This operator is delocalized between the ends of the Majorana wire.

Figure 4.26 illustrates this state.

Since $[d^\dagger d, H] = 0$, this is a twofold degenerate system with delocalized fermion mode ($d^\dagger d = 0, 1$). We might as well add this state to the original Hamiltonian as a zero-energy mode. This is the reason why we call the edge Majorana modes as **Majorana zero modes**. Thus, this system exhibits ground state degeneracy with the edge states where the two Majorana zero modes arise due to the delocalized fermion modes. At the sites where the fermions are localized, the ground states are unique and have finite energy, a characteristic of the SPT state. The ends of the Majorana wires should indeed be viewed as defects which bind anyons.

We must note that the end of the wire can behave like a **domain wall** between the topological phase and the vacuum, which is the trivial phase. Therefore, a domain wall can host a Majorana zero mode. The domain wall is illustrated in Fig. 4.27.

Now, lets us equate this with the Ising anyons. The trivial particle state **1** represents the ground state of the topological phase in which there are neither excitations nor Majorana zero modes. We can assume the Majorana zero modes as the σ anyons. The intrinsic excitations created by the operator $d_j^\dagger$ are the fermionic quasiparticles, and they should be considered as the ψ particles. In a superconductor, the ground states are the condensate of the Cooper pairs formed from fermions. The elementary excitations can be created by breaking up the pairs by spending energy of 2Δ. This is equal to the energy gap between a pair of ψ. As per the fusion rules, these two fermions pair up to become a

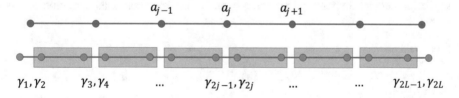

Fig. 4.26 Case #2, Topological phase arising from the pairing of fermion modes with adjacent sites

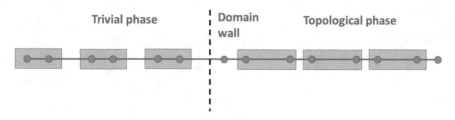

Fig. 4.27 Domain wall between topological and trivial phases hosting a Majorana zero mode

Cooper pair and dissolve into the condensate. This action preserves the parity of ψ. Thus, we can assume that the system is comprised only of σ anyons. From our previous discussions, the state $d^\dagger d = 0$ can be identified with the fusion rule $\sigma \times \sigma = 1$, and we can say that the state $d^\dagger d = 1$ can be identified with $\sigma \times \sigma = \psi$.

The two degenerate states of the ground state can be now written as the orthogonal fusion channels:

$$i\gamma_1 \gamma_{2L} |\sigma\sigma, 1\rangle = -|\sigma\sigma, 1\rangle$$
$$i\gamma_1 \gamma_{2L} |\sigma\sigma, \psi\rangle = +|\sigma\sigma, \psi\rangle \ ,$$
(4.38)

where $|\sigma\sigma, \psi\rangle = d^\dagger |\sigma\sigma, 1\rangle$. However, we cannot use this to construct a qubit, as they belong to two separate charge sectors. To construct a qubit, we need two nanowires.

4.6.4.4 Majorana Qubit

With a pair of Majorana wires, we have four Majorana zero modes. With this, we can create a ground state which has two charge states in each sector. By choosing the odd parity scheme, we can write the computational basis as:

$$|0\rangle \equiv |\sigma\sigma, 1\rangle |\sigma\sigma, 1\rangle$$
$$|1\rangle \equiv |\sigma\sigma, \psi\rangle |\sigma\sigma, \psi\rangle.$$
(4.39)

Quantum gate operations are performed by braiding, and it requires moving the anyons. We can achieve this by adiabatically tuning the chemical potential (μ). By adjusting the chemical potential, we can even make one nanowire behave like multiple segments of Majorana wires. By locally tuning the chemical potential, we can extend or contract the topological domains and move the domain wall that binds the Majorana modes along the wire. Figure 4.28 illustrates how we can implement a Clifford gate with the help of a T junction.

The $\frac{\pi}{8}$-phase gate can be implemented by adjusting the chemical potential in such a way that two domain walls are brought nearby and allowed to dephase for a specific time. This function can ensure the universality of this modality.

To perform a measurement, the Majorana fermions are brought close to each other adiabatically by shrinking the topological domain and allowing them to fuse together. This operation results in the

Fig. 4.28 (a) Braiding operation demonstrating a NOT gate. The dark circles are the Majorana zero modes. The dotted lines along the "T" junction are the trivial phase and the solid lines are the topological phase. (b) Worldlines of these anyons. (Ref: Jiannis K. Pachos)

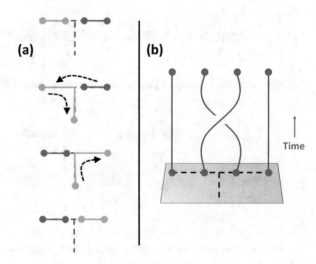

Table 4.3 List of research institutes and companies working on topological qubits

Chalmers University of Technology	Microsoft	QuTech
Neils Bohr Institute	Nokia Bell Labs	University of California, Santa Barbara
Joint Quantum Institute/Institute of Maryland		

system state to either vacuum (**1**) or the fermion ψ. Since the energy associated with the fermion Δ, is relatively high, the differentiation between the states can be easily measured.

A good number of research institutes and companies are working on this technology [26, 27], and Table 4.3 summarizes them:

4.6.4.5 Nobel Prize—2016

David J. Thouless, University of Washington, Seattle, **F. Duncan M. Haldane**, Princeton University, and **J. Michael Kosterlitz**, Brown University won the Nobel Prize for Physics in 2016 for their work on topological phases of matter and topological phase transitions.

4.6.5   Electron Spin/Quantum Dot

Quantum dots take advantage of our decades of accumulated knowledge of making semiconductor devices. In 1997 **Daniel Loss** and **David P. DiVincenzo** proposed [21] a scalable model of making quantum computers by confining the spin-$\frac{1}{2}$ states of electrons inside semiconductor devices, called quantum dots (QDs) In quantum dots, the tunnel barrier between neighboring dots can be changed by altering the gate voltage using electrical circuits. By altering the tunnel barrier, we can enable time-dependent electron–electron interactions that modify the spin states through the Heisenberg exchange coupling. Quantum information could be encoded in the spin states of the quantum dots. This qubit modality has the potential for universal and scalable quantum computing.

There are at least three types of quantum dots under research. In the first category of self-assembled QDs, the InGaAs layers are grown on top of GaAs, where islands form spontaneously due to mismatch in lattice constants. After a critical layer height, the structure is encapsulated in further layers of GaAs. Electron confinement happens due to the differences in the conductance and valence bands of the materials used. The second category of lateral QDs utilizes 2DEG (two-dimensional electron gas) formed between heterogeneous layers of semiconductors. The 2DEG is confined along the growth direction and can be depleted using metal gates mounted at the top of the fabrication. In the third category, semiconductor nanowires are used as the confinement due to their nanosizes, which can be shrouded with other layers to provide the barrier. QDs are also made of colloids, or carbon-based materials such as graphene and are used in displays.

In this book, we examine the second category, the quantum dots.

4.6.5.1 Qubit Physics

In the previous chapter, we learned about the spin-$\frac{1}{2}$ states and how they make a two-state system. Electrons can spin upward $|\uparrow\rangle$ or downward $|\downarrow\rangle$ depending upon the external magnetic field. In this qubit modality, we trap electrons inside quantum dots, fabricated as metal gate electrodes on top of a semiconductor heterostructure outlined in Fig. 4.29.

The semiconductor heterostructure is fabricated with layers of silicon doped gallium arsenide (GaAs) and aluminum gallium arsenide (AlGaAs.) The layers of GaAs and AlGaAs are grown on top of each other using a technique known as **molecular beam epitaxy** (MBE), which produces near-

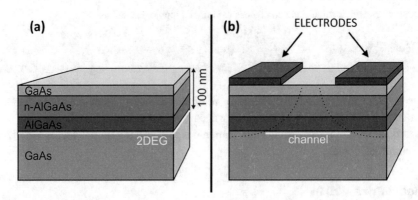

Fig. 4.29 The quantum dot. (**a**) Heterostructure of the semiconductor device with 2DEG between the GaAs and AlGaAs layers. (**b**) The 2DEG is depleted by applying a negative voltage on the electrodes. (Credits: Laurens Willems Van Beveren, with permission to reuse)

perfect crystalline layers of extreme purity with an atomically sharp transition between the layers [22].

In standard semiconductors, the energy levels of the electrons are so close that we can assume that they are *continuous*. Some of the energy levels are forbidden to electrons, and this region is called the **bandgap**. While most of the electrons occupy the energy levels below the bandgap called the **valence band**, some of them may be found in the energy levels above the bandgap called the **conduction band**. Electrons in the valence band can be made to jump to the conduction band by imparting some energy, the minimum of which is equivalent to the bandgap. An electron that leaves the valence band creates a hole, and the pair formed by the excited electron and the hole is called an **exciton**. However, the excited electrons fall back to the valence band in a short time.

In a quantum dot, the energy levels of the electrons are discrete due to the quantum confinement. Hence, quantum dots are also called "artificial atoms." Besides, the bandgap can be modified by the geometry of the quantum dot and by choosing the right materials. One method to increase the bandgap is to replace a few of the Ga atoms with Al atoms. This changes the crystalline structure from GaAs to AlGaAs, and increases the bandgap ΔE_g from 1.42 to 1.79 eV. When layers of GaAs and AlGaAs are stacked up, it creates a bandgap mismatch ΔE_c, due to the difference in the electron affinity between GaAs and AlGaAs. The discontinuity in the valence band is $\Delta E_V = \Delta E_c - \Delta E_g$. These energy levels are shown in Fig. 4.30.

Si atoms are added as dopants in the AlGaAs layer marking it as a donor region (n-AlGaAs). This arrangement causes free electrons to accumulate at the heterojunction in a triangular-shaped quantum well at about 90 nm below the surface of the heterostructure. The electrons in this quantum well are strongly confined in the growth direction, forming discrete states. The electrons can move only on the plane of the interface, forming **two-dimensional electron gas** (2DEG.)

The 2DEG is separated from the donor region by an undoped AlGaAs layer. This separation reduces the scattering of electrons by the Si donors, and electrons with high mobility are concentrated in the 2DEG layer. The heterostructure is to be kept in a dilution fridge. At a temperature of 4.2 K, the 2DEG has mobility of 10^5–10^6 cm^2/V s and a density of ~3×10^{15}/m^2. Compare this with Si at room temperature (300 K): the electron mobility is 1400 cm^2/V s , and density is 9.65×10^9/cm^3. Due to the high mobility of the 2DEG, the GaAs–AlGaAs heterostructure is often called high electron mobility transistor (HEMT) wafer [22]. The electron density is relatively lower. Hence, they can be depleted by applying a potential.

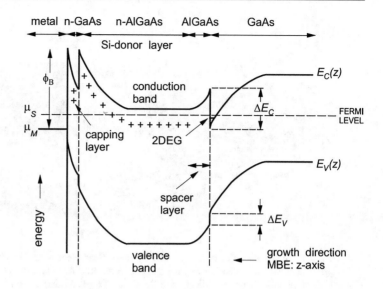

Fig. 4.30 Energy band diagram of the heterostructure. The triangular well is in the conduction band, where we can find the 2DEG. (Source: Laurens Willems Van Beveren, with permission to reuse)

Metal electrodes (~50 nm) are formed at the top of the heterostructure using lithographic techniques. By applying a negative voltage through the electrodes, the 2DEG can be depleted, and we can confine the 2DEG to form a dot of electrons. The gate voltage is defined by the chemical potential of the metal, and the semiconductor: $eV_g = \mu_M - \mu_S$. Φ_B represents the Schottky barrier formed at the metal-semiconductor junction, and it ensures that charges do not leak into the heterostructure. If we examine Fig. 4.30, we can see that the conduction band has a steep slope at the metal–semiconductor junction.

The steep slope of the conduction band is due to the high charge density at the surface, but donors balance this. Thus, most of the donor electrons are required to balance the Schottky barrier, while a fraction makes it to the 2DEG. The 2DEG remains isolated from the surface.

If the gate voltage is higher, it pushes the conduction band below the **Fermi level** (energy of the highest occupied single-particle state in a quantum system) and creates parallel conduction [22]. If the parallel conduction interferes with the qubit operation, the qubit may become unusable. The qubit is usually characterized to avoid this.

The "+" signs in Fig. 4.30, denote the Si doping region. The Fermi level ensures that no net charge transport can happen in the growth direction (Fig. 4.31).

The quantum dot is coupled via tunnel barriers into the source and drain reservoirs, as shown in Fig. 4.32. The quantum dot is coupled with gate electrodes V_g, which can tune the dot's electrostatic potential with respect to the reservoirs. By measuring the voltage V_{SD} between the reservoirs and current I in the circuit, the electronic properties of the dots can be estimated. A **quantum point contact** (QPC) device implemented close to the dots serves as a sensitive electrometer.

Quantum point contact is a narrow constriction in a 2DEG formed by electrostatic gating. The conductance through the QPC channel is quantized in integer multiples of $\frac{2e^2}{h}$, where e is the elementary charge. The transition between the quantized conductance is sensitive to the electrostatic environment, such as the number of electrons in the dot [24].

Figure 4.33 illustrates the quantum dot equipped with QPCs. The dotted circles are the two quantum dots. The gates T, M, R define the quantum dot on the right, and the gates T, M, L define the quantum dot on the left. The gates $R, Q - R$, and $L, Q - L$ for the QPCs on the right and left, respectively. The gates P_R and P_L are connected to pulse sources to perform qubit rotations.

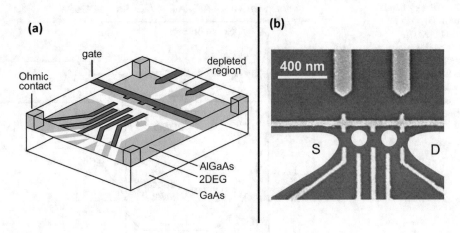

Fig. 4.31 Lateral quantum dot device formed by the surface electrodes. (a) Device schematics showing depleted 2DEG. (b) The two white circles are the quantum dots connected via tunable tunnel barriers to the source (S) and drain (D) reservoirs. (Source: Laurens Willems Van Beveren, with permission to use)

Fig. 4.32 A schematic view of the quantum dot. (Source: Laurens Willems Van Beveren, with permission to use)

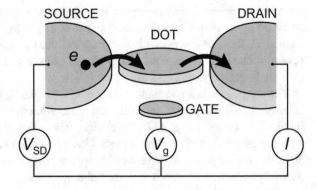

Fig. 4.33 Scanning electron microscope image of the quantum dot. (Source: Laurens Willems Van Beveren, with permission to use)

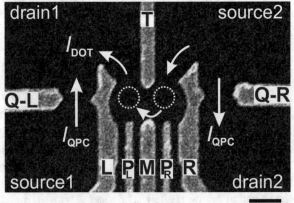

The arrows show the possible direction of the current. A bias voltage V_{DOT} applied between the source2 and drain1 is responsible for the current I_{DOT}. The bias voltages V_{VSD} between the sources and drains of the left and right quantum dots give rise to the respective currents I_{QPC}.

Returning to our discussions, for an electron to tunnel from the source to the dot and then to the drain, the electrochemical potential must be within the bias window: $\mu_s \geq \mu \geq \mu_D$ with $-|e| V_{SD} = \mu_s - \mu_D$, so that the electron does not lose or gain energy. The quantum dot is small, typically in the dimensions around 10–50 nm. Hence it forms a low-capacitance tunnel junction. At low temperatures, the junction resistance increases because the electron charge is discrete. This process is known as the **Coulomb blockade**. Due to the Coulomb blockade, the number of electrons in the dot is fixed, and no current flows through the dot.

By adjusting the gate voltage, the chemical potential of the reservoirs can be adjusted in such a way that only when an electron tunnels off into the drain; another electron can flow into the source. This current is known as a **single-electron tunneling current**. Thus, by tuning the gate voltage, we can control the number of electrons in the quantum dot. Since we can have a precise number of electrons, the quantum dot displays an atom-like shell structure associated with a two-dimensional harmonic potential. When a magnetic field is applied in a direction parallel to the tunneling current, the electrons pair into spin degenerate single-particle states.

In a single electron system ($N = 1$), the two possible spin orientations are $|\uparrow\rangle$ parallel to the magnetic field or $|\downarrow\rangle$ along the antiparallel direction to the magnetic field. The Zeeman splitting between these two basis states is given by $\Delta E_Z = \mu_B \cdot g \cdot \vec{B}$, where μ_B is the Bohr Magneton, $\vec{B}$ is the magnetic field, and g is the gyromagnetic ratio (or the g-factor) that describes the effects of nonpoint charge distribution. This system is equivalent to an artificial hydrogen atom.

When there are two electrons ($N = 2$) in a quantum dot (artificial Helium atom), the electron state is a product of orbital and spin states. Since the electrons are fermions, the state must be antisymmetric in particle exchange. If the orbital part is symmetric, the spin part is antisymmetric and vice versa. The antisymmetric two spin state is the spin-singlet $|S\rangle = \frac{|\uparrow\downarrow\rangle - |\downarrow\uparrow\rangle}{\sqrt{2}}$, with a total spin $S = 0$. This spin-singlet state is the ground state at zero magnetic fields. The symmetric two spin states are the spin-triplet states with a total spin $S = 1$ and a magnetic quantum $m_s = -1, 0, +1$. We can define this as $|T_+\rangle = |\uparrow\uparrow\rangle, |T_0\rangle = \frac{|\uparrow\downarrow\rangle + |\downarrow\uparrow\rangle}{\sqrt{2}}$, and $|T_-\rangle = |\downarrow\downarrow\rangle$. The triplet states are split by the Zeeman energy ΔE_Z. The spin-triplet states are also the lowest excited states.

Figure 4.34a shows the energy diagram for a fixed gate voltage. By changing the gate voltage, the single-electron state below the dotted horizontal line can jump upward or downward relative to the two-electron states. The vertical arrows show the six allowed transactions. Figure 4.34b shows the electrochemical potential difference between the one- and two-electron states. Note that the six transitions correspond to four different electrochemical potentials. By adjusting the gate voltage, the electron levels can be shifted upward or downward. These transitions can easily form a qubit system.

4.6.5.2 Qubit Initialization

The qubit can be initialized to the state $|\uparrow\rangle$ by waiting until the energy relaxation causes the spin to relax to the ground state $|\uparrow\rangle$. This method is robust, but time-consuming, and hence not suitable for quantum error correction. The fast way to initialize the qubit is to place the $|\uparrow\rangle$ level below the Fermi level and $|\downarrow\rangle$ above the Fermi level. If we initialize the qubit in this method, it causes a spin-up electron to stay in the dot, whereas a spin-down electron will tunnel out of the drain and replaced by a spin-up electron tunneling in from the source. After waiting for a few times—the tunnel time, there is a high probability that the dot is filled with a spin-up state.

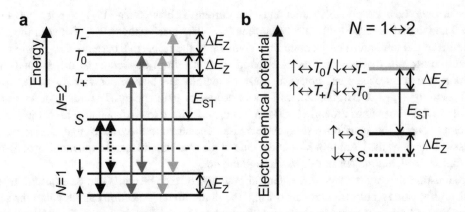

Fig. 4.34 Transitions of one and two electron states and the electrochemical potential. (Source: Laurens Willems Van Beveren, with permission to use)

4.6.5.3 Measurement

There are a few ways of performing measurements in this qubit methodology. One way to perform a single-shot readout of the spin state is to induce a spin-charge conversion using a Zeeman split with a strong magnetic field parallel to the 2DEG, and detecting single-electron tunneling events with the electrometer. Another method is to use an RF QPC tuned to a resonant frequency (~1 GHz) and measuring the reflection of the resonant carrier wave. In another method, the dot is raised above the Fermi level, which causes the electron to leave. The tunneling rate for the spin-up electron is higher than the tunneling rate of the spin-down electron. The spin state can be determined by measuring the charge on the dot within the spin relaxation time.

4.6.5.4 Gate Operations

A microwave magnetic field $\vec{B}_{ac}$ oscillating with a frequency $f = g\mu_B \vec{B}/h$ at a plane perpendicular to $\vec{B}$ causes the spin to transition between $|\!\uparrow\rangle$ and $|\!\downarrow\rangle$ states. Properly timed bursts of microwave power can rotate the spin state over a controlled angle, which helps implement the Pauli gates. The **Heisenberg exchange interaction** that arises due to the coupling between neighboring qubits can be used to build two-qubit gates. The Heisenberg exchange interaction is a quantum mechanical effect between identical particles, due to exchange symmetry.

The Hamiltonian of the exchange interaction between the electron spins $\vec{S}_1$ and $\vec{S}_2$ is: $H = J(t)\vec{S}_1 \cdot \vec{S}_2$. The coupling parameter $J(t)$ is a tunable parameter. The effect of the Hamiltonian is the application of a unitary operator $U(t) = \exp\left(-\frac{i}{\hbar}\vec{S}_1 \cdot \vec{S}_2 \int J(t)dt\right)$. For a certain time t_s, the spin–spin coupling is such that, $\int J(t)\,dt = J_0 t_s = \pi$, and the operator behaves like a swap operator [21]. By adjusting the coupling, $\int J(t)\,dt$, we can make the unitary perform a SWAP operation. If the neighboring qubits have identical spins, the interaction does not change their orientation. However, if the qubits start with opposite spins, then their states will be swapped after the time t_s. An interaction active for half this time performs the $\sqrt{\text{SWAP}}$ gate. This interaction, when combined with single-qubit operations, can implement universal quantum gates [23]. In Sect. 5.5.3, we use the $\sqrt{\text{SWAP}}$ gate to build a CNOT gate.

Table 4.4 List of research institutes and companies working on spin qubits

Archer Materials Limited	C12 Quantum Electronics	Intel
CEA-Leti/Inac	Silicon Quantum Computing	Photonic
RWH Aachen	Hitachi Cambridge Laboratory	Qutech
HRL Laboratories	Sandia National Laboratories	Equal1
Origin Quantum Computing	University of Wisconsin	Simon Fraser University
Quantum Motion Technologies	University of Basel	
Center for Quantum Computation & Communication Technology		

4.6.5.5 Summary

Ever since the first proposals were made, the quantum dots have become a subject of study at various research labs, and many improvements have been made. We now have the ability to manipulate the qubits precisely. The primary challenge with this modality is decoherence, which occurs due to nuclear spins and spin–orbit interactions. These aspects require further studies to channelize these effects. The other challenges are with cabling systems and measurement electronics. Scalability of exchange schemes, surface codes, long-distance spin–spin coupling, exchange interaction in triple quantum dots are some areas where current research is focused.

Quantum dots are a promising technology as they have applications beyond quantum computing. Because of their ability to be controlled purely through electrical circuits, they have a great appeal for the future since they can be fabricated like the regular semiconductor devices.

Table 4.4 lists the organizations and research institutes that are engaged in researching this modality [26].

4.7 A Note on the Dilution Refrigerator

The dilution refrigerator provides a low temperature (cryogenic) environment for hosting the qubits. The current design uses the **heat mixing** of ^{3}He and ^{4}He isotopes, and can produce temperatures as low as 2 mK. The dilution refrigerator [25] was first designed by the German-British physicist **Heinz London** in 1950 and experimentally verified in 1964 at the Leiden University.

Figure 4.35 shows the schematics of the dilution refrigerator.

4.7.1 How Does the Dilution Fridge Cool the Qubits?

The ^{3}He gas is first cooled and purified by liquid nitrogen at 77 K. It is then cooled to a temperature of 4.2 K using a ^{4}He bath. After this stage, the ^{3}He gas enters a vacuum chamber, where it is further cooled to a temperature of 1.2 – 1.5 K by the 1 K bath. Here a vacuum-pump decreases the pressure of ^{4}He, and its boiling point reduces to the 1 K range. At this stage, the ^{3}He gas becomes a liquid.

The liquefied ^{3}He passes through the main and secondary impedance capillary tubes, where a set of heat exchangers cool it to about 500–700 mK using the returning ^{3}He liquid from the main chamber.

In the mixing chamber, the two phases of the ^{3}He–^{4}He mixture are in equilibrium and separated by a phase boundary, as shown in Fig. 4.36. The concentrated phase consists purely of ^{3}He, and the dilute phase is a mixture of about 6.6% ^{3}He, and 93.4% ^{4}He.

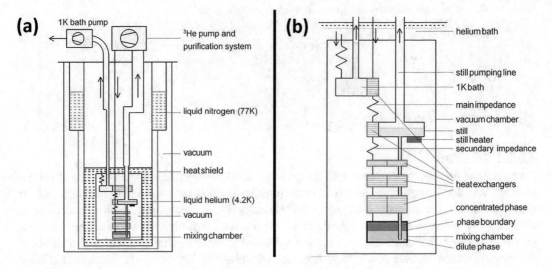

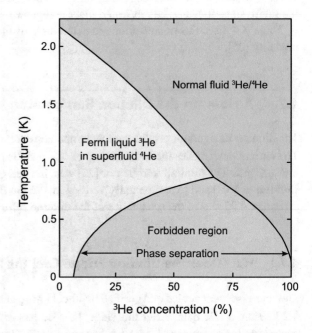

Fig. 4.35 (**a**) Schematic of the dilution refrigerator. (**b**) Schematic of the cold part of the dilution refrigerator. (Image courtesy of Wikipedia. Author: Adwaele, CC BY-SA3.0)

Fig. 4.36 The phase diagram of ^{3}He–^{4}He mixture. (Courtesy of Wikipedia Author: Mets501)

Inside the chamber, the ^{3}He flows through the phase boundary—from the concentrated phase to the dilute phase and gets diluted (hence the name **dilution refrigerator**.) This movement is endothermic and removes heat from the mixing chamber. This endothermic process provides the cooling power of the refrigerator.

The ^{3}He then leaves the mixing chamber, cools the downward following ^{3}He, and reaches the still. In the still, the ^{3}He flows through ^{4}He, which is in a superfluid state. Vacuum pumps keep the pressure in the still to about 10 Pa. The partial pressure difference between ^{3}He and ^{4}He in the still creates an osmotic pressure, which draws the ^{3}He upward from the mixing chamber. The temperature of the still is maintained to ensure a steady flow of ^{3}He. Vacuum pumps take out the ^{3}He from the still, purify, and pump it back at a pressure of a few hundred millibars. This action keeps the cycle flowing.

4.8 Summary

In this interesting chapter, we learned about the vocabulary of quantum computing and qubit metrics. We studied in detail how the qubits are constructed. We learned about the four leading qubit modalities – trapped ion qubits, superconducting transmons, topological qubits, and quantum dots. We ended this chapter after learning about how dilution refrigerators work.

The next chapter focuses on quantum circuits and quantum gates. This chapter also briefs the basic elements of quantum computing.

Practice Problems

1. The spin–orbit interaction of an electron is defined by the Hamiltonian $H = gL \cdot s$, where g is the spin–orbit coupling, L is the angular momentum operator, and s is the spin operator. If the electron is in p state (that is, $l = 1$), calculate the eigenvalues of H.

2. In the context of a superconductor, explain what does coherence length (ξ) mean?
3. Write the braid word for the braid shown in Fig. 4.37.
4. C. F. Gauss' famous four braids are given by the braid word $\sigma_3 \sigma_2^{-1} \sigma_2^{-1} \sigma_1 \sigma_3$. Draw the corresponding braid diagram.
5. **Yang-Baxter Equation** is given below:

$$(R \otimes I)(I \otimes R)(R \otimes I) = (I \otimes R)(R \otimes I)(I \otimes R). \tag{4.40}$$

Fig. 4.37 An example system of 4 braids

Draw a braid diagram showing the braids. Assume that the identity matrix I represents a straight strand and the solution matrix R represents two winding strands. Explain how matrix representation of the braid groups can be used to implement quantum gates.

6. According to definition, a braid word ω is considered to be reduced if it is null or if the main generator of ω occurs only positively or only negatively. For example, the braid word $\sigma_1\sigma_2\sigma_2^{-1}\sigma_1$ is not reduced as the generator σ_2 appears both positively and negatively. Whereas, the braid word $\sigma_1\sigma_2\sigma_3^{-1}\sigma_2$ is reduced. Implement Dehornoy's reduction algorithm [32] on braid words.

7. Develop a logic to swap the status of two Majoraja qubits.

8. How would you implement a CNOT gate on a trapped ion qubit?

9. Device a scheme for transferring qubit status to photons.

10. Define quantum volume. Compute the quantum volume of commercially available quantum computers, based on publicly available information.

11. Differentiate between quantum supremacy and quantum volume.

12. Write a comparative report on the various qubit modalities under research today.

References

1. https://en.wikipedia.org/wiki/Quantum_Turing_machine
2. Deutsch, David, 1985. Quantum theory, the Church-Turing principle and the universal quantum computer.
3. A Polynomial Quantum Algorithm for Approximating the Jones Polynomial, Aharonov et al, 2005, https://arxiv.org/pdf/quant-ph/0511096.pdf
4. Polynomial-Time Quantum Algorithms for Pell's Equation and the Principal Ideal Problem, Sean Hallgren, 2006, http://www.cse.psu.edu/~sjh26/pell.pdf
5. Quantum Computers, Jon Schiller.
6. For scholarly discussions on Quantum Supremacy, refer to these publications: (a) Characterizing Quantum Supremacy in Near-Term Devices, Boixo et al, 2016, https://arxiv.org/abs/1608.00263, (b) Quantum Computational Supremacy, Aram W Harrow, Ashley Montanaro, 2018, https://arxiv.org/abs/1809.07442, (c) How many qubits are needed for quantum computational supremacy?, Dalzell, Harrow et al, 2018, https://arxiv.org/abs/1805.05224
7. Validating quantum computers using randomized model circuits, Cross et al, 2019. https://arxiv.org/pdf/1811.12926.pdf
8. For further reading, Digital Computer Electronics by Jerald A Brown Albert P. Malvino and Electronic Principles by Albert Malvino and David Bates
9. Quantum Computing in the NISQ era and beyond, John Preskill, https://arxiv.org/abs/1801.00862
10. Spin relaxation and decoherence of two-level systems, Wang et al, 2005, https://arxiv.org/pdf/cond-mat/0509395.pdf
11. Superconducting Qubits: Dephasing and Quantum Chemistry, Peter James Joyce O'Malley, 2016, https://web.physics.ucsb.edu/martinisgroup/theses/OMalley2016.pdf
12. The Josephson Effect, Jakob Blomgren 1998, Per Magnelind 2005 , http://fy.chalmers.se/~delsing/LowTemp/Labbar/SQUIDlab-rev3.pdf
13. A Quantum Engineer's Guide to Superconducting Qubits, Krantz et al, 2019, https://arxiv.org/pdf/1904.06560.pdf
14. Charge insensitive qubit design derived from the Cooper pair box, Koch et al, 2007.
15. Lectures on tensor categories, Damien Calaque and Pavel Etingof, 2008, https://arxiv.org/pdf/math/0401246.pdf
16. A Short Introduction to Fibonacci Anyon Models, Simon Trebst et al, 2008, https://arxiv.org/abs/0902.3275
17. For further reading on Clifford group and universality of quantum computing interested readers can read: (a) Implementation of Clifford gates in the Ising-anyon topological quantum computer, Ahlbrecht et al, https://arxiv.org/pdf/0812.2338.pdf, (b) Clifford group, Maris Ozols, http://home.lu.lv/~sd20008/papers/essays/Clifford%20group%20[paper].pdf, (c) Gottesman–Knill theorem, https://en.wikipedia.org/wiki/Gottesman%E2%80%93Knill_theorem
18. Paired states of fermions in two dimensions with breaking of parity and time-reversal symmetries, and the fractional quantum Hall effect. N. Reed, Dimitry Green, 1999, https://arxiv.org/abs/cond-mat/9906453
19. Unpaired Majorana fermions in quantum wires. Alexei Yu Kitaev, 2000, Microsoft Corporation. https://arxiv.org/abs/cond-mat/0010440

20. Physics of p-wave spin-triplet pairing with the experimental examples of Strontium Ruthenate, Suk Bum Chung, Department of Physics, University of Illinois at Urbana-Champaign, Illinois 61801, USA. http://guava.physics.uiuc.edu/~nigel/courses/569/Essays_2004/files/chung.pdf

21. Quantum Computation with Quantum Dots, Daniel Loss, David P. DiVincenzo, 20-July-1997, https://arxiv.org/pdf/cond-mat/9701055.pdf

22. Electron Spins in few-electron lateral quantum dots, Laurens Henry Willems Van Beveren, 1-Sep-2015, https://repository.tudelft.nl/islandora/object/uuid:b33738cd-b8e5-49b6-aff2-573640012f98/datastream/OBJ/download

23. Electron spin qubits in quantum dots, R. Hanson et al, September 24, 2014, https://qutech.nl/wp-content/uploads/2019/01/2004_09_01419211.pdf

24. Real-time detection of single-electron tunneling using a quantum point contact, Vandersypen et al, 8-November-2004, http://kouwenhovenlab.tudelft.nl/wp-content/uploads/2011/10/149-real-time-detection.pdf

25. For further reading on the dilution fridge, interested readers can refer to: https://en.wikipedia.org/wiki/Dilution_refrigerator

26. List of organizations and research labs: https://quantumcomputingreport.com/qubit-technology/

27. For further reading on Topological Quantum Computing, readers may refer to the following publications: (a) Introduction to Topological Quantum Computation, Jiannis K. Pachos, (b) Non-Abelian Anyons and Topological Quantum Computation, Nayak et al, 2008, https://arxiv.org/abs/0707.1889, (c) A Short Introduction to Topological Quantum Computation Lahtinen et al, 2017. https://arxiv.org/abs/1705.04103, (d) A Pedagogical Overview On 2d And 3d Toric Codes And The Origin Of Their Topological Orders, Resente et al, 2019, https://arxiv.org/abs/1712.01258, (e) Introduction to abelian and non-abelian anyons, Sumathi Rao, 2016, https://arxiv.org/abs/1610.09260, (f) Designer non-Abelian anyon platforms: from Majorana to Fibonacci, Jason Alicea and Ady Stern, https://arxiv.org/pdf/1410.0359.pdf

28. Bifluxon: Fluxon-Parity-Protected Superconducting Qubit, Kalashnikov et al, 9-Oct-2019, https://arxiv.org/pdf/1910.03769.pdf

29. Not All Qubits Are Created Equal, A Case for Variability-Aware Policies for NISQ-Era Quantum Computers, Swamit S. Tannu, Moinuddin K. Qureshi, https://arxiv.org/ftp/arxiv/papers/1805/1805.10224.pdf#:~:text=A%20qubit%20in%20an%20high,retention%20time%20of%20DRAM%20cells).

30. Coherence properties of the 0-π qubit, Peter Groszkowski et al, 9-Aug-2017, https://arxiv.org/pdf/1708.02886.pdf

31. A Parity-Protected Superconductor-Semiconductor Qubit, T. W. Larsen et al, 8-April-2020, https://arxiv.org/pdf/2004.03975.pdf

32. A fast method for comparing braids, Patrick Dehornoy, https://dehornoy.users.lmno.cnrs.fr/Papers/Dfo.pdf

33. Protected gates for superconducting qubits, Peter Brooks, Alexei Kitaev, John Preskill, 17-Feb-2013, https://arxiv.org/abs/1302.4122v1

Quantum Circuits and DiVincenzo Criteria 5

"And how many hours a day did you do lessons?" said Alice, in a hurry to change the subject.
"Ten hours the first day," said the Mock Turtle: "nine the next, and so on."
"What a curious plan!" exclaimed Alice.
"That's the reason they're called lessons," the Gryphon remarked: "because they lessen from day to day."

— *Lewis Carroll, Alice's Adventures in Wonderland.*

The previous chapters provided an overview of the formulation of quantum mechanics. We built sufficient mathematical background and learned about the qubit modalities currently under research. We also learned how to represent the qubit as a Bloch sphere and perform Pauli rotations. With this knowledge, we are ready to write some quantum code. In the first part of this chapter, we set up our systems with **IBM Q** and **Microsoft Quantum Development Kit** and then proceed to learn about the quantum gates. In the second part of this chapter, we learn about constructing quantum circuits. We use Dirac's bra-ket notation to describe the circuits. We intermix theory with practice throughout this chapter.

▶ **Learning Objectives**
- Quantum programming languages
- Quantum circuits, transpilers, and optimization
- Single and multi-qubit gates, the universality of quantum gates
- Gottesman–Knill theorem
- Quantum entanglement and GHZ states
- Quantum teleportation
- No-cloning theorem
- Superdense coding
- Quantum parallelism
- Walsh–Hadamard transform
- Quantum interference
- Phase kickback
- DiVincenzo's criteria for quantum computing

V. Kasirajan, *Fundamentals of Quantum Computing*, https://doi.org/10.1007/978-3-030-63689-0_5

5.1   Setting Up the Development Environment

This section focuses on setting up the development system with the tools to build quantum circuits and perform experiments on IBM Q and Microsoft QDK.

5.1.1   Setting Up IBM Q

IBM Quantum Experience is a cloud-based platform for experimenting and deploying quantum applications in areas such as chemistry, optimization, finance, and AI [3–5, 9–11]. A visualizer helps in building the quantum circuits visually, while we edit the code side by side. **IBM Q** also provides a full-stack quantum development kit—**Qiskit**, available as a **Jupyter notebook**.

To sign up with IBM Q, launch the following website from your web browser and create your **IBMid**. If you already have an IBMid or a social account, you can log in using the login button or clicking the social account you want to use. The login link is provided here for your easy reference:

https://quantum-computing.ibm.com/login

After creating your IBMid, or signing in using a social account, accept the IBM Quantum End User Agreement and select "Accept & Continue." After this, fill in the survey by providing your name and your organization details. You are all set!

You can launch the **Circuit Composer** or the **Qiskit Notebook** from the page that opens up. You may want to bookmark this launcher page for easy access in the future. If you ever forget the link, it is provided here for your easy reference:

https://quantum-computing.ibm.com/

5.1.1.1 Installing Qiskit

For performing extensive experiments, the circuit composer is not enough. We need a development environment, and installing Qiskit is helpful. The following link generally documents the instructions for setting up Qiskit for development:

https://qiskit.org/documentation/install.html

If you are using a Windows system, follow these steps:

[1]. Install **Visual Studio 2017 runtime** from the link below:
https://go.microsoft.com/fwlink/?LinkId=746572
[2]. Download and install **Anaconda** from the following link:
https://www.anaconda.com/download/
[3]. Open **Anaconda Prompt** by searching it in the windows search in the taskbar. Launch the Anaconda Prompt in administer mode. Execute the following commands:

```
conda create -n ibmqiskit python=3.7
activate ibmqiskit
pip install qiskit
pip install qiskit[visualization]
```

These commands install the Qiskit by creating a new environment, "ibmqiskit." You can explore the wealth of information available on the Qiskit website and gain more insights into using IBM Q and quantum computing in general.

5.1.2   Installing Microsoft Quantum Development Kit

Microsoft Quantum Development Kit (QDK) comes in various flavors supporting C# and Python [6–8]. All flavors require us to write the quantum code in the Microsoft Q# language, which we shall get acquainted with by writing code. Installation instructions for **QDK** can be found in the link below or by googling.

https://docs.microsoft.com/en-us/quantum/install-guide/?view=qsharp-preview

In this book, we shall use the **Visual Studio Code**. Instructions for setting up Visual Studio Code can be found in the following link, or by searching the Internet.

https://docs.microsoft.com/en-us/quantum/install-guide/csinstall?view=qsharp-preview

You can install the QDK by following the given steps:

[1]. Install VS Code from the following link:

https://code.visualstudio.com/download

[2]. Install.Net Core SDK 3.1 or later from the following link:
https://www.microsoft.com/net/download
Remember to select the SDK and not the runtime!

[3]. Install the Quantum VS Code Extension:
https://marketplace.visualstudio.com/items?itemName=quantum.quantum-devkit-vscode

[4]. Install Q# Project Template
Launch VS Code now.
Go to **View -> Command Palette**
Select **Q#: Install project templates**

You should be all set now. The process requires the installation of additional extensions. Follow the instructions that pop-up on the screen.

5.1.3   Verifying the Installation

It is a good idea to verify the installation and troubleshoot any errors at this stage. Both these two development environments install several subcomponents. Some minor differences may exist as different groups develop these subcomponents in parallel and release with varying schedules and features. The general advice is to follow the installation instructions. Installing later versions of Python is usually problematic as the subcomponents do not support them readily.

5.1.3.1 Verifying Qiskit

An API token, associated with your account, is required to use Qiskit. You can generate the API token from your account. Launch the following website and login with your IBMid, if required.

https://quantum-computing.ibm.com/account

Select the user icon at the top right-hand corner and select "My Account." From the page that opens, select "Copy token." The API token associated with your account gets copied into the clipboard. Copy the API token into the given code.

To verify Qiskit, launch **Anaconda Navigator** (you can locate this application from the windows search bar). Ensure that you have selected "**ibmqiskit**" from the application channels. Selecting this ensures the usage of the right environment. Look for the **Jupyter notebook** from the list of applications in the home tab. Install the Jupyter notebook by selecting the "install" button.

Now launch Jupyter notebook. The Jupyter notebook opens in your default system browser.

In the browser tab that opens, select **New->Python 3** from the menu on the right-hand side panel. This selection opens a new browser window. In this browser window, enter the Python code from Listing 5.1 in the first cell.

```
import qiskit

# Save the API Token
MY_API_TOKEN = 'COPY YOUR TOKEN HERE'
from qiskit import IBMQ
IBMQ.save_account(MY_API_TOKEN)
```

Listing 5.1   Saving the IBM Q account token

The given Python code saves the API token in your local cache. Don't forget to copy your API token obtained earlier in line 3. In the next cell, type the code from Listing 5.2:

```
# Import the needed libraries
from qiskit import QuantumRegister, ClassicalRegister
from qiskit import QuantumCircuit, execute, Aer

# load the IBM Q account
provider = IBMQ.load_account()

# Define a quantum circuit with one-qubit and a classical register
q = QuantumRegister(1, 'q')
c = ClassicalRegister(1, 'c')
qc = QuantumCircuit(q, c)

# Apply an X gate
qc.x(q[0])

#  Project the qubit into a classical register
qc.measure(q, c)
```

Listing 5.2   Sample Qiskit code

The given code creates a quantum circuit with one qubit. It then applies an X-gate to the qubit and projects the qubit into a classical register. The final step is to plot the circuit. These steps become familiar in the next few sections. Let us continue writing some code to execute this on a simulator and plot the output. Type the code provided in Listing 5.3 in the third cell, and execute the same:

```
# Let's now draw the circuit
%matplotlib inline
qc.draw(output="mpl")
```

Listing 5.3 Drawing the circuit

Executing the given code plots in the following circuit (Fig. 5.1).

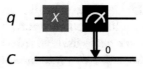

Fig. 5.1 Quantum circuit with one qubit and a X-gate

Now, type the code given in Listing 5.4 in a new cell and execute the same:

```
# Simulate the circuit with 1024 shots
simulator = Aer.get_backend('qasm_simulator')
job = execute(qc, backend=simulator, shots=1024)
result = job.result()

# Finally plot a histogram of the results
counts = result.get_counts(qc)
from qiskit.tools.visualization import import plot_histogram
plot_histogram(counts)
```

Listing 5.4 Executing the code in the Qiskit simulator

The given code fragment plots a histogram as shown in Fig. 5.2.

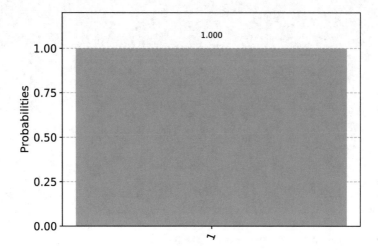

Fig. 5.2 Histogram of the X-gate

This quantum circuit illustrates the X-gate we implemented with one qubit q. The measurement operation projected the qubit into the classical register c. The histogram records an output "1," 100% of the times of the 1024 shots we took. Each shot is a repeat of the experiment. Note that your outcome may be slightly different from the histogram shown in Fig. 5.2. We shall discuss this outcome in subsequent sections. For now, you have successfully run a quantum program. Voila!

5.1.3.2 Verifying Microsoft Quantum Kit

From the windows search bar, search for "Visual Studio Code" and launch the application. Now select **View->Command Palette->Q#: Create new project** and create a Q# command-line application. You can create the project in any folder you may want and name the project as **QTest1**. A new project gets created, and the application reopens itself with the new project. The new project has two files, namely **Driver.cs** and **Program.qs**. The Driver.cs links with the Q# code contained in the file Program.qs. The file Program.qs contains the Q# code that implements the quantum circuit.

Now edit the file **Program.qs** and type code from Listing 5.5:

```
namespace QTest1 {
    open Microsoft.Quantum.Canon;
    open Microsoft.Quantum.Intrinsic;

    operation Xgate() : Result {

        using (qubit = Qubit()) {  // Define a single qubit

            X(qubit);                // Apply an X gate to the qubit
            let result = M(qubit);   // Now measure the same
            if (result == One )
            {
                Reset(qubit);        // It is a good practice to reset
                                     // the qubit
            }
            return result;
        }
    }

    operation RunQuantumMain() : Unit {
        mutable count = 0;

        for (i in 0..99) {
            set count += Xgate() == One ? 1 | 0;
        }
        Message($"The probability of getting 1 for the operation\
                X|0) is:{count} / 100");

    }
}
```

Listing 5.5 Sample code for the QDK. Program.qs, the Q# file

This code creates a single qubit, performs an X-gate, and measures the results of the X-gate. This step executes for about 100 times.

Edit the file **Driver.cs** and enter the code from Listing 5.6:

```
using System;
using Microsoft.Quantum.Simulation.Core;
using Microsoft.Quantum.Simulation.Simulators;


namespace QTest1
{
    class Driver
    {
        static void Main(string[] args)
        {
            using (var qsim = new QuantumSimulator())
            {
                RunQuantumMain.Run(qsim).Wait();
            }
        }
    }
}
```

Listing 5.6 The Driver.CS, C# accompanying file for the QDK program

This file creates an instance of the quantum simulator and invokes the method **RunQuantumMain** created in the Program.qs file.

Execute the program by selecting the menus **Debug->Start Debugging**. The program executes, and displays the following line in the Debug Console:

```
The probability of getting 1 for the operation X|0⟩ is: 100 /100
```

In the given example codes that we run on IBM Q and Microsoft Quantum Development Kit, we prepared a qubit in its default state $|0\rangle$ and applied a bit-flip gate—the X-gate—to flip the qubit to the state $|1\rangle$. We saw the bit-flip as a reliable operation, and we got the state $|1\rangle$ all the time.

5.1.4 Keeping the Installation Updated

The pip package gets upgraded frequently. It is a good idea to update pip. To upgrade pip, close all Jupyter notebooks and Anaconda instances. Execute the following command from the Anaconda command prompt "ibmqiskit."

```
python.exe -m pip install --upgrade pip
```

It is a good idea to keep qiskit and the visualization library updated. Execute the following commands from the Anaconda command prompt "ibmqiskit."

```
pip install qiskit --upgrade
pip install qiskit[visualization] --upgrade
```

5.2   Learning Quantum Programming Languages

The quantum programming languages referred to in this book are somewhat easy to learn. Anyone with a reasonable programming experience can learn them with ease. The quantum code for Microsoft QDK requires coding in the Q# programming language. The quantum code developed in Q# integrates with a C# or Python code to provide a user interface. Qiskit integrates well with the Python development environment. The OpenQASM is another quantum programming language. It somewhat resembles assembly language programming. In this section, we provide a quick tutorial to these three systems. Reference links provided at the end of this chapter can help our readers with further reading and practice.

5.2.1   Programming for OpenQASM

OpenQASM is a programming language that provides the interface to the IBM Quantum Experience. The Circuit Composer converts the QASM code into a quantum circuit, which we can edit using the graphical UI. Any change we make on the quantum circuit using the graphical UI automatically reflects in the QASM code. This graphical UI makes it easy to construct and verify quantum circuits with less circuit depth.

The QASM code snippet provided in Listing 5.7 declares two quantum and classical registers. A set of quantum gates are used to create entanglement between the two quantum registers (qubits), and the last segment of the code projects the qubits into the respective classical registers.

```
OPENQASM 2.0;                          File header and standard include file
include "qelib1.inc";


qreg q[2];                             Declare quantum and classical registers
creg c[2];


h q[0];                                Gate operations. H and CNOT gates are examples here.
cx q[0],q[1];


measure q[0] -> c[0];                  Measurement of qubits q[0] and q[1]
measure q[1] -> c[1];
```

Listing 5.7 The general structure of the OPENQASM code

```
OPENQASM 2.0;

include "qelib1.inc";


gate nG0(theta, phi) q  {        Subroutine nG0 with two parameters theta and phi and
                                 a quantum gate, implementing a U2 rotation.
  u2 theta,phi,q;

}



qreg q[2];

creg c[2];


nG0(0,pi) q[0];            ──────▶  Calling the subroutine nG0 with two parameters 0 and
                                    pi to operate on qubit q[0].
cx q[0],q[1];



measure q[0] -> c[0];

measure q[1] -> c[1];
```

Listing 5.8 Implementing a subroutine in the OPENQASM code

A QASM code generally follows the given convention. Besides, there is a provision to create a subroutine. An example is given in Listing 5.8.

5.2.2   QDK Using Q#

The Q# programming language is used to develop quantum algorithms in conjunction with the Microsoft QDK. The Q# code has the structure outlined in Listing 5.9.

The C# driver code creates an instance of the *QuantumSimulator* object and passes that as the argument to the *"Run"* method of the Quantum operation you want to invoke. Listing 5.10 contains the structure of the C# driver code.

Besides the intrinsic methods, the QDK provides a variety of functions for direct integration into the program. Libraries for Quantum Chemistry, Quantum Machine Learning, and Quantum Numerics contain the prework required for easy integration into projects.

```
namespace QTest1 {                          ──────────────►  The namespace of the project

    open Microsoft.Quantum.Canon;
    open Microsoft.Quantum.Intrinsic;    ──────────────►  Referenced modules

                                                            An operation
    operation NOTGate(input:Bool, shots:Int) : (Int) {──►  is a Q# function
        mutable c0 = 0;
        using (q0 = Qubit()) { ───────────────────────►  Declare the Qubits here
            for (i in 1..shots){
                Reset(q0);
                if ( true == input) {
                    X(q0);

                }
                                                          Gate operations

                X(q0);

                let result = M(q0); ──────────────►      Measurement

                if ( One == result ) {
                    set c0 += 1;

                }
            }
            Reset(q0); ───────────────────────────►      Reset the qubits before returning
            return c0;

        }
    }

                                                          The method that links with the
    operation RunQuantumMain() : Unit {  ──────────────►  C# code.
        mutable j = 0;

        set j = NOTGate(true, 100);
        Message($"We got 1 {j} times for NOTGate(true).");

    }
}
```

Listing 5.9 General structure of the QDK Q# code

```
using System;
using Microsoft.Quantum.Simulation.Core;  ──────▶  Import Microsoft Quantum
                                                     Core and Simulators
using
Microsoft.Quantum.Simulation.Simulators;


namespace QTest1
{
    class Driver
    {
        static void Main(string[] args)
        {
            Using(var qsim = new         ──────────▶  Quantum Simulator
QuantumSimulator())                                   instance
            {
                RunQuantumMain.Run(qsim).Wait();  ──▶ Call the main
            }                                         method in the Q#
        }                                             code
    }
}
```

Listing 5.10 The accompanying driver code for QDK implementing the main function

5.2.3 Qiskit Python

The Qiskit, when integrated with a Python development environment such as the Jupyter notebook, is an excellent tool for developing and experimenting with quantum computing. Like the OpenQASM, the Qiskit interfaces with IBM Q and provides the opportunity to work directly with real quantum processors (backends). Listing 5.11 outlines the structure of Python code using Qiskit.

The given example executes the code on the simulator. Make the changes provided in Listing 5.12 to execute the code on a backend.

Qiskit comes with four frameworks.

- **Qiskit Terra** is a software framework that provides visualization, transpilation, and handles user inputs. Qiskit Terra also interfaces with the Qiskit Aer simulator and other backends.
- **Qiskit Aer** provides the high-performance simulator functions.
- **Qiskit Ignis** provides the framework for understanding and mitigating noise in quantum circuits and systems.
- **Qiskit Aqua** is a library of algorithms we can experiment on chemistry, AI, optimization, and finance applications for near-term quantum computers.

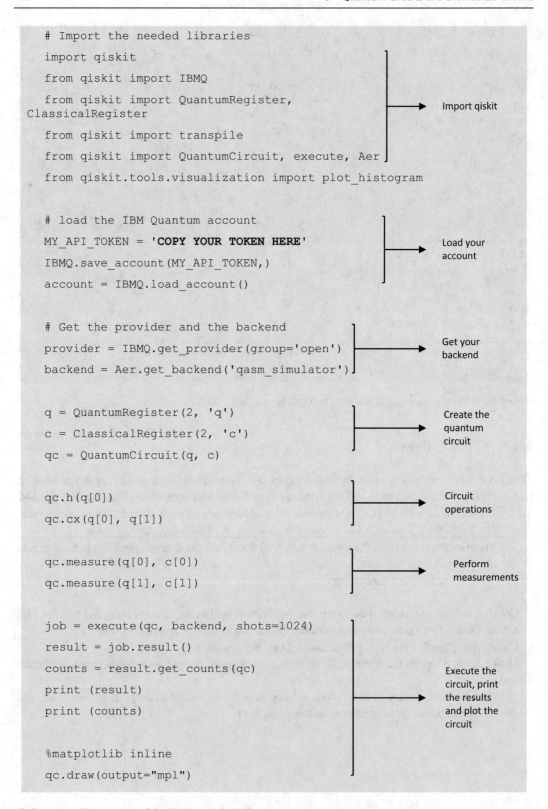

```
# Import the needed libraries
import qiskit
from qiskit import IBMQ
from qiskit import QuantumRegister,
ClassicalRegister                                    Import qiskit
from qiskit import transpile
from qiskit import QuantumCircuit, execute, Aer
from qiskit.tools.visualization import plot_histogram


# load the IBM Quantum account
MY_API_TOKEN = 'COPY YOUR TOKEN HERE'                Load your
IBMQ.save_account(MY_API_TOKEN,)                     account
account = IBMQ.load_account()


# Get the provider and the backend
provider = IBMQ.get_provider(group='open')          Get your
                                                     backend
backend = Aer.get_backend('qasm_simulator')


q = QuantumRegister(2, 'q')                          Create the
c = ClassicalRegister(2, 'c')                        quantum
qc = QuantumCircuit(q, c)                            circuit


qc.h(q[0])                                           Circuit
qc.cx(q[0], q[1])                                    operations


qc.measure(q[0], c[0])                              Perform
qc.measure(q[1], c[1])                              measurements


job = execute(qc, backend, shots=1024)
result = job.result()
counts = result.get_counts(qc)                       Execute the
print (result)                                       circuit, print
print (counts)                                       the results
                                                     and plot the
                                                     circuit
%matplotlib inline
qc.draw(output="mpl")
```

Listing 5.11 The structure of the Qiskit code in Python

```
provider = IBMQ.get_provider()                          Get a backend
                                                        that is free.
processor = least_busy(provider.backends())


# construct your quantum circuit here...

#...

#...

                                                        Execute the
# setup a job monitor                                   job in a
                                                        backend and
job = execute(qc, backend=processor, shots=1024)        setup a job
                                                        monitor
job_monitor(job)
```

Listing 5.12 Executing the code on the backend

5.2.4 Comparing the Development Platforms

For a comparison of the software development platforms at the gate level, interested readers can refer to Ryan LaRose [2].

5.3 Introducing Quantum Circuits

On a classical digital computer, any operation such as an arithmetic operation starts by preparing individual bits with an initial value. We then use a sequence of digital gates to perform the required arithmetic operation. We finally read out the bits to know the results of the operation deterministically. Any operation we do at the application level, boils down to a sequence of digital gate operations, irrespective of the complexity of the operation.

A quantum computer works similarly. Similar to the classical computer, we construct a quantum circuit to implement a specific quantum algorithm. The quantum circuit solves a specific problem by implementing unitary operations in the Hilbert space on a finite number of qubits. Quantum gates implement the unitary operations. The unitary operations also mean that the quantum circuits have the same number of inputs and outputs. Furthermore, the quantum circuits are **acyclic**, meaning there are no feedback circuits or loops.

We learned in previous chapters that the quantum evolution is unitary. Hence, every quantum circuit qc corresponds to a particular unitary operator U_{qc} in the Hilbert space, meeting the criteria: $U_{qc}U_{qc}^{\dagger} = U_{qc}^{\dagger}U_{qc} = 1$. From this equation, we can infer that U_{qc} has an inverse operator. Hence, we can say that the quantum circuit is **reversible**. In other words, if we start from the output and work backward, we should be able to retrieve the inputs. This process preserves the information. Therefore, the input data are fully recoverable. The reversibility is a specific property of quantum gates.

A quantum circuit consists of three stages. In the first stage, we prepare a certain number of qubits with the initial state. In the second stage, we perform quantum gate operations and use quantum mechanics to solve the problem. In the final stage, we measure the qubits by projecting them to a set of classical bits using a computational basis. The quantum circuit shown in Fig. 5.3 illustrates this.

In this quantum circuit, we start with a device having six quantum registers $q[0]. . q[5]$. There are also three classical registers (digital bits) $c[0]$, $c[1]$, $c[2]$. We start the qubits with an initial state of $|0\rangle$.

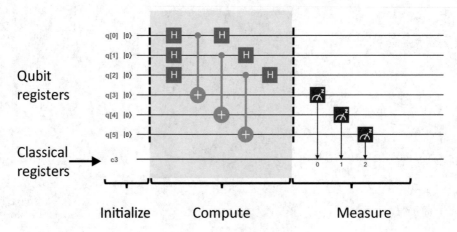

Fig. 5.3 A quantum circuit showing the three stages of execution

Note that in this example, we start the qubits with the state $|0\rangle$. Most algorithms require the qubits to be initialized to a superposition state and ancilla qubits to be set for phase kickback. As we progress through this chapter, these steps become clear.

Proceeding further with this circuit into the compute state, we apply H-gates to qubits $q[0]$, $q[1]$, $q[2]$. After this we apply a sequence of 3 controlled-NOT gates. The first CNOT gate is applied with $q[0]$ as the control and $q[3]$ as the target qubit. The second CNOT gate is applied with $q[1]$ as the control and $q[4]$ as the target qubit. The third CNOT gate is applied with $q[2]$ as the control and $q[5]$ as the target qubit. After this step, we apply H-gates one more time to qubits $q[0]$, $q[1]$, $q[2]$. We then finally measure the state of the qubits $q[3]$, $q[4]$, $q[5]$ by projecting them to classical registers $c[0]$, $c[1]$, $c[2]$.

You can construct this circuit on the IBM Q circuit composer and execute it by selecting the "Run" button. Select 1024 shots and "**ibmq_qasm_simulator**" as the back end. The execution queue appends the quantum experiment, and you may have to return after a few minutes to see the result. The dotted vertical bars indicate the barrier component, which separates circuit sections from being optimized by the transpiler. The transpiler is a pre-processing software that analyzes quantum circuits for optimization and translation to fit the topology of a given hardware platform. The barrier is introduced in this circuit to improve clarity. Its purpose can be safely ignored for now. We shall return to the barrier circuit in a later section. Executing this circuit generates a histogram, as shown in Fig. 5.4:

We can interpret the histogram as follows. The probability of projecting the state to $|000\rangle$ is 10.8%, the probability of projecting the state to $|001\rangle$ is 13%, the probability of projecting the state to $|010\rangle$ is 13.3%, and so on. Ideally, we expect an equal 12.5% distribution with this circuit, since we put the qubits $q[0]$, $q[1]$, $q[2]$ in an equal superposition state. The differences are due to the way the quantum systems work in nature. In a real device, it is challenging to perform an error-free measurement, and the quantum device is always subject to environmental noises and interactions. This process becomes clear as we progress through this chapter.

Note that the measured values of the classical registers $c0 . . c2$ are shown under the bars along the x-axis. The order of the individual bits is the same as that of digital bits. The least significant classical registers are on the right, and the most significant classical registers are on the left, that is, $... - c2 - c1 - c0$.

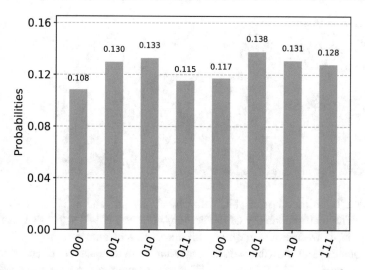

Fig. 5.4 Histogram resulting from executing the code in Fig. 5.3

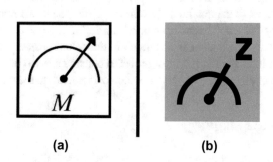

Fig. 5.5 Circuit symbol for measurement

5.3.1 On Quantum Measurements

We can describe the quantum state of a qubit in an arbitrarily unknown state using the quantum superposition principle as:

$$|\psi\rangle = a|0\rangle + b|1\rangle, \tag{5.1}$$

where the coefficients a and b are complex numbers representing probability amplitudes of finding the qubit in one of the states $|0\rangle$ or $|1\rangle$. When we measure a qubit in the computational basis, we project the qubit to the states $|0\rangle$ or $|1\rangle$. The probability of the outcome is determined by the coefficients a and b.

In the quantum circuits, the analog dial symbol shown in Fig. 5.5 represents a measurement operation. The alphabet "M" denotes the measurement. Some systems use "Z," as we measure the qubit in the z basis. The quantum devices are, in general, designed to perform z measurement. In the quantum circuit diagrams, the measurement operation projects the qubit that needs to be measured into the classical bit. The "Measure" stage of the circuit shown in Fig. 5.3 illustrates this.

Table 5.1 Possible measurements

c[2]	c[1]	c[0]	Outcome
0	0	0	10.8%
0	0	1	13.0%
0	1	0	13.3%
0	1	1	11.5%
1	0	0	11.7%
1	0	1	13.8%
1	1	0	13.1%
1	1	1	12.8%

We measure a $'0'$ on the classical bit, with probability $|a|^2$ and a $'1'$ in the classical bit, with probability $|b|^2$. The total probability of the system, however, is $|a|^2 + |b|^2 = 1$. At the end of the compute operation, if we measure n qubits, the measurement outcome projected on the classical register can be one of the 2^n possible values. For example, in the quantum circuit shown in Fig. 5.3, we measure three qubits, $q[0]$, $q[1]$, and $q[2]$. The measurement outcome could be one of the eight ($2^3 = 8$) possible values shown in Table 5.1.

The outcome can be different each time the experiment is run for the same circuit. This difference is because of the randomness of the quantum systems in nature. The measurement errors can only add to this. When we repeat the experiment a certain number of times, we get a distribution of the measurement outcome. This distribution is illustrated in the column named "Outcome" in Table 5.1. The distribution gives an indication of the most probable outcomes based on which inferences can be drawn. Note that the sum of all outcomes is 100%, which corresponds to the total probability being "1."

5.3.2   On the Compute Stage

Before understanding how the "compute stage" works, we must build a knowledge of the quantum gates. The following section takes us through this next step in our journey.

5.4   Quantum Gates

In the previous chapters, we learned that the qubits are two-state systems, implemented in the computational basis $\{|0\rangle, |1\rangle\}$. The following column matrices or ket vectors describe the two-states of the computational basis:

$$|0\rangle = \begin{bmatrix} 1 \\ 0 \end{bmatrix}$$

$$|1\rangle = \begin{bmatrix} 0 \\ 1 \end{bmatrix}$$

(5.2)

The quantum gates are unitary operations on the qubits and are reversible. We shall study the universality of quantum gates in detail in subsequent sections. If we start with a qubit in state $|\psi\rangle$ and

Fig. 5.6 The X-gate rotating the qubit from state $|0\rangle$ to $|1\rangle$

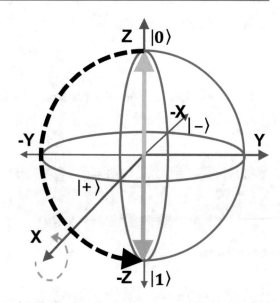

apply a unitary operation U, the final state of the qubit $|\psi'\rangle$ is as shown in the following transformation equation:

$$|\psi'\rangle = U\,|\psi\rangle \tag{5.3}$$

For single qubits the unitary operations can be defined using 2×2 unitary matrices.

5.4.1 Clifford Group Gates for Single Qubits

5.4.1.1 The Bit-Flip Gate or the NOT Gate or the Pauli X-Gate

The Pauli X-gate applies a $\frac{\pi}{2}$ rotational pulse along the x-axis. The action of this gate is to flip the qubit from state $|0\rangle$ to $|1\rangle$ and vice versa. The X-gate is illustrated in Fig. 5.6.

The matrix form of this gate is given by Eq. (5.4):

$$X = \begin{bmatrix} 0 & 1 \\ 1 & 0 \end{bmatrix} \tag{5.4}$$

The action of this gate on the basis states can be derived as follows:

$$X|0\rangle = \begin{bmatrix} 0 & 1 \\ 1 & 0 \end{bmatrix}\begin{bmatrix} 1 \\ 0 \end{bmatrix} = \begin{bmatrix} 0 \\ 1 \end{bmatrix} = |1\rangle$$
$$X|1\rangle = \begin{bmatrix} 0 & 1 \\ 1 & 0 \end{bmatrix}\begin{bmatrix} 0 \\ 1 \end{bmatrix} = \begin{bmatrix} 1 \\ 0 \end{bmatrix} = |0\rangle \tag{5.5}$$

Alternate form

$$X = U_3(\pi, 0, \pi) \tag{5.6}$$

U3 is a rotational gate described in Sect. 5.4.3.1.

Syntax

Circuit Composer	OpenQASM	Qiskit Python	Q#—MS QDK
X	x qubit|qreq; x q[0];	qc.x(qubit) qc.x(q[0])	operation X(qubit:Qubit):Unit X(q[0]);

Truth Table

Input	Output
$\|0\rangle$	$\|1\rangle$
$\|1\rangle$	$\|0\rangle$

5.4.1.2 The Identity Gate

The identity gate I does not perform any operation. It leaves the basis states $|0\rangle$ and $|1\rangle$ unmodified. In matrix form, it is represented by the identity matrix.

$$I = \begin{bmatrix} 1 & 0 \\ 0 & 1 \end{bmatrix} \tag{5.7}$$

The action of the identity gate on the basis states can be derived as follows:

$$I|0\rangle = \begin{bmatrix} 1 & 0 \\ 0 & 1 \end{bmatrix} \begin{bmatrix} 1 \\ 0 \end{bmatrix} = \begin{bmatrix} 1 \\ 0 \end{bmatrix} = |0\rangle$$
$$I|1\rangle = \begin{bmatrix} 1 & 0 \\ 0 & 1 \end{bmatrix} \begin{bmatrix} 0 \\ 1 \end{bmatrix} = \begin{bmatrix} 0 \\ 1 \end{bmatrix} = |1\rangle \tag{5.8}$$

Syntax

Circuit Composer	OpenQASM	Qiskit Python	Q#—MS QDK
Id	id qubit|qreq; id q[0];	qc.iden(q[0]) qc.iden(q[0]	operation I(qubit:Qubit):Unit I(q[0]);

Truth Table

Input	Output
$\|0\rangle$	$\|0\rangle$
$\|1\rangle$	$\|1\rangle$

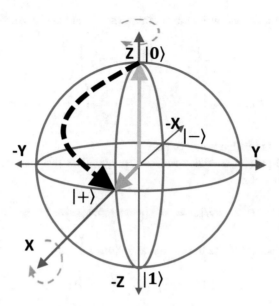

Fig. 5.7 The H-gate rotating the state $|0\rangle$ to $|+\rangle$

5.4.1.3 The Hadamard Gate—H-Gate

The Hadamard gate rotates the qubit by π degrees along an axis diagonal to the xy plane. It is equal to rotating the qubit by $\frac{\pi}{2}$ degrees along the y-axis and then by $\frac{\pi}{2}$ degrees along the x-axis. This rotation makes the z-axis to rotate to the x-axis and vice versa (Fig. 5.7).

Since the information of the qubit along the x-axis is rotated to the z-axis, the H-gate gives us a way of measuring the qubit in x basis. The Hadamard gate puts the qubit in an equal superposition state in the computational basis. The H-gate is defined by the following equation.

$$H = \frac{|0\rangle + |1\rangle}{\sqrt{2}} \langle 0| + \frac{|0\rangle - |1\rangle}{\sqrt{2}} \langle 1| \tag{5.9}$$

In the matrix form, the H-gate is defined as:

$$H = \frac{1}{\sqrt{2}} \begin{bmatrix} 1 & 1 \\ 1 & -1 \end{bmatrix} \tag{5.10}$$

The action of the H-gate on the basis states are as follows:

$$\begin{aligned} H|0\rangle &= \frac{|0\rangle + |1\rangle}{\sqrt{2}} \langle 0|0\rangle + \frac{|0\rangle - |1\rangle}{\sqrt{2}} \langle 1|0\rangle \\ &= \frac{|0\rangle + |1\rangle}{\sqrt{2}} = \frac{1}{\sqrt{2}} (|0\rangle + |1\rangle) \\ &= |+\rangle \end{aligned} \tag{5.11}$$

$$\begin{aligned} H|1\rangle &= \frac{|0\rangle + |1\rangle}{\sqrt{2}} \langle 0|1\rangle + \frac{|0\rangle - |1\rangle}{\sqrt{2}} \langle 1|1\rangle \\ &= \frac{|0\rangle - |1\rangle}{\sqrt{2}} = \frac{1}{\sqrt{2}} (|0\rangle - |1\rangle) \\ &= |-\rangle \end{aligned} \tag{5.12}$$

Using the matrix form, we can write the action of the H-gate on the basis states as:

$$H|0\rangle = \frac{1}{\sqrt{2}} \begin{bmatrix} 1 & 1 \\ 1 & -1 \end{bmatrix} \begin{bmatrix} 1 \\ 0 \end{bmatrix} = \frac{1}{\sqrt{2}} \begin{bmatrix} 1 \\ 1 \end{bmatrix} = |+\rangle$$

$$H|1\rangle = \frac{1}{\sqrt{2}} \begin{bmatrix} 1 & 1 \\ 1 & -1 \end{bmatrix} \begin{bmatrix} 0 \\ 1 \end{bmatrix} = \frac{1}{\sqrt{2}} \begin{bmatrix} 1 \\ -1 \end{bmatrix} = |-\rangle$$

(5.13)

The states $\{|+\rangle, |-\rangle\}$ are called **polar basis**.

We can write the H-gate in another form as follows. Note that the exponent evaluates to ± 1 depending upon the value of x.

$$H|x\rangle = \frac{1}{\sqrt{2}} \left(|0\rangle + e^{2\pi i \frac{x}{2}} |1\rangle \right)$$

(5.14)

In yet another form, we can write the H-gate as follows:

$$H|x\rangle = \frac{1}{\sqrt{2}} \left(|0\rangle + (-1)^x |1\rangle \right)$$

(5.15)

Recall that $(-1)^0 = 1$ and $(-1)^1 = -1$. Hence, this equation can be written as a summation.

$$H|x\rangle = \frac{1}{\sqrt{2}} \sum_{y=0}^{1} (-1)^{x \cdot y} |y\rangle$$

(5.16)

We shall use the representations shown in Eqs. (5.14) and (5.16) while developing quantum algorithms in the next few chapters.

Let us now try the H-gate. Construct the circuit, as shown in Table 5.2 using the circuit composer. You can also try this using the QDK or using the Qiskit.

Upon execution of this code, we get a histogram, as shown in Fig. 5.8.

The histogram shows a 49.8% probability of the qubit being in the state $|0\rangle$ and 50.2% probability in the state $|1\rangle$. Ideally, for an H-gate, we expect a 50–50% distribution. The differences are due to the

Table 5.2 The H-gate

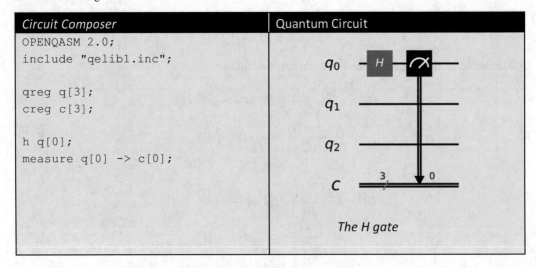

Fig. 5.8 The outcome of the experiment with the H-gate

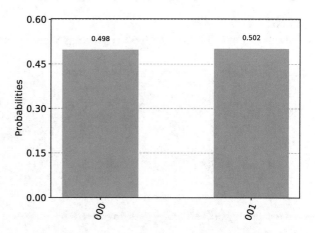

randomness in the underlying physics and measurement errors. Hence, each experiment could have a slightly different distribution.

Qiskit Python code:
The equivalent Qiskit code is provided in Listing 5.13.

```
# Import the needed libraries
from qiskit import QuantumRegister, ClassicalRegister
from qiskit import QuantumCircuit, execute, Aer
from qiskit.tools.visualization import import plot_histogram
provider = IBMQ.load_account()    # load the IBM Quantum account

q = QuantumRegister(1, 'q') # Define a quantum circuit with\
                            one-qubit
c = ClassicalRegister(1, 'c')     # and a classical register
qc = QuantumCircuit(q, c)

qc.h(q[0])                        # Apply a H gate
qc.measure(q, c)                  # Project the qubit into a\
                                  classical register
%matplotlib inline                # Let's now draw the circuit
qc.draw(output="mpl")

simulator = Aer.get_backend('qasm_simulator')
job = execute(qc, backend=simulator, shots=1024)
result = job.result()

counts = result.get_counts(qc)    # Finally plot a histogram of the\
                                  results
plot_histogram(counts)
```

Listing 5.13 Qiskit code with a single Hadamard gate

QDK:

To run this in QDK, edit the file **Program.qs** with the code from Listing 5.14.

```
namespace QTest1 {
    open Microsoft.Quantum.Canon;
    open Microsoft.Quantum.Intrinsic;
    operation Hgate() : Result {
        // Define a single qubit
        using (qubit = Qubit()) {
            // Apply a H gate to the qubit
            H(qubit);
            // Now measure the same
            let result = M(qubit);
            // It is a good practice to reset the qubit
            if (result == One )
            {
                Reset(qubit);
            }
            return result;
        }
    }


    operation RunQuantumMain() : Unit {
        mutable count = 0;
        for (i in 0..99) {
            set count += Hgate() == One ? 1 | 0;
        }
        Message($"The probability of getting 1 for the operation\
                H|0> is:{count} / 100");

    }
}
```

Listing 5.14   QDK code illustrating a Hadamard gate

The output of this code is:

```
The probability of getting 1 for the operation H|0⟩ is: 63 / 100
```

Recall the Eqs. (5.11) and (5.12). The probability amplitude vectors of the states $|+\rangle$ and $|-\rangle$ are $\left[\frac{1}{\sqrt{2}}, \frac{1}{\sqrt{2}}\right]$ and $\left[\frac{1}{\sqrt{2}}, -\frac{1}{\sqrt{2}}\right]$, respectively. However, the probability vector is the same, and it is $\left[\frac{1}{2}, \frac{1}{2}\right]$. The Hadamard transform is one example where two states can have different probability amplitudes but the same probability vector.

Alternate form

$$H = U_2(0, \pi) \tag{5.17}$$

The U2 rotation gate is described in Sect. 5.4.3.2.

Syntax

Circuit Composer	OpenQASM	Qiskit Python	Q#—MS QDK
H	h qubit|qreq; h q[0];	qc.h(qubit) qc.h(q[0]);	operation H(qubit:Qubit):Unit H (q[0]);

Truth Table

Input	Output
$\lvert 0 \rangle$	$\frac{1}{\sqrt{2}}(\lvert 0 \rangle + \lvert 1 \rangle)$
$\lvert 1 \rangle$	$\frac{1}{\sqrt{2}}(\lvert 0 \rangle - \lvert 1 \rangle)$

Let us cascade two H-gates, and study the system evolution. We can write the following equations for qubit states $\lvert 0 \rangle$ and $\lvert 1 \rangle$.

$$
\begin{aligned}
HH\lvert 0 \rangle = H\lvert + \rangle &= H(\frac{1}{\sqrt{2}}(\lvert 0 \rangle + \lvert 1 \rangle)) \\
&= \frac{1}{\sqrt{2}}(\frac{1}{\sqrt{2}}(\lvert 0 \rangle + \lvert 1 \rangle) + \frac{1}{\sqrt{2}}(\lvert 0 \rangle - \lvert 1 \rangle)) \\
&= \frac{1}{2}(\lvert 0 \rangle + \lvert 1 \rangle) + \frac{1}{2}(\lvert 0 \rangle - \lvert 1 \rangle) \\
&= \frac{1}{2}\lvert 0 \rangle + \frac{1}{2}\lvert 0 \rangle \\
&= \lvert 0 \rangle
\end{aligned} \tag{5.18}
$$

$$
\begin{aligned}
HH\lvert 1 \rangle = H\lvert - \rangle &= H(\frac{1}{\sqrt{2}}(\lvert 0 \rangle - \lvert 1 \rangle)) \\
&= \frac{1}{\sqrt{2}}(\frac{1}{\sqrt{2}}(\lvert 0 \rangle + \lvert 1 \rangle) - \frac{1}{\sqrt{2}}(\lvert 0 \rangle - \lvert 1 \rangle)) \\
&= \frac{1}{2}(\lvert 0 \rangle + \lvert 1 \rangle) - \frac{1}{2}(\lvert 0 \rangle - \lvert 1 \rangle) \\
&= \frac{1}{2}\lvert 1 \rangle + \frac{1}{2}\lvert 1 \rangle \\
&= \lvert 1 \rangle
\end{aligned} \tag{5.19}
$$

Thus, when we apply the H transform twice in succession, the qubit is returned to its original state. Let us verify this in IBM Q Circuit Composer. Construct the quantum circuit, as shown in Table 5.3.

Table 5.3 Successive application of H-gates

Circuit Composer	Quantum Circuit
OPENQASM 2.0; include "qelib1.inc"; qreg q[6]; creg c[5]; x q[0]; h q[0]; h q[0]; measure q[0] -> c[0];	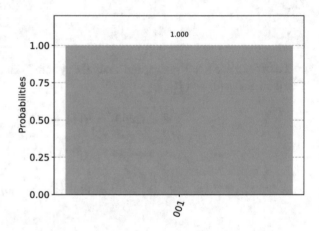 Two H gates in a sequence.

Fig. 5.9 The outcome of a circuit with an X-gate followed by two H-gates in succession

In the circuit shown in figure, we start with a state of $|0\rangle$ and apply an X-gate, which flips the qubit to $|1\rangle$. After this, we apply two H-gates in succession. Figure 5.9 shows the histogram we get upon executing the circuit.

The outcome is as expected. We got state $|1\rangle$, 100% of the time. The reader is advised to try this without the X-gate and verify that we get $|0\rangle$ all the time.

5.4.1.4 Pauli Y Gate

The Pauli Y gate rotates the qubit along the y-axis by π radians. In the matrix form the Pauli Y gate is as shown in the following equation:

$$Y = \begin{bmatrix} 0 & -i \\ i & 0 \end{bmatrix} \tag{5.20}$$

Now let us apply the Y gate on the basis states.

$$Y|0\rangle = \begin{bmatrix} 0 & -i \\ i & 0 \end{bmatrix} \begin{bmatrix} 1 \\ 0 \end{bmatrix} = \begin{bmatrix} 0 \\ i \end{bmatrix} = i \begin{bmatrix} 0 \\ 1 \end{bmatrix} = i|1\rangle$$

$$Y|1\rangle = \begin{bmatrix} 0 & -i \\ i & 0 \end{bmatrix} \begin{bmatrix} 0 \\ 1 \end{bmatrix} = \begin{bmatrix} -i \\ 0 \end{bmatrix} = -i \begin{bmatrix} 1 \\ 0 \end{bmatrix} = -i|0\rangle$$

(5.21)

Thus, the Y gate maps the state $|0\rangle$ to $i|1\rangle$ and the state $|1\rangle$ to $-i|0\rangle$.

Alternate form

$$Y = U_3\left(\pi, \frac{\pi}{2}, \frac{\pi}{2}\right)$$

(5.22)

Syntax

Circuit Composer	OpenQASM	Qiskit Python	Q#—MS QDK
Y	y qubit|qreq; y q[0];	qc.y(qubit) qc.y(q[0]	operation Y(qubit:Qubit):Unit Y(q[0]);

Truth Table

Input	Output
$|0\rangle$	$i|1\rangle$
$|1\rangle$	$-i|0\rangle$

5.4.1.5 Pauli Z Gate

The Pauli Z gate, rotates the qubit by π radians along the z-axis. The following equation describes the matrix form of the Pauli Z gate:

$$Z = \begin{bmatrix} 1 & 0 \\ 0 & -1 \end{bmatrix}$$

(5.23)

Applying the Z gate on the basis states, we get:

$$Z|0\rangle = \begin{bmatrix} 1 & 0 \\ 0 & -1 \end{bmatrix} \begin{bmatrix} 1 \\ 0 \end{bmatrix} = \begin{bmatrix} 1 \\ 0 \end{bmatrix} = |0\rangle$$

$$Z|1\rangle = \begin{bmatrix} 1 & 0 \\ 0 & -1 \end{bmatrix} \begin{bmatrix} 0 \\ 1 \end{bmatrix} = \begin{bmatrix} 0 \\ -1 \end{bmatrix} = -\begin{bmatrix} 0 \\ 1 \end{bmatrix} = -|1\rangle$$

(5.24)

Thus, the Z gate leaves the state $|0\rangle$ unchanged and changes the state $|1\rangle$ to $-|1\rangle$. For this reason, the Z gate is called the Phase-flip gate.

Alternate form

$$Z = U_1(\pi)$$

(5.25)

The U1 gate is described in Sect. 5.4.3.3 (Fig. 5.10).

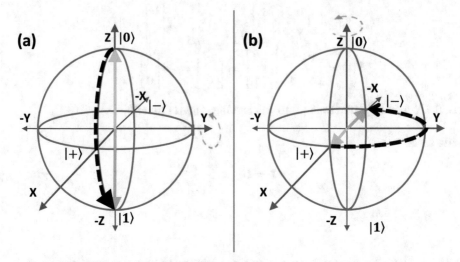

Fig. 5.10 Diagram illustrating (**a**) Pauli Y rotation from $|0\rangle$ to $|1\rangle$ and (**b**) Pauli Z rotation from $|+\rangle$ to $|-\rangle$

Syntax

Circuit Composer	OpenQASM	Qiskit Python	Q#—MS QDK
Z	z qubit|qreq; z q[0];	qc.z(qubit) qc.z(q[0])	operation Z(qubit:Qubit):Unit Z(q[0]);

Truth Table

Input	Output
$|0\rangle$	$|0\rangle$
$|1\rangle$	$-|1\rangle$

5.4.1.6 The *S* Gate or the Z90 Gate or the Phase Gate

The S gate rotates the qubit by $\frac{\pi}{2}$ radians along the z-axis. Some texts refer to this gate as a $\frac{\pi}{4}$ phase gate. The matrix form of this gate is as follows:

$$S = \begin{bmatrix} 1 & 0 \\ 0 & e^{\frac{i\pi}{2}} \end{bmatrix} = \begin{bmatrix} 1 & 0 \\ 0 & i \end{bmatrix} \tag{5.26}$$

The effect of this gate on the basis states can be derived as follows.

$$S|0\rangle = \begin{bmatrix} 1 & 0 \\ 0 & i \end{bmatrix} \begin{bmatrix} 1 \\ 0 \end{bmatrix} = \begin{bmatrix} 1 \\ 0 \end{bmatrix} = |0\rangle$$

$$S|1\rangle = \begin{bmatrix} 1 & 0 \\ 0 & i \end{bmatrix} \begin{bmatrix} 0 \\ 1 \end{bmatrix} = \begin{bmatrix} 0 \\ i \end{bmatrix} = i \begin{bmatrix} 0 \\ 1 \end{bmatrix} = i|1\rangle \tag{5.27}$$

Alternate form

$$S = U_1\left(\frac{\pi}{2}\right) \tag{5.28}$$

Syntax

Circuit Composer	OpenQASM	Qiskit Python	Q#—MS QDK
S	s qubit|qreq; s q[0];	qc.s(qubit) qc.s(q[0])	operation S(qubit:Qubit):Unit S(q[0]);

Truth Table

Input	Output
$\|0\rangle$	$\|0\rangle$
$\|1\rangle$	$i\|1\rangle$

5.4.1.7 The S† Gate or the Sdag Gate

The $S^\dagger$ gate is the conjugate transpose of the S gate. The following equation gives the matrix form of this gate.

$$S^\dagger = \begin{bmatrix} 1 & 0 \\ 0 & e^{\frac{-i\pi}{2}} \end{bmatrix} = \begin{bmatrix} 1 & 0 \\ 0 & -i \end{bmatrix} \tag{5.29}$$

The effect of this gate on the basis states are as follows:

$$S^\dagger|0\rangle = \begin{bmatrix} 1 & 0 \\ 0 & -i \end{bmatrix}\begin{bmatrix} 1 \\ 0 \end{bmatrix} = \begin{bmatrix} 1 \\ 0 \end{bmatrix} = |0\rangle$$

$$S^\dagger|1\rangle = \begin{bmatrix} 1 & 0 \\ 0 & -i \end{bmatrix}\begin{bmatrix} 0 \\ 1 \end{bmatrix} = \begin{bmatrix} 0 \\ -i \end{bmatrix} = -i\begin{bmatrix} 0 \\ 1 \end{bmatrix} = -i|1\rangle \tag{5.30}$$

Alternate form

$$S^\dagger = U_1\left(-\frac{\pi}{2}\right) \tag{5.31}$$

Syntax

Circuit Composer	OpenQASM	Qiskit Python	Q#—MS QDK
$S^\dagger$	sdg qubit|qreq; sdg q[0];	qc.sdg(qubit) qc.sdg(q[0])	To be implemented as a user gate.

Truth Table

Input	Output
$\|0\rangle$	$\|0\rangle$
$\|1\rangle$	$-i\|1\rangle$

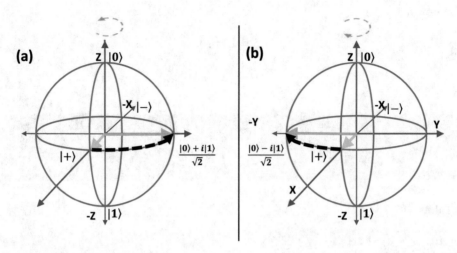

Fig. 5.11 Diagram illustrating S and S† gates

The gates S and $S^\dagger$ rotate the qubit between the x and y basis and are considered the two possible square roots of the Z gate. Figure 5.11 illustrates the S and S†gates.

5.4.2   Arbitrary Rotation of the Qubit

We can rotate the qubits arbitrarily along the x, y, or the z-axis. In this section, we learn about the R gates that can perform arbitrary rotation.

5.4.2.1 RX-Gate

The RX-gate rotates the qubit along the x-axis by θ radians. The matrix form of this gate is as follows:

$$R_x(\theta) = \begin{bmatrix} \cos\left(\dfrac{\theta}{2}\right) & -i\sin\left(\dfrac{\theta}{2}\right) \\ -i\sin\left(\dfrac{\theta}{2}\right) & \cos\left(\dfrac{\theta}{2}\right) \end{bmatrix} \tag{5.32}$$

We can write this rotation in terms of the Pauli matrix X.

$$R_x(\theta) = \cos\frac{\theta}{2} I - i\sin\frac{\theta}{2} X = e^{-i\frac{\theta}{2}X} \tag{5.33}$$

Alternate form

$$R_x(\theta) = U_3\left(\theta, -\frac{\pi}{2}, \frac{\pi}{2}\right) \tag{5.34}$$

Syntax

Circuit Composer	OpenQASM	Qiskit Python	Q#—MS QDK
R_x $\pi/2$	rx (theta) qubit|qreq; rx (pi/2) q[0];	qc.rx(theta, qubit) qc.rx(pi/2, q[0])	Operation Rx(theta:Double, qubit:Qubit): Unit let theta = 1.5707963268; Rx(theta, q[0]);

5.4.2.2 RY-Gate

The RY-gate rotates the qubit along the y-axis by θ radians. The following equation is the matrix form:

$$R_y(\theta) = \begin{bmatrix} cos\left(\dfrac{\theta}{2}\right) & -sin\left(\dfrac{\theta}{2}\right) \\ sin\left(\dfrac{\theta}{2}\right) & cos\left(\dfrac{\theta}{2}\right) \end{bmatrix} \tag{5.35}$$

We can describe $RY(\theta)$ in terms of Pauli rotation matrix Y as follows:

$$R_y(\theta) = cos\frac{\theta}{2} I - isin\frac{\theta}{2} Y = e^{-i\frac{\theta}{2}Y} \tag{5.36}$$

Alternate form

$$R_y(\theta) = U_3(\theta, 0, 0) \tag{5.37}$$

Syntax

Circuit Composer	OpenQASM	Qiskit Python	Q#—MS QDK
R_y $\pi/2$	ry (theta) qubit|qreq; ry (pi/2) q[0];	qc.ry(theta, qubit) qc.ry(pi/2, q[0])	Operation Ry(theta:Double, qubit:Qubit): Unit let theta = 1.5707963268; Ry(theta, q[0]);

5.4.2.3 RZ-Gate

The RZ-gate rotates the qubit along the $z-$axis by ϕ radians. The matrix form of this gate is as shown in the following equation:

$$R_z(\phi) = \begin{bmatrix} e^{-i\frac{\phi}{2}} & 0 \\ 0 & e^{i\frac{\phi}{2}} \end{bmatrix} \tag{5.38}$$

We can express $RZ(\phi)$ using the Pauli rotation matrix Z as follows:

$$R_z(\phi) = cos\frac{\phi}{2} I - isin\frac{\phi}{2} Z = e^{-i\frac{\phi}{2}z} \tag{5.39}$$

Alternate form

$$R_z(\phi) = e^{-i\frac{\phi}{2}}U_1(\phi) \tag{5.40}$$

Note that the RZ-gate is equivalent to a U1-gate, but with a global phase offset of $e^{-i\frac{\phi}{2}}$.

Syntax

Circuit Composer	OpenQASM	Qiskit Python	Q#—MS QDK
R_z $\pi/2$	rz (phi) qubit|qreq; rz (pi/2) q[0];	qc.rz(phi, qubit) qc.rz(pi/2, q[0])	Operation Rz(theta:Double, qubit:Qubit): Unit let theta =1.5707963268; Rz(theta, q[0]);

5.4.2.4 T Gate and the T† Gate or the $\sqrt{S}$ Gate

The T gate and the T† gates are cases of the R_z gate with $\theta = \pm\frac{\pi}{4}$. These two gates are the square root of the S gate. The matrix form of these two gates are as follows:

$$T = \begin{bmatrix} 1 & 0 \\ 0 & e^{i\frac{\pi}{4}} \end{bmatrix} = e^{i\frac{\pi}{8}} \begin{bmatrix} e^{-i\frac{\pi}{8}} & 0 \\ 0 & e^{i\frac{\pi}{8}} \end{bmatrix}$$

$$T^\dagger = \begin{bmatrix} 1 & 0 \\ 0 & e^{-i\frac{\pi}{4}} \end{bmatrix} \tag{5.41}$$

Alternate forms

$$T = U_1\left(\frac{\pi}{4}\right) \tag{5.42}$$

$$T^\dagger = U_1\left(-\frac{\pi}{4}\right) \tag{5.43}$$

Syntax

Circuit Composer	OpenQASM	Qiskit Python	Q#—MS QDK
T	t qubit|qreq; t q[0];	qc.t(qubit) qc.t(q[0])	Operation T(qubit:Qubit): Unit T(q[0]);
$T^\dagger$	tdg qubit|qreq; tdg q[0];	qc.tdg (qubit) qc.tdg(q[0])	To be implemented as a user gate.

Note that, historically, the T gate is sometimes referred to as the $\frac{\pi}{8}$ gate, because of the diagonal elements appearing in the form of a global phase of $e^{i\frac{\pi}{8}}$. The gate does indeed perform a $\frac{\pi}{4}$ rotation about the z-axis, as shown in the Fig. 5.12.

5.4.3 Physical Gates

The U gates are implemented physically in the quantum device. We can derive the single-qubit gates that we saw in the previous sections from the U gates.

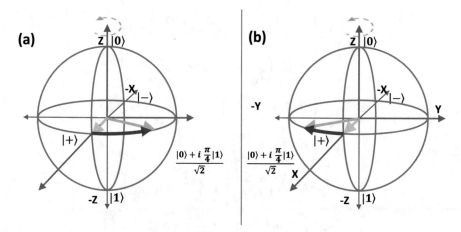

Fig. 5.12 Diagram illustrating T and the $T^{\dagger}$ gates

5.4.3.1 U3-Gate

Equation (5.44) gives the matrix form of the U3-gate.

$$U_3(\theta, \phi, \lambda) = \begin{bmatrix} cos\left(\dfrac{\theta}{2}\right) & -e^{i\lambda} sin\left(\dfrac{\theta}{2}\right) \\ e^{i\phi} sin\left(\dfrac{\theta}{2}\right) & e^{i\lambda+i\phi} cos\left(\dfrac{\theta}{2}\right) \end{bmatrix} \tag{5.44}$$

In this equation, θ and ϕ are the rotational angles in the Bloch sphere (see Fig. 3.10) and $e^{i\lambda}$ is a global phase. The matrix form of the U3-gate is also the general form of the unitary for single-qubit operations. Hence, we can derive the rest of the U gates from the U3-gate itself.

Let us now apply the U3-gate on the qubit states $|0\rangle$ and $|1\rangle$.

$$\begin{aligned} U_3|0\rangle &= \begin{bmatrix} \cos\left(\dfrac{\theta}{2}\right) & -e^{i\lambda}\sin\left(\dfrac{\theta}{2}\right) \\ e^{i\phi}\sin\left(\dfrac{\theta}{2}\right) & e^{i(\lambda+\phi)}\cos\left(\dfrac{\theta}{2}\right) \end{bmatrix} \begin{bmatrix} 1 \\ 0 \end{bmatrix} \\ &= \begin{bmatrix} \cos\left(\dfrac{\theta}{2}\right) \\ e^{i\phi}\sin\left(\dfrac{\theta}{2}\right) \end{bmatrix} = \cos\left(\dfrac{\theta}{2}\right)\begin{bmatrix} 1 \\ 0 \end{bmatrix} + e^{i\phi}\sin\left(\dfrac{\theta}{2}\right)\begin{bmatrix} 0 \\ 1 \end{bmatrix} \\ &= \cos\left(\dfrac{\theta}{2}\right)|0\rangle + e^{i\phi}\sin\left(\dfrac{\theta}{2}\right)|1\rangle \end{aligned} \tag{5.45}$$

$$U_3|1\rangle = \begin{bmatrix} \cos\left(\dfrac{\theta}{2}\right) & -e^{i\lambda}\sin\left(\dfrac{\theta}{2}\right) \\ e^{i\phi}\sin\left(\dfrac{\theta}{2}\right) & e^{i(\lambda+\phi)}\cos\left(\dfrac{\theta}{2}\right) \end{bmatrix} \begin{bmatrix} 0 \\ 1 \end{bmatrix}$$

$$= \begin{bmatrix} -e^{i\lambda}\sin\left(\dfrac{\theta}{2}\right) \\ e^{i(\lambda+\phi)}\cos\left(\dfrac{\theta}{2}\right) \end{bmatrix} \tag{5.46}$$

$$= -e^{i\lambda}\sin\left(\dfrac{\theta}{2}\right)\begin{bmatrix} 1 \\ 0 \end{bmatrix} + e^{i(\lambda+\phi)}\cos\left(\dfrac{\theta}{2}\right)\begin{bmatrix} 0 \\ 1 \end{bmatrix}$$

$$= e^{i\lambda}\left(-\sin\left(\dfrac{\theta}{2}\right)|0\rangle + e^{i\phi}\cos\left(\dfrac{\theta}{2}\right)|1\rangle\right)$$

We see that the U3-gate rotates the qubit into an arbitrary superposition state as determined by the angle of rotation.

Syntax

Circuit Composer	OpenQASM	Qiskit Python	Q#—MS QDK
U_3 π/2, π/2, π/2	u3 (theta, phi, lambda) qubit|qreq; u3 (pi/2, pi/2, pi/2) q[0];	qc.u3(theta, phi, lambda, qubit) qc.u3(pi/2, pi/2, pi/2, q[0])	See Appendix

5.4.3.2 U2-Gate
The following equation defines the U2-gate:

$$U_2(\phi, \lambda) = U_3\left(\frac{\pi}{2}, \phi, \lambda\right) = \frac{1}{\sqrt{2}}\begin{bmatrix} 1 & -e^{i\lambda} \\ e^{i\phi} & e^{i(\phi+\lambda)} \end{bmatrix} \tag{5.47}$$

Syntax

Circuit Composer	OpenQASM	Qiskit Python	Q#—MS QDK
U_2 π/2, π/2	u2 (phi, lambda) qubit|qreq; u2 (pi/2, pi/2) q[0];	qc. u2 (phi, lambda, qubit) qc. u2 (pi/2, pi/2, q[0])	See Appendix

5.4.3.3 U1-Gate
We can derive the U1-gate from the U3-gate by assuming θ and ϕ are 0.

$$U_1(\lambda) = U_3(0, 0, \lambda) = \begin{bmatrix} 1 & 0 \\ 0 & e^{i\lambda} \end{bmatrix} \tag{5.48}$$

In the Dirac's bra-ket notation, we can write this gate as follows:

$$U_1(\lambda)|x\rangle = e^{i\lambda \cdot x}|x\rangle \tag{5.49}$$

The U1-gate is equivalent to the Z gate.

Syntax

Circuit Composer	OpenQASM	Qiskit Python	Q#—MS QDK
U_1 $\pi/2$	u1 (lambda) qubit|qreq; u1 (pi/2) q[0];	qc. u1 (lambda, qubit) qc. u1 (pi/2, q[0])	See Appendix

5.4.4 Multiqubit Gates

In this section, we shall learn how to represent a system of two or more qubits and the gate operations, which are performed on two or more qubits.

5.4.5 Representing a Multi Qubit State

A system of n-qubits has 2^n orthonormal computational basis states, simultaneously. Collectively this is denoted by the ket $|x_1 x_2 \ldots x_n\rangle; x \in \{0, 1\}^n$. The quantum state of a n-qubit system is a superposition state with 2^n probability amplitudes $(a_x.)$ These statements are summarized in the following equation:

$$|\psi\rangle = \sum_{x \in \{0,1\}^n} a_x|x\rangle, \text{where} \sum_{x \in \{0,1\}^n} |a_x|^2 = 1 \tag{5.50}$$

When we measure this system, we get x, with a probability $|a_x|^2 = 1$. Let us study this in detail with a two-qubit example.

Recollect the tensor math, we learned in Chap. 2. We use the tensor product to represent the collective state of two or more qubits. Assume that we have two qubits $|\psi_1\rangle = a_1|0\rangle + b_1|1\rangle$ and $|\psi_2\rangle = a_2|0\rangle + b_2|1\rangle$. The two-qubit state describing them can be written as a tensor product described in the following equation:

$$\begin{aligned}|\psi\rangle &= |\psi_1\rangle \otimes |\psi_2\rangle \\ &= a_1 a_2|00\rangle + a_1 b_2|01\rangle + b_1 a_2|10\rangle + b_1 b_2|11\rangle\end{aligned} \tag{5.51}$$

We require this state to be orthonormal, that is, $\langle\psi|\psi\rangle = 1$. Hence, we infer that $|a_1 a_2|^2 + |a_1 b_2|^2 + |b_1 a_2|^2 + |b_1 b_2|^2 = 1$ from previous chapters.

When we measure the two-qubit state, we get two bits of information $x \in \{0, 1\}^2$. The four possible eigenvalues (λ_i) are as follows:

$$|00\rangle = \begin{bmatrix} 1 \\ 0 \end{bmatrix} \otimes \begin{bmatrix} 1 \\ 0 \end{bmatrix} = \begin{bmatrix} 1 \\ 0 \\ 0 \\ 0 \end{bmatrix}, \text{when we measure a '00'}$$

$$|01\rangle = \begin{bmatrix} 1 \\ 0 \end{bmatrix} \otimes \begin{bmatrix} 0 \\ 1 \end{bmatrix} = \begin{bmatrix} 0 \\ 1 \\ 0 \\ 0 \end{bmatrix}, \text{when we measure a '01'}$$

$$\hspace{10cm} (5.52)$$

$$|10\rangle = \begin{bmatrix} 0 \\ 1 \end{bmatrix} \otimes \begin{bmatrix} 1 \\ 0 \end{bmatrix} = \begin{bmatrix} 0 \\ 0 \\ 1 \\ 0 \end{bmatrix}, \text{when we measure a '10'}$$

$$|11\rangle = \begin{bmatrix} 0 \\ 1 \end{bmatrix} \otimes \begin{bmatrix} 0 \\ 1 \end{bmatrix} = \begin{bmatrix} 0 \\ 0 \\ 0 \\ 1 \end{bmatrix}, \text{when we measure a '11'}$$

Note that, the notation $|00\rangle$ is also used to represent an unsigned binar encoding of "00" into the qubits, and the usage of the tensor operator $\otimes$ is implicit. The alternate forms of representing the multi-qubit system are outlined here.

$$|00\rangle = \begin{bmatrix} 1 \\ 0 \end{bmatrix} \otimes \begin{bmatrix} 1 \\ 0 \end{bmatrix} = |0\rangle \otimes |0\rangle = |0\rangle^{\otimes 2} = |0\rangle \, |0\rangle \hspace{2cm} (5.53)$$

In a two or more qubit system, the qubits are correlated, and information is stored across all the qubits simultaneously. Joint measurement of the qubits is needed to reveal the information.

As in the case of single qubit, the probability of measuring an eigenvalue is given by the Born rule.

$$P_i^{zz} = |\langle \lambda_i | \psi \rangle|^2, \hspace{4cm} (5.54)$$

where z indicates the direction of measurement, and λ_i is one of the possible eigenvalues described in Eq. (5.52).

It is also possible to measure just one of the qubits separately. In that situation, the state of that qubit alone collapses into the computational basis. The other qubits continue to be in a superposition state.

The probability of measuring a 0 in the first qubit (this book uses a zero based indexing of qubits) is given by:

$$p_1(0) = P(00) + P(01) = |a_1 a_2|^2 + |a_1 b_2|^2 \tag{5.55}$$

Such a **partial measurement** of the system results in a new superposition state. We can calculate the new superposition state by removing the terms, which are no longer applicable. Moreover, we must normalize the new superposition state, since the sum of the square of the amplitudes is no longer 1.

$$|\psi\rangle = \frac{a_1 a_2 |00\rangle + a_1 b_2 |01\rangle}{\sqrt{|a_1 a_2|^2 + |a_1 b_2|^2}} \tag{5.56}$$

Similarly, the probability of measuring a 1 in the first qubit is:

$$p_1(1) = P(10) + P(11) = |b_1 a_2|^2 + |b_1 b_2|^2 \tag{5.57}$$

The following equation describes the new superposition state.

$$|\psi\rangle = \frac{b_1 a_2 |10\rangle + b_1 b_2 |11\rangle}{\sqrt{|b_1 a_2|^2 + |b_1 b_2|^2}} \tag{5.58}$$

We can now proceed to learn about multi-qubit gate operations.

5.4.5.1 The SWAP Gate

The SWAP gate is a two-qubit gate. It swaps the status of the two qubits. The matrix form of the SWAP gate is as follows (Fig. 5.13):

$$SWAP = \begin{bmatrix} 1 & 0 & 0 & 0 \\ 0 & 0 & 1 & 0 \\ 0 & 1 & 0 & 0 \\ 0 & 0 & 0 & 1 \end{bmatrix} \tag{5.59}$$

The SWAP gate performs the following mapping of the basis vectors of the target qubits:

$$\begin{aligned} |00\rangle &\mapsto |00\rangle \\ |01\rangle &\mapsto |10\rangle \\ |10\rangle &\mapsto |01\rangle \\ |11\rangle &\mapsto |11\rangle \end{aligned} \tag{5.60}$$

We can implement the SWAP gate using three CNOT gates. Section 5.5.1 describes the steps.

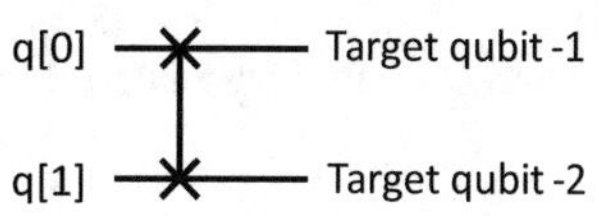

Fig. 5.13 The SWAP gate

q[0] ──✕──── Target qubit -1

q[1] ──✕──── Target qubit -2

Syntax

Circuit Composer	OpenQASM	Qiskit Python	Q#—MS QDK
q_0 ⨯ q_1 ⨯	swap qubit|qreq, qubit|qreq;	qc.swap (qubit, qubit)	Operation SWAP(qubit1:Qubit, qubit2: Qubit): Unit
	swap q[0], q[1];	qc.swap (q[0], q [1])	SWAP(q[0], q[1]);

Truth Table

Input	Output
$|00\rangle$	$|00\rangle$
$|01\rangle$	$|10\rangle$
$|10\rangle$	$|01\rangle$
$|11\rangle$	$|11\rangle$

5.4.5.2 Square Root of SWAP Gate

The Square root of SWAP or the $\sqrt{SWAP}$ gate performs half of the SWAP gate. This gate is easily implemented by applying half of the pulse needed for a Heisenberg interaction between two adjacent electron spin qubits. The $\sqrt{SWAP}$ gate is universal. $\sqrt{SWAP}$, along with a combination of rotational gates, can implement any other gate. In a matrix form, the $\sqrt{SWAP}$ gate is described as follows:

$$\sqrt{SWAP} = \begin{bmatrix} 1 & 0 & 0 & 0 \\ 0 & \frac{1+i}{2} & \frac{1-i}{2} & 0 \\ 0 & \frac{1-i}{2} & \frac{1+i}{2} & 0 \\ 0 & 0 & 0 & 1 \end{bmatrix} \tag{5.61}$$

5.4.6 Controlled U Gates

The general form of the Controlled-U gates (where U is the unitary matrix pertaining to the gate operation) is given by the following matrix:

$$C(U) = \begin{bmatrix} 1 & 0 & 0 & 0 \\ 0 & 1 & 0 & 0 \\ 0 & 0 & u_{00} & u_{01} \\ 0 & 0 & u_{10} & u_{11} \end{bmatrix}, \tag{5.62}$$

where, the unitary matrix $U = \begin{bmatrix} u_{00} & u_{01} \\ u_{10} & u_{11} \end{bmatrix}$ is one of the Pauli matrices.

Fig. 5.14 The CNOT gate

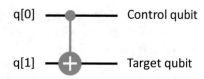

The Controlled-U gates perform the following mapping:

$$|00\rangle \mapsto |00\rangle$$
$$|01\rangle \mapsto |01\rangle$$
$$|10\rangle \mapsto |1\rangle \otimes U |0\rangle \qquad (5.63)$$
$$|11\rangle \mapsto |1\rangle \otimes U |1\rangle,$$

where the qubit on the left-hand side of the tensor product serves as the control qubit, and the qubit on the right-hand side of the tensor product serves as the target qubit.

5.4.6.1 The Controlled NOT (CNOT) or the cX Gate

The controlled-NOT gate operates on two qubits. Out of the two qubits, the first qubit serves as a control qubit. The second qubit is the target qubit. The CNOT gate applies an X-gate to the target qubit if the control qubit is in state $|1\rangle$. The target qubit is unaltered if the control qubit is in state $|0\rangle$. The control qubit remains unaltered during this process (Fig. 5.14).

The CNOT or the cX gate is described in the following matrix form:

$$\text{CNOT} = cX = \begin{bmatrix} 1 & 0 & 0 & 0 \\ 0 & 1 & 0 & 0 \\ 0 & 0 & 0 & 1 \\ 0 & 0 & 1 & 0 \end{bmatrix} \qquad (5.64)$$

The CNOT or the cX gate performs the following mapping:

$$|q_0, q_1\rangle \mapsto |q_0, q_0 \oplus q_1\rangle, \qquad (5.65)$$

where $\oplus$ is an XOR operation.

Syntax

Circuit Composer	OpenQASM	Qiskit Python	Q#—MS QDK
q_0 —•— q_1 —⊕—	cx qubit|qreq, qubit|qreq; cx q[0], q[1];	qc. cx (qubit, qubit) qc. cx (q[0], q[1])	Operation CNOT(control:Qubit, target:Qubit): Unit CNOT(q[0], q[1]);

Truth Table

Input	Output
$|00\rangle$	$|00\rangle$
$|01\rangle$	$|10\rangle$
$|10\rangle$	$|11\rangle$
$|11\rangle$	$|10\rangle$

5.4.6.2 cY and cZ (CPHASE) Gates

The cY and cZ gates operate in a fashion similar to the cX gate. The matrix representations for the cY and cZ gates are as follows:

$$
cY = \begin{bmatrix} 1 & 0 & 0 & 0 \\ 0 & 1 & 0 & 0 \\ 0 & 0 & 0 & -i \\ 0 & 0 & i & 0 \end{bmatrix}, \; cZ = \begin{bmatrix} 1 & 0 & 0 & 0 \\ 0 & 1 & 0 & 0 \\ 0 & 0 & 1 & 0 \\ 0 & 0 & 0 & -1 \end{bmatrix} \tag{5.66}
$$

The cY and cZ gates apply the unitary transform (Pauli Y or the Pauli Z gates) to the target qubit, only if the control qubit is in state $|1\rangle$. The target qubit is unaltered, if the control qubit is in state $|0\rangle$

Syntax

Circuit Composer	OpenQASM	Qiskit Python	Q#—MS QDK
q_0 q_1 Y	cy control|qreq, target|qreq; cy q[0], q[0];	qc. cy (control, target) qc. cy (q[0], q[1])	Operation CY(control:Qubit, target:Qubit): Unit CY(q[0], q[1]);
q_0 q_1 Z	cz control|qreq, target|qreq; cz q[0], q[1];	qc. cz (control, target) qc. cz (q[0], q[1])	Operation CZ(control:Qubit, target:Qubit): Unit CZ(q[0], q[1]);

Truth Table

Input	cY	cZ
$|00\rangle$	$|0\rangle \otimes |0\rangle$	$|0\rangle \otimes |0\rangle$
$|01\rangle$	$|0\rangle \otimes |1\rangle$	$|0\rangle \otimes |1\rangle$
$|10\rangle$	$|1\rangle \otimes i|1\rangle$	$|1\rangle \otimes |0\rangle$
$|11\rangle$	$|1\rangle \otimes -i|0\rangle$	$|1\rangle \otimes -|1\rangle$

The cZ gate is also called **CPHASE** gate.

5.4.6.3 Controlled Hadamard Gate or the cH Gate

The controlled H-gate operates on two qubits: the control qubit and the target qubit. The cH gate applies an H-gate to the target qubit when the control qubit is in state $|1\rangle$. When the control qubit is in state $|0\rangle$, the target qubit is unaltered. The state of the control qubit remains the same. The cH gate performs the following mapping:

$$
\begin{aligned}
|00\rangle &\mapsto |00\rangle \\
|01\rangle &\mapsto |01\rangle \\
|10\rangle &\mapsto |1\rangle \otimes H |0\rangle \\
|11\rangle &\mapsto |1\rangle \otimes H |1\rangle
\end{aligned} \tag{5.67}
$$

Syntax

Circuit Composer	OpenQASM	Qiskit Python	Q#—MS QDK
q_0 ——●—— q_1 ——H——	ch control|qreq, target|qreq; ch q[0], q[1];	qc. ch (control, target) qc. ch (q[0], q[1])	See Appendix

Truth Table

Input	Output
$\|00\rangle$	$\|00\rangle$
$\|01\rangle$	$\|01\rangle$
$\|10\rangle$	$\frac{1}{\sqrt{2}}(\|10\rangle + \|11\rangle)$
$\|11\rangle$	$\frac{1}{\sqrt{2}}(\|10\rangle - \|11\rangle)$

5.4.6.4 Controlled RZ-Gate or the cRz Gate

The controlled RZ-gate operates on two qubits: the control qubit, and the target qubit. The cRz gate applies a RZ-gate to the target qubit, when the control qubit is in state $|1\rangle$. When the control qubit is in state $|0\rangle$, the target qubit is unaltered. The state of the control qubit remains the same. The cRz gate performs the following mapping:

$$
\begin{aligned}
|00\rangle &\mapsto |00\rangle \\
|01\rangle &\mapsto |01\rangle \\
|10\rangle &\mapsto |1\rangle \otimes R_z(\theta)|0\rangle \\
|11\rangle &\mapsto |1\rangle \otimes R_z(\theta)|1\rangle
\end{aligned}
\tag{5.68}
$$

Syntax

Circuit Composer	OpenQASM	Qiskit Python	Q#—MS QDK
q_0 ——●—— q_1 ——R_z——	crz (theta) control|qreq, target|qreq; crz (pi/2) q[0], q[1];	qc. crz (theta, control, target) qc. crx (pi/2, q[0], q[1])	See Appendix

5.4.6.5 Controlled U3-Gate or the cU3 Gate

The controlled U3-gate operates on two qubits: the control qubit and the target qubit. The cU3 gate applies a U3-gate to the target qubit when the control qubit is in state $|1\rangle$. When the control qubit is in state $|0\rangle$, the target qubit is unaltered. The state of the control qubit remains the same. The cU3 gate performs the following mapping:

$$
\begin{aligned}
|00\rangle &\mapsto |00\rangle \\
|01\rangle &\mapsto |01\rangle \\
|10\rangle &\mapsto |1\rangle \otimes U_3(\theta, \phi, \lambda)|0\rangle \\
|11\rangle &\mapsto |1\rangle \otimes U_3(\theta, \phi, \lambda)|1\rangle
\end{aligned}
\tag{5.69}
$$

Syntax

Circuit Composer	OpenQASM	Qiskit Python	Q#—MS QDK
q_0 —•— q_1 —U_3—	cu3 (theta, phi, lambda) control|qreq, target|qreq; cu3 (pi/2, pi/2, pi/2) q[0], q[1];	qc. cu3 (theta, phi, lambda, control, target) qc. cu3 (pi/2, pi/2, pi/2, q[0], q[1])	See Appendix

5.4.6.6 Controlled U1-Gate or the cU1 Gate

The controlled U1-gate operates on two qubits: the control qubit, and the target qubit. The cU1 gate applies a U1-gate to the target qubit when the control qubit is in state $|1\rangle$. When the control qubit is in state $|0\rangle$, the target qubit is unaltered. The state of the control qubit remains the same. The cU1 gate performs the following mapping:

$$\begin{aligned}
|00\rangle &\mapsto |00\rangle \\
|01\rangle &\mapsto |01\rangle \\
|10\rangle &\mapsto |1\rangle \otimes U_1(\lambda) |0\rangle \\
|11\rangle &\mapsto |1\rangle \otimes U_1(\lambda) |1\rangle
\end{aligned} \tag{5.70}$$

Syntax:

Circuit Composer	OpenQASM	Qiskit Python	Q#—MS QDK
q_0 —•— q_1 —U_1—	cu1 (lambda) control|qreq, target|qreq; cu1 (pi/2) q[0], q[1];	qc. cu1 (lambda, control, target) qc. cu1 (pi/2, q[0], q[1])	See Appendix

5.4.7   Extended Gates

The Toffoli gate is a quantum gate with two control qubits and one target qubit. The Fredkin gate has one control qubit, but it operates on two target qubits. We study them as extended gates in this section.

5.4.7.1 The Toffoli or the CCNOT Gate or the ccX Gate

The Toffoli gate is a three-qubit gate. This gate has two control qubits and one target qubit. It applies the NOT (or the X-gate) gate to the target qubit, if both the control qubits are in state $|1\rangle$. We can write the operation of this gate as follows.

$$|q_0, q_1, q_2\rangle \mapsto |q_0, q_1, q_2 \oplus q_0 \wedge q_1\rangle, \tag{5.71}$$

where $\oplus$ is an XOR operation and $\wedge$ is an AND operation.

The classical version of the Toffoli gate is universal. But the quantum version is not universal as such. Together with a Hadamard gate, the Toffoli gate is universal in quantum computing [1][13]. The Toffoli gate is also called Deutsch $\frac{\pi}{2}$ gate (Fig. 5.15).

Syntax

Circuit Composer	OpenQASM	Qiskit Python	Q#—MS QDK
q_0 q_1 q_2	ccx control1|qreq, control2|qreq, target|qreq; ccx q[0], q[1], q[2];	qc. ccx (control1, control2, target) qc. ccx (q[0], q[1], q[2])	Operation CCNOT(control1:Qubit, control1:Qubit, target:Qubit): Unit CCNOT(q[0], q[1],q[2]);

The following permutation matrix describes the Toffoli gate.

$$
\begin{bmatrix}
1 & 0 & 0 & 0 & 0 & 0 & 0 & 0 \\
0 & 1 & 0 & 0 & 0 & 0 & 0 & 0 \\
0 & 0 & 1 & 0 & 0 & 0 & 0 & 0 \\
0 & 0 & 0 & 1 & 0 & 0 & 0 & 0 \\
0 & 0 & 0 & 0 & 1 & 0 & 0 & 0 \\
0 & 0 & 0 & 0 & 0 & 1 & 0 & 0 \\
0 & 0 & 0 & 0 & 0 & 0 & 0 & 1 \\
0 & 0 & 0 & 0 & 0 & 0 & 1 & 0
\end{bmatrix}
\tag{5.72}
$$

Truth Table

Input	Output	Input	Output
$|000\rangle$	$|000\rangle$	$|100\rangle$	$|100\rangle$
$|001\rangle$	$|001\rangle$	$|101\rangle$	$|101\rangle$
$|010\rangle$	$|010\rangle$	$|110\rangle$	$|111\rangle$
$|011\rangle$	$|011\rangle$	$|111\rangle$	$|110\rangle$

5.4.7.2 The Fredkin or the CSWAP or the cS Gate

The Fredkin gate performs a controlled SWAP of two target qubits. Hence, the Fredkin gate is a three-qubit gate. The CSWAP gate performs the following mapping:

$$
\begin{aligned}
CSWAP|0, q_1, q_2\rangle &\mapsto |0, q_1, q_2\rangle \\
CSWAP|1, q_1, q_2\rangle &\mapsto |1, q_2, q_1\rangle
\end{aligned}
\tag{5.73}
$$

Syntax

Circuit Composer	OpenQASM	Qiskit Python	Q#—MS QDK
q_0 q_1 q_2	cswap control|qreq, target1|qreq, target2|qreq; cswap q[0], q[1], q[2];	qc. ccx (control, target1, target2) qc. cswap (q[0], q[1], q[2])	To be implemented as a user gate

The following permutation matrix describes the Fredkin gate (Fig. 5.15).

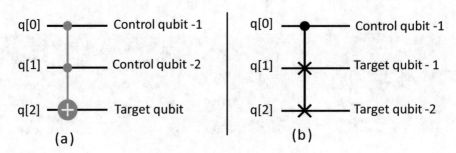

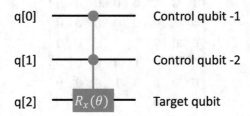

Fig. 5.15 Diagram showing: (**a**) Toffoli (CCNOT/CCX) and (**b**) Fredkin (CSWAP) gates

Fig. 5.16 The Deutsch gate

$$
\begin{bmatrix}
1 & 0 & 0 & 0 & 0 & 0 & 0 & 0 \\
0 & 1 & 0 & 0 & 0 & 0 & 0 & 0 \\
0 & 0 & 1 & 0 & 0 & 0 & 0 & 0 \\
0 & 0 & 0 & 1 & 0 & 0 & 0 & 0 \\
0 & 0 & 0 & 0 & 1 & 0 & 0 & 0 \\
0 & 0 & 0 & 0 & 0 & 0 & 1 & 0 \\
0 & 0 & 0 & 0 & 0 & 1 & 0 & 0 \\
0 & 0 & 0 & 0 & 0 & 0 & 0 & 1
\end{bmatrix}
\tag{5.74}
$$

Truth Table

Input	Output	Input	Output
$|000\rangle$	$|000\rangle$	$|100\rangle$	$|100\rangle$
$|001\rangle$	$|001\rangle$	$|101\rangle$	$|110\rangle$
$|010\rangle$	$|010\rangle$	$|110\rangle$	$|101\rangle$
$|011\rangle$	$|011\rangle$	$|111\rangle$	$|111\rangle$

5.4.7.3 The Deutsch Gate $D(\theta)$

The Deutsch gate is a three-qubit gate with two control qubits and a target qubit. When the two control qubits are in state $|1\rangle$, the Deutsch gate performs an $R_x(\theta)$ operation on the target qubit (Fig. 5.16).

The following mapping summarizes the Deutsch gate.

$$|q_0, q_1, q_2\rangle \mapsto \begin{cases} icos(\theta)|q_0, q_1, q_2\rangle + sin(\theta)|q_0, q_1, 1 - q_2\rangle \text{ if } q_0 \neq q_1 \neq 1 \\ |q_0, q_1, q_2\rangle \text{ otherwise} \end{cases} \qquad (5.75)$$

The specific example of the Deutsch gate with $\theta = \frac{\pi}{2}$ functions as the Toffoli gate.

5.4.8 Universality of Quantum Gates

5.4.8.1 Global Phase and Its Irrelevance

Recall that we can describe a qubit in its superposition state as a linear combination of the basis states.

$$|\psi\rangle = a|0\rangle + b|1\rangle, \qquad (5.76)$$

where a and b are complex numbers representing the probability amplitudes of finding the qubit in state $|0\rangle$ or $|1\rangle$. If we can write these complex numbers in exponential form, by drawing from Eq. (2.3), we get:

$$|\psi\rangle = r_1 e^{i\theta_1}|0\rangle + r_2 e^{i\theta_2}|1\rangle \qquad (5.77)$$

By rearranging this slightly, we get:

$$|\psi\rangle = e^{i\theta_1}\left(r_1|0\rangle + r_2 e^{i(\theta_2 - \theta_1)}|1\rangle\right) \qquad (5.78)$$

If we calculate the probability amplitude $|\psi|^2$, by drawing an analogy with Eq. (2.7), we can readily say that the factor $e^{i\theta_1}$, vanishes. This factor is called **global phase**, and it does not affect the quantum computation. The quantity $(\theta_2 - \theta_1)$ is called **relative phase** and it is an observable. To illustrate this further consider the action of the X-gate on the state $|-\rangle$. From Eq. (5.12),

$$|-\rangle = \frac{1}{\sqrt{2}}(|0\rangle - |1\rangle) \qquad (5.79)$$

Applying X-gate, we get:

$$\begin{aligned} X|-\rangle &= \frac{1}{\sqrt{2}}(X|0\rangle - X|1\rangle) = \frac{1}{\sqrt{2}}(|1\rangle - |0\rangle) \\ &= -1\left(\frac{1}{\sqrt{2}}(|0\rangle - |1\rangle)\right) \\ &= -|-\rangle \end{aligned} \qquad (5.80)$$

We see that this operation adds a global phase -1. But this phase is not observable. Now, consider the circuit shown in Fig. 5.17:

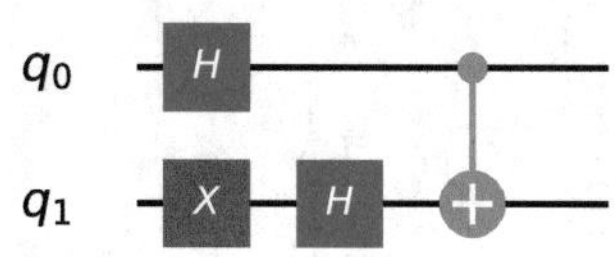

Fig. 5.17 Circuit illustrating relative phase

We can work out the system state of this circuit as follows. We begin the system with the qubits q[0] and q[1] initialized to state $|0\rangle$.

$$|\psi\rangle = |0\rangle|0\rangle \tag{5.81}$$

We then apply a H-gate to qubit q[0] and an X-gate to q[1].

$$|\psi\rangle = \frac{1}{\sqrt{2}}(|0\rangle + |1\rangle)|1\rangle \tag{5.82}$$

The next step is to apply a H-gate to q[1]. The new system state is as follows:

$$
\begin{aligned}
|\psi\rangle &= \frac{1}{\sqrt{2}}(|0\rangle + |1\rangle) \otimes \frac{1}{\sqrt{2}}(|0\rangle - |1\rangle) \\
&= \frac{1}{2}(|00\rangle - |01\rangle + |10\rangle - |11\rangle)
\end{aligned} \tag{5.83}
$$

Now we apply a CNOT gate, with the qubit q[0] as the control qubit and q[1] as the target qubit.

$$
\begin{aligned}
|\psi\rangle &= CNOT\left(\frac{1}{2}(|00\rangle - |01\rangle + |10\rangle - |11\rangle)\right) \\
&= \frac{1}{2}(|00\rangle - |01\rangle + |11\rangle - |10\rangle) \\
&= \frac{1}{2}(|00\rangle - |01\rangle - (|10\rangle - |11\rangle))
\end{aligned} \tag{5.84}
$$

Comparing Eqs. (5.83) and (5.84), we find that a global phase is added to the target qubit[1]. We can separate this state, into the states of two separate qubits.

$$= \frac{1}{\sqrt{2}}(|0\rangle - |1\rangle) \otimes \frac{1}{\sqrt{2}}(|0\rangle - |1\rangle) \tag{5.85}$$

Note that this operation has resulted in a relative phase to the control qubit. This process of transferring the global phase occurring in a qubit into a relative phase of another qubit using a controlled gate is called **phase kickback** or simply **kickback**.

The phase kickback method is used extensively in quantum algorithms. We shall see some of them in Chap. 7.

5.4.8.2 Sequence of Gates

In a quantum circuit, we employ a sequence of quantum gates to solve a problem. This sequence comprises of single and multi-gate operations. We can combine the unitary operations to draw some inferences. Let us study the sequence of HXH as an example. Using the corresponding matrices, we can write this sequence as:

$$
\begin{aligned}
HXH &= \frac{1}{\sqrt{2}}\begin{bmatrix} 1 & 1 \\ 1 & -1 \end{bmatrix} \cdot \begin{bmatrix} 0 & 1 \\ 1 & 0 \end{bmatrix} \cdot \frac{1}{\sqrt{2}}\begin{bmatrix} 1 & 1 \\ 1 & -1 \end{bmatrix} \\
&= \frac{1}{2}\begin{bmatrix} 1 & 1 \\ 1 & -1 \end{bmatrix} \cdot \begin{bmatrix} 1 & -1 \\ 1 & 1 \end{bmatrix} \\
&= \frac{1}{2}\begin{bmatrix} 2 & 0 \\ 0 & -2 \end{bmatrix} = \begin{bmatrix} 1 & 0 \\ 0 & -1 \end{bmatrix} = Z
\end{aligned} \tag{5.86}
$$

This sequence reduces to a Z gate. Similarly, we can prove the following identities:

HXH	Z
HYH	-Y
HZH	X

ZXZ	-X
ZYZ	-Y
ZZZ	Z

SXS†	Y
SYS†	-X
SZS†	Z

YXY	-X
YYY	Y
YZY	-Z

XXX	X
XYX	Y
XZX	Z

These identities mean that we can make composite gates, which can transform Pauli operators into other Pauli operators, which forms the basis for universality [12].

5.4.8.3 Universality

In the world of classical computers, we saw that every operation at the application level boils down to a set of digital gate operations at the hardware level. AND, OR, and NOT gates are the basic digital gates with which we can develop Boolean functions solving problems in Boolean algebra. If we have to find optimal gates that serve as the building blocks to solve any problem in Boolean algebra, these basic gates are probably not the right choices, for example, the AND gate can be constructed using OR and NOT gates. However, the two gates OR and NOT do not form a complete logical set by themselves. The choices are with NAND and NOR gates, as they are complete by themselves, and they can be used to construct all other logic. Hence, in the world of digital electronics, the NAND and NOR gates are called **Universal Logic Gates**.

Similarly, we can define **Universal Quantum Gates**. Universal Quantum Gates are a set of gates to which any operation possible on a quantum computer can be reduced. In other words, any unitary operation can be expressed as a finite sequence of gates from the set of Universal Quantum Gates.

If such a set of gates existed then, for example, we may be able to implement a gate $R(\theta), \forall \theta \in \Re$. Since this is a wide range, it may seem an impossibility to implement universality with a finite number of gates. So, we must resort to some approximations. The **Solovay–Kitaev theorem** proposes that approximations to any desired gate can be created using fairly short sequences of gates from the generating set. In short, we can say that the unitary U' simulates the required unitary U to an accuracy within ε if $\|U - U'\| \leq \varepsilon$.

According to Solovay–Kitaev theorem, we can say that a set of G quantum gates are universal if,

$$\forall U, \forall \varepsilon > 0, \exists g_1, g_2, \ldots, g_n \in G : \left\| U - U_{g1}, U_{g2}, \ldots, U_{gn} \right\| \leq \varepsilon. \tag{5.87}$$

One set of common Universal Quantum Gates available today are the Clifford Group and the T gates.

5.4.9 Circuit Optimization

5.4.9.1 Hardware Topology

In Sect. 5.4.8.2, we learned that a sequence of gates often reduces to single-qubit operations, which means there is some scope for circuit-level optimizations. Circuit level optimizations are essential for the current generation of NISQ processors. They have many design constraints due to the complexity of the underlying physics and the control electronics required to operate the qubits. Especially, qubit to qubit interaction channels and multi-qubit gate control are often limited. Due to these constraints, the NISQ processors have a specific configuration (called **Hardware Topology**) with a limited set of basis gates and qubits connections. For example, multi-qubit gates such as CNOT and SWAP operations can only be performed between certain qubit combinations. Hence, quantum circuits need to be optimized and translated into those basis gates and circuit combinations to suit specific hardware design constraints.

Let us explore these constraints a bit. The first constraint we shall explore is the hardware topology. Type the code from Listing 5.15 in your Jupyter notebook and execute the same.

Figure 5.18 shows the output of the program for the processor "ibmq_athens." Note that IBM Quantum Experience announces new backends, and old backends are taken offline from time to time. The current list of backends is found at this link: https://quantum-computing.ibm.com/systems? skip=0&system. if the backend "ibmq_athens" is not available at the time of your experimentation, please try another backend. You can run the code with the other processors, and get some familiarity with their hardware topology. Note that, in this processor, there is no connection between qubits 0 & 2, 0 & 3, 0 & 4, and 2 & 4. Does this mean, we cannot perform a multi-qubit operation between these

```
# Import the needed libraries
import qiskit
from qiskit import IBMQ
from qiskit.visualization import plot_histogram, plot_gate_map
from qiskit.visualization import plot_circuit_layout
from qiskit import QuantumRegister, ClassicalRegister
from qiskit import QuantumCircuit, execute, Aer


# load the IBM Quantum account
account = IBMQ.load_account()


# list the basis gates for a specific backend
provider = IBMQ.get_provider(group='open')
backend = provider.get_backend('imbq_athens')
plot_gate_map(backend)
```

Listing 5.15 Plotting the hardware topology

Fig. 5.18 Hardware
Topology of the backend
"ibmq_athens"

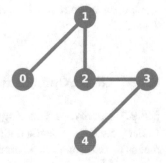

qubits? The answer is both yes and no. At the physical level, it is not possible. However, at the circuit level, we can construct a multi-qubit gate operation between these qubits. We shall learn how this works in a subsequent section.

Type the following code in a new cell and execute the same.

```
backend.configuration().basis_gates
```

Executing this code generates the list of basis gates for the backend "ibmq_athens." We get the following list upon execution of this code:

```
['id', 'rz', 'sx', 'x', 'cx', 'reset']
```

As the list shows, the backend "ibmq_athens" implements only the given basis gates at the hardware level! If we have expected all the gates outlined in the previous sections are supported at the hardware level, we are a little surprised by this information. Most backends implement the basis gates at the hardware level.

All other gates we outlined in previous sections can be constructed from combinations of the basis gates. Besides, gates such as Z, T, S, $T^\dagger$, and $S^\dagger$ are not performed on the hardware. These gates are implemented in software as "**virtual gates**." The other name for virtual gates is "**frame changes**." Frame changes execute in zero time and have no associated error. In other words, we regard them as **free gates** on hardware. To illustrate this, type the following code in a new cell and execute the same in your notebook (Listing 5.16).

```
# Define a quantum circuit with three qubits and a classical\
  register
q = QuantumRegister(3, 'q')
c = ClassicalRegister(1, 'c')
qc = QuantumCircuit(q, c)

qc.h(q[1])
qc.cx(q[0],q[1])
qc.cx(q[2],q[1])
qc.ccx(q[1],q[0],q[2])

qc.measure(q[1], c[0])
qc.draw(output='mpl')
```

Listing 5.16 Sample code to demonstrate circuit decomposition

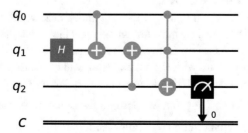

Fig. 5.19 A Test circuit to demonstrate circuit decomposition

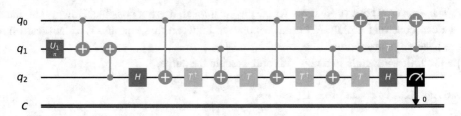

Fig. 5.20 Decomposition of the circuit in Fig. 5.19

Executing the given code generates the circuit shown in Fig. 5.19. We start with all qubits prepared at state $|0\rangle$. We then apply an H-gate to qubit $q[1]$, and a CNOT gate to qubits $q[0] \rightarrow q[1]$. After this step, we apply another CNOT gate to qubits $q[2] \rightarrow q[1]$. The next step is to apply a CCX gate to qubit $q[2]$, with qubits $q[0]$ and $q[1]$ as the control qubits. We finally measure qubit $q[2]$. This code does not do anything creative. We shall use this for illustration.

Enter the following code in a new cell and execute it in your notebook. These two lines of code decompose the circuit into its constituents.

```
qc_basis = qc.decompose()
qc_basis.draw(output='mpl')
```

Executing the code fragment generates a decomposed circuit as shown in Fig. 5.20.

The decomposition of our original circuit should not be surprising as we have learned that advanced gates are composed of basis gates!

We can try some math to prove that Figs. 5.19, 5.20 are the same. The question is: How did this happen?

5.4.9.2 Transpilation

We must also note that it is not efficient to write code for each hardware topology. As much research is happening in this space, we can expect frequent announcements of new backends with an increased number of qubits and hardware configuration. The older generation of backends is outdated and removed from the execution pipeline. So, it is advantageous if we can have software convert the given quantum circuit to suit any backend. When we submit a quantum circuit for execution, a process known as **transpilation** takes place first. The transpiler converts the input circuit to suit the target backend and optimizes it for execution. Note that the transpilers are improved as new optimization schemes are invented. So your version of the decomposed circuit may not be exactly similar to Fig. 5.20, but it should be quantum mechanically equivalent to Fig. 5.19.

5.4.9.3 Circuit Depth, Width, and Size

The **depth** of a quantum circuit is a measure of the quantum gates, which can be executed in parallel to complete the entire circuit. Hence, we can alternatively say that the depth of the circuit is a measure of the time taken to execute the circuit. Gate operations, such as the CCX gate used in the previous example, get converted into a circuit with significant depth and hence may be prone to gate errors and noises. We can use the circuit depth to check whether the quantum circuit can run on a backend by comparing it with the Gate Time $\left(\frac{T_2}{T_\varnothing}\right)$ and predicting how much noisy the circuit can be.

Let us examine the depth of our circuit now. Enter the following code in a new cell and execute the same.

```
qc.depth()
```

We should get an output of 5. Note that the depth calculations include the measurement operations. Execute the following code to determine the **width** of the circuit.

```
qc.width()
```

The width of the quantum circuit is the total number of qubits and classical registers. In our case, it is 4.

Now, let us check the size of the circuit with the following code:

```
qc.size()
```

The **size** of a quantum circuit is a measure of the raw operations. It is 3 for our circuit.

To check the total number of circuit operations, execute the following code.

```
qc.count_ops()
```

This command generates the list of gates used in the circuit.

```
OrderedDict([('cx', 2), ('h', 1), ('ccx', 1), ('measure', 1)])
```

Compare this with the circuit we constructed!

5.4.9.4 Barrier Gate

In a quantum circuit, a dotted vertical bar denotes a barrier gate. The barrier gate is not an actual gate. The purpose of the barrier gate is to prevent the transpiler from merging the gate operations. Let us illustrate this with an example.

Consider the code from Listing 5.17:

```
# Define a quantum circuit with a qubit and a classical register
q = QuantumRegister(1, 'q')
c = ClassicalRegister(1, 'c')
qc = QuantumCircuit(q, c)

qc.u3(math.pi,0,math.pi,q[0])
qc.u3(math.pi,0,math.pi,q[0])
qc.measure(q[0], c[0])
qc.draw(output='mpl')
```

Listing 5.17 Test code to illustrate transpilation and barrier gates

Upon executing the given code, we get the diagram shown in Fig. 5.21(a).

Now, execute the code provided in Listing 5.18.

The circuit is transpiled, and we get the diagram shown in Fig. 5.21(b).

In the given example, the $U3(pi, 0, pi)$ operation is equivalent to an X-gate. Two successive X-gates, bring the qubit to the original state. Hence, the transpilation process has merged the $U3$ gates into a no-operation state. The transpilation results in a circuit, which measures the qubit. To prevent

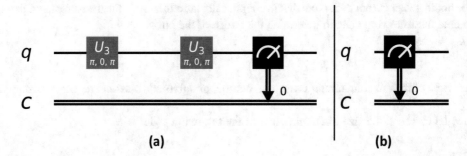

Fig. 5.21 Transpilation Example (**a**) Circuit with two successive U3 gates which are functionally equivalent to X-gates. (**b**) Circuit after transpilation

```
from qiskit import transpile

qc = transpile(qc)
qc_basis = qc.decompose()
qc_basis.draw(output='mpl')
```

Listing 5.18 Code illustrating the transpilation step

```
# Define a quantum circuit with a qubit and a classical register
q = QuantumRegister(1, 'q')
c = ClassicalRegister(1, 'c')
qc = QuantumCircuit(q, c)

qc.u3(math.pi,0,math.pi,q[0])
qc.barrier(q[0])
qc.u3(math.pi,0,math.pi,q[0])
qc.measure(q[0], c[0])
  qc.draw(output='mpl')
```

Listing 5.19 Test code with the barrier gate inserted between the U3 gates

Fig. 5.22 Circuit diagram with the barrier gate in between the U3 gates. This circuit is unaltered by the transpiler

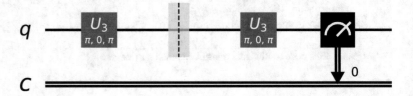

the transpiler from merging gates, we should introduce a barrier gate, as shown in the given code (Listing 5.19).

Upon execution of the given code, we get the circuit diagram, as shown in Fig. 5.22. The transpilation process does not alter this quantum circuit. A barrier separates the two X-gates.

We shall use the barrier gate in our future examples, appropriately.

5.4.9.5 The Reset Gate

It is possible to reset a qubit in the middle of an operation. The reset operation is not a physical gate. Hence, it is not reversible. The reset operation sets the qubit to state $|0\rangle$, irrespective of its previous state.

Syntax

Circuit Composer	QASM	Qiskit Python	Q#—MS QDK
$|0\rangle$	reset q[0]	qc.reset(q[0])	Reset(qubit)

5.4.10 State Visualization

In Listing 5.4, we used the plot_histogram function. In this section, we shall see a couple of other methods, which may be useful.

5.4.10.1 Plot_Bloch_Multivector

After executing a quantum circuit, the Bloch vector of the resultant state can be plotted. Be sure to use "statevector_simulator" as the backend (Listing 5.20).

Executing this code produces the image, as shown in Fig. 5.23. For the given circuit, the Bloch vector may point to different directions, each time the code is executed. The reason is that the qubits are in a superposition state.

```
from qiskit import QuantumCircuit, Aer, execute
from qiskit.visualization import plot_bloch_multivector,\
    plot_state_city
from qiskit.quantum_info import Statevector

q = QuantumRegister(2, 'q')
c = ClassicalRegister(2, 'c')
qc = QuantumCircuit(q, c)

qc.h(q[0])
qc.cx(q[0], q[1])
qc.measure(q[0], c[0])
qc.measure(q[1], c[1])

backend = Aer.get_backend('statevector_simulator')
job = execute(qc, backend).result()

plot_bloch_multivector(job.get_statevector(qc), title="My Bloch\
    vector")
```

Listing 5.20 Plotting the Bloch vector

Fig. 5.23 Bloch vector

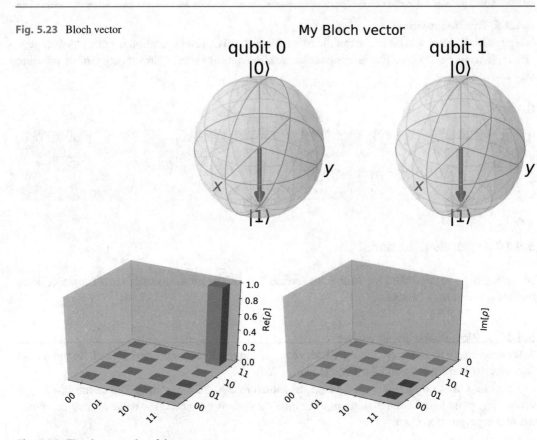

Fig. 5.24 The cityscape plot of the quantum state

5.4.10.2 Plot_State_City

In a new cell in the Jupyter notebook, execute the following statement.

```
plot_state_city(job.get_statevector(qc))
```

The above instruction produces the image shown in Fig. 5.24.

5.4.11 Gottesman–Knill Theorem

In the previous sections, we learned that the quantum circuit evolves unitarily. Assuming U is unitary in group $\mathbb{U}(2^n)$, and n is the number of qubits, the following equation describes the unitary evolution.

$$\left|\psi_{final}\right\rangle = U\left|\psi_{initial}\right\rangle = U|0_1, 0_2, \ldots, 0_n\rangle \tag{5.88}$$

According to the **Gottesman–Knill theorem**, a quantum unitary evolution that uses the following operations can be simulated effectively on a classical computer.

- Preparation of the qubit states in the computational basis
- Quantum evolution using the unitaries from the Clifford group (H-gate, CNOT-gate, Phase gates)
- Measurement in the computational basis

Daniel Gottesman and **Emanual Knill** proposed this theorem. This theorem proves that on a classical computer, we can simulate quantum circuits constructed from the Clifford group in polynomial time. Therefore, quantum circuits that rely on entanglement created by H and CNOT gates alone cannot create a quantum advantage.

Gottesman made some important contributions to quantum error correction and stabilizer circuits. In the last chapter of this book, we shall return to Gottesman–Knill theorem briefly. Several references related to Gottesman's work are provided in the reference section of that chapter.

5.5   The Compute Stage

With the knowledge we built so far on quantum gates, quantum circuits, and circuit optimizations, we are now ready to dive deep into the compute stage. In the following sections, we start by exploring simple circuits such as cascading of two CNOT gates and constructing digital logic equivalents. We then move on with some advanced topics such as quantum entanglement and quantum teleportation.

5.5.1   Experimenting with the CNOT Gate

Consider the circuit shown in Fig. 5.25:

This circuit is a cascade of two CNOT gates. Let us calculate the resultant state of the system. The starting state of the system is the tensor product of $|A\rangle$ and $|B\rangle$:

$$|\psi\rangle = |A\rangle|B\rangle \tag{5.89}$$

We now apply the first CNOT gate to qubit B with qubit A as the control. The equation becomes:

$$|\psi\rangle = |A\rangle|A \oplus B\rangle, \tag{5.90}$$

where $\oplus$ represents an XOR operation.

After this step, we apply the second CNOT gate to qubit B, with qubit A as the control. The equation now becomes:

$$|\psi\rangle = |A\rangle|A \oplus A \oplus B\rangle \tag{5.91}$$

We know that $A \oplus A = 0$. Hence, the given equation becomes:

$$|\psi\rangle = |A\rangle|0 \oplus B\rangle = |A\rangle|B\rangle \tag{5.92}$$

We infer that two CNOT gates return the system to the original state. Now consider the circuit shown in Fig. 5.26.

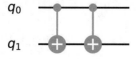

Fig. 5.25 Cascade of two CNOT gates

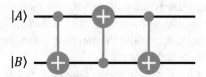

Fig. 5.26   Quantum circuit with three CNOT gates, performing a SWAP operation

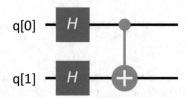

Fig. 5.27   CNOT with two qubits in superposition state

In this circuit, a CNOT gate in the opposite direction is sandwiched between two cascaded CNOT gates. Let us calculate the final state of the system. The starting state of the system is the same as the previous example:

$$|\psi\rangle = |A\rangle|B\rangle \tag{5.93}$$

We now apply the first CNOT gate to qubit B with qubit A as the control. The equation becomes:

$$|\psi\rangle = |A\rangle|A \oplus B\rangle \tag{5.94}$$

The next step is to apply a CNOT gate to qubit A with qubit B as the control. The system state now changes to (refer to previous example):

$$|\psi\rangle = |A \oplus A \oplus B\rangle|A \oplus B\rangle = |B\rangle|A \oplus B\rangle \tag{5.95}$$

The final step is to apply another CNOT gate to qubit B with qubit A as the control. The system now transforms into:

$$|\psi\rangle = |B\rangle|B \oplus A \oplus B\rangle = |B\rangle|A\rangle \tag{5.96}$$

The final state is the SWAP of the input states. Hence, we can infer that this sequence of CNOT gates functions as a SWAP gate.

Now, consider the circuit shown in Fig. 5.27.

In this circuit, we apply a CNOT gate to two qubits in equal superposition state. The system state can be derived as follows. We begin the system with the qubits q[0] and q[1] initialized to state $|0\rangle$.

$$|\psi\rangle = |0\rangle|0\rangle \tag{5.97}$$

The next step is to apply H-gates to q[0] and q[1]. The new system state is as follows:

$$
\begin{aligned}
|\psi\rangle &\to \frac{1}{\sqrt{2}}(|0\rangle + |1\rangle) \otimes \frac{1}{\sqrt{2}}(|0\rangle + |1\rangle) \\
&\to \frac{1}{2}(|00\rangle + |01\rangle + |10\rangle + |11\rangle)
\end{aligned} \tag{5.98}
$$

Now we apply a CNOT gate, with the qubit q[0] as the control qubit and q[1] as the target qubit.

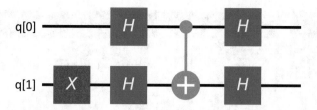

Fig. 5.28 CNOT gate surrounded by H-gates

$$\xrightarrow{CX\ q[0],q[1]} \frac{1}{2}\left(|00\rangle + |01\rangle + |10\rangle + |11\rangle\right)$$

$$\rightarrow \frac{1}{2}\left(|00\rangle + |01\rangle + |11\rangle + |10\rangle\right) \tag{5.99}$$

$$\rightarrow \frac{1}{\sqrt{2}}(|0\rangle + |1\rangle) \otimes \frac{1}{\sqrt{2}}(|0\rangle + |1\rangle)$$

Note that this leaves the system state unchanged. Consider a variant of this circuit which adds two H-gates after the CNOT, with the target qubit starting in state $|1\rangle$ (Fig. 5.28).

Comparing with Fig. 5.17, Eq. (5.85), we can write the state of the system upto the CNOT gate as follows:

$$|\psi\rangle = \frac{1}{\sqrt{2}}(|0\rangle - |1\rangle) \otimes \frac{1}{\sqrt{2}}(|0\rangle - |1\rangle) \tag{5.100}$$

Now, applying final set of H-gates to both the qubits, by referring to Eqs. (5.18) and (5.19) we can simplify this equation as:

$$|\psi\rangle = |1\rangle|1\rangle \tag{5.101}$$

We see that the control qubit has flipped its state. This also means the roles of the control qubit and target qubit are exchanged! Test this with other initial states!

5.5.2 Implementing Digital Logic Gates Using Quantum Circuits

We can now proceed with the next steps by constructing the quantum equivalent of the conventional digital logic gates. Starting at this level shall help us to understand the complex quantum circuits in the upcoming chapters. Of course, the quantum equivalent of digital gates is not significantly useful, beyond helping us to understand the quantum circuits better.

5.5.2.1 Quantum NOT Gate

The NOT gate is perhaps the easiest to construct. All we must do is to use a bit-flip gate to flip the state of the qubit. The code fragments in Listings 5.21 and 5.22 illustrate this.

Qiskit	QDK
```def NOTGate(input, backend, shots): q = QuantumRegister(1, 'q') c = ClassicalRegister(1, 'c') qc = QuantumCircuit(q, c) if (1 == input): qc.x(q) qc.x(q) qc.measure(q, c) job = execute(qc, backend, shots=shots) result = job.result() counts = result.get_counts(qc) return counts```	```operation NOTGate(input:Bool, shots:Int) : (Int) { mutable c0 = 0; using (q0 = Qubit()) { for (i in 1..shots){ Reset(q0); if (true == input) { X(q0); } X(q0); let result = M(q0); if (One == result) { set c0 += 1; } } Reset(q0); return c0; } }```

Listing 5.21   The NOT gate

Circuit Editor	Circuit Composer
```OPENQASM 2.0; include "qelib1.inc"; qreg q[1]; creg c[1]; x q[0]; measure q[0] -> c[0];```	q[0] — X — ◿ c[0] ═══════ 0 *The NOT gate.*

Listing 5.22   NOT gate

Truth Table—NOT Gate

Input	Output
$|0\rangle$	$|1\rangle$
$|1\rangle$	$|0\rangle$

5.5.2.2 Quantum OR Gate

The OR gate outputs a 1 when either of the inputs is in state 1. We can construct an OR gate using the logic outlined in Listing 5.23.

To construct a generic OR gate, we need to test the inputs. We can implement "**if**" statements on the classical registers. The Circuit Composer, at this time of writing, does not support setting the classical registers directly. The only way the classical registers can be set is by performing a measurement. So, we cannot implement the OR gate by using the classical registers as the input. Hence, we are taking an alternate approach by setting the qubits q[0] and q[1] to the desired input states. We then project the qubits to the classical registers and use the "if" statement to determine the inputs (Listing 5.24).

Qiskit	QDK
``def ORGate (Input1, Input2, backend,
shots):

    q = QuantumRegister(3, 'q')

    c = ClassicalRegister(1, 'c')

    qc = QuantumCircuit(q, c)

    if (0 == Input1):

        qc.x(q[0])

    if (0 == Input2):

        qc.x(q[1])

    qc.x(q[2])

    qc.ccx(q[0], q[1], q[2])

    qc.measure(q[2], c)

    job = execute(qc, backend,
shots=shots)

    result = job.result()

    counts = result.get_counts(qc)

    return counts``` | ```operation ORGate(Input1:Int, Input2:Int,
shots:Int): Int {

    mutable c0 = 0;

    using(q = Qubit[3]){

        if (0 == Input1){

            X(q[0]);

        }

        if (0 == Input2){

            X(q[1]);

        }

        for (i in 1..shots){

            X(q[2]);

            CCNOT(q[0], q[1], q[2]);

            let r = M(q[2]);

            if (One == r) {

                set c0 += 1;

            }

            Reset(q[2]);

        }

        Reset(q[0]);

        Reset(q[1]);

        return c0;

    }

}``` |

Listing 5.23 The OR gate

Circuit Editor	Circuit Composer
``OPENQASM 2.0;
include "qelib1.inc";

qreg q[3];
creg c0[1];
creg c1[1];
creg c2[1];

x q[0];
x q[1];
barrier q[1],q[0];
measure q[0] -> c0[0];
reset q[0];
measure q[1] -> c1[0];
reset q[1];
if (c0==0) x q[0];
if (c1==0) x q[1];
x q[2];
ccx q[0],q[1],q[2];
measure q[2] -> c2[0];``` | 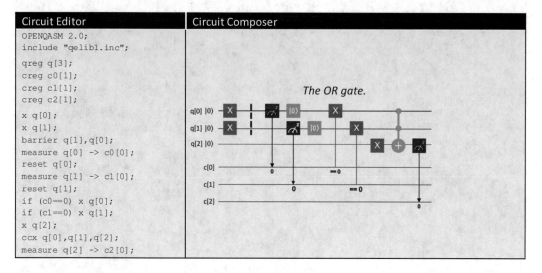 |

Listing 5.24 Constructing the OR gate using the circuit editor

Experiment with the circuit by changing the X-gates before the barrier gate and verify all four combinations of the inputs, as shown in the Truth Table.

Truth Table—OR Gate

Input1	Input2	Output
$|0\rangle$	$|0\rangle$	$|0\rangle$
$|0\rangle$	$|1\rangle$	$|1\rangle$
$|1\rangle$	$|0\rangle$	$|1\rangle$
$|1\rangle$	$|1\rangle$	$|1\rangle$

5.5.2.3 Quantum AND Gate

An AND gate outputs a "1" if all the inputs are "1." The logic for implementing the AND gate is similar to the OR gate. The source code for implementing the AND gate is provided in Listings 5.25 and 5.26.

Truth Table—AND Gate

Input1	Input2	Output
$|0\rangle$	$|0\rangle$	$|0\rangle$
$|0\rangle$	$|1\rangle$	$|0\rangle$
$|1\rangle$	$|0\rangle$	$|0\rangle$
$|1\rangle$	$|1\rangle$	$|1\rangle$

Qiskit	QDK
| ```python
def ANDGate (Input1, Input2, backend,
shots):
    q = QuantumRegister(3, 'q')
    c = ClassicalRegister(1, 'c')
    qc = QuantumCircuit(q, c)

    if ( 1 == Input1 ):
        qc.x(q[0]
    if ( 1 == Input2 ):
        qc.x(q[1])

    qc.ccx(q[0], q[1], q[2])
    qc.measure(q[2], c)

    job = execute(qc, backend,
shots=shots)
    result = job.result()
    counts = result.get_counts(qc)

    return counts
``` | ```
operation ANDGate( Input1 : Int, Input2 :
Int, shots : Int ) : Int {
    mutable c0 = 0;
    using (q = Qubit[3]) {
        for ( i in 1..shots ) {
            if (1 == Input1) {
                X(q[0]);
            }
            if ( 1 == Input2) {
                X(q[1]);
            }
            CCNOT(q[0], q[1], q[2]);
            let r = M(q[2]);
            if ( r == One) {
                set c0 += 1;
            }
            Reset(q[0]);
            Reset(q[1]);
            Reset(q[2]);
        }
    return c0;
    }
}
``` |

Listing 5.25 The AND gate

Circuit Editor	Circuit Composer

```OPENQASM 2.0;```
```include "qelib1.inc";```
 	
```qreg q[3];```
```creg c0[1];```
```creg c1[1];```
```creg c2[1];```
 	
```x q[0];```
```x q[1];```
```barrier q[1],q[0];```
```measure q[0] -> c0[0];```
```reset q[0];```
```measure q[1] -> c1[0];```
```reset q[1];```
```if (c0==1) x q[0];```
```if (c1==1) x q[1];```
```ccx q[0],q[1],q[2];```
```measure q[2] -> c2[0];```
	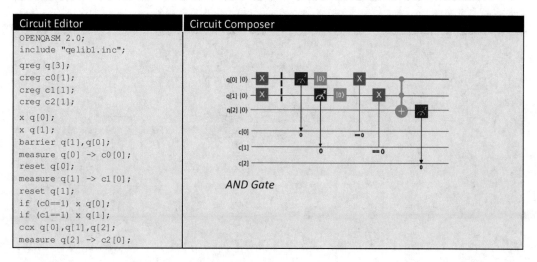
	
	AND Gate

Listing 5.26 Constructing the AND gate using the circuit editor

Qiskit	QDK

```def XORGate (Input1, Input2, backend, shots):```
  ```q = QuantumRegister(2, 'q')```
  ```c = ClassicalRegister(1, 'c')```
  ```qc = QuantumCircuit(q, c)```
 	
  ```if (1 == Input1):```
    ```qc.x(q[0])```
 	
  ```if (1 == Input2):```
    ```qc.x(q[1])```
 	
  ```qc.cx(q[0], q[1])```
  ```qc.measure(q[1], c)```
 	
  ```job = execute(qc, backend, shots=shots)```
  ```result = job.result()```
  ```counts = result.get_counts(qc)```
  ```return counts```
	```operation XORGate(Input1 : Int, Input2 : Int, shots : Int) : Int {```
	  ```mutable c0 = 0;```
	  ```using (q = Qubit[2]){```
	
	    ```for (i in 1..shots) {```
	      ```if (1 == Input1) {```
	        ```X(q[0]);```
	      ```}```
	      ```if (1 == Input2) {```
	        ```X(q[1]);```
	      ```}```
	      ```CNOT(q[0], q[1]);```
	      ```let r = M(q[1]);```
	      ```if (r == One) {```
	        ```set c0 += 1;```
	      ```}```
	      ```Reset(q[0]);```
	      ```Reset(q[1]);```
	    ```}```
	    ```return c0;```
	  ```}```
	```}```

Listing 5.27 The XOR gate

5.5.2.4 Quantum XOR Gate

The XOR gate is similar to the OR gate. But it outputs a 0 if all the inputs are 1. The logic for the XOR gate is outlined in Listings 5.27 and 5.28.

Circuit Editor	Circuit Composer

``OPENQASM 2.0;
include "qelib1.inc";
qreg q[2];
creg c0[1];
creg c1[1];
creg c2[1];
 	
x q[0];
x q[1];
barrier q[1],q[0];
measure q[0] -> c0[0];
reset q[0];
measure q[1] -> c1[0];
reset q[1];
if (c0==1) x q[0];
if (c1==1) x q[1];
cx q[0],q[1];
measure q[1] -> c2[0];```
	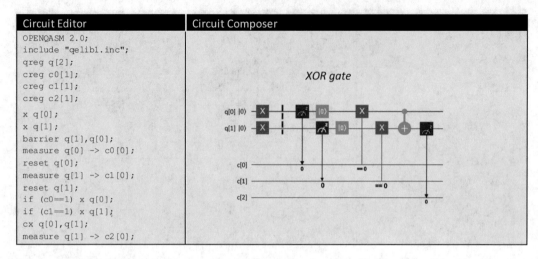

Listing 5.28 Constructing the XOR gate using the circuit editor

Truth Table—XOR Gate

Input1	Input2	Output
$\lvert 0\rangle$	$\lvert 0\rangle$	$\lvert 0\rangle$
$\lvert 0\rangle$	$\lvert 1\rangle$	$\lvert 1\rangle$
$\lvert 1\rangle$	$\lvert 0\rangle$	$\lvert 1\rangle$
$\lvert 1\rangle$	$\lvert 1\rangle$	$\lvert 0\rangle$

Exercise 5.1

1. *Write unit test code to prove that the given circuits work the way we meant.*
2. *Construct NOR, XNOR and NAND gates using the examples.*

5.5.2.5 Quantum Full Adder Circuit

In digital electronics, the full adder circuit forms an essential element in arithmetic operations and used in microprocessors and other integrated circuits. There are two kinds of adder circuits. The **half adder** adds two digital bits (A and B) and produces a sum (S) and a carry-out (C_{OUT}). In addition to the two digital bits (A and B), the **full adder** circuit can add a carry-in (C_{IN}) bit. It produces a sum (S) and a carry-out (C_{OUT}). Two half adder circuits are used to implement a full adder circuit. The full adder circuit is illustrated in Fig. 5.29.

The Sum (S) in the given circuit is as follows:

$$S = A \oplus B \oplus C_{IN} \tag{5.102}$$

We can calculate the Carry out (C_{OUT}) as follows:

$$C_{OUT} = (A \wedge B) + (A \oplus B) \wedge C_{IN} \tag{5.103}$$

Here, we use the following symbols to denote the gate operations:

Fig. 5.29 Digital full adder circuit, obtained by cascading two half adder circuits

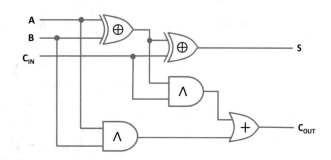

$$\oplus - XOR, \wedge -AND, + -OR$$

An XOR gate can replace the OR gate at the last stage of computing C_{OUT}. This is apparent from Table 5.4.

Truth Table—Full adder

Table 5.4 The truth table for the full adder

A	B	C_{IN}	$A \oplus B$	$A \oplus B \oplus C_{IN}$	$A \wedge B$	$(A \oplus B) \wedge C_{IN}$	$(A \wedge B) + (A \oplus B) \wedge C_{IN}$	$(A \wedge B) \oplus (A \oplus B) \wedge C_{IN}$
0	0	0	0	0	0	0	0	0
0	1	0	1	1	0	0	0	0
1	0	0	1	1	0	0	0	0
1	1	0	0	0	1	0	1	1
0	0	1	0	1	0	0	0	0
0	1	1	1	0	0	1	1	1
1	0	1	1	0	0	1	1	1
1	1	1	0	1	1	0	1	1

The last two columns of this table are the same. Therefore, we can rewrite Eq. (5.103) as:

$$C_{OUT} = (A \wedge B) \oplus (A \oplus B) \wedge C_{IN} \tag{5.104}$$

We can cascade a sequence of full adder circuits to add two n-bit numbers. To perform the full addition, the C_{OUT} of the nth bit needs to connect with the C_{IN} of n+1th bit. Such a circuit is known as a **"ripple adder"** as the carry ripples through all the adders. We shall implement a quantum ripple adder in Chap. 7. Let us return to our discussions on the full adder circuit and implement a quantum version of it.

Consider the quantum circuit shown in Fig. 5.30. This circuit uses four qubits. Qubits A and B are the inputs to be added. The qubit C_{IN} is the carry-in. When the circuit completes, this qubit also contains the sum (S). The qubit AUX holds information temporarily. However, when the circuit completes, it holds the carry-out (C_{OUT}).

Let us now derive the math to show how this circuit works.

The circuit starts with qubit AUX in state $|0\rangle$. The inputs A, B, and C_{IN} can be in state $|0\rangle$ or $|1\rangle$. The following equation describes the starting state of the system.

Fig. 5.30 A full adder
quantum circuit

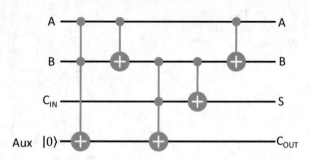

$$|\psi\rangle = |A\rangle|B\rangle|C_{IN}\rangle|0\rangle \tag{5.105}$$

We then apply a Toffoli gate to AUX qubit, with A and B serving as the control qubits. By referring to Eq. (5.71), we can rewrite the above equation.

$$= |A\rangle|B\rangle|C_{IN}\rangle|0 \oplus A \wedge B\rangle, \tag{5.106}$$

where $\oplus$ is an XOR operation, and $\wedge$ is an AND operation.

In the given equation $0 \oplus A \wedge B = A \wedge B$, and hence,

$$= |A\rangle|B\rangle|C_{IN}\rangle|A \wedge B\rangle \tag{5.107}$$

We then apply a CNOT gate to B, with A as the control gate. By referring to Eq. (5.65), we can rewrite the given equation as follows:

$$= |A\rangle|A \oplus B\rangle|C_{IN}\rangle|A \wedge B\rangle \tag{5.108}$$

After this operation, we perform a Toffoli gate with AUX qubit, with B and C_{IN} as the control qubits. The system state is as follows:

$$= |A\rangle|A \oplus B\rangle|C_{IN}\rangle|(A \wedge B) \oplus (A \oplus B) \wedge C_{IN}\rangle \tag{5.109}$$

The next step is to perform a CNOT on C_{IN}, with B as the control gate.

$$= |A\rangle|A \oplus B\rangle|A \oplus B \oplus C_{IN}\rangle|(A \wedge B) \oplus (A \oplus B) \wedge C_{IN}\rangle \tag{5.110}$$

Comparing the previous equation with Eqs. (5.102) and (5.104), we infer the following:

$$= |A\rangle|A \oplus B\rangle|S\rangle|C_{OUT}\rangle \tag{5.111}$$

It is optional to retrieve A and B from the circuit. To do so, we have to apply a CNOT operation to qubit B with A as the control. The above equation now becomes:

$$= |A\rangle|A \oplus A \oplus B\rangle|S\rangle|C_{OUT}\rangle \tag{5.112}$$

From our calculations that lead to Eq. (5.92), we can simplify this equation as:

$$= |A\rangle|B\rangle|S\rangle|C_{OUT}\rangle \tag{5.113}$$

This equation preserves the input, at the same time has the elements representing the Sum (S) and Carry-out (C_{OUT}).

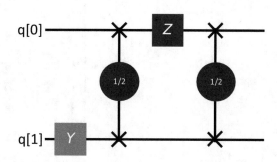

Fig. 5.31 CNOT gate using square root of SWAP gates

Exercise 5.2

Construct the full adder circuit using Qiskit and QDK and prove that it works by developing unit testing code.

5.5.3 Implementing a CNOT Gate with a $\sqrt{SWAP}$ gate

In section 5.4.5.2, we learned about the $\sqrt{SWAP}$ gate with a note that it is a universal quantum gate. In this section, we shall implement a CNOT gate using two $\sqrt{SWAP}$ gates and a couple of rotational gates. Consider the quantum circuit shown in Fig. 5.31. Let us assume we started the system in the state $|1\rangle|1\rangle$. To the second qubit, a Y gate is applied. We can write the system state as follows:

$$|\psi\rangle \rightarrow |1\rangle\, Y|1\rangle \rightarrow \begin{bmatrix} 0 \\ 1 \end{bmatrix} \begin{bmatrix} 0 & -i \\ 0 & 0 \end{bmatrix} \begin{bmatrix} 0 \\ 1 \end{bmatrix} \rightarrow \begin{bmatrix} 0 \\ 1 \end{bmatrix} \begin{bmatrix} -i \\ 0 \end{bmatrix} \rightarrow \begin{bmatrix} 0 \\ 0 \\ -i \\ 0 \end{bmatrix} \tag{5.114}$$

Note that we used tensor multiplication in the last step of the given equation. After the Y gate, $\sqrt{SWAP}$ is applied. The new system state can be derived as follows:

$$\rightarrow \begin{bmatrix} 1 & 0 & 0 & 0 \\ 0 & \frac{1+i}{2} & \frac{1-i}{2} & 0 \\ 0 & \frac{1-i}{2} & \frac{1+i}{2} & 0 \\ 0 & 0 & 0 & 1 \end{bmatrix} \begin{bmatrix} 0 \\ 0 \\ -i \\ 0 \end{bmatrix} = \begin{bmatrix} 0 \\ \frac{1+i}{2} \\ \frac{-1+i}{2} \\ 0 \end{bmatrix} \tag{5.115}$$

A Z gate on the first qubit transforms the system state as:

$$\rightarrow \begin{bmatrix} 1 & 0 & 0 & 0 \\ 0 & -1 & 0 & 0 \\ 0 & 0 & 1 & 0 \\ 0 & 0 & 0 & 1 \end{bmatrix} \begin{bmatrix} 0 \\ \frac{1+i}{2} \\ \frac{-1+i}{2} \\ 0 \end{bmatrix} \rightarrow \begin{bmatrix} 0 \\ \frac{-1-i}{2} \\ \frac{-1+i}{2} \\ 0 \end{bmatrix} \tag{5.116}$$

Finally, we use the square root of the swap gate.

$$\rightarrow \begin{bmatrix} 1 & 0 & 0 & 0 \\ 0 & \dfrac{1+i}{2} & \dfrac{1-i}{2} & 0 \\ 0 & \dfrac{1-i}{2} & \dfrac{1+i}{2} & 0 \\ 0 & 0 & 0 & 1 \end{bmatrix} \begin{bmatrix} 0 \\ \dfrac{-1-i}{2} \\ \dfrac{-1+i}{2} \\ 0 \end{bmatrix}$$

$$\rightarrow \begin{bmatrix} 0 \\ 0 \\ -1 \\ 0 \end{bmatrix} \tag{5.117}$$

$$\rightarrow \begin{bmatrix} 0 \\ -1 \end{bmatrix} \begin{bmatrix} 1 \\ 0 \end{bmatrix}$$

$$\rightarrow -|1\rangle|0\rangle$$

We see that up to a global phase; this circuit works like a CNOT gate. As an exercise, this sequence can be tried with possible input combinations.

5.6   Quantum Entanglement

In the previous sections, we learned that we represent a quantum system with n qubits as a complex vector in the Hilbert space with 2^n dimensions. We use the tensor product to represent such a composite system. For example, a system of two qubits $|\psi_0\rangle$ and $|\psi_1\rangle$ is written as a tensor product $|\psi\rangle = |\psi_0\rangle \otimes |\psi_1\rangle$. This composite system is a collection of individual qubits. There is no correlation between the qubits. If we can somehow find a correlation between the qubits and describe the system as a single system, we say that the qubits are **entangled**.

Entanglement is a unique property of quantum systems, and we use it, especially in quantum communications. Consider the circuit shown in Fig. 5.32, which we can construct equally in Qiskit/QDK.

In this circuit, we prepare two qubits in the default state of $|0\rangle$. We then apply an H-gate to qubit q[0], and perform a CNOT operation to the target qubit q[1], with the q[0] being the control qubit. Finally, we measure the qubits q[0] and q[1].

We can derive the final state of the quantum circuit as follows:

Fig. 5.32 Circuit demonstrating quantum entanglement

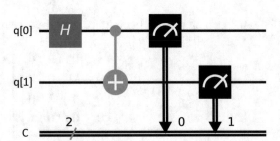

Step 1   Apply H-gate $|00\rangle \overset{H[0]}{\rightarrow} \frac{1}{\sqrt{2}}(|0\rangle + |1\rangle) \otimes |0\rangle$

$$\rightarrow \frac{1}{\sqrt{2}}\left(\begin{bmatrix}1\\0\end{bmatrix} + \begin{bmatrix}0\\1\end{bmatrix}\right) \otimes \begin{bmatrix}1\\0\end{bmatrix}$$

$$\rightarrow \frac{1}{\sqrt{2}}\left(\begin{bmatrix}1\\0\\0\\0\end{bmatrix} + \begin{bmatrix}0\\0\\1\\0\end{bmatrix}\right) \tag{5.118}$$

Step 2   Apply CNOT gate $\overset{CX\ q[0],q[1]}{\rightarrow} \frac{1}{\sqrt{2}}$

$$\times \left(\begin{bmatrix}1\\0\\0\\0\end{bmatrix} + \begin{bmatrix}0\\0\\1\\0\end{bmatrix}\right) \begin{bmatrix}1&0&0&0\\0&1&0&0\\0&0&0&1\\0&0&1&0\end{bmatrix} \tag{5.119}$$

Final State : $\rightarrow \frac{1}{\sqrt{2}}\left(\begin{bmatrix}1\\0\\0\\0\end{bmatrix} + \begin{bmatrix}0\\0\\0\\1\end{bmatrix}\right) = \frac{1}{\sqrt{2}}(|00\rangle + |11\rangle) = |\phi^+\rangle$ (5.120)

The final state $|\phi^+\rangle = \frac{1}{\sqrt{2}}(|00\rangle + |11\rangle)$ cannot be expressed separately in terms of q[0] and q[1]. We call this resultant state an entangled state, and the qubits are said to be in a **maximally entangled** state with each other. The entangled qubits are called **EPR Pair.** $|\Phi^+\rangle$ is one of the **Bell States**. We can create the remaining Bell states by changing the initial state of the qubits. The following set of equations describes the four Bell states.

$$|00\rangle \rightarrow \frac{1}{\sqrt{2}}(|00\rangle + |11\rangle) = |\phi^+\rangle \tag{5.121}$$

$$|01\rangle \rightarrow \frac{1}{\sqrt{2}}(|00\rangle - |11\rangle) = |\phi^-\rangle \tag{5.122}$$

$$|10\rangle \rightarrow \frac{1}{\sqrt{2}}(|01\rangle + |10\rangle) = |\Psi^+\rangle \tag{5.123}$$

$$|11\rangle \rightarrow \frac{1}{\sqrt{2}}(|01\rangle - |10\rangle) = |\Psi^-\rangle \tag{5.124}$$

Readers can calculate the four Bell states to gain more familiarity. Execute the Qiskit code shown in the Listing 5.29 to try the Bell state $|\phi^+\rangle$.

A plot of this code should get almost equal measurements for both the qubits, as shown in Fig. 5.33. The differences, as we know, are due to the underlying physics.

The equivalent code in QDK is given in Listing 5.30.

```
q = QuantumRegister(2, 'q')
c = ClassicalRegister(2, 'c')
qc = QuantumCircuit(q, c)

qc.h(q[0])
qc.cx(q[0], q[1])

qc.measure(q[0], c[0])
qc.measure(q[1], c[1])
```

Listing 5.29 Creating quantum entanglement

Fig. 5.33 Outcome of the quantum entanglement experiment

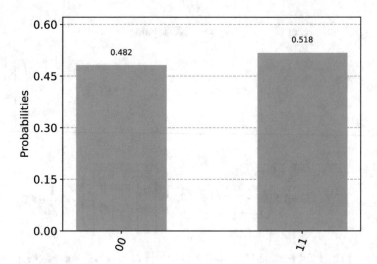

Upon executing the code, we can verify that both the qubits measure the same.

We shall return to the Bell States in the next chapter and explore its applications in quantum communications.

Exercise 5.3

Create the four Bell States and verify that the four Bell states form an orthonormal basis.

5.7   No-Cloning Theorem

The no-cloning theorem is an essential observation in quantum computing. According to this theorem, we cannot copy qubits in an arbitrarily unknown state. This theorem is attributed to **Wootters, Zurek,** and **Dieks,** who published this in 1982. Quantum entanglement is not cloning. The cloning process should produce separable states that are identical.

Assume we are given a qubit in an arbitrarily unknown quantum state, and the task is to guess whether it is in state $|\phi\rangle$ or $|\psi\rangle$. We can measure the qubit to determine whether it is in state $|\phi\rangle$ or $|\psi\rangle$. However, unless the states $|\phi\rangle$ and $|\psi\rangle$ are mutually orthogonal, no measurement can perfectly distinguish them. The reason is, there is always a probability of error. One thing we can probably do is to create multiple copies of the qubit and make optimal measurements to minimize the error.

```
operation BellState( ) : (Int) {
    mutable c0 = 0;
   mutable c1 = 0;

   using ( (q0, q1) = (Qubit(), Qubit())) {
       for ( i in 1..99)
       {
           Reset(q0);
           Reset(q1);

           H(q0);                     // Apply a H gate to the qubit
           CNOT(q0, q1);              // Apply a CNOT gate
           let result1 = M(q0);
           let result2 = M(q1);

           if ( result1 == One) {
               set c0 += 1;
           }
           if ( result2 == One){
               set c1 += 1;
           }
       }
       Message($"We got: {c0} ,  {c1} ");
       return (0);
   }
}
```

Listing 5.30 The Bell state using QDK

According to the no-cloning theorem, this is not physically possible. We can only copy mutually orthogonal states.

Assume we have a cloning operator C, which can copy states $|\phi\rangle$ or $|\psi\rangle$. We start with two qubits. The first qubit is in an arbitrarily unknown state. The second qubit is set to state $|0\rangle$. We expect the cloning operation to copy the information in the first qubit to the second qubit. After the cloning operation, we expect both the qubits to have the same information.

$$|\phi\rangle \otimes |0\rangle \xrightarrow{C} |\phi\rangle \otimes |\phi\rangle$$
$$|\psi\rangle \otimes |0\rangle \xrightarrow{C} |\psi\rangle \otimes |\psi\rangle$$

(5.125)

Now, notice that:

$$\langle\phi|\psi\rangle = \langle\phi|\psi\rangle \langle0|0\rangle$$

(5.126)

$$= ((\langle\phi| \otimes \langle0|) C^\dagger C (|\psi\rangle \otimes |0\rangle))$$

(5.127)

$$= ((\langle\phi| \otimes \langle\phi|)\,(|\psi\rangle \otimes |\psi\rangle)) \tag{5.128}$$

$$= \langle\phi|\psi\rangle\langle\phi|\psi\rangle \tag{5.129}$$

$$= \langle\phi|\psi\rangle^2 \tag{5.130}$$

This equality is possible only if C is a unitary and $\langle\phi|\psi\rangle = 0$ or 1, that is, $|\phi\rangle$ and $|\psi\rangle$ are mutually orthogonal.

5.8   Quantum Teleportation

The No-Cloning theorem forbids us from making copies of qubits in arbitrary quantum states. Of course, we can make a copy of an original qubit by destroying its state and transmitting the copied qubit's state over vast distances. Consider the quantum circuit shown in Fig. 5.34, which can be constructed by typing the code in Listing 5.31 in the Circuit Editor.

In this circuit, Alice and Bob[1] share two maximally entangled qubits q[1] and [q2]. Alice has access to qubit q[1], and Bob has access to qubit q[2].

Now, Alice has access to a data qubit q[0], which is in an arbitrarily unknown state $|\psi\rangle = a|0\rangle + b|1\rangle$. Alice first creates entanglement between her qubits q[0] and q[1]. Alice then measures her qubit q[1] and sends it status to Bob using a classical channel. If Bob receives a 1 on the classical channel, he performs a bit flip (X-gate) on his qubit q[2].

Alice then measures her data qubit q[0] and sends the status to Bob. If Bob receives a 1 on the classical channel, he performs a phase flip (Z gate) on his qubit q[2].

In this process, when Alice first measures her qubit q[1], its state collapses and the entanglement is between q[0] and q[2]. When the qubit q[0] is measured, information is encoded only in q[2]. The X and Z gates are correction operations. Let us look at the math to make this concept clear.

The initial state of the system is as follows:

$$|\psi\rangle \otimes |0\rangle \otimes |0\rangle \tag{5.131}$$

After applying H-gate to q[2], we get

Fig. 5.34 Circuit diagram illustrating Quantum Teleportation. This circuit transfers the state of qubit q[0] to qubit q[2]

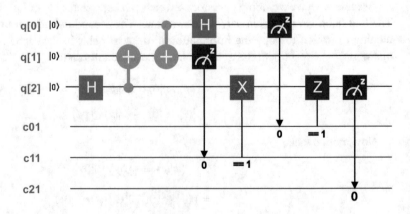

---

[1] Alice and Bob are fictional characters used in cryptography and other thought experiments. http://cryptocouple.com/

```
OPENQASM 2.0;
include "qelib1.inc";
qreg q[3];
creg c0[1];
creg c1[1];
creg c2[1];

h q[2];
cx q[2],q[1];
cx q[0],q[1];
h q[0];

measure q[1] -> c1[0];
if (c1==1) x q[2];

measure q[0] -> c0[0];
if (c0==1) z q[2];
measure q[2] -> c2[0];
```

Listing 5.31 Circuit for Quantum Teleportation

$$|\psi\rangle \otimes |0\rangle \otimes H|0\rangle$$

$$\rightarrow (a|0\rangle + b|1\rangle) \otimes |0\rangle \otimes \frac{1}{\sqrt{2}}(|0\rangle + |1\rangle) \tag{5.132}$$

Then, we apply a CNOT gate to qubit q[1]. This operation puts q[1] and q[2] in a maximally entangled state. By using the Bell state Eq. (5.121), we get the following:

$$(a|0\rangle + b|1\rangle) \otimes CNOT \left(|0\rangle \otimes \frac{1}{\sqrt{2}}(|0\rangle + |1\rangle) \right)$$

$$\rightarrow (a|0\rangle + b|1\rangle) \otimes \frac{1}{\sqrt{2}}(|00\rangle + |11\rangle) \tag{5.133}$$

$$\rightarrow \frac{1}{\sqrt{2}}(a|000\rangle + a|011\rangle + b|100\rangle + b|111\rangle)$$

We can see that the use of Dirac's bra-ket notation helps us to write the tensor products easily. The next step is to apply a CNOT gate, with q[0] as the control qubit and q[1] as the target qubit. The state of the system now becomes:

$$\rightarrow \frac{1}{\sqrt{2}}(a|000\rangle + a|011\rangle + b|110\rangle + b|101\rangle) \tag{5.134}$$

In this equation, we just flipped the state of q[1] depending upon the state of q[0]. Now, we apply an H-gate to q[0]. Applying H-gate to q[0] in the given equation may be a little difficult to understand. To simplify our understanding of this step, we can temporarily **separate** the kets, apply the H-gate to q[0] and then recombine the kets. Separable kets are a property of tensor products.

To begin with, let us separate the kets. We can rewrite the given equation as:

$$\to \frac{1}{\sqrt{2}}(a|0\rangle|00\rangle + a|0\rangle|11\rangle + b|1\rangle|10\rangle + b|1\rangle|01\rangle) \tag{5.135}$$

Now we can apply the H-gate to q[0].

$$\to \frac{1}{\sqrt{2}}(aH|0\rangle|00\rangle + aH|0\rangle|11\rangle + bH|1\rangle|10\rangle + bH|1\rangle|01\rangle) \tag{5.136}$$

Application of the H-gate expands the above equation as follows:

$$\to \frac{1}{\sqrt{2}}(a\frac{1}{\sqrt{2}}(|0\rangle + |1\rangle)|00\rangle + a\frac{1}{\sqrt{2}}(|0\rangle + |1\rangle)|11\rangle + b\frac{1}{\sqrt{2}}(|0\rangle - |1\rangle)|10\rangle + b\frac{1}{\sqrt{2}}(|0\rangle - |1\rangle)|01\rangle)$$
$$\tag{5.137}$$

$$\to \frac{1}{\sqrt{2}}\frac{1}{\sqrt{2}}(a|0\rangle|00\rangle + a|1\rangle|00\rangle + a|0\rangle|11\rangle + a|1\rangle|11\rangle + b|0\rangle|10\rangle - b|1\rangle|10\rangle + b|0\rangle|01\rangle - b|1\rangle|01\rangle)$$
$$\tag{5.138}$$

We can now recombine the kets to get the final form of the quantum circuit at this stage.

$$\to \frac{1}{2}(a|000\rangle + a|100\rangle + a|011\rangle + a|111\rangle + b|010\rangle - b|110\rangle + b|001\rangle - b|101\rangle) \tag{5.139}$$

We can now examine the outcome of the possible measurements.

Case 1:
If we measured a $|0\rangle$ in q[1], the system collapses to the following state with qubits q[0] and q[2]:

$$\left|\psi_{q[0]q[2]}\right\rangle \to \frac{1}{2}(a|00\rangle + a|10\rangle + b|01\rangle - b|11\rangle) \tag{5.140}$$

Since we measured a $|0\rangle$, we do not perform the X-gate with q[2].

Case 2:
If we measured a $|1\rangle$ in q[1], the system collapses to the following state:

$$\left|\psi_{q[0]q[2]}\right\rangle \to \frac{1}{2}(a|01\rangle + a|11\rangle + b|00\rangle - b|10\rangle) \tag{5.141}$$

Since we measured a $|1\rangle$, we perform the X-gate with q[2]. This results in a new state given by:

$$\left|\psi_{q[0]q[2]}\right\rangle \to \frac{1}{2}(a|00\rangle + a|10\rangle + b|01\rangle - b|11\rangle) \tag{5.142}$$

We see that Eqs. (5.140) and (5.142) are the same. The correction operation X has ensured that we have the same system state at this point. We now proceed with measuring q[0]. There can be two possible outcomes now.

Subcase 1:
If we measured a $|0\rangle$ with q[0], the system collapses into the following state for q[2]:

$$\left|\psi_{q[2]}\right\rangle \to \frac{1}{2}(a|0\rangle + b|1\rangle) \tag{5.143}$$

Since a and b are complex numbers, we can factor in $\frac{1}{2}$ into these constants and we can rewrite the above equation as:

$$\left|\psi_{q[2]}\right\rangle \rightarrow a|0\rangle + b|1\rangle = |\psi\rangle \tag{5.144}$$

Subcase 2:
If we measured a $|1\rangle$ with q[0], the system collapses into the following state for q[2]:

$$\left|\psi_{q[2]}\right\rangle \rightarrow \frac{1}{2}\left(a|0\rangle - b|1\rangle\right) \tag{5.145}$$

In this case, we apply a Z gate to q[2] and the result of this operation, after considering the logic in Eq. (5.144), we get:

$$\begin{aligned}\left|\psi_{q[2]}\right\rangle &\rightarrow Z\left(\frac{1}{2}\left(a|0\rangle - b|1\rangle\right)\right) \\ &\rightarrow \frac{1}{2}\left(a|0\rangle + b|1\rangle\right) \\ &\rightarrow a|0\rangle + b|1\rangle = |\psi\rangle\end{aligned} \tag{5.146}$$

From Eqs. (5.144) and (5.146), we can conclude that the resultant state of q[2] is $|\psi\rangle = a|0\rangle + b|1\rangle$. This superposition state is where Alice started with her data qubit q[0]. We can, therefore, say that the quantum state of q[0] is copied into q[2]. This operation is known as "**Quantum Teleportation**." The status of q[0] is, however, destroyed. The reason for calling this operation as "teleportation" is because the qubits q[1] and q[2] can be far apart. We shall explore the concept of two entangled qubits being far apart in the next chapter.

Upon executing this circuit in the simulator for 1024 times, we get the histogram shown in Fig. 5.35.

In the histogram, the values of the classical registers are shown along the x-axis under the bars. Recall that we projected q[2] into c2. It must be easy to figure out that the state of the classical register c2 is 0. This can mean that the state $|0\rangle$ of qubit q[0] was copied to qubit q[2]. To confirm this, construct the following circuit, by flipping the state of q[0] to $|1\rangle$ (Fig. 5.36).

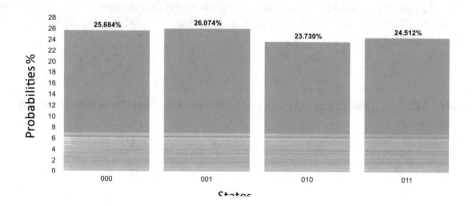

Fig. 5.35 Histogram obtained after teleporting qubit q[0] in state |0⟩ to qubit q[2]

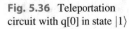

Fig. 5.36 Teleportation
circuit with q[0] in state $|1\rangle$

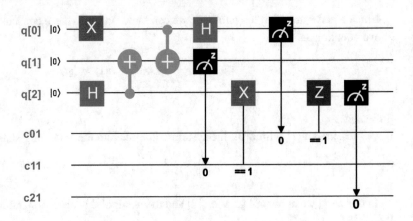

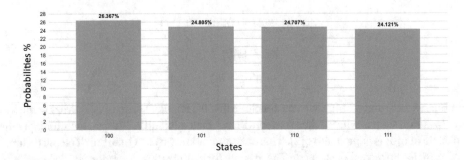

Fig. 5.37 Histogram for the teleportation experiment with qubit q[0] in state

We can flip the qubit q[0] by adding the instruction $x\ q[0]$ above the instruction that performs H-gate on q[2]. Upon executing the given circuit, we get the histogram shown in Fig. 5.37.

If we observe the values of the classical registers below the bars along the x-axis, we note that the classical register c2 is always 1.

By comparing both the experiments, we can conclude that we teleported qubit q[0] into qubit q[2].

Exercise 5.4

1. *Perform the teleportation experiment using Qiskit using Python code and using QDK in Q# code.*
2. *Execute the experiment on a Quantum Processor on the IBM Q, compare the results with that of the simulator. What do you conclude? Is your Quantum Processor able to produce the same results as that of the simulator?*

Hints:
To execute the code on a real Quantum Processor, use the modifications shown in the Listing 5.32 into your Python code.

```
# To run this on the emulator
provider = IBMQ.get_provider()
processor = Aer.get_backend('qasm_simulator')

# To run this on the processor.
# Get the least busy backend
provider = IBMQ.get_provider()
processor = least_busy(provider.backends())

# construct your quantum circuit here...
#...
#...

# It is a good idea to setup a job monitor while executing on a\
  backend
job = execute(qc, backend=processor, shots=1024)
job_monitor(job)
```

Listing 5.32 Executing the quantum circuit on a backend

◀

5.9 Superdense Coding

Holevo's theorem puts an upper bound on how much information we can transmit using qubits. According to this theorem, it is not possible to transmit more than n classical bits of information by transmitting n qubits [14]. Let us say Alice has got some information $x \in \{0, 1\}^n$ that she wishes to transmit to Bob. She does that by creating a quantum state $|\psi_x\rangle$ in the Hilbert space $\mathbb{H} \in \mathbb{C}^d$ and transmits that to Bob over a quantum communication channel. Bob receives the qubits and measures their quantum state to retrieve the input $x \in \{0, 1\}^n$ faithfully. This transmission is possible only if the dimensionality of the state $d = 2^n$. In our discussion on the no-cloning theorem, we learned that non-orthogonal states could not be precisely measured. In order to have a dimensionality of $d = 2^n$, we need to have n qubits. In other words, in order to transmit n bits of classical information, we need n qubits.

Superdense coding comes handy here. Using this protocol, Alice can send two classical bits of information by just sending one qubit to Bob, provided they both share an entangled qubit.

Assume that both Alice and Bob have access to a source that creates entangled pairs of qubits and sends one qubit each to Alice and Bob. Assume that the qubits are prepared in the Bell state $|\phi^+\rangle = \frac{1}{\sqrt{2}}(|00\rangle + |11\rangle)$. Alice has the qubit A, and Bob has the qubit B. There are four scenarios now.

• If Alice wants to send the message "00" to Bob, she does not perform any operation on her qubit.
  If Alice wants to send the message "01" to Bob, she performs an X-gate on her qubit A. The state of the entanglement now becomes $|\psi^+\rangle$ as shown in the following steps.

$$\frac{1}{\sqrt{2}}(X|0\rangle|0\rangle + X|1\rangle|1\rangle)$$

$$= \frac{1}{\sqrt{2}}(|1\rangle|0\rangle + |0\rangle|1\rangle) \tag{5.147}$$

$$= \frac{1}{\sqrt{2}}(|10\rangle + |01\rangle) = |\psi^+\rangle$$

- If Alice wants to send the message "10" to Bob, she performs a Z-gate on her qubit A. The state of the entanglement becomes $|\phi^-\rangle$ as shown in the following derivation:

$$\frac{1}{\sqrt{2}}(Z|0\rangle|0\rangle + Z|1\rangle|1\rangle)$$

$$= \frac{1}{\sqrt{2}}(|0\rangle|0\rangle - |1\rangle|1\rangle) \tag{5.148}$$

$$= \frac{1}{\sqrt{2}}(|00\rangle - |11\rangle) = |\phi^-\rangle$$

- If Alice wants to send the message "11" to Bob, she performs an X-gate and then a Z-gate to her qubit A. The new entangled state is $|\psi^-\rangle$.

$$\frac{1}{\sqrt{2}}(ZX|0\rangle|0\rangle + ZX|1\rangle|1\rangle)$$

$$= \frac{1}{\sqrt{2}}(Z|1\rangle|0\rangle + Z|0\rangle|1\rangle)$$

$$= \frac{1}{\sqrt{2}}(-|1\rangle|0\rangle + |0\rangle|1\rangle) = \frac{1}{\sqrt{2}}(|0\rangle|1\rangle - |1\rangle|0\rangle)) \tag{5.149}$$

$$= \frac{1}{\sqrt{2}}(|01\rangle - |10\rangle) = |\psi^-\rangle$$

After encoding the information using the above scheme, Alice then sends her qubit A to Bob. Bob now has both the qubits A and B, entangled in one of the four mutually orthogonal Bell states. To retrieve the encoded information, Bob must perform a couple of operations. Bob first performs a CNOT on the qubit B, with qubit A as the control. The following set of equations illustrate how each of the Bell states transforms after the CNOT operation.

$$|\phi^+\rangle = \frac{1}{\sqrt{2}}(|00\rangle + |11\rangle) \stackrel{CNOT_{A,B}}{\rightarrow} \frac{1}{\sqrt{2}}(|00\rangle + |10\rangle) \tag{5.150}$$

$$|\psi^+\rangle = \frac{1}{\sqrt{2}}(|10\rangle + |01\rangle) \stackrel{CNOT_{A,B}}{\rightarrow} \frac{1}{\sqrt{2}}(|11\rangle + |01\rangle) \tag{5.151}$$

$$|\phi^-\rangle = \frac{1}{\sqrt{2}}(|00\rangle - |11\rangle) \stackrel{CNOT_{A,B}}{\rightarrow} \frac{1}{\sqrt{2}}(|00\rangle - |10\rangle) \tag{5.152}$$

$$|\psi^-\rangle = \frac{1}{\sqrt{2}}(|01\rangle - |10\rangle) \stackrel{CNOT_{A,B}}{\rightarrow} \frac{1}{\sqrt{2}}(|01\rangle - |11\rangle) \tag{5.153}$$

After performing the CNOT, Bob performs an H-gate on qubit A. By referring to Eqs. (5.11) and (5.12) we can write the transforms as follows:

$$\text{Case} \quad \#1 \qquad \frac{1}{\sqrt{2}}(|00\rangle + |10\rangle) \rightarrow \frac{1}{\sqrt{2}}(H|0\rangle|0\rangle + H|1\rangle|0\rangle)$$

$$= \frac{1}{\sqrt{2}}\left(\frac{1}{\sqrt{2}}(|0\rangle + |1\rangle)|0\rangle + \frac{1}{\sqrt{2}}(|0\rangle - |1\rangle)|0\rangle\right) \qquad (5.154)$$

$$= \frac{1}{2}(|0\rangle|0\rangle + |1\rangle|0\rangle + |0\rangle|0\rangle - |1\rangle|0\rangle)$$

$$= |0\rangle|0\rangle = |00\rangle$$

$$\text{Case} \quad \#2 \qquad \frac{1}{\sqrt{2}}(|11\rangle + |01\rangle) \rightarrow \frac{1}{\sqrt{2}}(H|1\rangle|1\rangle + H|0\rangle|1\rangle)$$

$$= \frac{1}{\sqrt{2}}\left(H\frac{1}{\sqrt{2}}(|0\rangle - |1\rangle)|1\rangle + \frac{1}{\sqrt{2}}(|0\rangle + |1\rangle)|1\rangle\right) \qquad (5.155)$$

$$= \frac{1}{2}(|0\rangle|1\rangle + |1\rangle|1\rangle + |0\rangle|1\rangle - |1\rangle|1\rangle)$$

$$= |0\rangle|1\rangle = |01\rangle$$

$$\text{Case} \quad \#3 \qquad \frac{1}{\sqrt{2}}(|00\rangle - |10\rangle) \rightarrow \frac{1}{\sqrt{2}}(H|0\rangle|0\rangle - H|1\rangle|0\rangle)$$

$$= \frac{1}{\sqrt{2}}\left(\frac{1}{\sqrt{2}}(|0\rangle + |1\rangle)|0\rangle - \frac{1}{\sqrt{2}}(|0\rangle - |1\rangle)|0\rangle\right) \qquad (5.156)$$

$$= \frac{1}{2}(|0\rangle|0\rangle + |1\rangle|0\rangle - |0\rangle|0\rangle + |1\rangle|0\rangle)$$

$$= |1\rangle|0\rangle = |10\rangle$$

$$\text{Case} \quad \#4 \qquad \frac{1}{\sqrt{2}}(|01\rangle - |11\rangle) \rightarrow \frac{1}{\sqrt{2}}(H|0\rangle|1\rangle - H|1\rangle|1\rangle)$$

$$= \frac{1}{\sqrt{2}}\left(\frac{1}{\sqrt{2}}(|0\rangle + |1\rangle)|1\rangle - \frac{1}{\sqrt{2}}(|0\rangle - |1\rangle)|1\rangle\right) \qquad (5.157)$$

$$= \frac{1}{2}(|0\rangle|1\rangle + |1\rangle|1\rangle - |0\rangle|1\rangle + |1\rangle|1\rangle)$$

$$= |1\rangle|1\rangle = |11\rangle$$

By performing a Z measurement on the qubits A and B, Bob can retrieve the encoded information as "00" or "01" or "10" or "11."

In 1992, **Charles Henry Bennett** (IBM Research) and **Stephen J. Wiesner** (Columbia University) proposed the superdense coding protocol. Four years later, it was verified using entangled photon pairs.

5.10 Greenberger–Horne–Zeilinger State (GHZ State)

Our discussion on entanglement will be incomplete without describing the Greenberger–Horne–Zeilinger state (**GHZ State**). Consider the circuit shown in Fig. 5.38.

This circuit creates a maximally entangled quantum state of three qubits, called **tripartite entanglement**. This entanglement has peculiar properties, and it is non-biseparable. A quantum state $|\psi\rangle$ of n parties is **biseparable**, if we can partition the parties into two disjoint subsets A and B (with $A \cup B = \{1, 2, ..., n\}$), such that $|\psi\rangle = |\psi_A\rangle \otimes |\psi_B\rangle$. We can calculate the GHZ state of the three parties (qubits) as follows:

Fig. 5.38 Circuit
describing the GHZ State

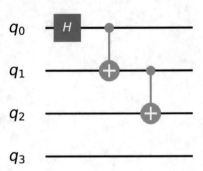

$$|GHZ\rangle = \frac{1}{\sqrt{2}}(|000\rangle + |111\rangle) \tag{5.158}$$

The GHZ state is a special case of the cat state $\frac{1}{\sqrt{2}}\left(|0\rangle^{\otimes n} + |1\rangle^{\otimes n}\right)$ with $n = 3$.

Let us assume Alice, Bob, and Charlie each share one of the qubits. If Charlie measures his qubit in the computational basis, the system collapses to an unentangled pure state of $|00\rangle$ or $|11\rangle$, depending upon what Charlie measured—$|0\rangle$ or $|1\rangle$. Now the correlation between Alice's qubit and Bob's qubit is trivial. If Alice's qubit is $|0\rangle$, then Bob's qubit is also $|0\rangle$. If Alice's qubit is $|1\rangle$, then Bob's qubit is also $|1\rangle$. This is true, if Charlie didn't measure his qubit, and either Alice or Bob did the measurement. Therefore, in order to have non-trivial correlations, we need to determine the joint statistics of Alice, Bob, and Charlie. The entanglement of the GHZ state lies purely with the collective measurement of the three parties!

There is another state called **W-state**, which also exhibits **multipartite entanglement**. The W-state for a three-party system has the following definition.

$$|W\rangle = \frac{1}{\sqrt{3}}(|001\rangle + |010\rangle + |100\rangle) \tag{5.159}$$

The W-state is quite different from the GHZ state. If one of the qubits is measured or lost, the rest of the system still stays entangled. In the GHZ state, the entanglement is lost, if one of the qubits is measured. Because of this property, the W-state has applications in implementing quantum memory. In a W-state, if one of the states is lost, the system can be reconstructed.

Exercise 5.5

Construct a quantum circuit demonstrating W-state.

◀

5.11   Walsh–Hadamard Transform

The **Walsh–Hadamard transform** is a reversible randomization operation and a tensor power of the Hadamard transform. Given n qubits, the Walsh–Hadamard transform produces a quantum mechanical system of $N = 2^n$ states. This collective state forms the basis of **Quantum Parallelism** and creates a single large superposition state of all possible register values. Let us explore this concept in detail (Fig. 5.39).

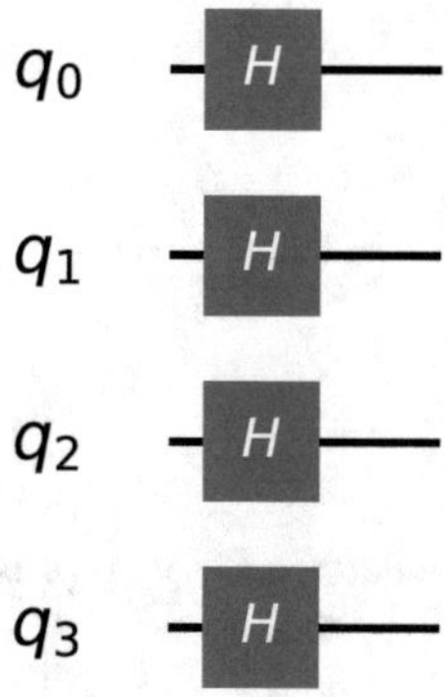

Fig. 5.39 The Walsh–Hadamard transform of a four-qubit register

For the sake of simplicity, consider a two-qubit register (that is, $n = 2$). The concept can be scaled to any number of qubits. Let us assume that the qubits are prepared in state $|0\rangle$. The Walsh–Hadamard transform on this two-qubit register can be written as follows:

$$H^{\otimes 2}|00\rangle = H|0\rangle \otimes H|0\rangle$$
$$= |+\rangle^{\otimes 2} \tag{5.160}$$

$$= \left[\frac{1}{\sqrt{2}}\left(|0\rangle + |1\rangle\right)\right]^{\otimes 2} \tag{5.161}$$

$$= \left[\frac{1}{\sqrt{2}}\left(|0\rangle + |1\rangle\right)\right] \otimes \left[\frac{1}{\sqrt{2}}\left(|0\rangle + |1\rangle\right)\right] \tag{5.162}$$

$$= \left(\frac{1}{\sqrt{2}}\right)^2 (|0\rangle + |1\rangle) \otimes (|0\rangle + |1\rangle) \tag{5.163}$$

$$= \frac{1}{2}\left(|00\rangle + |01\rangle + |10\rangle + |11\rangle\right) \tag{5.164}$$

Note that this is a superposition of all possible states with a two-qubit register. By referring to Eq. (5.163), we can generalize this superposition state as follows, for a $n-$qubit system:

$$H^{\otimes n}|0\rangle^{\otimes n} = \left(\frac{1}{\sqrt{2^n}}\right)\left((|0\rangle + |1\rangle)_1 \otimes (|0\rangle + |1\rangle)_2 \otimes \ldots \otimes (|0\rangle + |1\rangle)_n\right) \tag{5.165}$$

We can expand the tensor product.

$$H^{\otimes n}|0\rangle^{\otimes n} = \frac{1}{\sqrt{2^n}}\left(|000\ldots 0\rangle + |000\ldots 1\rangle + |000\ldots 10\rangle + \ldots + |111\ldots 1\rangle\right) \tag{5.166}$$

In this equation, the term $(|000\ldots 0\rangle + |000\ldots 1\rangle + |000\ldots 10\rangle + \ldots |111\ldots 1\rangle)$ represents a single large superposition of 2^n components, that is, a component with n-qubits in state $|000\ldots 0\rangle$ to a component of n-qubits in state $|111\ldots 1\rangle$, and all states in between. The given equation can be written as a summation:

$$H^{\otimes n}|0\rangle^{\otimes n} = \frac{1}{\sqrt{2^n}} \sum_{x=0}^{2^n-1} |x\rangle^{\otimes n} \tag{5.167}$$

We can also write this equation in a general form:

$$H^{\otimes n}|x\rangle^{\otimes n} = \frac{1}{\sqrt{2^n}} \sum_{y=0}^{2^n-1} (-1)^{x\cdot y}|y\rangle^{\otimes n}, \tag{5.168}$$

where $x \cdot y$ represents the bitwise inner product *mod* 2 defined as $x \cdot y = x_0 y_0 \oplus x_1 y_1 \oplus x_2 y_2 \oplus \ldots \oplus x_{n-1} y_{n-1}$ *mod* 2.

When we make an operation to this large superposition state, it applies simultaneously to all 2^n possible states. This simultaneous operation readily gives us **exponential parallelism**; an advantage quantum computers provide over classical computers.

Shor's algorithm and Grover's algorithm use this method to create a superposition of all inputs during the initialization phase.

5.12 Quantum Interference

In Eq. (5.13), we derived the action of the Hadamard gate on a qubit already put in an equal superposition state by another Hadamard gate (that is, cascading of two Hadamard gates). We used the matrix method to prove that this operation reverses the state of the qubit to its initial state. Let us now derive the same operation using the Dirac's bra-ket notation. Assume the initial state of the qubit as follows.

$$|\psi\rangle = H|0\rangle = \frac{1}{\sqrt{2}}(|0\rangle + |1\rangle) \tag{5.169}$$

Applying a Hadamard gate to this initial quantum state, we get the following:

$$\begin{aligned} H|\psi\rangle &= H\left(\frac{1}{\sqrt{2}}(|0\rangle + |1\rangle)\right) = \frac{1}{\sqrt{2}}(H|0\rangle + H|1\rangle) \\ &= \frac{1}{2}(|0\rangle + |1\rangle + |0\rangle - |1\rangle) \\ &= \frac{1}{2}(|0\rangle + |0\rangle) = |0\rangle \end{aligned} \tag{5.170}$$

Upon studying Eq. (5.170), we find that the probability amplitudes of $|1\rangle$ had the same magnitude, but were opposite in sign. Therefore, the probability amplitudes of the state $|1\rangle$ have interfered destructively, and hence, removed from the system. Therefore, the resultant state $H|\psi\rangle$ does not have a component along the direction $|1\rangle$ anymore. The probability amplitudes of the state $|0\rangle$ have interfered constructively, and made it to the final state of the system.

We use this property in quantum computing. Using circuit elements, we can create constructive and destructive **Quantum Interference** to systematically evolve the most probable system states, which can be the candidate solutions to a given problem.

We can explain the concept of quantum interference (which is similar to the interference of light, we saw in Chap. 1) using a lab experiment using the Mach–Zehnder Interferometer. Consider the setup shown in Fig. 5.40.

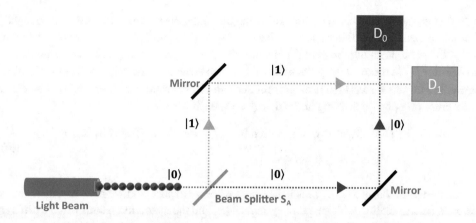

Fig. 5.40 The Mach–Zehnder Interferometer

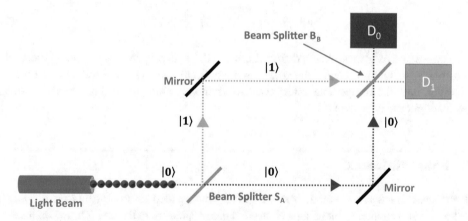

Fig. 5.41 The Mach–Zehnder Interferometer with the second beam splitter

In this experimental setup, we use laser equipment that emits one photon at a time. Using a set of attenuators, we can filter out light from the laser and ensure that only one photon emerges from the apparatus at a time. This light is then passed through a beam splitter S_A, which splits this light into two components $|0\rangle$ and $|1\rangle$ with equal probabilities. Let us assume that P_0 be the probability for the photon to be split along the path $|0\rangle$, and P_1 be the probability for the photon to be split along the path $|1\rangle$. The beam with component $|0\rangle$ passes horizontally, gets deflected by a mirror, and reaches the detector D_0. The beam with component $|1\rangle$ passes vertically, get deflected by a mirror, and reaches the detector D_1. In this setup, the detectors D_0 and D_1 receive an equal number of photons. The beam splitter, indeed, behaves like a Hadamard gate, creating equal superposition of light.

If we assume that the light emitted by the laser apparatus was prepared in the ground state $|0\rangle$, we can write the action of the beam splitter as shown in the following equation:

$$|0\rangle \xrightarrow{S_A} \frac{1}{\sqrt{2}}(|0\rangle + |1\rangle) \tag{5.171}$$

Now, let us modify the setup by adding another beam splitter at the intersecting point before the detectors (Fig. 5.41).

In this setup, the second beam splitter splits the incoming beams into two new components. The photons taking path $|0\rangle$ can directly pass to detector D_0, or it can be deflected and sent to detector D_1. Similarly, the photons taking the path $|1\rangle$ can directly pass to detector D_1, or it can be deflected and sent to detector D_0. We expect the probability of a photon arriving at the detector D_0 or D_1 to be still $\frac{1}{2}$. The reasoning for this assumption is clear from our knowledge of probabilities, and we can calculate the probability of a photon arriving at the detector D_0 as follows:

$$P(D_0) = P_0 \times P(\text{not deflected by } S_B) + P_1 \times P(\text{deflected by } S_B)$$
$$= \frac{1}{2} \times \frac{1}{2} + \frac{1}{2} \times \frac{1}{2} = \frac{1}{2} \qquad (5.172)$$

We can derive similarly for D_1. This behavior is what we expect, and we have no challenges with it. However, we are dealing with quantum systems here, and the quantum systems behave differently. At the second beam splitter, the photons from the two incoming paths undergo interference. The new state can be obtained by extending Eq. (5.171) and by referring to Eq. (5.170) as follows:

$$|0\rangle \xrightarrow{S_A} \frac{1}{\sqrt{2}}(|0\rangle + |1\rangle) \xrightarrow{S_B} |0\rangle \qquad (5.173)$$

This is the same result we obtained from Eq. (5.170). This property means the photon was in a superposition state, and it travelled through both the paths $|0\rangle$ and $|1\rangle$ simultaneously. At the second beam splitter, both the components interfered, constructively along $|0\rangle$ and destructively along $|1\rangle$. The resultant photon is in state $|0\rangle$.

5.13   Phase Kickback

Recall that, **Kickback** is a method, where a global phase added to a qubit by a gate operation is "kicked back" into another qubit using a controlled gate operation. In this section, we shall see how a cU1 gate can be used to demonstrate this concept.

Recall the matrix form of the U_1 gate from Eq. (5.48). Applying this on state $|1\rangle$, we get,

$$U_1\left(\frac{\pi}{4}\right) = U_3\left(0, 0, \frac{\pi}{4}\right) = \begin{bmatrix} 1 & 0 \\ 0 & e^{i\frac{\pi}{4}} \end{bmatrix}$$
$$U_1\left(\frac{\pi}{4}\right)|1\rangle = \begin{bmatrix} 1 & 0 \\ 0 & e^{i\frac{\pi}{4}} \end{bmatrix} \begin{bmatrix} 0 \\ 1 \end{bmatrix} = \begin{bmatrix} 0 \\ e^{i\frac{\pi}{4}} \end{bmatrix} \qquad (5.174)$$
$$= e^{i\frac{\pi}{4}}|1\rangle$$

The operation of the U_1 gate adds a global phase $e^{i\frac{\pi}{4}}$ to the state $|1\rangle$. This global phase cannot be observed. Note that the $U_1\left(\frac{\pi}{4}\right)$ gate is equivalent to a T gate. Consider the circuit shown in Fig. 5.42, in which a $cU_1\left(\frac{\pi}{4}\right)$ gate operates on an equal superposition state.

Let us derive the system state. The system starts with the qubits in state $|0\rangle$. The first step is to apply an H-gate to qubit q[0] and an X-gate to the qubit q[1]. The resultant state of the system can be written as:

Fig. 5.42 Phase kickback
with a cU_1 gate

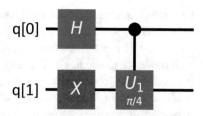

$$|\psi\rangle = \frac{1}{\sqrt{2}}(|0\rangle + |1\rangle)\,|1\rangle = \frac{1}{\sqrt{2}}(|01\rangle + |11\rangle) \tag{5.175}$$

The next step is to apply the $cU_1\left(\frac{\pi}{4}\right)$ gate. The new system state is as follows:

$$
\begin{aligned}
cU_1\left(\frac{\pi}{4}\right)|\psi\rangle &= \frac{1}{\sqrt{2}}\left(|01\rangle + e^{i\frac{\pi}{4}}|11\rangle\right)\\
&= \frac{1}{\sqrt{2}}\left(|0\rangle + e^{i\frac{\pi}{4}}|1\rangle\right)|1\rangle
\end{aligned} \tag{5.176}
$$

Note that this has added a relative phase to the qubit q[0].

5.14 DiVincenzo's criteria for Quantum Computation

We can conclude Chap. 5 with a brief discussion on DiVincenzo's criteria for quantum computation. Ever since the early pioneers—the Russian physicist **Yuri Manin** and the American physicist **Richard Feynman** proposed the concept of a quantum computer, several technologies are under development to create one on a working scale. In Chap. 4, we learned about the leading qubit modalities. Some of them have the potential to create a scalable system to run sophisticated quantum algorithms, such as Shor's algorithm. When we build such a quantum computer, what are its characteristics? DiVincenzo's criteria provide answers to this question by setting up the basic expectations of a quantum computer. In 2000, the American physicist **David P. DiVincenzo** proposed the seven criteria any quantum computer should meet. The first five criteria are about the quantum computation itself. The remaining two criteria describe the needs of quantum communication. Here are the seven criteria!

1	A scalable physical system with well characterized qubit
2	The ability to initialize the state of the qubits to a simple fiducial state
3	Long relevant decoherence times
4	A "universal" set of quantum gates
5	A qubit-specific measurement capability
6	The ability to interconvert stationary and flying qubits
7	The ability to faithfully transmit flying qubits between specified locations

5.14.1 A Scalable Physical System with Well-Characterized Qubit

In the previous chapter, we learned that a qubit is a two-state system. In a well-characterized qubit, the two states are distinct, and the measurement systems are well-calibrated. Well-characterized qubits are somewhat possible with the current qubit modalities. The problem is, however, with scalability. A multi-qubit system must have phase channels and coupling between arbitrary qubits.

Quantum tunneling between the qubits and the presence of stray particles should not create a mixed state. Cooling and electromagnetic isolation systems for such a vast array of qubits are engineering challenges we need to resolve. Most qubit modalities we are experimenting with today operate under high power magnetic fields. Targeting the magnetic flux lines for specific qubits is another engineering challenge. Error correction is another challenge. For community adaption, error-corrected logical qubits are a must. These are some challenges associated with scalability. The more the number of qubits embedded in a device, the more these problems are.

The requirement here is that the qubits in large-scale systems are well- characterized and error-corrected. In such environments, it shall be possible to run real-time mission-critical algorithms.

5.14.2 The Ability to Initialize the State of the Qubits to a Simple Fiducial State

The quantum circuits help the unitary evolution of the system. The unitary evolution starts with an initial state. Hence, preparing the qubits to the initial state of a well-defined $|0\rangle$ is essential. Furthermore, it should be possible to initialize the qubits to the state $|0\rangle$ rapidly, before the decoherence sets in. Fast initialization is essential, especially in large-scale systems, so that all qubits are relatively in the same T_2 state. Parallel initialization of the qubits is essential to minimize the latency from initializing q[0] to initializing q[n]. Otherwise, the qubits initialized first would have started to decohere and lose their initialized state. Partial decoherence of the qubits makes the circuit unusable after a few steps.

5.14.3 Long Relevant Decoherence Times

For effectively running a quantum algorithm, the qubits must have a longer decoherence time than the gate time. Longer decoherence time also helps in building error correction schemes. Qubit modalities that have high response times (T_1), and high fidelity may quickly interact with the environment and start to decohere. We need to make a trade-off between these characteristics in large-scale systems.

5.14.4 A "Universal" Set of Quantum Gates

To implement most quantum algorithms, we need a minimum set of universal quantum gates. We can, today, expect that the industry is aware of this requirement, and most experimental systems offer a universal set of quantum gates.

5.14.5 A Qubit-Specific Measurement Capability

The ability to measure the qubit's state is an essential step in any quantum computation. The measurement scheme is quite complicated, and it is specific to the qubit modality. We learned about some of that in Chap. 4. A strong measurement is needed to project the qubit's state faithfully into classical registers. However, weak measurements and non-destructive measurements are reported quite often and may help improve circuit reliability, error resilience coding, and may even help invent new quantum algorithms.

5.14.6 The Ability to Interconvert Stationary and Flying Qubits

When we create an entanglement of the qubits, the state is within the quantum device. We must transfer qubit states to other quantum devices to enable quantum communications, but it is not easy. A quantum communications channel requires us to send the quantum state as a flying qubit (photon), essentially encoding information by polarizing photons and transmitting them to the receiving device using fiber optics. The receiving station must decode the information in the photons and transfer the state to the local quantum device. Some research labs are already working on establishing this capability.

5.14.7 The Ability to Faithfully Transmit Flying Qubits between Specified Locations

Even if we can encode information in photons, decoherence is another challenge. The photons start interacting with the environment and lose their state. Research labs are using optical repeaters to transfer quantum information over several miles with some success [15, 16].

5.15   Summary

This chapter was another milestone in our learning. We configured our systems with the required tools for quantum algorithm development, and learned about quantum circuits and quantum gate operations. We learned about the essential elements of quantum computing—quantum entanglement, no-cloning theorem, quantum teleportation, superdense coding, quantum interference, and phase kickback.

We concluded this chapter by reading about DiVincenzo's criteria for quantum computing.

The next chapter focuses on quantum communications.

Practice Problems

1. Prove that a controlled unitary matrix can be written as $|0\rangle \langle 0| \otimes I + |1\rangle\langle 1| \otimes U$.

2. The state of a three-qubit system is described as $\frac{1}{\sqrt{2^3}}(|0\rangle + |0\rangle + |1\rangle)$. Verify whether this is a valid quantum state. (Hints: The norm of a valid quantum state is 1.)

3. The state of a qubit is described by $|\psi\rangle = a|0\rangle + b|1\rangle$, where $a = \left(\frac{1}{3} + \frac{i}{3}\right)$ and $b = \left(\frac{\sqrt{3.5}}{3} + \frac{i\sqrt{3.5}}{3}\right)$. Check whether the wavefunction is normalized. If it is normalized, evaluate the probability of measuring the qubit in a state of "0" or "1."

4. When we discussed global and relative phases, it became clear that $|1\rangle$ and $-|1\rangle$ are the same state, excepting up to a phase, which cannot be measured. Explain why the same argument does not hold good for the states $\frac{1}{\sqrt{2}}(|0\rangle + |1\rangle)$ and $\frac{1}{\sqrt{2}}(|0\rangle - |1\rangle)$.

5. A quantum gate performs the transformation $|-\rangle \rightarrow |0\rangle$ and $|+\rangle \rightarrow |1\rangle$. Write the corresponding unitary matrix.

6. Verify whether the CCNOT gate is universal.

7. If two states $|\phi\rangle$ and $|\psi\rangle$ can be cloned, prove that $\langle \phi| \psi \rangle = 0$.

Fig. 5.43 Example
quantum circuits with
CNOT gates surrounded by
H-gates

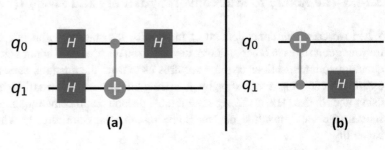

Fig. 5.44 An example
quantum circuit with a
network of CNOT gates
with multiple controls

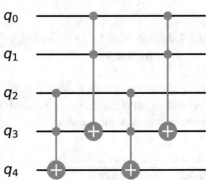

Fig. 5.45 An example
quantum circuit illustrating
phase kickback and
reversed CNOT gates with
multiple controls

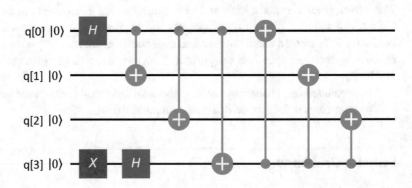

8. Create a unitary matrix to generate GHZ state.
9. Use the Walsh–Hadamard transform in a quantum circuit to generate random numbers.
10. Verify whether the circuits shown in Fig. 5.43 are equivalent.
11. What is the output of the quantum circuit shown in Fig. 5.44, assuming all qubits are initialized to the default state $|0\rangle$? Calculate the circuit's width and depth. Evaluate the circuit for other combinations of inputs and produce a truth table.
12. What is the output of the quantum circuit illustrated in Fig. 5.45? Calculate the circuit's width and depth.
13. Assume that you are a systems engineer developing the software for the backend control systems. Which quantum gates would you implement in software? Why?
14. Design a quantum comparator circuit that compares the status of two qubits in an arbitrary state. (Hints: Use the swap test from Sect. 7.10.1.)

15. The square root of the X-gate and its Hermitian conjugate are called the V and $V^\dagger$ gates. The matrices in Eq. 5.177 define them. Develop quantum circuits to emulate them. Derive the math and establish that the circuit works.

$$V = \frac{1}{2}\begin{bmatrix} 1+i & 1-i \\ 1-i & 1+i \end{bmatrix}$$
$$V^\dagger = \frac{1}{2}\begin{bmatrix} 1-i & 1+i \\ 1-i & 1-i \end{bmatrix}$$

(5.177)

References

1. Quantum Computation and Quantum Information, 10th Anniversary Edition, Michael A. Nielsen, Isaac I. Chuang.
2. Overview and Comparison of Gate Level Quantum Software Platforms Ryan LaRose, March 22, 2019, https://arxiv.org/pdf/1807.02500.pdf
3. Open Quantum Assembly Language, Andrew, Lev, John, Jay, January 10, 2017 https://arxiv.org/pdf/1707.03429.pdf
4. OpenQASM language reference: https://github.com/Qiskit/openqasm/blob/master/spec-human/qasm2.pdf
5. Github for OpenQASM: https://github.com/Qiskit/openqasm
6. Q# Programming language reference: https://docs.microsoft.com/en-us/quantum/language/
7. Microsoft Quantum Intrinsic reference: https://docs.microsoft.com/en-us/qsharp/api/qsharp/microsoft.quantum.intrinsic
8. https://github.com/microsoft/Quantum
9. Qiskit: An opensource framework for quantum computing, 2019. H'ector Abraham et al, https://github.com/Qiskit/qiskit-terra
10. Qiskit Textbook: https://qiskit.org/textbook/preface.html
11. Qiskit API Documentation: https://qiskit.org/documentation/
12. Equivalent Quantum Circuits, Juan Carlos and Pedro Chamorro, October 14, 2011, https://arxiv.org/pdf/1110.2998.pdf
13. A Simple Proof that Toffoli and Hadamard are Quantum Universal, Dorit Aharonov, 2003, https://arxiv.org/pdf/quant-ph/0301040.pdf
14. Limits on the ability of quantum states to convey classical messages. Ashwin Nayak, Julia Salzman, January 2006, http://www.math.uwaterloo.ca/~anayak/papers/NayakS06.pdf
15. Quantum Communications and Network Project, NIST: https://www.nist.gov/programs-projects/quantum-communications-and-networks
16. Argonne National Laboratory reports Quantum Teleportation across 52 miles. http://news.uchicago.edu/story/argonne-uchicago-scientists-take-important-step-developing-national-quantum-internet

Quantum Communications

6

"Cheshire Puss," she began, rather timidly, as she did not at all know whether it would like the name: however, it only grinned a little wider. "Come, it's pleased so far," thought Alice, and she went on. "Would you tell me, please, which way I ought to go from here?"
"That depends a good deal on where you want to get to," said the Cat.
"I don't much care where—" said Alice.
"Then it doesn't matter which way you go," said the Cat.
"—so long as I get somewhere," Alice added as an explanation.
"Oh, you're sure to do that," said the Cat, "if you only walk long enough."
— Lewis Carroll, Alice's Adventures in Wonderland.

Quantum Information Theory is a growing subject of specialization. Perhaps quantum communication is the first quantum technology the industry shall adopt. This chapter is a short introduction to some of the basic concepts of this focus area. This chapter begins with a brief introduction to the Einstein–Podolsky–Rosen (EPR) Paradox. Measurement of the Bell states, explanation of the local realism, and proof of Clauser, Horne, Shimony, and Holt (CHSH) inequality are the first few topics we explore in this chapter. The concept of density matrices, differences between pure and mixed states, a procedure for validating entanglement, and the no-communication theorem are introduced next. The final sections of this chapter discuss quantum communications with entangled photons, Quantum Key Distribution protocols, and RSA security. For further reading, a list of references is provided at the end of this chapter.

▶ **Learning Objectives**
- Bell measurement
- Local realism and proof of CHSH inequality
- Density matrices formalism
- Pure and mixed states
- No-communication theorem
- Von Neumann entropy
- Schumacher compression
- Schmidt decomposition and degree of entanglement
- Quantum communications with entangled photons
- Quantum Key Distribution protocols
- RSA security

© The Author(s), under exclusive license to Springer Nature Switzerland AG 2021
V. Sivasubramanian, *Fundamentals of Quantum Computing*, https://doi.org/10.1007/978-3-030-63689-0_6

6.1   EPR Paradox

A couple of properties of quantum mechanics are a bit of a deviation from classical theories. The first property is Heisenberg's principle of uncertainty (Sect. 1.6.3). The uncertainty principle forbids certain sets of observables to be measured with accuracy simultaneously. As an example, the position and momentum of a quantum system cannot be measured accurately at the same time. The second property is entanglement. Assume that Alice and Bob are far apart, but they have access to a source that produces entangled photons (Sect. 6.4.2). The source creates a pair of entangled photons and sends one entangled photon to both Alice and Bob. Since the photons are entangled, they share the same wavefunction, even after being far apart. If Alice measures her photon and finds it vertically polarized ↕, Bob will find his photon polarized in the horizontal ↔ direction and vice versa. Measuring one of the photons determines the other's state, despite the distance and no-communication between the photons. This behavior is called a non-local phenomenon.

Einstein called this "spooky action at a distance," as it seemed to violate relativity. In 1935, Einstein, Boris Podolsky, and Nathan Rosen proposed a thought experiment called the "EPR Paradox." Einstein proposed that hidden variables and certain unknown properties of quantum systems are necessary to make quantum mechanics a complete theory so that the spooky action is not necessary.

In 1964, the Irish physicist **John Stewart Bell** developed a set of inequalities that represent how two particles' measurements would distribute if they weren't entangled [1]. There is a lot of experimental evidence that the Bell inequalities are violated, which means that quantum entanglement does indeed occur.

In this section, we examine this subject!

6.1.1   Bell Measurement

In Chap. 5, we derived the four Bell states given by Eqs. (5.121), (5.122), (5.123), and (5.124). The general form of the Bell state equations is stated by the following equation.

$$|\mathcal{B}_{xy}\rangle = \frac{1}{\sqrt{2}} (|0y\rangle (-1)^x |1\bar{y}\rangle), \text{ where } \bar{y} \text{ is the complement of } y \tag{6.1}$$

Supposing, if we are given a Bell state $|\mathcal{B}_{xy}\rangle$, we can use the circuit described in Fig. 6.1 to measure it.

Fig. 6.1 Bell measurement

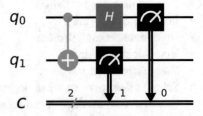

In this quantum circuit, the application of the CNOT gate converts the system into a "separable" state as follows:

$$\overset{CX\ q0,q1}{\rightarrow} \frac{1}{\sqrt{2}} \left(|0y\rangle \, (-1)^x \, |1y\rangle \right)$$

$$\rightarrow \frac{1}{\sqrt{2}} \left(|0\rangle \, (-1)^x \, |1\rangle \right) |y\rangle \tag{6.2}$$

Upon applying the H gate to the qubit q0, from Eqs. (5.18) and (5.19),

$$\overset{H\ q0}{\rightarrow} \frac{1}{\sqrt{2}} \left(|0\rangle \, (-1)^x \, |1\rangle \right) |y\rangle \rightarrow |x\rangle |y\rangle \tag{6.3}$$

This gives the state of the system before entanglement.

Exercise 6.1

If the input to this circuit is not a Bell state, calculate the probability of measuring a certain x and y.

6.1.2 Local Realism

In classical physics, **realism** means that the properties of objects have definite values before measurement, whether we measure them or not, that is, measurement just revealed the already existing property. Also, the measurement of property A of the system does not interfere with the measurement of another property B. If we measure B simultaneously, the results do not change, that is, the measurement of A does not depend on the context of the measurement. Therefore, we can say that the properties of the object exist independently of the experimental setup, especially the property we choose to measure.

Furthermore, consider two objects A and B, which are spatially separated. In this setup, if we measure object A, it does not affect the other object B instantaneously, that is, $\left| \vec{r}_A - \vec{r}_B \right| > ct$, where c is the speed of light, and t is the time between measuring A and B. The second principle is called **locality**. This classical scenario defined by Einstein–Podolsky–Rosen is known as **local realism**. Let us now check how this principle relates to quantum mechanics.

Assume we are given a qubit in state $|\psi\rangle = \frac{1}{\sqrt{2}}(|0\rangle + |1\rangle)$. Measuring this qubit can give a value 0 or 1, each with a 50% probability. We know this property from the Born rule, and from the histogram shown in Fig. 5.8. With this data, we cannot infer that, before measurement, the qubit was in state 0 or 1 with a 50% probability.

The reason is quite simple. If we apply an H gate, we get $H|\psi\rangle = |0\rangle$. If the qubit was in state 0 or 1 with a 50% probability, we should have got either $\frac{1}{\sqrt{2}}(|0\rangle + |1\rangle)$ or $\frac{1}{\sqrt{2}}(|0\rangle - |1\rangle)$, each with a 50% probability.

Hence in quantum mechanics, we cannot assume the state of the system before measurement.

In quantum mechanics, simultaneous measurement is possible only if the system is in a simultaneous eigenstate of A and B. Therefore, in general, the outcome of measurement in quantum mechanics depends upon the other commuting measurements.

Let us study these conclusions that led to the EPR Paradox briefly.

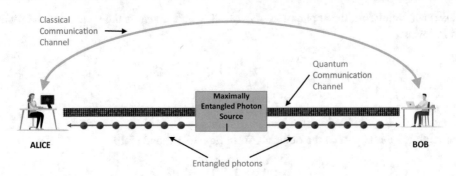

Fig. 6.2 Alice and Bob communicating on a quantum communication channel. A photon source sends them a series of entangled photons. Note that Alice and Bob may have a classical communication channel as well

Assume that Alice and Bob have a source that produces singlet (entangled) states $|\Psi^-\rangle = \frac{1}{\sqrt{2}}(|01\rangle - |10\rangle)$. The source sends Alice one of the entangled qubits, and Bob the other qubit that is entangled with the qubit sent to Alice. Assume that Alice and Bob are far apart in the opposite ends of the communication, as shown in Fig. 6.2. Let us also assume that both Alice and Bob measure their qubits in the computational basis $\{|0\rangle, |1\rangle\}$.

Based on what Alice measures, we can predict what Bob shall measure. If Alice measures her qubit in state $|0\rangle$ (with eigenvalue +1), Bob shall measure his qubit to be in state $|1\rangle$ (with eigenvalue −1), and vice versa. Recall from our discussions in Chap. 3, Eq. (3.34), that this is true for any basis that Alice and Bob may use. The only condition is that they both must use the same basis to measure their qubits. Therefore, we can say that there is a strong correlation between the measurements.

Note that both Alice and Bob can perform rotational operations on their qubits, changing the entangled state. For example, Z gate (Eq. (5.24)) can flip $|\phi^+\rangle$ to $|\phi^-\rangle$, and vice versa. Application of X gate flips $|\phi^+\rangle$ to $|\psi^+\rangle$, and vice versa.

Hence, we find that entangled states are inseparable. Even if the constituent states are separated by a distance, the constituent states are affected by the action on each other. It looks as though they can communicate instantaneously (but we cannot harness this communication to send information at speeds higher than the speed of light).

These were the contradictions in the 1935 paper made by **Einstein–Podolsky–Rosen**, now called the EPR Paradox. According to the paper, quantum mechanics is incomplete, and there must be some hidden variables. The hidden variables should reveal the state of the system before measurement.

If Bob decides to measure his qubit in the polar basis $\{|+\rangle, |-\rangle\}$, his measurement outcome shall be one of the states in the polar basis with a 50% probability in his reality.

There are a couple of assumptions here—realism and locality. The first assumption is that, at the time when Alice measures her qubit, Bob's system is in a reality that is independent of Alice. The second assumption is that nothing changes Bob's system when Alice measures her qubit.

Let us assume that the two qubits have some hidden variables associated with them. Since the qubits are entangled and show a strong correlation, we should assume that these hidden variables are related to each other. We shall use this to create a bound in the next section.

To continue with our experiment, the source keeps sending the entangled qubits to Alice and Bob. Both Alice and Bob measure each of the qubits they receive on some basis and record the time, the measurement basis used, and the outcome of the measurements. Later, when they both meet, they can exchange notes and discover the correlation between their measurements.

6.1.3 Bell's Inequalities

Bell found that the EPR Paradox can be expressed in terms of assumptions that lead to falsifiable predictions known as Bell's inequalities. Bell's inequalities enable the EPR Paradox to be experimentally verified. With technological advancements, several experiments have been conducted on these inequalities. Today we readily accept quantum entanglement and the non-locality of the quantum phenomenon. In this section, we examine one of the inequalities.

Let us assume that Alice measures her qubits a and b using measurement operators $Z = \sigma_z$ and $X = \sigma_x$. She measures her observables on the σ_z basis if she applied a Z gate and on the σ_x basis if she applied an X gate. Similarly, Bob measures his qubits c and d using measurement operators $C = \frac{1}{\sqrt{2}}(-\sigma_z - \sigma_x)$ and $D = \frac{1}{\sqrt{2}}(\sigma_z - \sigma_x)$. There can be four possible combinations of experiments we can perform with the qubits a, b, c, d. The respective expectation values of the outcomes of these experiments are given in the following equations.

$$\langle \Psi^- | ZC | \Psi^- \rangle = \langle ac \rangle = \frac{1}{\sqrt{2}} \tag{6.4}$$

$$\langle \Psi^- | ZD | \Psi^- \rangle = \langle ad \rangle = -\frac{1}{\sqrt{2}} \tag{6.5}$$

$$\langle \Psi^- | XC | \Psi^- \rangle = \langle bc \rangle = \frac{1}{\sqrt{2}} \tag{6.6}$$

$$\langle \Psi^- | XD | \Psi^- \rangle = \langle bd \rangle = \frac{1}{\sqrt{2}} \tag{6.7}$$

These expectation values can be derived as follows. We can easily show that: $(-\sigma_z - \sigma_x) = \begin{bmatrix} -1 & -1 \\ -1 & 1 \end{bmatrix}$ and $(\sigma_z - \sigma_x) = \begin{bmatrix} 1 & 1 \\ 1 & -1 \end{bmatrix}$. By applying this,

$$ZC = \sigma_z \otimes \frac{1}{\sqrt{2}}(-\sigma_z - \sigma_x) = \frac{1}{\sqrt{2}} \begin{bmatrix} 1 & 0 \\ 0 & -1 \end{bmatrix} \otimes \begin{bmatrix} -1 & -1 \\ -1 & 1 \end{bmatrix}$$

$$= \frac{1}{\sqrt{2}} \begin{bmatrix} -1 & -1 & 0 & 0 \\ -1 & 1 & 0 & 0 \\ 0 & 0 & 1 & 1 \\ 0 & 0 & 1 & -1 \end{bmatrix} \tag{6.8}$$

By referring to Eq. (2.133), we can derive the expectation value $\langle ac \rangle = \langle \Psi^- | ZC | \Psi^- \rangle$. For the sake of simplicity, we can derive this in two steps using matrix methods.

$$ZC|\Psi^-\rangle = \frac{1}{\sqrt{2}} \begin{bmatrix} -1 & -1 & 0 & 0 \\ -1 & 1 & 0 & 0 \\ 0 & 0 & 1 & 1 \\ 0 & 0 & 1 & -1 \end{bmatrix} \frac{1}{\sqrt{2}} (|01\rangle - |10\rangle)$$

$$= \frac{1}{\sqrt{2}} \begin{bmatrix} -1 & -1 & 0 & 0 \\ -1 & 1 & 0 & 0 \\ 0 & 0 & 1 & 1 \\ 0 & 0 & 1 & -1 \end{bmatrix} \cdot \frac{1}{\sqrt{2}} \begin{bmatrix} 0 \\ 1 \\ -1 \\ 0 \end{bmatrix} = \frac{1}{2} \begin{bmatrix} -1 \\ 1 \\ -1 \\ -1 \end{bmatrix}$$

(6.9)

$$\langle \Psi^- | ZC | \Psi^- \rangle = \frac{1}{\sqrt{2}} \begin{bmatrix} 0 & 1 & -1 & 0 \end{bmatrix} \cdot \frac{1}{2} \begin{bmatrix} -1 \\ 1 \\ -1 \\ -1 \end{bmatrix} = \frac{1}{\sqrt{2}}$$
(6.10)

Similarly, we can derive the other expectation values and arrive at Eqs. (6.5), (6.6), and (6.7). The sum of the expectation values is:

$$\langle ac \rangle + \langle bc \rangle + \langle bd \rangle - \langle ad \rangle = \frac{1}{\sqrt{2}} + \frac{1}{\sqrt{2}} + \frac{1}{\sqrt{2}} + \frac{1}{\sqrt{2}}$$
$$= 2\sqrt{2}$$
(6.11)

We shall return to this value in a while.

Let λ be the hidden variable. Let us assume that λ contains enough information for Alice and Bob to calculate the probabilities $p(\lambda)$ of the outcomes of their measurements. Let $Z(\lambda)$ be the probability associated with the measurement of the observable a, $X(\lambda)$ be the probability associated with the measurement of the observable b. Similarly, let $C(\lambda)$ and $D(\lambda)$ be the probabilities associated with Bob's observables c and d. Let us now study the outcome of the experiments we can do with various combinations of the observables. Consider the following outcome:

$$Z(\lambda)C(\lambda) + X(\lambda)C(\lambda) + X(\lambda)D(\lambda) - Z(\lambda)D(\lambda)$$
(6.12)

We can now integrate this over all such hidden variables

$$\int p(\lambda)[Z(\lambda)C(\lambda) + X(\lambda)C(\lambda) + X(\lambda)D(\lambda) - Z(\lambda)D(\lambda)]\, d\lambda$$
(6.13)

This equation can be rewritten by grouping the elements:

$$\int p(\lambda)[X(\lambda)(C(\lambda) + D(\lambda)) + Z(\lambda)(C(\lambda) - D(\lambda))]\, d\lambda$$
(6.14)

Since the measurement outcomes can take only values ± 1, the terms in the sum can only take values +1 and −1 for the hidden variable λ. Therefore, we see that $C(\lambda) + D(\lambda) = 0$ shows zero correlation and $C(\lambda) - D(\lambda) = 2$ shows a strong correlation. With this assumption, we can bind this integral:

$$\left| \int p(\lambda)\left[X(\lambda)\left(C(\lambda) + D(\lambda) \right) + Z(\lambda)(C(\lambda) - D(\lambda)) \right] d\lambda \right| \leq 2 \tag{6.15}$$

This inequality is known as CHSH inequality, named after **Clauser**, **Horne**, **Shimony**, and **Holt**. CHSH inequality is one of the Bells Inequalities.

By comparing Eqs. (6.15) and (6.11), we find that quantum mechanics violates the bound we derived for the local hidden variable. Therefore, it is clear that local hidden variable theories cannot explain the statistics of certain quantum mechanical experiments.

The CHSH inequalities are the simplest and widely studied. There could be many other Bell's inequalities. The inequalities assume a local hidden value model to describe the correlation between the measurement outcomes at different locations. In multi-partite systems, the inequalities are bound in the region of experimentally observable probabilities accessible by the local variable models. We can characterize this space if we find a minimal set of Bell's inequalities that are complete. The completeness here means iff the correlations permit a local hidden variable model. This is a problem we can readily solve in lower dimensions. However, it is generally a problem that is computationally hard.

Let us assume that we set up thousands of computers isolated from each other to emulate the Bell experiments classically. We then setup the different experiments the computers should emulate. Collectively the classical computers cannot simulate the quantum experiment! This inability is one proof that quantum computers can do certain things that classical computers cannot do.

Note that the correlations produced by quantum entanglement are not correlations that can be used to signal parties like that in a classical communication channel. If we must explain the correlations classically, we need to have communication to establish the correlations. We shall explore this topic shortly.

6.2   Density Matrix Formalism

So far, we have been assuming **pure states**, which are isolated quantum systems. The quantum systems are usually interacting with other systems, which could be larger, with many degrees of freedom. It shall be advantageous if we could describe a quantum system without going into many details of the larger system. Such systems are in **mixed states**. The density matrix representation is something we could use here. The density matrix representation also helps us understand why Bob cannot learn anything about the basis Alice used to measure her qubits. This formalism can also explain why Alice cannot use entanglement to transmit information to Bob at speeds greater than the speed of light c.

Consider a system of photons emanating from a table lamp. These photons are polarized in random directions. This is because of the random orientation of atoms in the lamp's filament. For example, if 25% of the photons are polarized vertically, 25% of the photons are polarized diagonally to the left, 25% polarized diagonally to the right, and 25% of the photons are polarized horizontally, we cannot describe them as $0.25|\updownarrow\rangle + 0.25|\leftrightarrow\rangle + 0.25|\nwarrow\rangle + 0.25|\nearrow\rangle$. The reason is the equal distribution of 25% represents the probabilities and not the probability amplitudes. Here, the probability p_i represents the occurrence of the state $|\psi_i\rangle$ in the *ensemble*.

This ensemble of photons is an excellent example of a **mixed state** which can be described as an ordered pair $\{(|\psi_1\rangle, p_1), (|\psi_2\rangle, p_2), \ldots\}$, where the probabilities $\sum_i p_i = 1$. We can think of the pure state as a degenerate case of a mixed state with probability 0 or 1. Note that the different probability distribution of the pure states gives rise to the same mixed state.

Recall that every vector $|\psi\rangle \in \mathbb{H}^n$ can be represented as a matrix $A \in \mathbb{C}^{n \times n}$, which is an inner product space. We can describe this mapping as a **density matrix** defined by the following equation for mixed states.

$$\rho = \sum_{i=1}^{n} p_i |\psi_i\rangle \langle \psi_i| \tag{6.16}$$

The density matrix is the sum of the projections on to the vector space spawned by $|\psi_i\rangle$ weighted by the corresponding probabilities p_i. The density matrix of a mixed state is a Hermitian operator, and its trace is 1. Note that, if the vectors $|\psi_i\rangle$ do not form an orthonormal basis, then we must find the orthonormal basis vectors before calculating the trace.

The density matrix for a pure state $|\psi\rangle = \sum_i a_i |v_i\rangle$ is given by:

$$\rho = \sum_{i=1}^{n} p_i |\psi\rangle\langle\psi| = \begin{bmatrix} a_1 a_1^* & a_1 a_2^* & \cdots & a_1 a_n^* \\ a_2 a_1^* & a_2 a_2^* & \cdots & a_2 a_n^* \\ \cdots & \vdots & \vdots & \vdots \\ a_n a_1^* & a_2 a_2^* & \cdots & a_n a_n^* \end{bmatrix} \tag{6.17}$$

This is a Hermitian matrix with a trace of 1. The density matrix for a pure state is simply a projection operator $\rho = |\psi\rangle\langle\psi|$.

Qiskit provides a class library to work with the density matrices.
https://qiskit.org/documentation/stubs/qiskit.quantum_info.DensityMatrix.html
We can create a density matrix based on numerical values or from a state vector. The calculated density matrix can be used to initialize the qubits in a quantum circuit. Listing 6.1 contains a sample implementation.

6.2.1 Maximally Mixed States

Now, let us look at some example states. The density matrices of basis states are:

$$|0\rangle\langle 0| = \begin{bmatrix} 1 & 0 \\ 0 & 0 \end{bmatrix} \text{ and } |1\rangle\langle 1| = \begin{bmatrix} 0 & 0 \\ 0 & 1 \end{bmatrix} \tag{6.18}$$

If we have an equal mixture (that is, $p = \frac{1}{2}$) of them, then the density matrix is:

$$\frac{|0\rangle\langle 0| + |1\rangle\langle 1|}{2} = \begin{bmatrix} \frac{1}{2} & 0 \\ 0 & \frac{1}{2} \end{bmatrix} = \frac{1}{2} I \tag{6.19}$$

Similarly, the density matrices of the pure states are:

$$|+\rangle\langle +| = \begin{bmatrix} \frac{1}{2} & \frac{1}{2} \\ \frac{1}{2} & \frac{1}{2} \end{bmatrix} \text{ and } |-\rangle\langle -| = \begin{bmatrix} -\frac{1}{2} & \frac{1}{2} \\ \frac{1}{2} & -\frac{1}{2} \end{bmatrix} \tag{6.20}$$

If we have an equal mixture of the pure states, then the density matrix of that state is:

$$\frac{|+\rangle\langle+| + |-\rangle\langle-|}{2} = \begin{bmatrix} \frac{1}{2} & 0 \\ 0 & \frac{1}{2} \end{bmatrix} = \frac{1}{2}I \qquad (6.21)$$

Note that the density matrix is the same for the equal mixture of states ($|0\rangle$, $|1\rangle$) and ($|+\rangle$, $|-\rangle$). The density matrix should be the same for any orthogonal basis Alice chooses, which is an important observation. It follows that, if Alice is changing the measurement basis, it does not affect Bob's density matrix!

States with the density matrix $\frac{1}{2}I$ are also known as a **maximally mixed state**.

Note that measuring ρ on a given basis, gives an outcome $|i\rangle$ on that basis with a probability given by the following equation.

$$Pr(|i\rangle) = \rho_{ii} = \langle i|\rho|i\rangle \qquad (6.22)$$

This equation means the diagonal elements of the density matrices represent probabilities.

6.2.2   Unitary Transformations on Density Matrices

The following equation defines the action of a unitary U on the density matrix ρ.

$$
\begin{aligned}
U\rho &= U\sum_i p_i |\psi_i\rangle\langle\psi_i| = \sum_i p_i \left(U|\psi_i\rangle\right)(U|\psi_i\rangle)^\dagger \\
&= \sum_i p_i \left(U|\psi_i\rangle\right)\langle\psi_i| U^\dagger \qquad (6.23) \\
&= U\rho U^\dagger
\end{aligned}
$$

6.2.3   Eigen Decomposition

Recall from Eq. (2.146) that the density matrix ρ can be written as:

$$\rho|\psi\rangle = \lambda|\psi\rangle, \text{ where } |\psi\rangle \text{ are eigenvectors and } \lambda \text{ is some eigenvalue.} \qquad (6.24)$$

The density matrix can then be written as the spectral decomposition:

$$\rho = \sum_i \lambda_i |\psi_i\rangle\langle\psi_i| \qquad (6.25)$$

Therefore,

$$\langle\psi_i|\rho|\psi_i\rangle = \lambda_i \qquad (6.26)$$

Here the λ_i sum up to the trace of the density matrix, which is 1. This way, we can derive the eigenvalues and eigenvectors.

6.2.4 Verification of Entanglement

We can use density matrices to verify whether a state is entangled or whether it is a product state. Let us start with a simple product state of two qubits A and B denoted by the following equation, which is familiar to us now.

$$|\psi\rangle = |\psi_A\rangle|\psi_B\rangle \tag{6.27}$$

For the qubit A, one of the eigenvalues of the density matrix is 1, and the rest are all 0. If more than one of the eigenvalues are positive, then this represents a mixed state. The other way to check this is to take the density matrix square and check its trace.

$$Tr\left(\rho^A\right)^2 = \sum_i \lambda_i^2, \tag{6.28}$$

where ρ^A is the mathematical notation that denotes the density matrix of A, $0 \le \lambda_i$ and $\sum_i \lambda_i = 1$.

Hence $\sum_i \lambda_i^2 \le 1$. Now we have two cases. The first case is when one of the λ_i is equal to 1, and the rest is equal to 0. The second case is not restricted, and the λ_i must be positive and in the range.

$$Tr\left(\rho^A\right)^2 \rightarrow \begin{cases} = 1 \text{ for product states} \\ < 1 \text{ for entangled states} \end{cases} \tag{6.29}$$

6.2.5 Reduced Density Matrices

Let A and B be two qubits in the Hilbert spaces H_A and H_B. The composite state of these two quantum systems is described by the state $|\psi\rangle = H_A \otimes H_B$. The density matrix of the composite system is $\rho^{AB} = |\psi\rangle\langle\psi|$. From this, there is no direct way of deriving the pure state for subsystem A, but we can define a reduced density matrix ρ^A as a partial trace over B. This is done as follows.

$$\rho^A = \sum_i \langle i_B|(|\psi\rangle\langle\psi|)|i_B\rangle = Tr_B\left(\rho^{AB}\right) \tag{6.30}$$

Let us apply this logic to a Bell state. The density matrix for the Bell state $|\phi^+\rangle$ described in Eq. (5.121) is as follows.

$$\rho^{AB} = \frac{(|0\rangle|0\rangle + |1\rangle|1\rangle)\,(\langle 0|\langle 0| + \langle 1|\langle 1|)}{2} \tag{6.31}$$

$$= \frac{|00\rangle\langle 00| + |11\rangle\langle 00| + |00\rangle\langle 11| + |11\rangle\langle 11|}{2} \tag{6.32}$$

$$= \frac{1}{2}\left((|0\rangle\langle 0| \otimes |0\rangle\langle 0|) + (|1\rangle\langle 0| \otimes |1\rangle\langle 0|) + (|0\rangle\langle 1| \otimes |0\rangle\langle 1|) + (|1\rangle\langle 1| \otimes |1\rangle\langle 1|)\right) \tag{6.33}$$

From this, we trace over the Hilbert space for the qubit B. To do this, we keep the terms $|0\rangle\langle 0|$ and $|1\rangle\langle 1|$ for the qubit B and discard the rest of the terms. Then we take the terms corresponding to the qubit A.

$$\rho^A = \frac{1}{2}(|0\rangle\langle 0| + |1\rangle\langle 1|) = \frac{1}{2}\,I \tag{6.34}$$

Similarly,

$$\rho^B = \frac{1}{2}(|0\rangle\langle 0| + |1\rangle\langle 1|) = \frac{1}{2} I \tag{6.35}$$

The reduced density matrices again prove that, say if we measure qubit A, we shall measure $|0\rangle$ with a probability of $\frac{1}{2}$ and $|1\rangle$ with a probability of $\frac{1}{2}$, irrespective of the basis in which the qubit is measured. Due to this randomness, by simply measuring the qubits A or B locally, we gain no information about the state they were prepared.

6.2.6 No-Communication Theorem

According to the **no-communication theorem** [3, 4], if Alice and Bob share an entangled state, any operation performed by Alice does not change the density matrix of Bob. Therefore, it is not possible to send classical information using the entangled state instantaneously.

Assume that Alice and Bob share an entangled state $|\phi^+\rangle$. The probability of Alice measuring a $|0\rangle$ is $P(|0\rangle) = |0\rangle\langle 0| = \begin{bmatrix} 1 & 0 \\ 0 & 0 \end{bmatrix}$ and $P(|1\rangle) = |1\rangle\langle 1| = \begin{bmatrix} 0 & 0 \\ 0 & 1 \end{bmatrix}$.

In this experimental setup, the density matrix of the qubit shared with Bob is given by the Eq. (6.35).

Alice now measures her qubit. This act collapses the system. The density matrix of Bob's qubit in the post-measurement scenario can be calculated as follows.

$$\rho'^B = \sum_i P(|i\rangle) \, \rho^B P(|i\rangle) \tag{6.36}$$

$$= \frac{1}{2}\begin{bmatrix} 1 & 0 \\ 0 & 0 \end{bmatrix}\begin{bmatrix} 1 & 0 \\ 0 & 1 \end{bmatrix}\begin{bmatrix} 1 & 0 \\ 0 & 0 \end{bmatrix} + \frac{1}{2}\begin{bmatrix} 0 & 0 \\ 0 & 1 \end{bmatrix}\begin{bmatrix} 1 & 0 \\ 0 & 1 \end{bmatrix}\begin{bmatrix} 0 & 0 \\ 0 & 1 \end{bmatrix} \tag{6.37}$$

This equation simplifies to:

$$= \frac{1}{2}\begin{bmatrix} 1 & 0 \\ 0 & 0 \end{bmatrix} + \frac{1}{2}\begin{bmatrix} 0 & 0 \\ 0 & 1 \end{bmatrix} = \frac{1}{2}I = \frac{1}{2}(|0\rangle\langle 0| + |1\rangle\langle 1|) = \rho^B \tag{6.38}$$

We find that the density matrix of Bob's qubit is not altered by the measurement of Alice's qubit. Hence, the information one of them changes is something the other person cannot read.

Exercise 6.2

Prove that the density matrix of Bob does not change if Alice performs a unitary on her qubit.

Some points to remember
- The eigenvalues are positive because the density matrix is positive semidefinite (refer Sect. 2.10)
- The eigenvalues are real because the density matrix is Hermitian.
- If the rank of the density matrix is 1, then it represents a pure state.
- The density matrices encode observables.
- Anything that Alice does on her entangled qubit does not change the density matrix of Bob.
- The basis states are in the computational basis of $|0\rangle$ and $|1\rangle$.
- The pure states are superposition states of the basis states $|+\rangle$ and $|-\rangle$.

- The mixed states are probability distribution over pure states.
- $\rho^2 = \rho$ for pure states.
- The expectation value of an observable A is given by $\langle A \rangle = Tr(\rho A)$.

6.3   Von Neumann Entropy

The term **entropy** was coined by the German physicist **Rudolf Clausius**, from the Greek words "en," and "tropē" meaning *change within*. Entropy is the measure of the energy contained in a physical system that could not be used to do work. It is also interpreted as the measure of disorder or randomness in the system. Clausius postulated that the entropy of a closed system cannot decrease, which is the second law of thermodynamics. However, Clausius did not provide an explanation. **Ludwig Boltzmann** provided an interpretation of this postulate using statistical mechanics in later years.

Consider a large isolated system of N identical particles. Each of these particles occupies one of the energy levels $E_1 < E_2 < \ldots < E_m$. The particles exchange energy among themselves. Hence the system achieves a thermal equilibrium over time. If there are N_1 particles with energy E_1, N_2 particles with energy E_2, and so on, the total number of particles is $N = \sum_i N_i$, and the total energy of the system is $E = \sum_i N_i E_i$. The N_i are also referred as the occupancy numbers in literature.

Supposing that if we partition the system and put the $N_1, N_2, \ldots, N_m$ particles into m microstates corresponding to the energy levels. The number of ways the N particles can be arranged is

$$\text{number of microstates} = \frac{N!}{N_1! N_2! \ldots N_m!} \tag{6.39}$$

This is the number of ways in which the microstates can be put together to realize the large system. It is also the probability of the microstates being equal. The most probable microscopic state can be obtained by maximizing Eq. (6.39) for E.

The probability of the microstate to be in state i can be defined as: $p_i = \frac{N_i}{N}$. Using Stirling's approximation[1] of the factorials,

$$\frac{1}{N} \log \left(\frac{N!}{N_1! N_2! \ldots N_i!} \right) = H(p_i) + O(N^{-1} \log N), \tag{6.40}$$

where the entropy function,

$$H(p_i) = -\sum_i p_i \log p_i \tag{6.41}$$

According to Shannon, Eq. (6.41) is the measure of ignorance before the experiment. It is also the information gathered by experimenting. The outcome of the experiment is maximum if p_i are equal.

In quantum mechanics, we associate statistical operators with system states. Von Neumann introduced the density matrix or the density operator formalism to represent the statistical state of quantum systems. The statistical description of quantum systems is useful in describing system evolution and entanglement. In these problems, each subsystem can be represented as density matrices, even though the entire system can be in a pure state.

Based on the principles of thermodynamics, von Neumann associated entropy with a statistical operator D. Von Neumann assumed a system of gas with N molecules. If the system can be compared

---

[1] In the limit $N \to \infty$, Stirling's asymptotic expansion is: $\ln N! \approx N \ln N - N + O(N)$.

to a quantum system, it can be described by a statistical operator D. Assume that the gas is a statistical mixture $\lambda|\varphi_1\rangle\langle\varphi_1| + (1 - \lambda)|\varphi_2\rangle\langle\varphi_2|)$, where λN molecules are in pure state φ_1 and $(1 - \lambda)N$ molecules are in pure state φ_2. Assuming φ_1 and φ_2 are orthogonal, von Neumann described a thought experiment in which the molecules φ_1 and φ_2 can be separated adiabatically without any work being done. The entropy of the original gas described by D must be equal to the sum of the entropies of φ_1 and φ_2. If the gases φ_1 and φ_2 are compressed to volumes λV and $(1 - \lambda)V$, keeping the temperature of the system T constant, the change in the entropy is $k\lambda N \log \lambda$ and $k(1 - \lambda)N \log \lambda$, respectively, where k is Boltzmann's constant[2]. The following relation must hold good:

$$S_0(\varphi_1, \lambda N) + S_0(\varphi_2, (1 - \lambda)N) = S_0(D, N) + k\lambda N \log \lambda + k(1 - \lambda)N \log \lambda, \tag{6.42}$$

where S_0 is the entropy function of a given gas. If S_0 is proportional to N, then dividing the equation by N, we get:

$$\lambda S(\varphi_1) + (1 - \lambda)S(\varphi_2) = S(D) + k\lambda \log \lambda + k(1 - \lambda) \log \lambda, \tag{6.43}$$

where S is a thermodynamic entropy. Instead of a mixture of two components φ_1 and φ_2, if we assume an infinite mixture, von Neumann concluded that:

$$S\left(\sum_i \lambda_i|\varphi_i\rangle\langle\varphi_i|\right) = \sum_i \lambda_i S(|\varphi_i\rangle\langle\varphi_i|) - k\sum_i \lambda_i \log \lambda_i \tag{6.44}$$

Recall that the density matrix of a quantum system in a pure state has only one element in its diagonal. The value of this element is 1, and the values of the rest of the diagonal elements are 0. The square of the trace of the density matrix is 1. The eigenvalues of the density matrix are the probabilities. A quantum state ρ can be written as a spectral decomposition. From Eq. (6.25),

$$\rho = \sum_i \lambda_i|\psi_i\rangle\langle\psi_i|, \tag{6.45}$$

where λ_i is an eigenvalue of ρ and $|\psi_i\rangle\langle\psi_i|$ is the projection of λ_i. If λ_i is degenerate, the spectral decomposition can be decomposed into one-dimensional projections called **Schatten Decomposition**,

$$\rho = \sum_i \lambda_i|\varphi_i\rangle\langle\varphi_i|, \tag{6.46}$$

where $|\varphi_i\rangle\langle\varphi_i|$ is the one-dimensional projections of λ_i. The degeneracy of λ_i equals the dimension of $|\varphi_i\rangle\langle\varphi_i|$. The Schatten decomposition is not unique if λ_i is degenerate. If λ_i is degenerate, then the corresponding eigenvectors can be chosen in several ways. If the entropy $S(D)$ is to be independent of the Schatten Decomposition, then $S(|\varphi_i\rangle\langle\varphi_i|)$ must be independent of the state vector $|\varphi_i\rangle$. With this assumption, the von Neumann entropy for a pure state can be written as a function of the density matrix as follows:

$$S\left(\sum_i \lambda_i|\varphi_i\rangle\langle\varphi_i|\right) = -Tr(\rho \log \rho) = -\sum_i \lambda_i \log \lambda_i, \tag{6.47}$$

where λ_i are the eigenvalues of the density matrix ρ. The logarithm is taken to base n, the dimension of the Hilbert space.

The eigenvalues (or the probabilities) do not change if we change the basis. Hence, the entropy itself is invariant. From Eq. (6.16),

```
# von Neumann entropy

from qiskit import QuantumCircuit, Aer, execute
from qiskit.quantum_info import Statevector
from qiskit.quantum_info import DensityMatrix
from qiskit.quantum_info import entropy
import numpy as np
import matplotlib.pyplot as plt
%matplotlib inline

q = QuantumRegister(2, 'q')
qc = QuantumCircuit(q)

#
# Setup some initial states,
# but ideally we start with |0>

backend = Aer.get_backend('statevector_simulator')
job = execute(qc, backend).result()
qc_state = job.get_statevector(qc)
dm_start = DensityMatrix(qc_state)

q = QuantumRegister(2, 'q')
qc = QuantumCircuit(q)

qc.initialize(qc_state, q)

qc.x(q[1])
qc.h(q[0])
qc.h(q[1])
qc.cx(q[0], q[1])
qc.h(q[0])
qc.h(q[1])

job = execute(qc, backend).result()
dm_end = DensityMatrix(job.get_statevector(qc))
time_range = np.arange(0.0, 1, 0.01)

# Create list of von Neumann entropies
entropy_list = list()
for time in time_range:
    element = time * dm_start + (1 - time) * dm_end
    entropy_list.append(entropy(element))

plt.xlabel('Time')
plt.ylabel('Entropy')
plt.title('von Neumann Entropy')
plt.plot(time_range, entropy_list, color='black')
plt.show()
```

Listing 6.1. Plotting von Neumann entropy

$$\rho = \sum_{i=1}^{n} p_i \, |\psi_i\rangle \, \langle\psi_i| \tag{6.48}$$

We can rewrite the previous equation as:

$$S(\rho) = -\,Tr\left[\sum_{i=1}^{n} p_i \, |\psi_i\rangle \, \langle\psi_i| \sum_{i=1}^{n} \log p_i |\psi_i\rangle \, \langle\psi_i| \right]$$
$$= -\sum_{i=1}^{n} p_i \log p_i \tag{6.49}$$

Note that von Neumann's entropy equals Shannon entropy. Consider the following definition of a density matrix for a **Bernoulli distribution** with probabilities $\{p, 1-p\}$ of a quantum system being in state $|0\rangle$ or $|1\rangle$.

$$\rho(p) = p|0\rangle \, \langle 0| + (1-p)|1\rangle \, \langle 1| \tag{6.50}$$

The von Neumann entropy is,

$$S(\rho(p)) = -p \, log \, p + (1-p) \, log \, (1-p) \tag{6.51}$$

When $p = 0$, $S(\rho(0)) = \, log \, (1) = 0$, similarly,
When $p = 1$, $S(\rho(1)) = \, log \, (1) = 0$
Von Neumann entropy is > 0, for the range $0 < p < 1$. We can plot this as a graph for a system evolution in this time range for the circuit shown in Fig. 5.28. Type in the code in Listing 6.1 after the regular import sections and portions of code loading the IBM account from the earlier listings.

In this code, we started with two qubits in the default state. The density matrix at this stage is the starting density matrix at $t = 0$. We can get the density matrix by executing the circuit and fetching the resultant state vector. We then use the starting density matrix to initialize a two-qubit circuit and build it as per Fig. 5.28. We determine the ending density matrix at $t = 1$ by executing this circuit. Von Neumann entropy is calculated for the points in the range 0 to 1 with a step value of 0.01. Finally, we plot the entropy calculated for each point. Figure 6.3 contains the plot obtained from executing this circuit.

Fig. 6.3 Plot of the von Neumann entropy

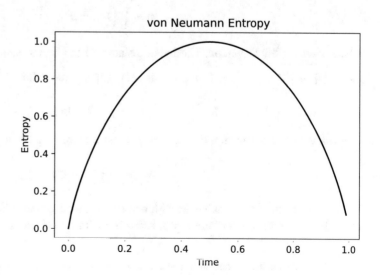

The von Neumann entropy describes the randomness of a quantum state. Von Neumann entropy quantifies the amount of information present in a system. It also signifies the amount of correlations between quantum systems.

6.3.1 Schumacher's Compression Theorem

Von Neumann entropy is used in the concept of quantum compression. According to Schumacher's compression theorem, if ρ is the density operator of the quantum information source, then the von Neumann entropy $S(\rho)$ is the smallest achievable rate R for quantum data compression: $\inf^3$ {R: R is achievable} $= S(\rho)$

Assume that Alice sends Bob a message of n letters. She chooses a random set of pure states, which are not necessarily orthogonal. The density matrix of the n letters is given by:

$$\rho = \rho^1 \otimes \rho^2 \otimes \ldots \otimes \rho^n \tag{6.52}$$

The best possible compression we can achieve using Schumacher's compression theorem, assuming $n \to \infty$, is compression to Hilbert's state $\mathcal{H}$:

$$\log\left(\dim(\mathcal{H})\right) = nS(\rho) \tag{6.53}$$

If we use n photons to send the message, we can compress the message to $m = nS(\rho)$.

6.3.2 Bloch Sphere

In this subsection, we shall return to the Bloch sphere briefly. Recall from Eq. (2.83), that any matrix can be written as a linear combination of the identity matrix and the Pauli matrices. We can apply that logic to density matrices as well:

$$\rho = \frac{1}{2}\left(I + x\sigma_x + y\sigma_y + z\sigma_x\right) \tag{6.54}$$

$$\rho = \frac{1}{2}\left(I + \vec{r}\,\vec{\sigma}\right), \tag{6.55}$$

where $\vec{r}$ is the **Bloch vector**. The radial distance of the Bloch vector from the origin is given by $r = \sqrt{|x|^2 + |z|^2 + |z|^2}$. The Bloch vector is the expectation value of the Pauli matrices.

$$\vec{r} = \langle\vec{\sigma}\rangle = Tr\left(\rho\vec{\sigma}\right) \tag{6.56}$$

Also, the determinant of the density matrix has a relation to the radius of the Bloch vector,

$$\det(\rho) = \frac{1}{4}\left(1 - r^2\right) \tag{6.57}$$

The density matrices of qubits lie within or on the Bloch sphere, whose radius is $\vec{r} = 1$. The length of the Bloch vector is 1 for pure states, and it is <1 for mixed states.

---

[3] Inf – the infimum is the greatest lower bound of a set. In a set $A = \{1, 5, 3, 7\}$, the infimum is 1.

We shall now derive an alternate interpretation of the von Neumann entropy in terms of the Bloch vector for single qubits. Let λ_1, λ_2 be the eigenvalues for the density matrix ρ. Since the trace of the density matrix is 1, we can say that $\lambda_2 = 1 - \lambda_1$. Also, $\det(\rho) = \lambda_1 \lambda_2$. Connecting these values with the above equation, we can derive that:

$$\lambda_1 = \frac{1+r}{2} \text{ and } \lambda_2 = \frac{1-r}{2} \tag{6.58}$$

With this information, we can define the von Neumann entropy as a function of r:

$$S(\rho) = -\left[\frac{1+r}{2} \log \frac{1+r}{2} + \frac{1-r}{2} \log \frac{1-r}{2} \right] \tag{6.59}$$

6.3.3 Schmidt Decomposition and Degree of Entanglement

A pure quantum system state $|\psi\rangle$ can be written as a separable tensor product decomposition $|\psi\rangle = |\psi_1\rangle \otimes |\psi_2\rangle \otimes \ldots \otimes |\psi_n\rangle$. If this is not possible, then the state $|\psi\rangle$ is entangled. For example, the state $|\psi\rangle = \frac{1}{\sqrt{2}} (|00\rangle + |11\rangle)$ is maximally entangled. From Eqs. (6.34) and (6.35), we know that the trace over the density matrix of each qubit is a mixed state $\frac{1}{2} I$. Notably, this state has the highest von Neumann entropy between two qubits. There is no correlation between the qubits in an unentangled state, and the von Neumann entropy is zero. The trace over gives a pure state. Thus, we can use von Neumann entropy as a measure of entanglement between two qubits.

Schmidt's theorem decomposes a vector as a tensor product of two inner produce spaces. Hence, Schmidt decomposition is a good tool in analyzing bipartite pure states. It can be used to decompose bipartite pure states into the superposition of constituent states.

Schmidt's Theorem A bipartite pure state $|\psi\rangle_{AB} \in \mathcal{H}_A \otimes \mathcal{H}_B$ can be expressed as follows:

$$|\psi\rangle_{AB} \equiv \sum_{i=0}^{d-1} \lambda_i |i\rangle_A \otimes |i\rangle_B, \tag{6.60}$$

where the **Schmidt coefficients** λ_i are strictly positive valued real numbers and normalized (that is, $\sum_i \lambda_i = 1$) constants. The states $|i\rangle_A$ and $|i\rangle_B$ form the orthonormal basis for the states A and B. d, the **Schmidt rank** or the **Schmidt number**, is the number of Schmidt coefficients counted with multiplicity. $d \leq \min\left(\dim(\mathcal{H}_A), \dim(\mathcal{H}_A)\right)$.

Proof

In an orthonormal basis $|j\rangle_A$ and $|k\rangle_B$, we can write $|\psi\rangle_{AB}$ as:

$$|\psi\rangle_{AB} = \sum_{j=0}^{d_A-1} \sum_{k=0}^{d_B-1} \alpha_{j,k} |j\rangle_A \otimes |k\rangle_B, \tag{6.61}$$

where $\alpha_{j,k}$ are some amplitudes corresponding to the orthonormal basis. d_A and d_B are the dimensions of the respective Hilbert spaces. The matrix formed by the coefficients $\alpha_{j,k}$ can be written as follows.

$$\alpha_{j,k} = [A]_{j,k} \tag{6.62}$$

The matrix A can be decomposed in terms of unitary matrices U, Σ, V, whose dimensions are mentioned as subscripts in the following equation.

$$A = U_{d_A \times d_A} \Sigma_{d_A \times d_B} V_{d_B \times d_B} \tag{6.63}$$

Assuming Σ is a diagonal matrix of dimension $d \times d$, with λ_i as its diagonal elements, we can write $\alpha_{j,k}$ as follows:

$$\alpha_{j,k} = \sum_{i=0}^{d-1} u_{j,i} \lambda_i v_{i,k} \tag{6.64}$$

Plugging this into Eq. (6.61),

$$|\psi\rangle_{AB} = \sum_{j=0}^{d_A-1} \sum_{k=0}^{d_B-1} \sum_{i=0}^{d-1} u_{j,i} \lambda_i v_{i,k} |j\rangle_A \otimes |k\rangle_B \tag{6.65}$$

The above equation can be rewritten as:

$$|\psi\rangle_{AB} = \sum_{i=0}^{d-1} \lambda_i \left(\sum_{j=0}^{d_A-1} u_{j,i} |j\rangle_A \right) \otimes \left(\sum_{k=0}^{d_B-1} v_{k,i} |k\rangle_B \right) \tag{6.66}$$

If $\left(\sum_{j=0}^{d_A-1} u_{j,i} |j\rangle_A \right)$ and $\left(\sum_{k=0}^{d_B-1} v_{k,i} |k\rangle_B \right)$ form an orthonormal basis, we can rewrite the equation as follows, proving the theorem.

$$|\psi\rangle_{AB} = \sum_{j=0}^{d_A-1} \sum_{k=0}^{d_B-1} \alpha_{j,k} |j\rangle_A \otimes |k\rangle_B \tag{6.67}$$

Returning to the bipartite pure systems, let us assume that $|\psi\rangle_{AB}$ is a bipartite system $A \otimes B$. The density matrix for $|\psi\rangle_{AB}$ can be written in terms of the Schmidt decomposition as follows.

$$\rho = |\psi\rangle_{AB} \langle\psi|_{AB} \equiv \sum_{i=0}^{d-1} \sum_{j=0}^{d-1} \lambda_i \lambda_j |i\rangle_A \langle j|_A \otimes |i\rangle_B \langle j|_B \tag{6.68}$$

Tracing over the terms for the qubits B and A, we can write the density matrices for the individual qubits as:

$$\rho^A = \sum_{i=0}^{d-1} \lambda_i^2 |i\rangle_A \langle i|_A$$

$$\rho^B = \sum_{i=0}^{d-1} \lambda_i^2 |i\rangle_B \langle i|_B \tag{6.69}$$

The von Neumann entropy can be derived from Eq. (6.47) as follows.

$$S(\rho^A) = -\sum_{i=0}^{d-1} \lambda_i^2 \log \lambda_i^2$$

$$S(\rho^B) = -\sum_{i=0}^{d-1} \lambda_i^2 \log \lambda_i^2$$

(6.70)

We find that $S(\rho^A) = S(\rho^B)$, which also gives the degree of entanglement between the bipartite pure states.

6.3.4 Purification

Consider the system state given in Eq. (6.67). Assume that the two qubits A and B are shared by Alice and Bob, respectively. If Alice measures her qubit, the outcome of the measurement j with probability $P(j) = |\alpha_{j,k}|^2$. The state for Bob is $\frac{\alpha_{j,k}|k\rangle}{\sqrt{P(k)}}$. Bob's density matrix is:

$$\rho^B = \sum_{j=0}^{d_A-1} P(k) \frac{\sum_{k=0}^{d_B-1} \alpha_{j,k}|k\rangle}{\sqrt{P(k)}} \frac{\sum_{k'=0}^{d_B-1} \alpha^*_{j,k'}\langle k'|}{\sqrt{P(k)}}$$

$$= \sum_{k=0}^{d_B-1} \sum_{k'=0}^{d_B-1} \sum_{j=0}^{d_A-1} \alpha_{j,k}\alpha^*_{j,k'} |k\rangle\langle k'| = (\alpha^\dagger \alpha)^T$$

(6.71)

It is easy to prove that, if Bob measures first, the density matrix of Alice will be $(\alpha\alpha^\dagger)$. So we can say $Tr(\rho^A) = Tr(\alpha\alpha^\dagger) = \sum_{j,k}|\alpha_{j,k}|^2 = 1$. Such states $|\psi\rangle_{AB}$ are called the purification of the density matrix.

Example 6.1

A system of two qubits is described by $|\psi\rangle = \frac{3}{10}|00\rangle + \frac{1}{10}|11\rangle$. Calculate the density matrices of the qubits. What are the coordinates of the qubits on the Bloch sphere? Also, calculate the von Neumann entropy of the system.

Solution:
The density matrix is given by:

$$\rho = \left(\frac{6}{8}|00\rangle + \frac{\sqrt{28}}{8}|11\rangle\right) \otimes \left(\frac{6}{8}\langle00| + \frac{\sqrt{28}}{8}\langle11|\right)$$

$$= \frac{36}{64}|00\rangle\langle00| + \frac{6\sqrt{28}}{64}|00\rangle\langle11| + \frac{6\sqrt{28}}{64}|11\rangle\langle00| + \frac{28}{64}|11\rangle\langle11|$$

$$= \frac{36}{64}\left(|0\rangle\langle0| \otimes |0\rangle\langle0|\right) + \frac{6\sqrt{28}}{64}\left(|0\rangle\langle1| \otimes |0\rangle\langle1|\right) + \frac{6\sqrt{28}}{64}\left(|1\rangle\langle0| \otimes |1\rangle\langle0|\right) + \frac{28}{64}\left(|1\rangle\langle1| \otimes |1\rangle\langle1|\right)$$

Tracing over B for A,

$$\rho^A = \frac{36}{64}|0\rangle\langle 0| + \frac{28}{64}|1\rangle\langle 1| = \begin{bmatrix} \frac{36}{64} & 0 \\ 0 & \frac{28}{64} \end{bmatrix}$$

Using Eq. (2.63),

$$\rho^A = \frac{1}{2}\left(I + \frac{1}{8}Z\right)$$

Hence, the coordinates of the qubit A are $\left(0, 0, \frac{1}{8}\right)$. We can similarly derive for the qubit B.
Von Neumann entropy is:

$$S(\rho^A) = -\left[\frac{36}{64}\log\frac{36}{64} + \frac{28}{64}\log\frac{28}{64}\right]$$

◄

Some points to remember
- Von Neumann entropy is ≥ 0.
- For a pure state, the trace of the density matrix is 1, and $\log\rho^A = 0$. Hence the von Neumann entropy is 0.
- For maximally mixed states with all N pure states in equal proportion, $p_i = \frac{1}{N}$ and von Neumann entropy is $\log N$.
- For a composite system AB, if it is in a pure state, then $S(A) = S(B)$.
- For a pure state, the Bloch vector starts from the origin and ends at a point on the surface of the Bloch sphere. For a mixed state, the Bloch vector ends at a point inside the Bloch sphere. Von Neumann entropy of a mixed state is measured by the distance from the point inside the sphere to the surface of the sphere.

6.4   Photons

Photons are the means of quantum communication. In this section, we examine how polarized light can encode quantum information. We also learn about the process of creating entangled photons using nonlinear optics.

6.4.1   Polarization

Bell's inequality experiments are usually done with photons. Photons are spin-1 particles. Hence, we use polarization to measure their states. The spin angular momentum of circularly polarized photons is $\pm\hbar$. It is positive for left polarized photons and negative for right polarized photons by convention. The spin of the photons is measured along the direction of propagation z called **helicity**.

Photons have some other interesting properties. They have zero mass, and stable mean lifetime. Photon–photon interaction can happen only with large energies. Hence, they can be safely transferred over vast distances using fiber optics. The fiber-optic cables use a glass fiber core and cladding with different refractive indices. These two materials are chosen in such a way that the light gets bent at a certain angle due to total internal reflection. The photons reflect off the core and cladding in a series of zig-zag bounces and zip along the fiber's length. Today, telecommunication and internet backbone are on fiber optics. Fiber optics is a proven technology we can use in quantum communications.

The light emanating from lamps vibrates in random directions. This light is unpolarized. Using polaroid filters, we can filter out vibrations in directions that are not needed. The polaroid filters use long-chain molecules that absorb components of light vibrating in the direction they are stretched. We can also change the direction of polarization by passing light through half-wave and quarter-wave plates. To measure the polarization of light, we pass the light through a polaroid filter and then use a photodetector to measure the intensity of light coming out of the filter.

Vertically polarized photons $|\updownarrow\rangle$ correspond to the state $|0\rangle$ in the computational basis. Likewise, the horizontally polarized photons $|\leftrightarrow\rangle$ correspond to the state $|1\rangle$ in the computational basis. Photons can also be polarized diagonally. The following equation describes the representation of diagonally polarized photons.

$$|\nearrow\rangle = \frac{1}{\sqrt{2}}(|\updownarrow\rangle + |\leftrightarrow\rangle), \text{and} \ |\searrow\rangle = \frac{1}{\sqrt{2}}(|\updownarrow\rangle - |\leftrightarrow\rangle) \tag{6.72}$$

A right-hand circular (RHC) polarized light has equal amplitudes polarized in vertical or horizontal directions, which are $90°$ apart, which also applies for the left-hand circular (LHC) polarized light. In the polar basis, the state of the photons in RHC or LHC polarization can be written in equal superposition states as:

$$\frac{1}{\sqrt{2}}(|\updownarrow\rangle + i|\leftrightarrow\rangle) = |+\rangle, \text{and} \ \frac{1}{\sqrt{2}}(|\updownarrow\rangle - i|\leftrightarrow\rangle) = |-\rangle \tag{6.73}$$

6.4.2 Entangled Photons

6.4.2.1 Cascade Experiments

In Atomic Radiation Cascade experiments, we can excite an atom to emit two polarization-correlated photons that move in opposite directions ($+z$ and $-z$). This happens when two electrons from the same upper orbital switch to the same lower orbital. When calcium atoms are excited [2] from the 4^1S_0 ground state to the 6^1P_1 state, two polarization-correlated photons γ_1 and γ_2 are emitted in succession. This process conserves parity and total angular momentum. Let $|x_1\rangle$ and $|y_1\rangle$ be the photon's polarization states γ_1 along the x and y axes, and $|x_2\rangle$ and $|y_2\rangle$ be that of the photon γ_2, then the state of entanglement between the photons can be proved to be one of the Bell states:

$$\frac{1}{\sqrt{2}}(|x_1\rangle|x_2\rangle + |y_1\rangle|y_2\rangle) = |\phi^+\rangle \tag{6.74}$$

This method is one way of producing entangled photons and widely used in EPR experiments. The other method we shall brief in this book uses nonlinear optics.

6.4.2.2 Spontaneous Parametric Down Conversion

Spontaneous Parametric Down Conversion is a process that can happen in nonlinear crystals. This process converts incoming single photons (called **pump**) into a pair of outgoing photons (called **signal** and **idler**) using an optical property called **birefringence**. Of course, this process should preserve the law of conservation of energy and the law of conservation of momentum. Hence, the two photons produced in this process have energies lower than the pump photon (Fig. 6.4).

Birefringence is a property of crystals whose refractive index changes with the polarization, and the direction of propagation of light. Recall that the refractive index changes with the frequency. So,

Fig. 6.4 Type-I,
spontaneous parametric
down conversion

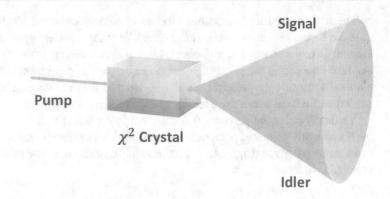

Fig. 6.5 Type-II SPDC
using a BBO crystal

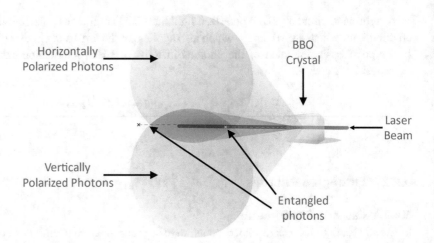

to conserve energy and momentum, we must "**phase match**," and only certain triplet frequencies work. Phase matching is typically done in nonlinear crystals. There are three types of Spontaneous Parametric Down Conversion (SPDC).

- If the pump, signal, and idler photons have the same polarization, it is type-0.
- If the signal and idler photons have the same polarization and orthogonal to the pump photon's polarization, it is type-I.
- If the signal and idler photons are polarized in orthogonal directions, it is type-II.

Beta-barium borate (BBO) crystals exhibit type-II behavior. When we direct a laser beam (pump) on a BBO crystal, most photons pass through. However, some photons undergo spontaneous parametric down conversion. Their trajectories are constrained along the edges of the signal and idler cones. The crystal is cut in such a way that the cones are symmetrical to the incident beam. Since momentum and energy are to be conserved, the correlated photons always move along the lines where the two cones intersect. Figure 6.5 illustrates the type-II SPDC.

6.5 Quantum Communication

Quantum communication is perhaps the first quantum computing technology to be adopted by the industry. The necessary pieces of technology required for quantum communication exist today. It is just a matter of time before commercial-grade systems are built. A reasonable understanding of quantum communications can be useful. The quantum communication system has three parts. There must be a source (usually Alice) preparing the qubits (usually photons) and encoding information on them. This is the first part. The second part is the transmission medium, such as the fiber optics. The third part is the recipient (Bob), who receives the photons and decodes them to retrieve the information.

Photons are prepared using one of the methods described in the previous section, and information is encoded on them by rotating the direction of polarization. At the receiving end, Bob uses polaroid filters and photodetectors to identify the direction of polarization of the photons.

Superdense coding, quantum entanglement, and quantum teleportation are some methods used in quantum communications. There are some error correction schemes available to detect and correct channel errors.

There could be some eavesdroppers (Eve) in the system who want to tap into the quantum channel and steal information. Eavesdropping on the quantum system requires measurement of the qubits, and it is an irreversible transformation. Hence, we can detect this with a high probability due to the way the quantum systems work.

Quantum communication offers additional security. Say, for example, Alice sends her qubit A to Bob. The transmitted qubit has the density matrix $\rho^A = \frac{1}{2}I$. It carries no other information. The information is encoded in the correlation between the qubits A and B. Hence, unless Eve gains control over both the qubits, she cannot gather knowledge of the transmission. Hence, if Alice encodes a secret key on a series of entangled photons and sends them over to Bob in a quantum communication channel, Eve cannot learn anything about the secret key.

6.6 The Quantum Channel

Quantum Channels transmit classical or quantum information from point A to B. The input to the **quantum channel** is the system state before transmission, and the output to the quantum channel is the system state after transmission. The quantum channel under discussion in this text is discrete and memoryless. It is discrete in the sense that it transforms a quantum system whose state vector is in a finite-dimensional Hilbert space. The memoryless property means that the output of the quantum channel is determined by the current input. We measure the quality of a noisy quantum channel by two parameters: channel capacity and channel fidelity.

Figure 6.6 illustrates communication through a quantum channel. The input to the communication block is an ensemble of n qubits in the pure state $|\varphi\rangle$ with the density matrix $\rho = |\varphi\rangle\langle\varphi|$. The encoder maps the qubits into $m > n$ intermediate states, which may or may not be qubits. This forms the input to the m quantum channels characterized by the superoperator ε, a trace-preserving completely positive linear map. The superoperator maps the input states with the density matrix ρ into the output states with the density matrix $\rho' = \varepsilon(\rho)$. At the receiving end, the decoder transforms the mixed states into n qubits defined by the density matrix ρ''. The fidelity of the quantum communication is $F = \langle\varphi|\rho''|\varphi\rangle$.

The quantum channel capacity Q is the highest rate at which quantum information can be transmitted from Alice to Bob over a noisy quantum channel. It can be defined as the largest number

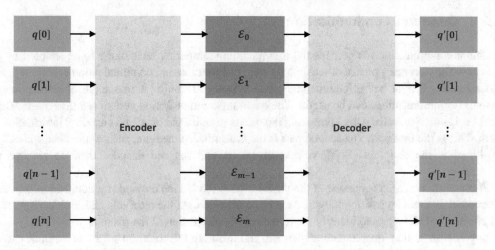

Fig. 6.6 Diagram illustrating communication through a quantum channel, from Dan C. Marinescu and Gabriela M. Marinescu

Q such that: the transmission rate $R < Q$, the error term $\epsilon > 0$, and the fidelity $F > (1 - \epsilon)$. The quantum channel can transmit both classical and quantum information. Depending upon the nature of the information, the **channel capacity** is defined as follows:

- The capacity to transmit classical information is defined as C.
- The capacity to transmit quantum states is defined as Q, which is less than C (that is, $Q < C$).
- The capacity to transmit quantum states with the assistance of a two-way classical channel is defined as Q_2, and it is in the range $Q \leq Q_2 \leq C$.
- Entanglement-assisted classical capacity C_E is defined as $Q \leq C \leq C_E$.

Channel fidelity is defined as the probability that the output of the quantum channel is the same as the input. If ρ^A and ρ^B be the density matrices of the input and output states of a quantum communication channel, then the channel fidelity $F(\rho^A, \rho^B)$ is defined as follows:

For pure states $|\varphi_A\rangle$ and $|\varphi_B\rangle$:

$$F\left(\rho^A, \rho^B\right) = |\langle \varphi_A | \varphi_B \rangle|^2 \tag{6.75}$$

In the case of mixed states,

$$F\left(\rho^A, \rho^B\right) = Tr\left(\sqrt{(\rho^A)^{1/2}\, \rho^B (\rho^A)^{1/2}}\right) \tag{6.76}$$

6.7   Quantum Communication Protocols

Quantum Key Distribution (QKD) is the protocol used to exchange security keys and other important information that needs protection. In 1984, **Charles Henry Bennett** and **Gilles Brassard** published the first QKD protocol called **BB84**. Since its publication, a number of QKD protocols have been announced, and most of them are built upon on the concepts of BB84. **E91** or **Ekert 91** is a protocol published by **Artur Ekert** in 1991 and uses quantum entanglement.

Under the QKD protocols, Alice transmits a secret key to Bob by encoding the key in photons' polarization. Eve cannot trap these photons and transmit them back to Bob without changing the photon's state. The no-cloning theorem prevents the photons from being cloned. Also, measuring the photon's state and creating a new photon in the measured state is impossible unless Eve knows the basis in which Alice prepared the photons. Alice and Bob can detect noise in the channel by introducing enough check-bits in the string, indicating a weak channel or the presence of Eve.

In this book, we shall study BB84 and Ekert 91 protocols from the theoretical perspective. Implementing these protocols in software using any of the quantum computing frameworks available today is a good exercise.

6.7.1 BB84 Protocol

Assume that Alice and Bob are two parties in communication. Alice intends to send a security key of length m bits of a large prime number N. Alice and Bob share a quantum communication channel. They also have a classical communication channel. Eve is an eavesdropper who tries to tap into the channel to learn about the security key transmitted. The BB84 protocol can be used to securely exchange the key. The m bits are embedded inside the string a described below.

The BB84 protocol is outlined in the following steps.

1. Alice chooses two strings a and b, each of length $(4 + \delta)n$ classical bits. The string a embeds the secret key that is to be transferred. The string b is randomly chosen. She then encodes them in blocks of $(4 + \delta)n$ photons. The resultant state is a tensor product of $(4 + \delta)n$ states, with each state representing the ith bit of a and b in the string.

$$|\psi\rangle = |\psi_{a\,b}\rangle^{\otimes(4+\delta)n} \qquad (6.77)$$

Each of the photons are encoded in one of the following states.

$$\begin{aligned}
|\psi_{00}\rangle &= |0\rangle; \quad a_i = 0, \ b_i = 0 \\
|\psi_{01}\rangle &= |1\rangle; \quad a_i = 0, \ b_i = 1 \\
|\psi_{10}\rangle &= |+\rangle; \quad a_i = 1, \ b_i = 0 \\
|\psi_{11}\rangle &= |-\rangle; \quad a_i = 1, \ b_i = 1
\end{aligned} \qquad (6.78)$$

This encodes the photons a based on b, either on the computational basis (Z) or on the Hadamard basis (X). With this encoding, the photons cannot be identified without the knowledge of b. Alice now sends the photons $|\psi\rangle$ to Bob over the quantum channel. Photons encoded in the X basis are polarized in vertical or horizontal directions. The photons encoded in the Z basis are polarized in diagonal or antidiagonal directions, as shown in the fragment below.

2. Let ε represent the quantum channel, which may include the channel noises and the effect of intruders like Eve. In that case, the set of photons received by Bob is $\varepsilon(|\psi\rangle\langle\psi|)$. At this step, Bob constructs a random bit string b' of length $(4 + \delta)n$ and measures the photons in the X or Z basis depending upon the value of b' for each corresponding bit position and stores the measured results in a'.

Basis:

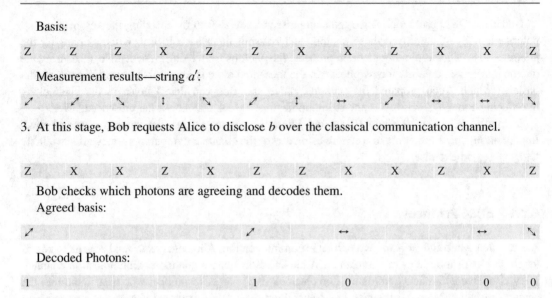

| Z | Z | Z | X | Z | Z | X | X | Z | X | X | Z |

Measurement results—string a':

3. At this stage, Bob requests Alice to disclose b over the classical communication channel.

| Z | X | X | Z | X | Z | Z | X | X | Z | X | Z |

Bob checks which photons are agreeing and decodes them.
Agreed basis:

Decoded Photons:

| 1 | | | | | 1 | | 0 | | | 0 | 0 |

At this point, Alice and Bob require at least $2n$ bits to be kept. By choosing a large value for δ, this is possible. If there are less than $2n$ bits, the protocol is aborted and repeated for a different value of b.

4. Now, Alice and Bob must perform several tests to confirm the channel noise and any eavesdropping on the channel. They divide the $2n$ bits into two subsets of n bits each. The first subset serves as the check-bits, and the second subset contains the secret key that is exchanged. Alice announces the position and value of the check-bits over the classical communication channel.

| | | | | | **1** | | **0** | | | **0** | |

Bob compares the check-bits, and if there are more than t bits that do not agree, the protocol is aborted and retried.

5. Alice and Bob perform information reconciliation and privacy amplification [5] to arrive at the final list of m bits, representing the secret key.

The BB84 protocol is well studied, and the proof of security has been established [6].

6.7.2 Ekert 91 Protocol

The Ekert 91 protocol uses a workflow similar to the BB84 protocol. Instead of a stream of photons encoded in Z or X basis, the E91 protocol uses entangled photons generated by a third-party source. Figure 6.2, illustrates this setup.

Since this protocol is based on entanglement, the correlation is not affected if Alice and Bob use an orthogonal basis for measurement. The E91 protocol requires Alice to perform the measurement in Z basis but with one of the rotational angles $0°$, $45°$, $90°$. Bob is also required to measure the photons in the same but with one of the rotational angles $45°$, $90°$, $135°$.

The protocol requires the photons to be measured in two ways. The first set of photons is measured on the same basis by Alice and Bob. The second set of photons are measured in random. Eavesdropping can be detected by computing the correlation coefficients between the measurement basis used by Alice and Bob. Recall from Eq. (6.11) that the correlation coefficient is $2\sqrt{2}$ for maximally entangled photons. If not, we can conclude that either the photons were not entangled or Eve has introduced local realism to the system, violating Bell's Theorem. If the protocol is successful, the first set of photons can be used to arrive at the keys.

For a detailed description of this protocol, refer to Nikolina Ilic's article [7].

6.8   RSA Security

Encryption is the science of converting information into a form that is readable only by the recipient. Humans have always been interested in protecting intellectual property: from protecting the art of making pottery and silk to military secrets. With the advent of the internet and social media, this has extended into protecting the privacy and personal information. Today, we do not do any transaction that is not protected. From the period (1900 BC), when the scribes in Egypt used non-standard hieroglyphs to record secret information, cryptography has come a long way. The notable ones in the history of cryptography are the shift by three substitution ciphers used by Julius Caesar in the 1st century BC, the substitution cipher created by Gabrieli di Lavinde in 1379, the first cipher disk by Leon Battista Alberti in 1466, and so on. A good number of treatises on cryptography were also written during the medieval period.

In 1585, the French diplomat Blaise de Vigenère wrote a book on ciphers. For the first time, this book described a plaintext and ciphertext based autokey system. Nevertheless, who shall forget "The Enigma Machine" used during World War II? In the United States, the confederation, and union armies used "Cipher Disks" to protect communication (Fig. 6.7).

In modern times, the **Data Encryption Standard** (DES) was invented by IBM in 1976. It was broken in 1997. DES used a 56-bit **symmetric-key algorithm**, which means the same key is used to encrypt the *plaintext* and decrypt the *cyphertext*. As the internet bloomed and e-commerce got popular, DES was replaced by **Advanced Encryption Standard** (AES) in 2001. AES is a block cipher algorithm, which can be implemented efficiently in hardware or software. The algorithm uses a **substitution–permutation network,** which is a series of linked mathematical operations. Decryption is performed by reversing the order of the network. The standard we shall explore in this section is the

Fig. 6.7 Cipher disk used to protect Union Army communications. This disk functioned by aligning the letters on the upper two disks with the numbers on the two lower disks. National Archives, Records of the Office of the Chief Signal Officer

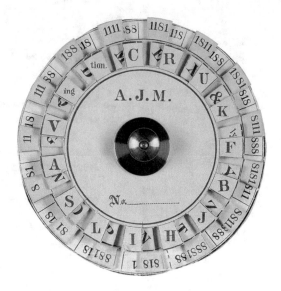

RSA algorithm, which was invented in 1977 by **Ron Rivest**, **Adi Shamir**, and **Leonard Adleman**. RSA is an asymmetric block cipher, and it is a popular public-key based cryptographic system. RSA is a crucial component of the **public key infrastructure** on which most transactions on the internet are done.

The public-key cryptographic systems help reliably transmit information on the network, which may be prone to attack. It is also used to authenticate entities using digital certificates. Being asymmetric, RSA has two dissimilar keys—the private key and the public key. The public key can be published. RSA remains secure, as long as the parties can keep the private key safe. Using RSA is simple—Alice encrypts her message, the plaintext m using the public key, and sends the encrypted message, the cyphertext c to Bob. Bob uses the private key to decipher the message.

The premise for the security of RSA comes from the fact that given a very large composite integer number N, it is considered a hard problem determining its prime factors (p and q). The **General Number Field Sieve** (GNFS) algorithm is the fastest well-known classical integer factorization algorithm. The GNFS algorithm, when performed on an AMD 2.2 GHz Opteron Processor with 2 GB of RAM takes the following time to factorize integers:

- 768 Bit Number (232 Digits, RSA-768) [8]—1500 years
- 1024 Bit Number—1.5 Million years
- DigiCert 2048-bit SSL certificate 617 digit number [9]—6.4 Quadrillion years

The steps outlined in Fig. 6.8 illustrates the workflow for the RSA algorithm.

6.8.1 Modular Exponentiation

Assume a large composite integer number N with two distinct primes p and q, such that $N = p * q$. Then, according to the **number theory**, there exists a **periodic function** $\mathcal{F}(x)$ with period r, such that, $\mathcal{F}(x) = a^x \bmod N$, where a is an integer, $a \neq N$, coprime with p or q and coprime with N. The period $r = \text{lcm}((p-1),(q-1))$ is defined by **Carmichael's Totient function**. Here lcm stands for least common multiple.

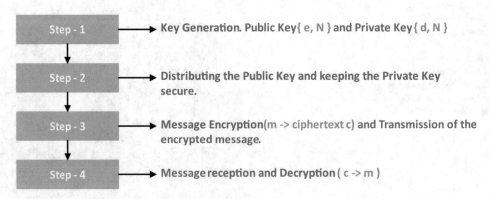

Fig. 6.8 The RSA algorithm

Table 6.1 The periodic function

$f(x_0)$	$f(x_1)$	$f(x_2)$
$f(1) = 5^1 \ mod \ 21 = 5$	$f(7) = 5^7 \ mod \ 21 = 5$	$f(13) = 5^{13} \ mod \ 21 = 5$
$f(2) = 5^2 \ mod \ 21 = 4$	$f(8) = 5^8 \ mod \ 21 = 4$	$f(14) = 5^{14} \ mod \ 21 = 4$
$f(3) = 5^3 \ mod \ 21 = 20$	$f(9) = 5^9 \ mod \ 21 = 20$	$f(15) = 5^{15} \ mod \ 21 = 20$
$f(4) = 5^4 \ mod \ 21 = 16$	$f(10) = 5^{10} \ mod \ 21 = 16$	$f(16) = 5^{16} \ mod \ 21 = 16$
$f(5) = 5^5 \ mod \ 21 = 17$	$f(11) = 5^{11} \ mod \ 21 = 17$	$f(17) = 5^{17} \ mod \ 21 = 17$
$f(6) = 5^6 \ mod \ 21 = 1$	$f(12) = 5^{12} \ mod \ 21 = 1$	$f(18) = 5^{18} \ mod \ 21 = 1$

Let us illustrate this with an example.

$$Let \ N = 21, \quad p = 7, \quad q = 3$$
$$Then, \quad N = p \times q = 7 \times 3 = 21$$
$$r = \text{lcm}((7 - 1), (3 - 1)) = \text{lcm}(6, 2) = 6 \quad\quad (6.79)$$
$$Assume, \quad a = 5$$

For the above values, Table 6.1 lists the first three periods of the periodic function. This table can be verified easily in WolframAlpha.

From the table, the period $r = 6$, as we calculated in Eq. (6.79). Now that we are clear about the periodic function, let us perform the inner workings of the RSA by taking slightly large numbers.

1. Select p and q.	Select two large prime numbers in random.	$p = 23$ $q = 31$
2. Calculate N	Calculate N, the composite number. $N = p \times q$	$N = 23 \times 31 = 713$
3. Find r The period	Calculate period r using Carmichael's totient function: $r = lcm\,((\,p - 1\,)\,(\,q - 1\,))$	$r = lcm\,((23 - 1\,),\,(31 - 1))$ $= lcm\,(\,22,\,30)$ $= 330$
4. Find e Public exponent	Calculate the public exponent e as follows: e is coprime to r $e < r$ $gcd\,(\,e, r\,) = 1$	Assume $e = 19$ $gcd\,(\,19,\,339\,) = 1$
5. Find d Private exponent	Calculate the private exponent d using modular inverse: $d \times e \ mod \ r = 1$ $d = e^{-1} \ mod \ r$	$19^{-1} \ mod \ 330 = 139$ $d = 139$

Now, we have the public exponent $e = 19$, and the private exponent $d = 139$. With this, we can try to perform an encryption and decryption operation of the alphabet "A."

Encryption	Public Key { e, N } = { 19, 713 } m = 65 → ASCII code for letter "A"	$c = m^e \ mod \ N$ $c = 65^{19} \ mod \ 713$ $c = 198$
Decryption	Private Key { d, N } = { 139, 713 } c = 198	$m = c^d \ mod \ N$ $m = 198^{139} \ mod \ 713$ $m = 65$

6.8.2 Period Estimation

- If p and q are known, it is easy to find r. But if only N is known, it is a hard problem. The hard problem of finding the period r of $\mathcal{F}(x) = a^x \bmod N$ ensures the security of RSA.
- Once r is known, it is easy to calculate the Private Exponent d using the formula $d = e^{-1} \bmod r$, and the RSA security is broken (where e is the Public Exponent. It is part of the Public Key $\{\ e, N\ \}$, and usually 65537.)
- Finding the value of r is called *order finding or period estimation*. Shor's algorithm (c. 1994) is a Quantum version of period estimation. We shall study Shor's algorithm in the next chapter.

6.9   Summary

This chapter focused on quantum communications. We learned about EPR Paradox, density matrix formalism, von Neumann entropy, creation of entangled photons, quantum communication protocols, and RSA security.

In this chapter, some concepts like Bell measurement, Bell's inequalities, verification of entanglement, reduced density matrices, and no-communication theorem were explored. This chapter also focused on Schumacher's compression and Schmidt decomposition.

We shall return to RSA security in the following chapter, which focuses on quantum algorithms.

Practice Problems

1. Differentiate between pure and mixed states. How are the density matrices defined for pure and mixed states?
2. Establish that the density operator and the partial trace are independent of the orthonormal basis used to calculate them.
3. Prove that the tensor product of two density operators is also a density operator.
4. Develop a quantum circuit to verify the GHZ state.
5. Explain whether the Bell state is a mixed state or a pure state? Using reduced density operators, establish that one of the qubits is in a mixed state.
6. What is the Schmidt number of a product state? Explain your answer.
7. Calculate the degree of entanglement between two qubits described by the quantum state $|\psi\rangle = \frac{1}{4}|00\rangle + \frac{1}{4}|01\rangle + \frac{1}{4}|10\rangle + \frac{1}{4}|11\rangle$.
8. A qubit in state $|\psi\rangle = a|0\rangle + b|1\rangle$ starts to decohere. If τ is its decoherence time, what is its density matrix.
9. Explain quantum teleportation. How is it different from sending classical bits over large distances on the internet? Can information be sent at speeds higher than the speed of light?
10. Simulate a quantum communication channel by implementing the BB84 Protocol as a hybrid software that runs partially on a classical computer and partially on a quantum computer.
11. In the above experiment, assume that Eve tries to listen to the communication between Alice and Bob. Eve uses a fixed basis to measure the qubits in transmission. She then prepares new qubits in the measured state and puts them in the transmission channel. How much of information can Eve retrieve at each stage in the protocol? How does this disrupt the communication between Alice and Bob?

12. Factorize 4331.
13. Write an article on post-quantum cryptographic methods under consideration.
14. Can we create a quantum channel that can remember the past information? Does such a "cache" even matter in quantum communications? Explain your answer.

References

1. Physics: Bell's theorem still reverberates, https://www.nature.com/news/physics-bell-s-theorem-still-reverberates-1. 15435
2. Polarization Correlation of Photons Emitted in an Atomic Cascade, Kocher et al, Physical Review Letters, 10 April 1967.
3. The Non-Signalling theorem in generalizations of Bell's theorem. J Walleczek and G Grössing 2014 J. Phys.: Conf. Ser. 504 012001. https://iopscience.iop.org/article/10.1088/1742-6596/504/1/012001/pdf
4. https://en.wikipedia.org/wiki/No-communication_theorem
5. https://en.wikipedia.org/wiki/Quantum_key_distribution#Information_reconciliation_and_privacy_amplification
6. Simple Proof of Security of the BB84 Quantum Key Distribution Protocol, Peter W. Shor and John Preskill, Physical Review Letters, 10 July 2000, http://www.theory.caltech.edu/~preskill/pubs/preskill-2000-proof.pdf
7. The Ekert Protocol, Nikolina Ilic, Journal of Physics, July 22, 2007, https://www.ux1.eiu.edu/~nilic/Nina's-article. pdf
8. Factorization of a 768-bit RSA modulus, Version 1.4, February 18, 2010, Thorsten Kleinjung et al, https://eprint.iacr. org/2010/006.pdf
9. https://www.digicert.com/blog/moving-beyond-1024-bit-encryption/

Quantum Algorithms

7

> *"Why," said the Dodo, "the best way to explain it is to do it."*
> — *Lewis Carroll, Alice's Adventures in Wonderland*

This chapter focuses on exploring some of the fundamental quantum algorithms. We start this chapter by building a simple ripple adder circuit, an extension to the full adder circuit we learned in Chap. 5. The purpose of introducing the ripple adder circuit is to increase our exposure in building layered quantum circuits. The underlying math work helps us understand how to apply unitary transformations in each stage—qubit by qubit, and evolve the final quantum state.

After completing this exercise, we learn about the quantum Fourier transform (QFT), an essential milestone in our learning. After that, we learn about oracle functions and study a few oracle-based algorithms such as the Deutsch–Jozsa algorithm, Bernstein–Vazirani problem, and Simon's algorithm. We then return to QFT and apply it to perform arithmetic! Grover's search algorithm is an interesting concept, which employs amplitude magnification. After exploring Grover's search, we learn about Shor's algorithm and factor the number 15 using a fixed quantum circuit. We then build a quantum version of the k-means algorithm, widely used in unsupervised machine learning. Another concept introduced in this chapter is quantum phase estimation and its application in solving linear equations.

Throughout this chapter, we mix theory, math, and practice. Step-by-step derivation of the math is provided to help improve understanding. Working codes accompany the algorithms explained in this chapter. Experiment with the code, analyze, optimize, measure, and find bugs!

We complete this chapter by looking at the complexity theory.

▶ Learning Objectives

- Quantum ripple adder
- Oracle-based algorithms: Deutsch–Jozsa, Bernstein–Vazirani, and Simon's problem
- Quantum Fourier transform and its applications in performing arithmetic in the Fourier basis
- Amplitude magnification and Grover's search algorithm
- Modular exponentiation and integer factorization using Shor's algorithm
- Quantum k-means algorithm
- Quantum phase estimation and HHL algorithm for solving linear system of equations
- Quantum complexity theory

© The Author(s), under exclusive license to Springer Nature Switzerland AG 2021
V. Kasirajan, *Fundamentals of Quantum Computing*, https://doi.org/10.1007/978-3-030-63689-0_7

7.1   Quantum Ripple Adder Circuit

In Chap. 5, we implemented a quantum full adder circuit. In this section, we shall extend that concept, and implement a reusable in-place ripple adder that adds two 2-bit positive integers. The logic outlined in this section [1] can be extended to add two integers of any size. The total number of qubits required is $2n + 2$, where n is the number of bits required to represent the unsigned integers. We are not using any sign bits in this exercise. Implementing sign bits is left as an exercise. Adding numbers is not the best use of quantum computers. We shall use this as an exercise to learn the construction of complex circuits.

Let us assume two n-bit positive integers with the binary representation, $a = a_{n-1} \ldots a_2 a_1 a_0$ and $b = b_{n-1} \ldots b_2 b_1 b_0$. Let us also define the carry $c = c_{n-1} \ldots c_2 c_1 c_0$. At a given level i, the circuit adds the corresponding bits a_i, b_i, c_i and produces the carry c_{i+1}, and the sum s_i. The carry bit generated at level i is an input to the next level $i + 1$. This cycle repeats through all the bits. Hence, this circuit is called the "**ripple adder**." From Chap. 5, Eqs. (5.102) and (5.103), we can define the sum and carry as follows:

$$s_i = a_i \oplus b_i \oplus c_i \tag{7.1}$$

$$c_{i+1} = (a_i \wedge b_i) + (a_i \oplus b_i) \wedge c_i \tag{7.2}$$

Let us define two functions MAJ and UMS.

The majority function (MAJ) is implemented, as shown in Fig. 7.1. It uses two CNOT gates, and one Toffoli gate. Let us calculate the output of this function.

The initial state of the circuit can be written as follows:

$$\psi = |c_i\rangle |b_i\rangle |a_i\rangle \tag{7.3}$$

At the start of the circuit, we apply a CNOT gate to the target qubit b_i with a_i as the control qubit. The new state is:

$$\psi = |c_i\rangle |a_i \oplus b_i\rangle |a_i\rangle \tag{7.4}$$

The next step is to apply a CNOT gate to the target qubit c_i with a_i as the control qubit.

$$\psi = |a_i \oplus c_i\rangle |a_i \oplus b_i\rangle |a_i\rangle \tag{7.5}$$

The final step is to apply a Toffoli gate to qubit a_i, with qubits c_i and b_i as the control qubits.

$$\psi = |a_i \oplus c_i\rangle |a_i \oplus b_i\rangle |((a_i \oplus c_i) \wedge (a_i \oplus b_i)) \oplus a_i\rangle \tag{7.6}$$

The third term in this equation can be written as (for brevity, we skip this step, but it can be easily worked out):

Fig. 7.1 The majority function

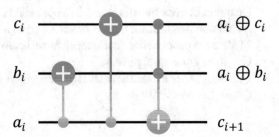

Table 7.1 Truth table of the third term in Eq. (7.7)

A	B	C	a ∧ b	a ∧ c	b ∧ c	(a ∧ b) ⊕(a ∧ c) ⊕ (b ∧ c)
0	0	0	0	0	0	0
0	1	0	0	0	0	0
1	0	0	0	0	0	0
1	1	0	1	0	0	1
0	0	1	0	0	0	0
0	1	1	0	0	1	1
1	0	1	0	1	0	1
1	1	1	1	1	1	1

Fig. 7.2 The unmajority and sum function

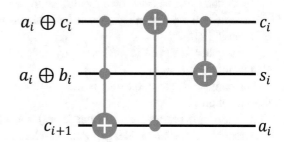

$$\psi = |a_i \oplus c_i\rangle \, |a_i \oplus b_i\rangle \, |a_i b_i \oplus a_i c_i \oplus b_i c_i\rangle \tag{7.7}$$

By comparing the last column of Table 7.1 with the last two columns of Table 5.4 the truth table of the full adder in Chap. 5, we can infer that the third term in Eq. (7.7) calculates the c_{i+1} of this step. We can, therefore, rewrite Eq. (7.7).

$$\psi = |a_i \oplus c_i\rangle \, |a_i \oplus b_i\rangle \, |c_{i+1}\rangle \tag{7.8}$$

The unmajority-sum (UMS) function performs the in-place addition, and it is implemented, as shown in Fig. 7.2.

The first step is to perform a Toffoli gate on the third qubit, with the first and second qubits as the control qubits. To make the math understandable, we must expand the third term in Eq. (7.7) and apply the Toffoli operation.

$$\psi = |a_i \oplus c_i\rangle|a_i \oplus b_i\rangle|((a_i \oplus c_i) \wedge (a_i \oplus b_i)) \oplus (a_i b_i \oplus a_i c_i \oplus b_i c_i). \tag{7.9}$$

This equation simplifies to (this can be worked out easily on an online Boolean expression evaluator [2]) the following:

$$\psi = |a_i \oplus c_i\rangle \, |a_i \oplus b_i\rangle \, |a_i\rangle \tag{7.10}$$

The second step is to perform the CNOT of the first qubit, with the third qubit as the control. We get,

$$\psi = |a_i \oplus (a_i \oplus c_i)\rangle \, |a_i \oplus b_i\rangle \, |a_i\rangle \tag{7.11}$$

The equation simplifies to:

$$\psi = |\, c_i\rangle \, |a_i \oplus b_i\rangle \, |a_i\rangle \tag{7.12}$$

The third step is to perform a CNOT of the second qubit, with the first qubit as the control. We can write Eq. (7.12) as:

$$\psi = |\, c_i\rangle\, |c_i \oplus (a_i \oplus b_i)\rangle\, |a_i\rangle \tag{7.13}$$

By comparing this equation with Eq. (5.102), we can write the next step:

$$\psi = |\, c_i\rangle\, |s_i\rangle\, |a_i\rangle \tag{7.14}$$

We note that this method calculates the sum s_i in place of b_i and retrieves c_i and a_i.

We can create a ripple adder by connecting the MAJ and UMS functions in a sequence. The bit c_0 can be set as "0" if there is no incoming carry. It can be set as "1," if there is an incoming carry. The outgoing carry can be calculated by performing a CNOT of the output qubit z with c_{n+1}.

Assume two 2-bit numbers A and B. Their binary representations are described by qubits q[1] q[2] and q[3]q[4].

For number A: the qubit q[2] is the least significant qubit ($i = 0$), and the qubit q[1] is the most significant qubit ($i = 1$).

For number B: the qubit q[4] is the least significant bit ($i = 0$), and the qubit q[3] is the most significant bit ($i = 1$).

Assume q[0] represents C_{IN} (that is, c_0) and q[5] represents C_{OUT}. Figure 7.3 illustrates the construction of the 2-bit adder using the MAJ and UMS functions.

This circuit can be implemented in any of the development platforms. Type the code fragment from Listing 7.1 in a new cell on your Jupyter Notebook for Qiskit. Copy the import sections from previous work. Use "qasm_simulator" as your backend.

In this code fragment, we are creating two composite gates. These two composite gates are implemented as independent quantum circuits. The 'to_instruction()' method helps the composite gates to be appended as instructions to the target circuits. This level of modularization helps us develop and manage code easily.

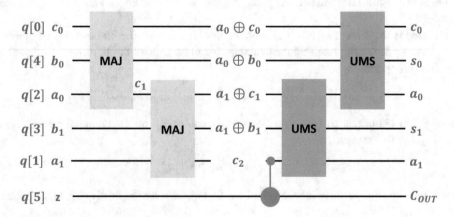

Fig. 7.3 The 2-bit ripple adder circuit with the MAJ and UMS functions

```
# Create a composite gate for the MAJ function
maj_q = QuantumRegister(3)
maj_circ = QuantumCircuit(maj_q, name='maj_func')
maj_circ.cx(maj_q[2], maj_q[1])
maj_circ.cx(maj_q[2], maj_q[0])
maj_circ.ccx(maj_q[0], maj_q[1], maj_q[2])

# Convert this to a composite gate
maj_inst = maj_circ.to_instruction()

# Create a composite gate for the UMS function
ums_q = QuantumRegister(3)
ums_circ = QuantumCircuit(ums_q, name='ums_func')
ums_circ.ccx(ums_q[0], ums_q[1], ums_q[2])
ums_circ.cx(ums_q[2], ums_q[0])
ums_circ.cx(ums_q[0], ums_q[1])

# Convert this to a composite gate
ums_inst = ums_circ.to_instruction()
```

Listing 7.1 The MAJ and UMS functions

Listing 7.2 implements the 2-bit adder circuit.

In this circuit, we create a 6-qubit quantum register and a 3-bit classical register. We use bitwise AND operation to set the initial conditions of the qubits. The classical registers are used to project the sum ($c[1]c[0]$) and carry $c[2]$. Note that this is a 2-bit adder. Hence, the inputs cannot be >3 (Fig. 7.4).

The equivalent QDK code can be found in Listing 7.3.

Exercise 7.1

1. *Write unit testing code to prove that this function works.*
2. *Extend this function to support 8-bit positive integers.*
3. *Extend this function to support negative integers.*
4. *What is the circuit depth?*

◀

```
# Implements a two-bit ripple adder
# Input1 and Input2 -> the two, 2-bit unsigned integers
# CarryIn - incoming carry
# backend - processor to run the circuit.
# shots    - number of times, the experiment is to be repeated.

def   TwoBitRippleAdder   (Input1,   Input2,   CarryIn,\
                           Backend, shots=1024):
    q = QuantumRegister(6, 'q')
    c = ClassicalRegister(3, 'c')
    qc = QuantumCircuit(q, c)

    if (CarryIn == 1):  qc.x(q[0])

    if (Input1 & 0x01): qc.x(q[2])
    if (Input1 & 0x02): qc.x(q[1])

    if (Input2 & 0x01):  qc.x(q[4])
    if (Input2 & 0x02):  qc.x(q[3])

    qc.append(maj_inst, [q[0], q[4],q[2]])
    qc.append(maj_inst, [q[2], q[3],q[1]])

    qc.cx(q[1], q[5])

    qc.append(ums_inst, [q[2], q[3],q[1]])
    qc.append(ums_inst, [q[0], q[4],q[2]])

    qc.measure(q[4], c[0])
    qc.measure(q[3], c[1])
    qc.measure(q[5], c[2])

    job = execute(qc, backend=backend, shots=shots)
    result = job.result()
    counts = result.get_counts(qc)

    return counts
```

Listing 7.2   2-bit quantum ripple adder

7.2   Quantum Fourier Transformation

7.2.1   Classical Fourier Transformation

The Fourier transform is a reversible linear transformation function. It converts a signal $g(t)$ in the time domain into a signal $G(f)$ in the frequency domain. Due to this property, the Fourier transform has several applications in signal and image processing. The Fourier transforms are key components

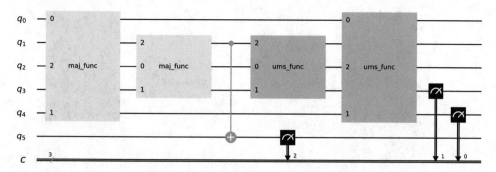

Fig. 7.4 The quantum circuit for the 2-bit adder, as constructed using Qiskit

in period finding (which we shall apply in a later section) application and RF signal receivers. Equation (7.15) describes the Fourier transform for **aperiodic signals**. Aperiodic signals are signals that don't repeat itself after a period of time, but they can also be signals with an infinite period, that is, a signal is periodic if $x(t) = x(t + T_0)$, where T_0 is the period.

$$X(\omega) = \mathcal{F}[x(t)] = \int_{-\infty}^{+\infty} x(t)e^{-i\omega t}\, dt, \tag{7.15}$$

where $\omega = 2\pi f$, f is the frequency, and t is the time. This equation is also known as the **forward transform**.

The inverse of the Fourier transform is described in the following equation: **inverse transform**.

$$x(t) = \mathcal{F}^{-1}[X(\omega)] = \int_{-\infty}^{+\infty} X(\omega)e^{i\omega t} d\omega \tag{7.16}$$

By comparing the exponential term with the Euler's formula[1] $e^{j\omega t} = \cos \omega t + j \sin \omega t$, we can conclude that complex sinusoidal functions are the basis functions of the signal. This term can also be considered as a unit vector rotating in a complex plane at a rate of ω radians per second. Figure 7.6 illustrates this concept. Note that $\cos\omega t$ and $\sin\omega t$ are projections of this vector along the real and imaginary axes. From these two equations, we can interpret $x(t)$ as the weighted sum $X(\omega)$ of $e^{j\omega t}$ components of frequencies ω.

7.2.2 Discrete Fourier Transform (DFT)

The Discrete Fourier Transform is equivalent to the Continuous Fourier Transform described in the previous section. However, it applies to N samples of the signals taken at T time intervals. Figure 7.5 illustrates the DFT of a sinusoidal function with a period t into its constituent frequency.

Let us assume $x(t)$ be the continuous signal under consideration. We take N samples of this signal $x[0]$, $x[1]$, $x[2]$, $\ldots$, $x[k]$, $\ldots$, $x[N-1]$. Each sample $x[k]$ is taken after a time T from the previous sample $x[k-1]$. The signal is periodic. Hence, we can assume that the data in the range

---

[1] Between literature, the unit imaginary number may be represented as i or j, interchangeably.

```
// Implement the Majority function
// q0 : Input c
// q1 : Input b
// q2 : Input a

operation MajFunction (q0:Qubit, q1:Qubit, q2:Qubit) : Unit {
   CNOT(q2, q1);
   CNOT(q2, q0);
   CCNOT(q0, q1,q2);
}


// Implement the UMS function
// q0 : Input c
// q1 : Input b
// q2 : Input a

operation UmsFunction (q0:Qubit, q1:Qubit, q2:Qubit) : Unit {
   CCNOT(q0, q1, q2);
   CNOT(q2, q0);
   CNOT(q0, q1);
}


// Implement the 2-bit adder
// Input1 and Input2 - the 2 bit unsigned integers to add
// Cin - Carry In
// Returns: the sum of the inputs.

operation TwoBitAdder( Input1: Int, Input2: Int, Cin:Int) : Int {
   mutable r = 0;
   using ( q = Qubit[6] )
   {
       // Setup initial conditions
       if ( Cin != 0 ) { X(q[0]); }
       if ( (Input1 &&& 0x01 ) == 0x01 ) { X(q[2]); }
       if ( (Input1 &&& 0x02 ) == 0x02 ) { X(q[1]); }
       if ( (Input2 &&& 0x01 ) == 0x01 ) { X(q[4]); }
       if ( (Input2 &&& 0x02 ) == 0x02 ) { X(q[3]); }

       // Setup the MAJ functions
       MajFunction( q[0], q[4], q[2] );
       MajFunction( q[2], q[3], q[1] );

       // The Carry-OUT bit
       CNOT(q[1], q[5]);

       // The Setup the UMS Functions
       UmsFunction( q[2], q[3], q[1] );
       UmsFunction( q[0], q[4], q[2] );
```

Listing 7.3 2-bit ripple adder using QDK.

```
            // Perform the measurements and order
            // the bits to the result
        if (One == M(q[4])) { set r = r ||| 0x01; }
        if (One == M(q[3])) { set r = r ||| 0x02; }
        if (One == M(q[5])) { set r = r ||| 0x04; }

        ResetAll(q);
    }

    Message($"We got {r} expected: {Input1 + Input2 + Cin}, " + \
            $"Data: Input1 {Input1} Input2 {Input2} Carry {Cin}.");

    return r;
}
```

Listing 7.3 (continued)

Fig. 7.5 The discrete Fourier transform. (**a**) A sine wave with a period "*t*" picturized in the time domain. (**b**) The Discrete Fourier transform of the sine wave picturized in the frequency domain

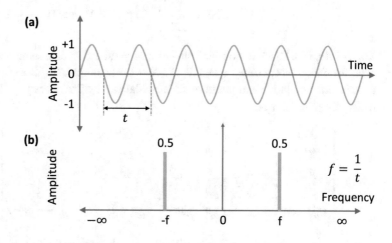

Fig. 7.6 The rotating vector $e^{j\omega t}$

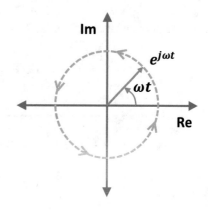

$x[0], \ldots, x[N-1]$ are the same as the data in the range $x[N], \ldots, x[2N-1]$. Since this is a sampled data, it exists only at specific times. Hence, we can write Eq. (7.15) as:

$$X(\omega) = \int_0^{(N-1)T} x(t)e^{-i\omega t}dt \tag{7.17}$$

$$= x[0]e^{-i0} + x[1]e^{-i\omega T} + \ldots + x[k]e^{-i\omega kT} + \ldots + x[N-1]e^{-i\omega(N-1)T} \tag{7.18}$$

This equation can be rewritten as a summation.

$$X(\omega) = \sum_{k=0}^{(N-1)} x[k]e^{-i\omega kT} \tag{7.19}$$

The sample data are periodic. Hence, we should calculate the DFT for the fundamental frequency $\frac{2\pi}{NT}$ rad/sec and its harmonics, that is, the values of ω we should evaluate are:

$$\omega = 0, \frac{2\pi}{NT} \times 1, \frac{2\pi}{NT} \times 2, \ldots, \frac{2\pi}{NT} \times n, \ldots, \frac{2\pi}{NT} \times (N-1) \tag{7.20}$$

Applying this in Eq. (7.19), we get after normalization,

$$X[n] = \frac{1}{\sqrt{N}} \sum_{k=0}^{(N-1)} x[k]e^{-i\frac{2\pi}{N}n\cdot k}, n = 0 : N-1 \tag{7.21}$$

$X[n]$ is called the Discrete Fourier Transform of the sampled data sequence $x[k]$. The sampling is done at uniform intervals of time $t = 0, 1, 2, \ldots, N-1$. The output $X[n]$ is a vector of complex numbers that encodes the amplitude and phase of a sinusoidal wave of frequency $\frac{n}{N}$. The DFT can be written in a matrix form as follows:

$$\begin{bmatrix} X[0] \\ X[1] \\ X[2] \\ \vdots \\ X[N-1] \end{bmatrix} = \frac{1}{\sqrt{N}} \begin{bmatrix} \omega^0 & \omega^0 & \omega^0 & \cdots & \omega^0 \\ \omega^0 & \omega^1 & \omega^2 & \cdots & \omega^{(N-1)} \\ \omega^0 & \omega^2 & \omega^4 & \cdots & \omega^{2(N-1)} \\ \omega^0 & \vdots & \vdots & \cdots & \vdots \\ \omega^0 & \omega^{(N-1)} & \omega^{2(N-1)} & \cdots & \omega^{(N-1)(N-1)} \end{bmatrix} \begin{bmatrix} x[0] \\ x[1] \\ x[2] \\ \vdots \\ x[N-1] \end{bmatrix}, \tag{7.22}$$

where $\omega = e^{-i\frac{2\pi}{N}}$. Note that $e^{-i\frac{\pi}{2}} = -i$. The term $\frac{1}{\sqrt{N}}$ is a normalization constant, and the DFT matrix is unitary.

The Inverse Discrete Fourier Transform is given by the following equation:

$$x[k] = \frac{1}{N} \sum_{k=0}^{(N-1)} X[n]e^{i\frac{2\pi}{N}nk}, n = 0 : N-1 \tag{7.23}$$

Example 7.1

A sine wave is sampled every 0.25 seconds. The samples are $\{0, 1, 0, -1\}$. Calculate the DFT.

◀

Solution

Using Eq. (7.22), we can write

$$
\begin{bmatrix} X[0] \\ X[1] \\ X[2] \\ X[3] \end{bmatrix} = \frac{1}{\sqrt{4}} \begin{bmatrix} 1 & 1 & 1 & 1 \\ 1 & -i & -1 & i \\ 1 & -1 & 1 & -1 \\ 1 & i & -1 & -i \end{bmatrix} \begin{bmatrix} 0 \\ 1 \\ 0 \\ -1. \end{bmatrix} = \frac{1}{2} \begin{bmatrix} 0 \\ -2i \\ 0 \\ 2i \end{bmatrix} = \begin{bmatrix} 0 \\ -i \\ 0 \\ i \end{bmatrix}
\tag{7.24}
$$

7.2.3 Quantum Fourier Transformation

Recall from our discussions in the earlier section that the Discrete Fourier Transform takes a vector $x[N] \in \mathbb{C}^N$ and maps it into a vector $X[N] \in \mathbb{C}^N$. Similarly, the Quantum Fourier Transform acts on a quantum state $|x\rangle = \sum_{i=0}^{N-1} x_i |i\rangle$ and maps it into a state $\sum_{i=0}^{N-1} X_i |i\rangle$. This transformation is given by the following equation:

$$
X_k = \frac{1}{\sqrt{N}} \sum_{n=0}^{(N-1)} x_n \, \omega_N^{kn}, k = 0 : N - 1,
\tag{7.25}
$$

where $\omega_N = e^{\frac{2\pi i}{N}}$, which is called the primitive N^{th} root of unity. Note that we are using a phase factor with a positive sign, and it is similar to the Inverse Discrete Fourier Transform. Assuming $|x\rangle$ is a basis state, the QFT can be written as the following mapping:

$$
|x\rangle = \frac{1}{\sqrt{N}} \sum_{y=0}^{(N-1)} \omega_N^{xy} |y\rangle
\tag{7.26}
$$

In the matrix form, the QFT can be written as a unitary matrix.

$$
U_{\text{QFT}} = \frac{1}{\sqrt{N}} \sum_{x=0}^{N-1} \sum_{y=0}^{N-1} \omega_N^{xy} |y\rangle \langle x|
$$

$$
= \frac{1}{\sqrt{N}} \begin{bmatrix} 1 & 1 & 1 & 1 & \cdots & 1 \\ 1 & \omega^1 & \omega^2 & \omega^3 & \cdots & \omega^{(N-1)} \\ 1 & \omega^2 & \omega^4 & \omega^6 & \cdots & \omega^{2(N-1)} \\ 1 & \omega^3 & \omega^6 & \omega^9 & \cdots & \omega^{3(N-1)} \\ \vdots & \vdots & \vdots & \vdots & \vdots & \vdots \\ 1 & \omega^{(N-1)} & \omega^{2(N-2)} & \omega^{3(N-1)} & \cdots & \omega^{(N-1)(N-1)} \end{bmatrix}
\tag{7.27}
$$

For a value of $N = 4$, we can show that $\omega^0 = 1, \omega^1 = i, \omega^2 = -1$, and $\omega^3 = -i$.

Now, let us assume a qubit in a state $|\psi\rangle = a \, |0\rangle + b \, |1\rangle$. For this qubit $x_0 = a$, and $x_1 = b$. From Eq. (7.25), we can calculate the QFT as follows:

$$X_0 = \frac{1}{\sqrt{2}}\left(a \cdot e^{2\pi i \frac{0\times 0}{2}} + a \cdot e^{2\pi i \frac{0\times 1}{2}}\right), k = 0, N = 2$$

$$= \frac{1}{\sqrt{2}}(a \cdot e^0 + b \cdot e^0) \tag{7.28}$$

$$= \frac{1}{\sqrt{2}}(a + b)$$

$$X_1 = \frac{1}{\sqrt{2}}\left(a \cdot e^{2\pi i \frac{1\times 0}{2}} + a \cdot e^{2\pi i \frac{1\times 1}{2}}\right), k = 1, N = 2$$

$$= \frac{1}{\sqrt{2}}(a \cdot e^0 + b \cdot e^{\pi i}), \text{ we know that } e^{\pi i} = -1, \ e^0 = 1. \tag{7.29}$$

$$= \frac{1}{\sqrt{2}}(a - b)$$

When the QFT is applied to the wavefunction, we get:

$$U_{QFT} |\psi\rangle = \frac{1}{\sqrt{2}}(a + b) |0\rangle + \frac{1}{\sqrt{2}}(a - b) |1\rangle \tag{7.30}$$

We can prove that this operation is equal to an H gate operation on the qubit. From Eq. (5.9), we can write the following operation:

$$H|\psi\rangle = \left(\frac{|0\rangle + |1\rangle}{\sqrt{2}} \langle 0| + \frac{|0\rangle - |1\rangle}{\sqrt{2}} \langle 1| \right) a |0\rangle + b |1\rangle$$

$$= \frac{1}{\sqrt{2}}(a(|0\rangle + |1\rangle) + b(|0\rangle - |1\rangle)) \tag{7.31}$$

$$= \frac{1}{\sqrt{2}}((a + b)|0\rangle + (a - b)|1\rangle)$$

By comparing the above two equations, we can conclude that the H gate performs a QFT on $N = 2$ states.

Recall that the H gate transforms the Z basis (the computational basis) into the X basis (the polar basis, or the $\{|+\rangle, |-\rangle\}$ basis). Similarly, we can define the QFT as the transformation from the computational basis to the **Fourier basis**. With this assumption, we can rewrite Eq. (7.30) as:

$$U_{QFT} |\psi\rangle \equiv \tilde{a} |0\rangle + \tilde{b} |1\rangle \tag{7.32}$$

We use the tilde accent ($\tilde{}$) to denote the Fourier basis.

Let us now learn how to calculate the QFT for large states $N = 2^n$. This will be useful in constructing multi-qubit QFT circuits. Assume the state $|x\rangle = |x_1 x_2 x_3 \ldots x_n\rangle$. In this convention, x_1 is the most significant qubit. We start from Eq. (7.26)

$$\text{QFT}_N|x\rangle = \frac{1}{\sqrt{N}} \sum_{y=0}^{(N-1)} \omega_N^{xy} |y\rangle$$

$$= \frac{1}{\sqrt{2^n}} \sum_{y=0}^{(N-1)} e^{2\pi i \frac{xy}{2^n}} |y\rangle \tag{7.33}$$

Expanding this for y in the computational basis $\{0, 1\}$, the range of the sum becomes $y_1, y_2 \ldots y_n \in \{0, 1\}^n$. Also, we can write y as a binary number:

$$y = y_1 2^{n-1} + y_2 2^{n-2} + y_3 2^{n-3} + \ldots + y_n 2^0.$$

Applying this logic, we get:

$$\text{QFT}_N |x\rangle = \frac{1}{\sqrt{2^n}} \sum_{y_1, y_2 \ldots y_n \in \{0,1\}} \exp\left(\frac{2\pi i}{2^n} \cdot x \cdot \sum_{k=1}^{n} 2^{n-k} y_k\right) |y_1 y_2 \ldots y_n\rangle \tag{7.34}$$

The exponential of the sum over k can be rewritten as the product of exponentials:

$$= \frac{1}{\sqrt{2^n}} \sum_{y_1, y_2 \ldots y_n \in \{0,1\}} \otimes_{k=1}^{n} \left(\exp\left(\frac{2\pi i x}{2^n} \cdot 2^{n-k} y_k\right) |y_k\rangle\right) \tag{7.35}$$

By rearranging the sum and product operators, we get:

$$= \frac{1}{\sqrt{2^n}} \otimes_{k=1}^{n} \left(\sum_{y_k \in \{0,1\}} \exp\left(\frac{2\pi i x}{2^n} \cdot 2^{n-k} y_k\right) |y_k\rangle\right) \tag{7.36}$$

Expanding the sum part of this equation,

$$= \frac{1}{\sqrt{2^n}} \otimes_{k=1}^{n} \left(\exp\left(2\pi i x \cdot \frac{0}{2^k}\right) |0\rangle + \exp\left(2\pi i x \cdot \frac{1}{2^k}\right) |1\rangle\right)$$

$$= \frac{1}{\sqrt{2^n}} \otimes_{k=1}^{n} \left(|0\rangle + \exp\left(\frac{2\pi i x}{2^k}\right) |1\rangle\right) \tag{7.37}$$

After this step, we can proceed to expand the tensor product as follows:

$$= \frac{1}{\sqrt{2^n}} \left(|0\rangle + \exp\left(2\pi i \frac{x}{2^1}\right) |1\rangle\right) \otimes$$

$$\left(|0\rangle + \exp\left(2\pi i \frac{x}{2^2}\right) |1\rangle\right) \otimes$$

$$\left(|0\rangle + \exp\left(2\pi i \frac{x}{2^3}\right) |1\rangle\right) \otimes \cdots \otimes \tag{7.38}$$

$$\left(|0\rangle + \exp\left(2\pi i \frac{x}{2^n}\right) |1\rangle\right)$$

Let us analyze the exponent part of Eq. (7.37) separately, by expanding x as a binary number.

$$\exp\left(2\pi i \frac{x}{2^k}\right) = \exp\left(2\pi i \frac{\sum_{l=1}^{n} 2^{n-l} x_l}{2^k}\right)$$

$$= \exp\left(2\pi i \sum_{l=1}^{n} 2^{n-k-l} x_l\right) \tag{7.39}$$

The summation on the exponential can be split into two parts. This helps us in separating the integer and fractional parts, and solve them separately.

$$= \exp\left(2\pi i \sum_{l=1}^{n-k} 2^{n-k-l}x_l + 2\pi i \sum_{l=n-k+1}^{n} 2^{n-k-l}x_l\right)$$

$$= \exp\left(2\pi i \sum_{l=1}^{n-k} 2^{n-k-l}x_l\right) \times \exp\left(2\pi i \sum_{l=n-k+1}^{n} 2^{n-k-l}x_l\right)$$

(7.40)

In the first exponent, we can easily find that $l \leq n - k$. This means $(n - k - l) \geq 0$. Therefore, we can say that $2^{n-k-l} \geq 0$. This implies that the first exponent is always 1. Equation (7.40) reduces to:

$$= \exp\left(2\pi i \sum_{l=n-k+1}^{n} 2^{n-k-l}x_l\right)$$

(7.41)

Expanding the summation, we get:

$$= \exp\left(2\pi i(2^{-1}x_{n-k+1} + 2^{-2}x_{n-k+2} + 2^{-3}x_{n-k+3} + \cdots + 2^{-n}x_n)\right)$$

$$= \exp\left(2\pi i\left(\frac{x_{n-k+1}}{2^1} + \frac{x_{n-k+2}}{2^2} + \frac{x_{n-k+2}}{2^2} + \cdots + \frac{x_n}{2^n}\right)\right)$$

(7.42)

Recall that a binary fraction can be expanded as follows:

$$0.x_1x_2x_3\ldots x_n = \frac{x_1}{2^1} + \frac{x_2}{2^2} + \frac{x_3}{2^3} + \ldots + \frac{x_n}{2^n}$$

(7.43)

Comparing the above two equations, we can rewrite the Eq. (7.37):

$$\text{QFT}_N|x\rangle = \frac{1}{\sqrt{2^n}} \otimes_{k=1}^{n}\left[\, |0\rangle + \exp\left(2\pi i\left(0.x_{n-k+1}x_{n-k+2}x_{n-k+3}\ldots x_n\right)\right)|1\rangle\right]$$

(7.44)

The final step is to expand the tensor product.

$$= \frac{1}{\sqrt{2^n}}\left[\begin{array}{c}(|0\rangle + e^{2\pi i\,(0.x_n)}|1\rangle) \otimes (|0\rangle + e^{2\pi i\,(0.x_{n-1}x_n)}|1\rangle) \\ \otimes (|0\rangle + e^{2\pi i\,(0.x_{n-2}x_{n-1}x_n)}|1\rangle) \otimes \ldots \otimes (|0\rangle + e^{2\pi i\,(0.x_1x_2\ldots x_n)}|1\rangle)\end{array}\right]$$

(7.45)

The exponents evaluate to 2-power roots of unity, and the terms are equivalent to the Hadamard transformation. Note that the last qubit depends upon all qubits in the system. As we progress towards the first qubit, the dependency reduces to itself. With this knowledge, we can construct the QFT circuit.

Consider the 4-qubit quantum circuit shown in Fig. 7.7. This circuit uses H gates and controlled rotations.

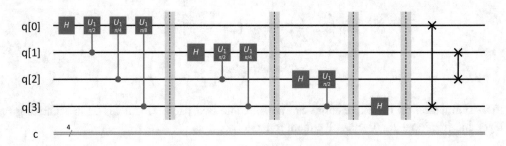

Fig. 7.7 Four-qubit QFT circuit

Recall from Eq. (5.14), that the H gate can be written as follows:

$$H|x\rangle = \frac{1}{\sqrt{2}} \left(|0\rangle + \exp\left(2\pi i \frac{x}{2^1}\right) |1\rangle \right) \tag{7.46}$$

Let us define a rotation operator R as follows:

$$R(k) = \begin{bmatrix} 1 & 0 \\ 0 & e^{\frac{2\pi i}{2^k}} \end{bmatrix} \tag{7.47}$$

Note that, in the quantum circuit, this rotation can be implemented as a U1 gate with $\lambda = \frac{2\pi}{2^k} = \frac{\pi}{2^{k-1}}$.

Let us define a controlled version of this R gate (CR). Assume a two-qubit state $|x_k x_j\rangle$ with x_k as the control qubit and x_j as the target qubit. The action of the CR gate on this 2-qubit state can be defined as follows:

$$\begin{aligned} CR\,|0\rangle|x_j\rangle &= |0\rangle|x_j\rangle \\ CR\,|1\rangle|x_j\rangle &= |1\rangle\left|e^{\frac{2\pi i}{2^k}x_j}\right\rangle \end{aligned} \tag{7.48}$$

With this definition, let us start working with our quantum circuit. The initial state of the system is as follows:

$$|\psi\rangle = |x_1\rangle \otimes |x_2\rangle \otimes |x_3\rangle \otimes |x_4\rangle \tag{7.49}$$

We begin the circuit by applying an H gate to qubit x_1. The resultant state can be written as follows:

$$\begin{aligned} |\psi\rangle &= H\,|x_1\rangle \otimes |x_2\rangle \otimes |x_3\rangle \otimes |x_4\rangle \\ &= \frac{1}{\sqrt{2}} \left(|0\rangle + \exp\left(2\pi i \frac{x_1}{2^1}\right) |1\rangle \right) \otimes |x_2\rangle \otimes |x_3\rangle \otimes |x_4\rangle \end{aligned} \tag{7.50}$$

We then apply a CR gate to the qubit x_1 with the qubit x_2 as the control qubit.

$$= \frac{1}{\sqrt{2}} \left(|0\rangle + \exp\left(2\pi i \frac{x_2}{2^2} + 2\pi i \frac{x_1}{2^1}\right) |1\rangle \right) \otimes |x_2\rangle \otimes |x_3\rangle \otimes |x_4\rangle \tag{7.51}$$

The next step is to apply a CR gate to the qubit x_1 with the qubit x_3 as the control qubit.

$$= \frac{1}{\sqrt{2}} \left(|0\rangle + \exp\left(2\pi i \frac{x_3}{2^3} + 2\pi i \frac{x_2}{2^2} + 2\pi i \frac{x_1}{2^1}\right) |1\rangle \right) \otimes |x_2\rangle \otimes |x_3\rangle \otimes |x_4\rangle \tag{7.52}$$

The final step with qubit x_1 is to perform a CR gate with it using the qubit x_4 as the control qubit.

$$= \frac{1}{\sqrt{2}} \left(|0\rangle + \exp\left(2\pi i \left(\frac{x_4}{2^4} + \frac{x_3}{2^3} + \frac{x_2}{2^2} + \frac{x_1}{2^1}\right)\right) |1\rangle \right) \otimes |x_2\rangle \otimes |x_3\rangle \otimes |x_4\rangle \tag{7.53}$$

We can now take the qubit x_2, apply an H gate first, a controlled rotation with qubit x_3 as the control qubit, and then another controlled rotation with qubit x_4 as the control qubit

$$= \frac{1}{\sqrt{2}} \left(|0\rangle + \exp\left(2\pi i \left(\frac{x_4}{2^4} + \frac{x_3}{2^3} + \frac{x_2}{2^2} + \frac{x_1}{2^1} \right) \right) |1\rangle \right) \otimes$$

$$\frac{1}{\sqrt{2}} \left(|0\rangle + \exp\left(2\pi i \frac{x_4}{2^3} + 2\pi i \frac{x_3}{2^2} + 2\pi i \frac{x_2}{2^1} \right) |1\rangle \right) \otimes \qquad (7.54)$$

$$|x_3\rangle \otimes |x_4\rangle$$

The next step is to take the qubit x_3, apply an H gate, and then a controlled rotation with x_4 as the control qubit.

$$= \frac{1}{\sqrt{2}} \left(|0\rangle + \exp\left(2\pi i \left(\frac{x_4}{2^4} + \frac{x_3}{2^3} + \frac{x_2}{2^2} + \frac{x_1}{2^1} \right) \right) |1\rangle \right) \otimes$$

$$\frac{1}{\sqrt{2}} \left(|0\rangle + \exp\left(2\pi i \left(\frac{x_4}{2^3} + \frac{x_3}{2^2} + \frac{x_2}{2^1} \right) \right) |1\rangle \right) \otimes \qquad (7.55)$$

$$\frac{1}{\sqrt{2}} \left(|0\rangle + \exp\left(2\pi i \frac{x_4}{2^2} + 2\pi i \frac{x_3}{2^1} \right) |1\rangle \right) \otimes |x_4$$

The final step is to apply an H gate with the qubit x_4 and rearranging the terms inside the exponents.

$$= \frac{1}{\sqrt{2}} \left(|0\rangle + \exp\left(2\pi i \left(\frac{x_1}{2^1} + \frac{x_2}{2^2} + \frac{x_3}{2^3} + \frac{x_4}{2^4} \right) \right) |1\rangle \right) \otimes$$

$$\frac{1}{\sqrt{2}} \left(|0\rangle + \exp\left(2\pi i \left(\frac{x_2}{2^1} + \frac{x_3}{2^2} + \frac{x_4}{2^3} \right) \right) |1\rangle \right) \otimes$$

$$\frac{1}{\sqrt{2}} \left(|0\rangle + \exp\left(2\pi i \left(\frac{x_3}{2^1} + \frac{x_4}{2^2} \right) \right) |1\rangle \right) \otimes \qquad (7.56)$$

$$\frac{1}{\sqrt{2}} \left(|0\rangle + \exp\left(2\pi i \frac{x_4}{2^1} \right) |1\rangle \right)$$

Comparing the fractional series inside the exponents with Eq. (7.43), we can rewrite the above equation:

$$= \frac{1}{\sqrt{2}} \left(\begin{array}{l} (|0\rangle + \exp(2\pi i\, (0.x_1x_2x_3x_4)) \, |1\rangle) \otimes (|0\rangle + \exp(2\pi i(0.x_2x_3x_4)) \, |1\rangle) \\ \otimes (|0\rangle + \exp(2\pi i(0.x_3x_4)) \, |1\rangle) \otimes (|0\rangle + \exp(2\pi i\, (0.x_4)) \, |1\rangle) \end{array} \right) \qquad (7.57)$$

By comparing this equation with Eq. (7.45), we can conclude that the 4-qubit circuit performs Quantum Fourier Transform. However, the order of the terms is reversed. This can be corrected by performing SWAP operations between the qubits pairs (x_1, x_4) and (x_2, x_3). The following Eq. (7.58) describes the final state of the system.

$$|\psi\rangle = \frac{1}{\sqrt{2}} \left((|0\rangle + \exp(2\pi i\, (0.x_4)) \, |1\rangle) \right.$$

$$\otimes (|0\rangle + \exp(2\pi i(0.x_3x_4)) \, |1\rangle) \qquad (7.58)$$

$$\otimes (|0\rangle + \exp(2\pi i(0.x_2x_3x_4)) \, |1\rangle)$$

$$\left. \otimes (|0\rangle + \exp(2\pi i\, (0.x_1x_2x_3x_4)) |1\rangle) \right)$$

This algorithm runs in polynomial time in n, that is, $O(n^2)$. The best estimate for classical Discrete Fourier Transform runs in an exponential time $O(n2^n)$. We can construct this circuit using $\frac{n(n+1)}{2}$ Hadamard and controlled U1 gates, and $\frac{n}{2}$ swap gates. The QFT does provide exponential speedup

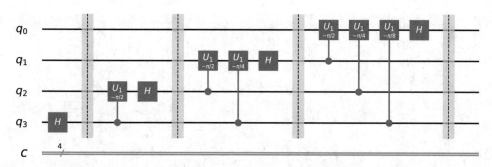

Fig. 7.8 Inverse Quantum Fourier Transform of 4-qubit registers. Note the order of the gates and the negative rotational angles

over classical DFT, but it works on the probability amplitudes of the quantum states. The classical Fourier transforms work on the complex vectors. Hence, not all applications that use DFT can benefit from QFT.

The inverse QFT is defined as follows:

$$x_n = \frac{1}{\sqrt{N}} \sum_{k=0}^{(N-1)} X_n \, \omega_N^{-kn}, n = 0 : N - 1 \tag{7.59}$$

The Inverse QFT $\left(\text{QFT}^{-1} \text{ or IQFT}\right)$ can be implemented by reversing the order of the gate operations and by negating the rotational angles. The circuit diagram shown in Fig. 7.8 illustrates the $U_{\text{QFT}}^{\dagger}$.

Note that the Quantum Fourier Transform is unitary, and it satisfies the relation $U_{\text{QFT}} U_{\text{QFT}}^{\dagger} = U_{\text{QFT}}^{\dagger} U_{\text{QFT}} = I$, where $U_{\text{QFT}}^{\dagger}$ is the Hermitian adjoint of U_{QFT} and $U_{\text{QFT}}^{\dagger} = U_{\text{QFT}}^{-1}$.

Let us develop the code that performs the forward and inverse QFT for an arbitrary size of qubits. Open your Jupyter notebook and enter the code from Listing 7.4 that imports the needed libraries and sets up the Qiskit.

We then develop two helper functions that swap the qubit registers and initialize the qubit registers based on the input. Enter the code for the two helper functions provided in Listing 7.5 in a new cell in your Jupyter notebook.

With these two helper functions available, we can create the code that constructs the forward QFT. We use two loops to create the circuit iteratively. The size of the qubit is taken directly from the quantum circuit. Hence, this function can create a QFT circuit of any size (Listing 7.6).

The inverse QFT must be defined backward and by using negative angles. Listing 7.7 demonstrates how to build the inverse QFT.

7.2.4   Testing the QFT

The QFT works on the amplitudes of the quantum states. Hence, we cannot compare QFT and DFT with a set of test data. To verify QFT, we construct a back to the back circuit of QFT and QFT^{-1}. Since QFT^{-1} is the Hermitian adjoint of QFT, cascading these two circuits should return the qubits to the initial state. The code snippet in Listing 7.8 illustrates this.

This circuit uses a 4-qubit QFT with an initial value of 7. Executing this circuit gets the histogram shown in Fig. 7.9. The circuit's output is the binary pattern "1110," and this is what we expected.

```
# Import the needed libraries
import qiskit
import math
from qiskit import *
from qiskit import IBMQ
from qiskit.visualization import plot_histogram, plot_gate_map
from qiskit.visualization import plot_circuit_layout
from qiskit.visualization import plot_bloch_multivector
from qiskit import QuantumRegister, ClassicalRegister, transpile
from qiskit import QuantumCircuit, execute, Aer
from qiskit.providers.ibmq import least_busy
from qiskit.tools.monitor import job_monitor
%config InlineBackend.figure_format = 'svg'

# load the IBM Quantum account
account = IBMQ.load_account()

# Get the provider and the backend
provider = IBMQ.get_provider(group='open')
backend = Aer.get_backend('qasm_simulator')
```

Listing 7.4 Import sections

```
# Swap the qubits
# qc - the quantum circuit

def swap_reg (qc):
    for qubit in range(qc.n_qubits // 2):
        qc.swap(qubit, qc.n_qubits - qubit - 1)

# Setup the qubits according to the bit pattern of the input number
# qc - the quantum circuit
# input - the input number whose binary pattern
#         is to be encoded in the qubits

def set_reg(qc, input):
    if (input.bit_length() > qc.n_qubits):
        print("Insufficient qubits")
        return False

    for i in range (0, input.bit_length()):
        if ((1 << i) & input ):
            qc.x(qc.n_qubits - i -1)
    return True
```

Listing 7.5 Two helper functions

```
# Construct the Forward QFT circuit
# qc - the quantum circuit


def qft_cct (qc):
    i = 0
    n = qc.n_qubits
    while ( n ):
        qc.h(i)
        n -= 1
        i += 1
        for qubit in range (n):
            qc.cu1(math.pi / 2** ( qubit + 1 ), qubit + i, i - 1)
        qc.barrier()


# Forward QFT
# qc - the quantum circuit
# input - the input number to be transformed.
# ensure sufficient number of qubits are allocated.
#

def qft_fwd (qc, input):
    if ( True == set_reg(qc,input) ):
        qft_cct(qc)
        swap_reg(qc)
        return True
    else:
        return False
```

Listing 7.6 Core performing forward QFT

```
# Inverse QFT
# qc - the quantum circuit
# do not forget the negative sign before math.pi

def qft_inv (qc):
    for i in range (qc.n_qubits - 1, -1, -1):
        for qubit in range (i, qc.n_qubits-1):
            # for inverse transform, we have to use negative angles
            qc.cu1(- math.pi/2** (qubit + 1 - i), qubit + 1, i)
        # the H transform should be done after the rotations
        qc.h(i)
        qc.barrier()
```

Listing 7.7 Inverse QFT

```
number = 7                      # The number for performing QFT and
                                  IQFT
numberofqubits = 4              # Required number of qubits
shots = 1024                    # Number of times the experiment is
                                  to be done.

q = QuantumRegister(numberofqubits , 'q')
c = ClassicalRegister(numberofqubits , 'c')
qc = QuantumCircuit(q, c)

qft_fwd(qc, number )
swap_reg(qc)
qft_inv(qc)

qc.measure_all()

backend = Aer.get_backend("qasm_simulator")
job = execute(qc, backend=backend, shots=shots)
counts = job.result().get_counts()
plot_histogram(counts)
```

Listing 7.8 Verifying the QFT

Fig. 7.9 Output of the
QFT verification code

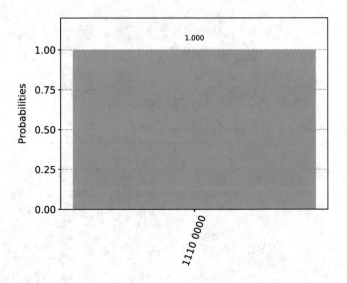

Exercise 7.2

1. *Try this code with all possible inputs (in the range of 0...15 for 4 qubits) and confirm the circuit works as intended on the simulator.*
2. *Try this code on a backend. Are you getting the same results? If not, where is the problem?*
3. *Reduce the circuit to using 3 qubits (change the inputs to the range 0...7) and retry the experiment on the backend. What do you observe now?*
4. *Compare this code with the QDK version. What differences do you see between the systems?*

Code example for QDK is included in the Appendix.

7.3 Deutsch–Jozsa Oracle

An oracle is a black-box function. It implements a complex algorithm to find an answer to a complex problem with a definitive "Yes," or "No" answer. We do not know how the black box works. However, we can build a circuit around it to learn about its properties. Oracles are interesting to quantum computing because quantum circuits are reversible, quantum computation is probabilistic, and the qubits are in superposition state. So, it is a natural tendency to study black-box functions and their properties. Deutsch–Jozsa oracle is one such function we study in this section.

Figure 7.10 illustrates the general form of an oracle function. The black box takes certain inputs $|x\rangle$ and $|y\rangle$. The oracle function is then calculated for all inputs and the outputs are set accordingly.

7.3.1 Deutsch Oracle

Given a function $f : \{0, 1\} \rightarrow \{0, 1\}$, the oracle determines whether the function is constant or balanced. The function is constant if $f(0) = f(1)$. It is balanced if $f(0) \neq f(1)$, that is, if the function returns the same value $\{0, 1\}$ irrespective of the input $\{0, 1\}$, it is a constant function. It is a balanced function if it returns different values.

Constant functions

$$f(0) = 0, \ f(1) = 0$$
$$f(0) = 1, \ f(1) = 1$$

Balanced functions

$$f(0) = 0, \ f(1) = 1$$
$$f(0) = 1, \ f(1) = 0$$

7.3.2 Deutsch–Jozsa Oracle

The Deutsch–Jozsa oracle is a generalization of the Deutsch oracle for a string of n bits. Given a function $f : \{0, 1\}^n \rightarrow \{0, 1\}$, the oracle determines whether the function is constant or balanced. The function is balanced if it returns 0 for half of the arguments and returns 1 for the other half of the arguments. For a function that is neither constant nor balanced, Deutsch-Jozsa oracle is not of much help.

Classically, it is easy to solve this problem. By repeatedly calling the function f with inputs $\{0, 1\}^n$, we can determine the nature of this function. We must call the function for half of the inputs and for an additional time to ascertain whether it is a constant or a balanced function. For half of the

Fig. 7.10 The general form of an oracle function

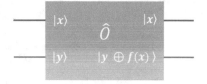

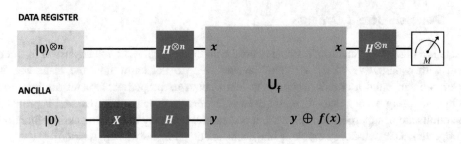

Fig. 7.11 The Deutsch–Jozsa algorithm

inputs, if we get a 0 (or a 1), then we need to call the function one more time to make sure it returns a 1 (or a 0) to be a balanced function. So we have to call the function $2^{n-1} + 1$ times to get 100% confidence.

Let us now look at the quantum version. Figure 7.11 illustrates how we construct the quantum circuit for the Deutsch–Jozsa algorithm.

This circuit uses an n qubit data register and an ancilla qubit. The black box U_f implements the oracle function $y \oplus f(x)$. We start the system with the data register and the ancilla qubit initialized to state $|0\rangle$.

$$|\psi\rangle = |0\rangle^{\otimes n}|0\rangle \tag{7.60}$$

The next step is to apply an X gate to the ancilla qubit. This gate flips its state to $|1\rangle$.

$$= |0\rangle^{\otimes n}|1\rangle \tag{7.61}$$

Recall from our discussion on Walsh–Hadamard Transform equation (5.167), that the simultaneous application of H gates on n qubits prepared at state $|0\rangle$ can be written as the following summation:

$$H^{\otimes n}|0\rangle^{\otimes n} = \frac{1}{\sqrt{2^n}} \sum_{x=0}^{2^n-1} |x\rangle \tag{7.62}$$

Getting back to our circuit, we apply H gates to both the data register and the ancilla qubit. The H gates put the system in a single-large-equal superposition state. By referring to the Eq. (7.62), we can write the system state as:

$$H^{\otimes n}|0\rangle^{\otimes n}H|1\rangle = \frac{1}{\sqrt{2^{n+1}}} \sum_{x=0}^{2^n-1} |x\rangle \left(|0\rangle - |1\rangle\right) \tag{7.63}$$

By applying the oracle function $y \oplus f(x)$ into the ancilla qubit, we get:

$$= \frac{1}{\sqrt{2^{n+1}}} \sum_{x=0}^{2^n-1} |x\rangle \left((|0\rangle - |1\rangle) \oplus f(x)\right)$$

$$= \frac{1}{\sqrt{2^{n+1}}} \sum_{x=0}^{2^n-1} |x\rangle \left(|0 \oplus f(x)\rangle - |1 \oplus f(x)\rangle\right) \tag{7.64}$$

$$= \frac{1}{\sqrt{2^{n+1}}} \sum_{x=0}^{2^n-1} |x\rangle \left(|f(x)\rangle - |1 \oplus f(x)\rangle\right)$$

Now, let us focus on the term $|f(x)\rangle - |1 \oplus f(x)\rangle$.

- If $f(x) = 0$, this term evaluates to $|0\rangle - |1 \oplus 0\rangle = |0\rangle - |1\rangle$.
- If $f(x) = 1$, this term evaluates to $|1\rangle - |1 \oplus 1\rangle = |1\rangle - |0\rangle$.

Therefore, we can write this term as $(-1)^{f(x)} (|0\rangle - |1\rangle)$.
Note that $(-1)^0 = 1$, and $(-1)^1 = -1$. Substituting this observation in Eq. (7.64), we get:

$$= \frac{1}{\sqrt{2^{n+1}}} \sum_{x=0}^{2^n-1} |x\rangle (-1)^{f(x)} (|0\rangle - |1\rangle) \tag{7.65}$$

We find that the state of the ancilla qubit remains unchanged after the application of the oracle function. We can drop the ancilla qubit from our discussion as we are not going to use this qubit anymore. The final step in the compute stage is to apply H gates to the data register. By referring to Eq. (5.16), we get:

$$= \frac{1}{\sqrt{2^n}} \sum_{x=0}^{2^n-1} (-1)^{f(x)} H^{\otimes n} |x\rangle$$

$$= \frac{1}{\sqrt{2^n}} \sum_{x=0}^{2^n-1} (-1)^{f(x)} \cdot \frac{1}{\sqrt{2^n}} \sum_{y=0}^{2^n-1} (-1)^{x \cdot y} |y\rangle \tag{7.66}$$

Here, $x \cdot y = x_0 y_0 \oplus x_1 y_1 \oplus x_2 y_2 \oplus \ldots \oplus x_{n-1} y_{n-1}$
This simplifies to:

$$= \sum_{y=0}^{2^n-1} \left(\frac{1}{2^n} \sum_{x=0}^{2^n-1} (-1)^{f(x)} \cdot (-1)^{x \cdot y} \right) |y\rangle \tag{7.67}$$

At this stage, we are ready to measure the data register. All we need to be sure of is that the data register projects into $|0\rangle^n$, that is, when $|y\rangle = |0\rangle^n$. For this case $y = 0$, therefore, $x \cdot y = 0$ Hence, the probability amplitude associated with this state is given by:

$$\frac{1}{2^n} \sum_{x=0}^{2^n-1} (-1)^{f(x)} \tag{7.68}$$

The probability of measuring $|0\rangle^n$ is:

$$\left| \frac{1}{2^n} \sum_{x=0}^{2^n-1} (-1)^{f(x)} \right|^2 = \begin{cases} 1 \text{ for constant function} \\ 0 \text{ for balanced function} \end{cases} \tag{7.69}$$

We can, therefore, confirm that the algorithm works. Let us verify this algorithm by implementing a quantum circuit with a 4-qubit data register and an oracle function.

Let us first examine the oracle function. A balanced function can be constructed using a set of CNOT gates. In this function, each qubit in the data register serves as the control qubit, and the ancilla qubit serves as the target qubit. This part of the circuit is given in Fig. 7.12. By referring to Sect. 5.4. 6.1 on the CNOT gate, we can write the truth table of this setup as shown in Table 7.2.

From the truth table, we can find that this function produces an equal number of $|0\rangle$s and $|1\rangle$s. Hence, we can say that this function is balanced.

It is relatively easy to construct a constant function. If we need the function to output a $|1\rangle$, we must apply an X gate to the ancilla qubit. If we need the function to output a $|0\rangle$, we do not do any operation. To try the Deutsch–Jozsa algorithm on the IBM Q, start with the Listing 7.9 which contains the import functions.

Fig. 7.12 Balanced function

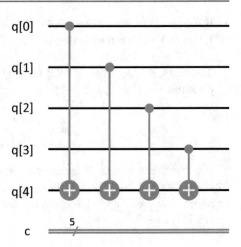

Table 7.2 Truth table for the balanced function

Input $\lvert q_0q_1q_2q_3\rangle$	Output $\lvert q_4\rangle$	Input $\lvert q_0q_1q_2q_3\rangle$	Output $\lvert q_4\rangle$
$\lvert 0000\rangle$	$\lvert 0\rangle$	$\lvert 1000\rangle$	$\lvert 1\rangle$
$\lvert 0001\rangle$	$\lvert 1\rangle$	$\lvert 1001\rangle$	$\lvert 0\rangle$
$\lvert 0010\rangle$	$\lvert 1\rangle$	$\lvert 1010\rangle$	$\lvert 0\rangle$
$\lvert 0011\rangle$	$\lvert 0\rangle$	$\lvert 1011\rangle$	$\lvert 1\rangle$
$\lvert 0100\rangle$	$\lvert 1\rangle$	$\lvert 1100\rangle$	$\lvert 0\rangle$
$\lvert 0101\rangle$	$\lvert 0\rangle$	$\lvert 1101\rangle$	$\lvert 1\rangle$
$\lvert 0110\rangle$	$\lvert 0\rangle$	$\lvert 1110\rangle$	$\lvert 1\rangle$
$\lvert 0111\rangle$	$\lvert 1\rangle$	$\lvert 1111\rangle$	$\lvert 0\rangle$

```
import qiskit
import math
from qiskit import *
from qiskit import IBMQ
from qiskit.visualization import plot_histogram, plot_gate_map
from qiskit.visualization import plot_circuit_layout
from qiskit.visualization import plot_bloch_multivector
from qiskit import QuantumRegister, ClassicalRegister, transpile
from qiskit import QuantumCircuit, execute, Aer
from qiskit.providers.ibmq import least_busy
from qiskit.tools.monitor import job_monitor
%matplotlib inline
%config InlineBackend.figure_format = 'svg'

# load the IBM Quantum account
account = IBMQ.load_account()

# Get the provider and the backend
provider = IBMQ.get_provider(group='open')
backend = Aer.get_backend('qasm_simulator')
```

Listing 7.9 The import functions

Listing 7.10 implements balanced and constant functions. It also includes the Deutsch–Jozsa algorithm.

```python
# The balanced function
# qc - the quantum circuit
# n - number of bits to use
# input - the input bit string

def balanced_func (qc, n, input):
    # first apply the input as a bitmap to the data register
    for i in range (0, input.bit_length()):
        if ((1 << i) & input ):
            qc.x(i)
    qc.barrier()
    #setup the CNOT gates
    for i in range (n):
        qc.cx(i, n)
    qc.barrier()
    # apply the input one more time to restore state
    for i in range (0, input.bit_length()):
        if ((1 << i) & input ):
            qc.x(i)

# The constant function
# qc - the quantum circuit
# n - number of bits to use
# constant - set this to true for constant functions.
#             False if not.

def constant_func (qc,n,constant=True):
    qc.barrier()
    # just apply a NOT gate to the ancilla qubit to output 1
    if (True == constant):
        qc.x(n)

# The Deutsch Jozsa algorithm
# qc - the quantum circuit
# n - the number of bits to use
# input - the input bit string
# function - set this to "constant" for constant functions.
# constant - set this to True for constant functions.

def deutsch_jozsa(qc, n, input, function, constant = True):
    # set the ancilla qubit to state 1.
    qc.x(n)
    #setup the H gates for both data register and ancilla qubit
    for i in range (n+1):
        qc.h(i)
```

Listing 7.10

```
    # call constant or balanced function as required
    if ("constant" == function ):
        constant_func(qc,n,constant)
    else:
        balanced_func(qc, n,input  )
    qc.barrier()
    #setup the H gates for the data register and setup measurement
    for i in range (n):
        qc.h(i)
    qc.barrier()
    #measure the data register
    for i in range (n):
        qc.measure(i,i)
```

Listing 7.10 (continued)

```
number = 15 # The hidden string
numberofqubits = 5
shots = 1024

q = QuantumRegister(numberofqubits , 'q')
c = ClassicalRegister(numberofqubits , 'c')
qc = QuantumCircuit(q, c)
deutsch_jozsa(qc, numberofqubits-1, number,"constant", True)
#qc.draw('mpl')

backend = Aer.get_backend("qasm_simulator")
job = execute(qc, backend=backend, shots=shots)
counts = job.result().get_counts()
plot_histogram(counts)
```

Listing 7.11   Test code to evaluate the Deutsch–Jozsa algorithm

Finally use Listing 7.11 to set up the quantum circuit, execute the algorithm, and measure the final state.

Executing this code produces the histograms shown in Fig. 7.13 for the balanced and constant functions. The function produces a "1111" with 100% confidence (or in other words measuring a "0000" is 0% probability) for the balanced function, and it produces a "0000" with 100% confidence for the constant function. This result is what we expected, and it aligns with Eq. (7.69) (Figs. 7.14 and 7.15).

The code that runs on the Q# QDK is given in Listing 7.12.

The helper functions that build the balance and constant functions are provided in Listing 7.13.

To make our discussions complete, we shall derive the math for the circuit we constructed. We begin the quantum system with the qubits initialized to the default state. We also skip the computation steps for the hidden string for the sake of simplicity. Verification of the hidden string is left as an exercise.

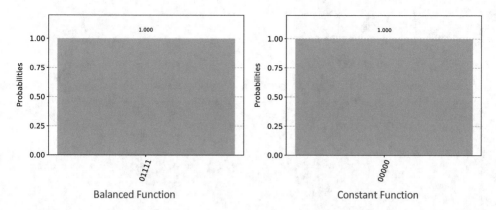

Fig. 7.13 The histogram for the balanced and constant functions, each producing 100% confidence

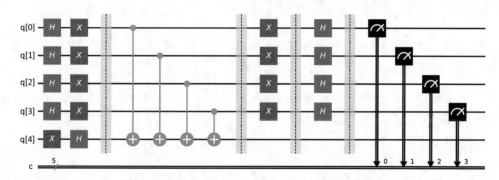

Fig. 7.14 Deutsch–Jozsa algorithm with a balanced function of 4-qubit data register (q0...q3) and an ancilla qubit (q4). The hidden string is applied before and after the oracle function

Fig. 7.15 Deutsch–Jozsa algorithm constant function outputting "1."

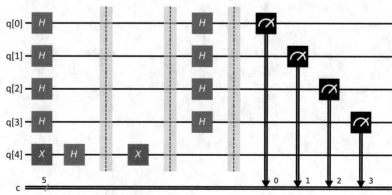

$$|\psi\rangle = |0000\rangle|0\rangle \tag{7.70}$$

We then apply an X gate to flip the ancilla qubit.

$$|\psi\rangle = |0000\rangle|1\rangle \tag{7.71}$$

H gates are applied to the system to put the qubits in a large superposition state.

```
/// # Summary
/// Implements the Deutsch-Jozsa Algorithm
///
/// # Inputs
/// input: a number whose' binary form represents a hidden string
/// functionType: 0 - Constant function, 1 - Balanced function
/// constant: {0,1} - The value the constant function need to\
    output
///
/// # Returns
/// The measured value denoting the probability of measuring all
    0s.
/// Should be returning all 1's or all 0's indicating the\
    probability
/// all 1's for balanced and all 0's for constant functions

operation DeutschJozsaAlgorithm ( input:Int, functionType:Int,
constant:Int ): Int {
    mutable n = 0;
    mutable r = 0;

    // determine how many bits are required to store the input
    // ignore the sign bit. We dont go that far
    for ( i in 0..63)
    {
        if (0 != (  (1 <<< i ) &&& input ) )
        {
            set n = i;
        }
    }

    using ( q = Qubit[n+2] )
    {
        // Set the ancilla qubit
        X(q[n+1]);

        // H transform for all
        ApplyToEach(H, q);

        if ( 0 == functionType)
        {
            // Constant function
            ConstantFunction(q, n, constant);
        }
        else
        {
            BalancedFunction(q, n,input);
        }
```

Listing 7.12 The Deutsch–Jozsa algorithm.

```
              // H transform for the data register
              ApplyToEach(H, Subarray( SequenceI(0,n), q));

              // Perform the measurements and order the bits to the
                 result
              for ( i in 0..n)
              {
                  if (One == M(q[i])) { set r = r ||| (1 <<< i); }
              }

              ResetAll(q);
          }

      Message($"The measured value is:  {r} ");

      return r;
  }
```

Listing 7.12 (continued)

$$= \frac{1}{\sqrt{2^4}} \left(|0000\rangle + |0001\rangle + \ldots + |1111\rangle \right) \frac{1}{\sqrt{2}} (|0\rangle - |1\rangle) \tag{7.72}$$

The next step is to apply the oracle function to the ancilla qubit. The oracle function $y \oplus f(x)$ is implemented using CNOT gates with the data registers as the control qubits and the ancilla qubit as the target. This step is an excellent example of how quantum parallelism works. Each state of the data register needs to be CNOT with the states of the ancilla qubit. The new state of the system can be written as:

$$= \frac{1}{\sqrt{2^4}\sqrt{2}} [|0000\rangle \left(|0 \oplus 0 \oplus 0 \oplus 0 \oplus 0\rangle - (|1 \oplus 0 \oplus 0 \oplus 0 \oplus 0\rangle) \right)$$

$$+ |0001\rangle(|0 \oplus 0 \oplus 0 \oplus 0 \oplus 1\rangle - (|1 \oplus 0 \oplus 0 \oplus 0 \oplus 1\rangle))$$

$$+ |0010\rangle \left(|0 \oplus 0 \oplus 0 \oplus 1 \oplus 0\rangle - (|1 \oplus 0 \oplus 0 \oplus 1 \oplus 0\rangle) \right)$$

$$+ |0011\rangle \left(|0 \oplus 0 \oplus 0 \oplus 1 \oplus 1\rangle - (|1 \oplus 0 \oplus 0 \oplus 1 \oplus 1\rangle) \right)$$

$$+ \ldots +$$

$$+ |1111\rangle \left(|0 \oplus 1 \oplus 1 \oplus 1 \oplus 1\rangle - (|1 \oplus 1 \oplus 1 \oplus 1 \oplus 1\rangle) \right)] \tag{7.73}$$

Applying XOR operations to the second term, this equation reduces to the following equation:

$$= \frac{1}{\sqrt{2^4}\sqrt{2}} |0000\rangle [(|0\rangle - |1\rangle) - |0001\rangle (|0\rangle - |1\rangle) - |0010\rangle(|0\rangle - |1\rangle)$$

$$+ |0011\rangle(|0\rangle - |1\rangle) + \ldots + |1111\rangle(|0\rangle - |1\rangle)] \tag{7.74}$$

```
// # Summary
/// **Implements the balanced function.**
///
/// # Inputs
/// q        - an array of qubits
/// n        - the total number of qubits in the data register.
///          The ancilla qubit is just next to
/// input    - a number whose binary form is the hidden string

operation BalancedFunction( q:Qubit[], n:Int, input:Int): Unit {
    for (i in 0..n) // Set the hidden string
    {
        if ( 0 != ( ( 1 <<< i) &&& input ) )
        {
            X(q[i]);
        }
    }

    for ( i in 0..n ) // The CNOT operation
    {
        CNOT(q[i], q[n+1]);
    }

    for (i in 0..n) // Unset the hidden string
    {
        if ( 0 != ( ( 1 <<< i) &&& input ) )
        {
            X(q[i]);
        }
    }
}


/// # Summary
/// **Implements the constant function.** Sets the ancilla qubit to 1,
/// if we have to output 1. No actions, otherwise.
///
/// # Inputs
/// q - an array of qubits
/// n - the total number of qubits in the data register.
///     The ancilla qubit is just next to this
/// constant - 0 - No actions, 1 - if we have to output 1.

operation ConstantFunction (q:Qubit[], n:Int, constant:Int): Unit
{
    if ( 1 == constant )
    {
        X(q[n+1]);
    }
}
```

Listing 7.13 The helper functions implementing the balanced and constant functions

$$= \frac{1}{\sqrt{2^4}} \left[|0000\rangle - |0001\rangle - |0010\rangle + |0011\rangle + \cdots + |1111\rangle \right] \frac{1}{\sqrt{2}} (|0\rangle - |1\rangle) \tag{7.75}$$

The kets belonging to the data register can be split, and this equation can be rewritten.

$$= \frac{1}{\sqrt{2}} (|0\rangle - |1\rangle) \otimes \frac{1}{\sqrt{2}} (|0\rangle - |1\rangle) \otimes \frac{1}{\sqrt{2}} (|0\rangle - |1\rangle) \otimes \frac{1}{\sqrt{2}} (|0\rangle - |1\rangle)$$
$$\otimes \frac{1}{\sqrt{2}} (|0\rangle - |1\rangle) \tag{7.76}$$

We finally apply H gates to the data register.

$$= H^{\otimes 4} \left(\frac{1}{\sqrt{2}} (|0\rangle - |1\rangle) \otimes \frac{1}{\sqrt{2}} (|0\rangle - |1\rangle) \otimes \frac{1}{\sqrt{2}} (|0\rangle - |1\rangle) \otimes \frac{1}{\sqrt{2}} (|0\rangle - |1\rangle) \right)$$
$$\otimes \frac{1}{\sqrt{2}} (|0\rangle - |1\rangle) \tag{7.77}$$

Application of the H gates reduces the system state to the following:

$$= |1\rangle \otimes |1\rangle \otimes |1\rangle \otimes |1\rangle \otimes \frac{1}{\sqrt{2}} (|0\rangle - |1\rangle) \tag{7.78}$$

When we measure the data register, it is projected to the state "1111" deterministically (see Fig. 7.13), indicating that it is a balanced function. We ignore the ancilla qubit.

This algorithm was first proposed in 1992 by the British physicist **David Deutsch** and the Australian mathematician **Richard Jozsa**. The Deutsch–Jozsa algorithm is of no practical use but is a good example where quantum parallelism is used. Besides, this algorithm is also an excellent example of a **deterministic quantum algorithm**. The classical method requires $2^{N-1} + 1$ evaluations, whereas the quantum method can perform the same evaluation with one evaluation. Deutsch–Jozsa algorithm demonstrates an exponential speedup that quantum computing can offer.

Exercise 7.3

1. *Try this code on a backend.*
2. *Derive the equations for a constant function.*
3. *Derive the equations with the hidden string in use and prove that it still works.*

◀

7.4   The Bernstein–Vazirani Oracle

Now that we have learned about the oracle functions, we can focus on a couple of more such problems. The next in line is the Bernstein–Vazirani Oracle. This oracle is similar to the Deutsch–Jozsa problem, and it is defined as follows:

Given a function $f : \{0, 1\}^n \to \{0, 1\}$, with the oracle function $f(x) = a \cdot x$, where a is a constant vector of $\{0, 1\}^n$. $a \cdot x$ is a scalar product of a and x. The goal of the oracle is to find the hidden bit string a.

A classical solution to this problem requires n steps. We can retrieve the hidden string a $(a_0 a_1 a_2 \ldots a_{n-1})$ by querying the oracle with a value of x, such that each query returns one of the bits in the hidden string. For example (note that the bits are reversed in the argument to the function):

$$f(0001) = a_0$$
$$f(0010) = a_1$$
$$f(0100) = a_2 \qquad (7.79)$$
$$f(1000) = a_3$$

After querying all the bits, we can stitch the results and obtain the hidden string $(a_0a_1a_2a_3)$.

There are a few ways to solve this problem on a quantum computer, and we can get the answer in one step. In this exercise, we solve this in a similar way to the Deutsch–Jozsa problem. The block diagram for the solution to the Bernstein–Vazirani oracle is given in Fig. 7.16.

We can implement the oracle function as a set of CNOT gates, with the ancilla qubit as the target and the data register qubits corresponding to the bit value of the hidden string a as the control qubits. Figure 7.17 illustrates this for a hidden string $10 = 1010b$. Upon executing this circuit, we get the histogram, as shown in Fig. 7.18.

Let us prove that the circuit works as intended.

Like the Deutsch–Jozsa algorithm, we start the system state with both the data register and the ancilla qubit initialized to $|0\rangle$.

$$|\psi\rangle = |0000\rangle|0\rangle \qquad (7.80)$$

We then apply an X gate to flip the ancilla qubit.

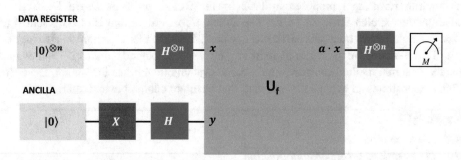

Fig. 7.16 Block diagram for Bernstein–Vazirani algorithm

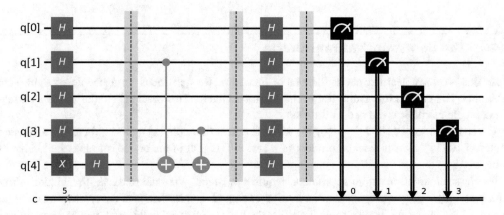

Fig. 7.17 4-Qubit Bernstein–Vazirani algorithm with a hidden string 1010b

Fig. 7.18 Histogram output of the 4 qubit Bernstein–Vazirani algorithm with a hidden string 1010b

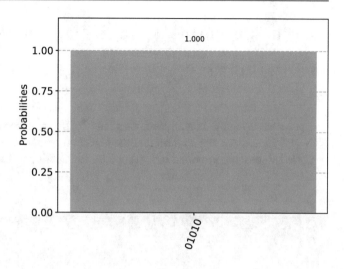

$$|\psi\rangle = |0000\rangle|1\rangle \tag{7.81}$$

H gates are applied to the system to put the qubits in a single large superposition state.

$$= \frac{1}{\sqrt{2}}(|0\rangle + |1\rangle) \otimes \frac{1}{\sqrt{2}}(|0\rangle + |1\rangle) \otimes \frac{1}{\sqrt{2}}(|0\rangle + |1\rangle) \otimes$$
$$\frac{1}{\sqrt{2}}(|0\rangle + |1\rangle) \ \otimes \frac{1}{\sqrt{2}}(|0\rangle - |1\rangle) \tag{7.82}$$

After this step, we apply the CNOT gates. But not for all qubits in the data register. We apply them for the qubits q[3] and q[1], representing the binary pattern 1010b, equivalent to decimal number 10. For easy understanding, let's first apply the CNOT between q[1] and the ancilla qubit. The system state becomes:

$$\xrightarrow{\text{CNOT}(1,anc)} \left[\frac{1}{\sqrt{2}}(|0\rangle + |1\rangle) \otimes \frac{1}{\sqrt{2}}(|0\rangle - |1\rangle) \right]$$
$$\xrightarrow{\text{CNOT}(1,anc)} \left[\frac{1}{2}(|00\rangle - |01\rangle + |10\rangle - |11\rangle) \right]$$
$$= \frac{1}{2}(|00\rangle - |01\rangle + |11\rangle - |10\rangle)$$
$$= \frac{1}{\sqrt{2}}(|0\rangle - |1\rangle) \otimes \frac{1}{\sqrt{2}}(|0\rangle - |1\rangle) \tag{7.83}$$

Note that the operation did not change the state of the ancilla qubit. We can write the next step after applying the same logic to the qubit q[3], and combining all states in the right order.

$$= \frac{1}{\sqrt{2}}(|0\rangle - |1\rangle) \otimes \frac{1}{\sqrt{2}}(|0\rangle + |1\rangle) \otimes \frac{1}{\sqrt{2}}(|0\rangle - |1\rangle) \otimes$$
$$\frac{1}{\sqrt{2}}(|0\rangle + |1\rangle) \ \otimes \frac{1}{\sqrt{2}}(|0\rangle - |1\rangle) \tag{7.84}$$

The final step is to apply H gates to the data register. We can safely ignore the ancilla qubit.

$$= H^{\otimes 4}\left(\frac{1}{\sqrt{2}}(|0\rangle - |1\rangle) \otimes \frac{1}{\sqrt{2}}(|0\rangle + |1\rangle) \otimes \frac{1}{\sqrt{2}}(|0\rangle - |1\rangle) \otimes \frac{1}{\sqrt{2}}(|0\rangle + |1\rangle)\right) \otimes \frac{1}{\sqrt{2}}(|0\rangle - |1\rangle)$$

$$= |1\rangle \otimes |0\rangle \otimes |1\rangle \otimes |0\rangle \otimes \frac{1}{\sqrt{2}}(|0\rangle - |1\rangle)$$

$$(7.85)$$

Upon measuring the data register, we get a "1010b" as expected, and it matches with the histogram Fig. 7.18. Hence, we can confirm that the algorithm works with one step.

Code snippets are provided in Listing 7.14 for Qiskit and Listing 7.15 for QDK.

```
# Implements the Bernstein-Vazirani problem
# qc - quantum circuit
# hiddenstring - the integer holding the binary bitstring
# ensure that sufficient qubits are allocated in qc
#
def bv_func (qc, hiddenstring):

    # set the ancilla qubit to state |1>
    qc.x(qc.n_qubits-1)

    # Set all qubits to an equal superposition state.
    for i in range (qc.n_qubits):
        qc.h(i)

    qc.barrier()

    # implement the hidden string - the oracle function
    for i in range (qc.n_qubits - 1):
        if ( 0!= ( hiddenstring & (1 << i))):
            qc.cx(I, qc.n_qubits-1)

    qc.barrier()

    # Apply H gates one more time
    for i in range (qc.n_qubits-1):
        qc.h(i)

    qc.barrier()

    # finally measure the data register
    for i in range(qc.n_qubits - 1):
        qc.measure(i,i)
```

Listing 7.14 Qiskit example code implementing the Bernstein–Vazirani algorithm

```
/// # Summary
/// Implements the Bernstein-Vaziranai Algorithm
///
/// # Input
/// hiddenString: a number whose' binary form represents the hidden
/// string
///
/// # Returns
/// The output of the algorithm. Should be equal to the hiddenString
///

operation BernsteinVaziraniAlgorithm ( hiddenString:Int):Int {
    mutable r = 0;
    mutable n = 0;

    // determine how many bits are required to store the input
    // ignore the sign bit. We dont go that far
    for ( i in 0..63)
    {
        if (0 != (  (1 <<< i ) &&& hiddenString ) )
        {
            set n = i;
        }
    }
    using ( q = Qubit[n+2] )
    {
        // Set the ancilla qubit
        X(q[n+1]);

        // H transform for all
        ApplyToEach(H, q);

        // Apply CNOT gates, ancilla qubit is the target.
        for ( i in 0..n )
        {
            if ( 0!= ((1 <<< i) &&& hiddenString))
            {
                CNOT(q[i], q[n+1]);
            }
        }
        // H transform for the data register
        ApplyToEach(H, Subarray( SequenceI(0,n), q));

        // Perform the measurements and order the bits to the result
        for ( i in 0..n)
        {
            if (One == M(q[i])) { set r = r ||| (1 <<< i); }
        }
```

Listing 7.15   Q#, QDK example of the Bernstein–Vazirani algorithm.

```
        ResetAll(q);
    }
    Message($"The measured value is:  {r} ");
    return r;
}
```

Listing 7.15 (continued)

Ethan Bernstein and **Umesh Vazirani** first proposed this algorithm in 1992. This algorithm is another example where we create a superposition of 2^n states. For the vector $|a\rangle$, the probability amplitudes interfered constructively. For all other values, the probability amplitudes interfered destructively, and their values are 0. The quantum version of this algorithm gets a **linear speed** up advantage.

7.5   Simon's Algorithm

It is believed that **Daniel Simon** was analyzing the reported quantum algorithms, and he was not satisfied that any of them gave an impressive speedup. Simon thought that there must be a theoretical limit that prevents anyone from getting the much-promised exponential speedup. In a quest to find that limit, in 1994, he ended up finding an oracle-based function with an exponential speedup [3]. The Quantum Fourier Transform and Shor's algorithm were invented around the same time.

Simon's algorithm uses quantum parallelism and quantum interference. Hence, a study of Simon's algorithm helps in understanding complex algorithms such as Shor's. Therefore, our next step is to explore Simon's algorithm. Here is the definition of Simon's problem, to start with:

Assume we have an oracle function that maps n-bits to n-bits, that is, $f : \{0, 1\}^n \mapsto \{0, 1\}^n$. Also, assume that the oracle function is guaranteed to satisfy the condition: $f(x) = f(y) \Leftrightarrow y = x \oplus a$. Simon's problem is then querying the oracle function and determining whether the hidden string $a = 0^n$ or $a \neq 0^n$.

If $a = 0^n$, then the oracle function is a *one-to-one* function (Table 7.3). If $a \neq 0^n$, then the function is a *two-to-one* function. A one-to-one function produces a unique output for each input. A two-to-one function produces a unique output for two inputs (Table 7.4).

Table 7.4 illustrates a two-to-one function. By examining this table, we can trace four distinct values: 000, 001, 010, 011. Each distinct value is repeated twice. Let us choose one of these distinct values: 011. The corresponding x values are 000 and 011. The hidden string $a = 000 \oplus 011 = 011$. We can verify this for other distinct values. From the table, we can recognize the following property.

Table 7.3   Example: One-to-one function:
A 3 qubit example of a one-to-one function

x	000	001	010	011	100	101	110	111
$f(x)$	111	110	101	100	011	010	001	000

Table 7.4   Example: Two-to-one function:
A 3 qubit example of a two-to-one function

x	000	001	010	011	100	101	110	111
$f(x)$	011	010	010	011	000	001	001	000

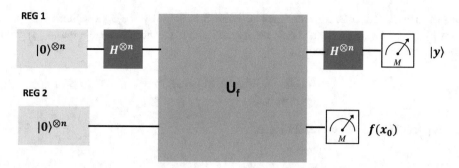

Fig. 7.19 The Simon's algorithm

$$f(x) = f(x \oplus a) \tag{7.86}$$

We know that $x \oplus a \oplus a = x$, we can rewrite the above equation.

$$f(x) = f(x \oplus a) = f(x \oplus a \oplus a) \tag{7.87}$$

This equation means $f(x)$ is periodic with the period a, in bitwise *mod* 2 operation. We shall use this property in later sections.

Solving Simon's problem classically is somewhat challenging. One way to determine the hidden string a, is to keep calling the oracle function until we find two repeat values. Let us assume that we had to call the oracle function i times to detect a repeat value. The number of pairs we may have compared is $\frac{i(i-1)}{2} \approx 2^n$. This is the minimum number of pairs that need to be examined before determining a. The complexity of this classical method is $O\left(2^{\frac{n}{2}}\right)$, exponential to n. Therefore, the number of times we call the oracle function grows exponentially with n, the number of bits in the hidden string a.

Figure 7.19 shows the block diagram of the quantum circuit that solves Simon's problem.

In this circuit, we use two sets of quantum registers: REG 1 and REG 2. We use a Walsh–Hadamard transform on REG 1 before it is connected to the oracle function. At the output section of the oracle function, we first measure REG 2. We apply a Walsh–Hadamard transform on REG 1 before it is measured.

Like the previous algorithms, we start the system with REG1 and REG2 in state $|0\rangle$.

$$|\psi\rangle = |0\rangle^{\otimes n} |0\rangle^{\otimes n} \tag{7.88}$$

We then apply a Walsh–Hadamard transform on REG1. Recall the Walsh–Hadamard transform we learned in Chap. 5. The system state can be written as follows:

$$|\psi\rangle = \frac{1}{\sqrt{2^n}} \sum_{x \in \{0,1\}^n} |x\rangle |0\rangle^{\otimes n}, \tag{7.89}$$

where $x \in \{0,1\}^n$ denotes an n-bit string. The summation is over 2^n such strings. The next step is to perform the oracle function on REG 2. This task is performed in one step using quantum parallelism.

$$
\begin{aligned}
|\psi\rangle &= \frac{1}{\sqrt{2^n}} \sum_{x \in \{0,1\}^n} |x\rangle |0 \oplus f(x)\rangle^{\otimes n} \\
&= \frac{1}{\sqrt{2^n}} \sum_{x \in \{0,1\}^n} |x\rangle |f(x)\rangle^{\otimes n}
\end{aligned} \tag{7.90}
$$

We then measure REG 2. Measurement of REG 2 is done by the application of 2^n projection operators $|i\rangle \langle i|$. In Chap. 2, Sect. 2.6.5, we learned about the projection operators. Applying the projection operator on REG 2, we get:

$$\frac{1}{\sqrt{2^n}} \sum_{x \in \{0,1\}^n} |x\rangle |i\rangle \langle i| |f(x)\rangle^{\otimes n} \tag{7.91}$$

From Eq. (2.125), we can write REG 2 as:

$$\sum_{x \in \{0,1\}^n} f_0(x) \tag{7.92}$$

Now, we have two options, if $a = 0^n$, then each value of $f_0(x)$ is unique. This means, the first register $|x\rangle$ has an equal probability of being in any state over $x \in \{0, 1\}^n$.

If $a \neq 0^n$, then from our definition of the problem, the function $f_0(x)$ can have two possible values of x, which we can denote as x_0 and $x_0 \oplus a$.

Thus, the act of measuring REG 2 has a side effect on REG 1. REG 1 collapses into a superposition state of $|x\rangle^{\otimes n}$ for which $f(x)$ has two solutions x_0 and $x_0 \oplus a$. This is an example of quantum interference. The terms which are not the probable solutions are canceled out due to destructive interference. The probability amplitudes of the states, which are probable solutions, are increased due to constructive interference. The new state of the system is as follows.

$$|\psi\rangle = \frac{1}{\sqrt{2}} \left(|x_0\rangle^{\otimes n} + |x_0 \oplus a\rangle^{\otimes n} \right) \tag{7.93}$$

This new state has an equal probability for all possible strings $x_0 \in \{0, 1\}^n$. Note that we have omitted REG 2 as it is measured, and we do not need it anymore. We proceed with our examination of this algorithm by taking up the second case. We shall return to the first case $a = 0^n$ towards the later stage of our discussion.

The next step is to apply a Walsh–Hadamard transform to REG 1.

$$H^{\otimes n} |x\rangle^{\otimes n} = \frac{1}{\sqrt{2^n}} \sum_{y \in \{0,1\}^n} (-1)^{x \cdot y} |y\rangle^{\otimes n} \tag{7.94}$$

In this equation, $y \in \{0, 1\}^n$ denotes an n-bit string. The summation is over all possible n-bit strings. $x \cdot y$ represents the bitwise inner product mod 2 defined as $x \cdot y = x_0 y_0 \oplus x_1 y_1 \oplus x_2 y_2 \oplus \cdots \oplus x_{n-1} y_{n-1}$.

Applying this in Eq. (7.93) and omitting REG 2, we get:

$$\begin{aligned}
|\psi\rangle &= \frac{1}{\sqrt{2}} H^{\otimes n} (|x_0\rangle^{\otimes n} + |x_0 \oplus a\rangle^{\otimes n}) \\
&= \frac{1}{\sqrt{2^{n+1}}} \sum_{y \in \{0,1\}^n} \left((-1)^{x_0 \cdot y} + (-1)^{(x_0 \oplus a) \cdot y} \right) |y\rangle^{\otimes n}
\end{aligned} \tag{7.95}$$

At this stage, let us focus on the term $\left((-1)^{x_0 \cdot y} + (-1)^{(x_0 \oplus a) \cdot y} \right)$. From Boolean algebra, recall that $(x_0 \oplus a) \cdot y = (x_0 \cdot y) \otimes (a \cdot y)$.

Applying this logic and rewriting the terms, we get:

$$|\psi\rangle = \frac{1}{\sqrt{2^{n+1}}} \sum_{y \in \{0,1\}^n} (-1)^{x_0 \cdot y} \left(1 + (-1)^{a \cdot y} \right) |y\rangle^{\otimes n} \tag{7.96}$$

By examining this equation, we can conclude a few points. The amplitude $(-1)^{x_0 \cdot y} \left(1 + (-1)^{a \cdot y}\right)$ is non-zero, only if $a \cdot y = 0$. Besides, each of these amplitudes has an equal magnitude $\frac{1}{\sqrt{2^{n+1}}}$. This means, when we measure REG 1, we uniformly get randomized bit strings y, such that $a \cdot y = 0$. Now, we have a set of strings y, such that $a \cdot y = 0$. The question is, how do we find a from this.

We run this function n times and get n strings: $y_1 y_2 y_3 \ldots y_n$. With these n strings, we have n equations: $a. y_1 = 0$; $a. y_2 = 0; \ldots a. y_n = 0$. If $y_1 y_2 y_3 \ldots y_n$ are linearly independent, then we can quickly solve this system of equations, and the total number of possible substrings is 2^n. Say, if we tried this algorithm for a few times more than n, then the probability of getting n linearly independent strings is:

$$P_r = \left(1 - \frac{1}{2^1}\right)\left(1 - \frac{1}{2^2}\right) \cdots \left(1 - \frac{1}{2^n}\right) > \frac{1}{4} \qquad (7.97)$$

If we are not finding the set of linearly independent string y_i, we can keep calling the function beyond the n attempts, until we can determine a. If we totally cannot get the required set of linearly independent strings, then it means $a = 0^n$. The probability of this occurrence is given by:

$$P_r = \left(1 - \frac{1}{4}\right)^{4x} < e^{-x} \qquad (7.98)$$

If we try this for $x = 10$ times, the error rate is approximately 1 in 10000. The advantage of this algorithm is that the number of times we need to try beyond n is not dependent on n. The classical algorithm requires $2^{\frac{n}{2}}$ steps, whereas the quantum version requires a little over n steps, which is an exponential advantage.

Figure 7.20 shows the histogram and Fig. 7.21 shows the quantum circuit that implements Simon's algorithm for the hidden string $011b = 3$.

Listing 7.16 implements Simon's algorithm in Qiskit. Listing 7.17 puts the code together and plots the output.

We can obtain the value of the hidden string from the output $a = 3 = 011b$. Let us conclude this section by deriving the math for the circuit, shown in Fig. 7.21. We begin the system with the registers REG 1 and REG 2 initialized to the default state.

$$|\psi\rangle = |000\rangle |000\rangle \qquad (7.99)$$

Fig. 7.20 Histogram of the Simon's algorithm with hidden string 011b

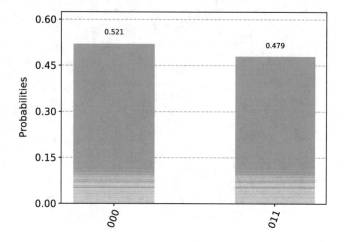

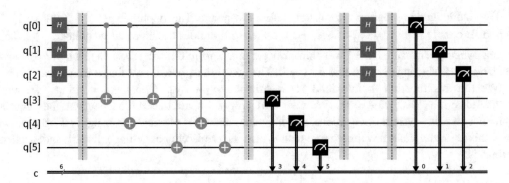

Fig. 7.21 The Simon's algorithm for the hidden string 011b

```
# Simons function
# qc - the quantum circuit
# a - the hidden string

def simons_func (qc, a):
    n = qc.n_qubits // 2

    # put the first half - REG 1 through H gate
    for i in range (n):
        qc.h(i)

    qc.barrier()

    # build the oracle function
    for i in range (n):
        if ( 0!= (( 1 << i) & a)):
            for j in range (n):
                qc.cx(q[i], q[j+n])
    qc.barrier()

    # measure the lower half REG 2
    for i in range (n, qc.n_qubits):
        qc.measure(i,i)
    qc.barrier()

    # Apply H transform to REG 1
    for i in range (n):
        qc.h(i)
    qc.barrier()

    # Finally measure the first half  REG 1
    for i in range (n):
        qc.measure(i,i)
```

Listing 7.16 Code implementing Simon's problem in Qiskit

```
a = 3                        # the hidden string
numberofqubits = 6           # number of qubits required.
shots = 1024                 # experiments

q = QuantumRegister(numberofqubits , 'q')
c = ClassicalRegister(numberofqubits , 'c')
qc = QuantumCircuit(q, c)

simons_func(qc, a)
qc.draw('mpl')

backend = Aer.get_backend("qasm_simulator")
job = execute(qc, backend=backend, shots=shots)
counts = job.result().get_counts()

# plot the histogram for REG 1 alone
res_plot = {}
for i in counts.keys():
    inp = i[numberofqubits // 2:]
    if inp in res_plot:
        res_plot[inp] += counts[i]
    else:
        res_plot[inp] = counts[i]
plot_histogram(res_plot)
```

Listing 7.17 Code fragment to plot the histogram for REG 1 alone

First, a Walsh–Hadamard transform is applied to REG 1.

$$|\psi\rangle = H^{\otimes 3}|000\rangle|000\rangle \tag{7.100}$$

$$= \frac{1}{\sqrt{2^3}} (|000\rangle + |001\rangle + |010\rangle + |011\rangle + \\ |100\rangle + |101\rangle + |110\rangle + |111\rangle) |000\rangle \tag{7.101}$$

$$= \frac{1}{\sqrt{2^3}} (|000\rangle|000\rangle + |001\rangle|000\rangle + |010\rangle|000\rangle + \\ |011\rangle|000\rangle + |100\rangle|000\rangle + |101\rangle|000\rangle + \\ |110\rangle|000\rangle + |111\rangle|000\rangle) \tag{7.102}$$

We now apply the oracle function to REG 2. This applies the value of $f(x)$ for the corresponding value of x, from Table 7.4 for the two-to-one function. The new system state is as follows.

$$= \frac{1}{\sqrt{2^3}} (|000\rangle|011\rangle + |001\rangle|010\rangle + |010\rangle|010\rangle + \\ |011\rangle|011\rangle + |100\rangle|000\rangle + |101\rangle|001\rangle + \\ |110\rangle|001\rangle + |111\rangle|000\rangle) \tag{7.103}$$

Let us analyze this state carefully. When we measure REG 2, it has an equal probability of measuring one of the following values: 000, 001, 010,011. If we assume that we measured REG 2 as 011, then the state of REG 1 collapses to the following state due to quantum interference.

$$= \frac{1}{\sqrt{2}} \left(|000\rangle + |011\rangle \right) \tag{7.104}$$

We can now apply the second Walsh–Hadamard transform to REG 1. To make our understanding easier, we can apply the H transforms qubit by qubit and expand the equation.

Applying the H gate on the first qubit, we get:

$$= \frac{1}{\sqrt{2^2}} \left(|000\rangle + |100\rangle + |011\rangle + |111\rangle \right) \tag{7.105}$$

We can now apply the H gate on the second qubit, expanding the system state as follows:

$$= \frac{1}{\sqrt{2^3}} \left(|000\rangle + |010\rangle + |100\rangle + |110\rangle + |001\rangle - |011\rangle + |101\rangle - |111\rangle \right) \tag{7.106}$$

Finally, we apply the H gate on the third qubit.

$$= \frac{1}{\sqrt{2^4}} (|000\rangle + |011\rangle + |010\rangle + |011\rangle +$$
$$|100\rangle + |101\rangle + |110\rangle + |111\rangle +$$
$$|000\rangle - |001\rangle - |010\rangle + |011\rangle +$$
$$|100\rangle - |101\rangle - |110\rangle + |111\rangle) \tag{7.107}$$

Due to quantum interference, opposite terms cancel out. This equation reduces to the following:

$$= \frac{1}{2} \left(|000\rangle + |011\rangle + |100\rangle + |111\rangle \right) \tag{7.108}$$

Let us rewind a bit to Eq. (7.103) and check what would happen if we measured other values in REG 2. For example, if we measured 001 in REG 2, REG 1 collapses as follows:

$$= \frac{1}{\sqrt{2}} \left(|101\rangle + |110\rangle \right) \tag{7.109}$$

We now apply the H transforms. Applying the H gate on the first qubit,

$$= \frac{1}{\sqrt{2^2}} \left(|001\rangle - |101\rangle + |010\rangle - |110\rangle \right) \tag{7.110}$$

Applying H gate on the second qubit,

$$= \frac{1}{\sqrt{2^3}} \left(|001\rangle + |011\rangle - |101\rangle - |111\rangle + |000\rangle - |010\rangle - |100\rangle + |110\rangle \right) \tag{7.111}$$

Applying H gate on the third qubit,

$$= \frac{1}{\sqrt{2^4}} (|000\rangle - |011\rangle + |010\rangle - |011\rangle$$
$$- |100\rangle + |101\rangle - |110\rangle + |111\rangle$$
$$+ |000\rangle + |001\rangle - |010\rangle - |011\rangle$$
$$- |100\rangle - |101\rangle + |110\rangle + |111\rangle \tag{7.112}$$

This equation reduces to the following. We can safely ignore the negative phase factors.

$$= \frac{1}{2} \left(|000\rangle - |011\rangle - |100\rangle + |111\rangle \right) \tag{7.113}$$

Therefore, when we measure REG 1, it can be projected to one of the following values with equal probability: $000, 011, 100, 111$. We can readily solve these values for $a = 011$ using classical methods. Code example for QDK is included in the Appendix.

7.6 Quantum Arithmetic Using QFT

In Sect. 7.1, we learned about constructing a simple ripple adder. In this section, we shall learn about using QFT to do arithmetic [7]. When we compare with classical methods, these are not efficient ways of doing arithmetic. However, these methods are used in modular exponentiation and similar algorithms requiring number work with superposition states. The first circuit we shall build is a simple QFT adder, followed by a controlled QFT adder, and a QFT-based quantum multiplier circuit.

7.6.1 Simple QFT Adder

The algorithm for the QFT-based adder was introduced by **Draper** in 2000 and optimized the number of qubits required to perform the addition. The algorithm takes two numbers, a and b, as the inputs and computes $QFT(b)$ as the first step. The number a is added to this by the application of a set of rotations, moving the system state to $QFT(a + b)$ in the **Fourier basis**. Finally, an inverse QFT transforms the system state to $(a + b)$, which is the desired output in the computational basis.

Recall the 4-qubit QFT shown in Fig. 7.22. We have just swapped the order of the transform, to avoid the additional swap operations. Referring to Eq. (7.58), we can write the system's state with a bit of shorthand.

$$QFT|b\rangle = |\phi(b)\rangle = \frac{1}{\sqrt{2}} \left(\left(|0\rangle + e(0.b_4) |1\rangle \right) \otimes \right.$$
$$\left(|0\rangle + e(0.b_3b_4) |1\rangle \right) \otimes \left(|0\rangle + e(0.b_2b_3b_4) |1\rangle \right) \otimes \tag{7.114}$$
$$\left. \left(|0\rangle + e(0.b_1b_2b_3b_4)|1\rangle \right) \right)$$

Now, consider the adder circuit shown in Fig. 7.23.

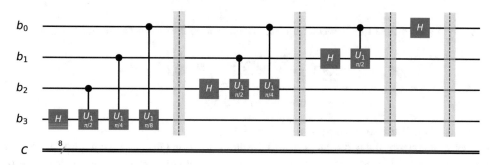

Fig. 7.22 Four-qubit QFT circuit, drawn in a swapped order

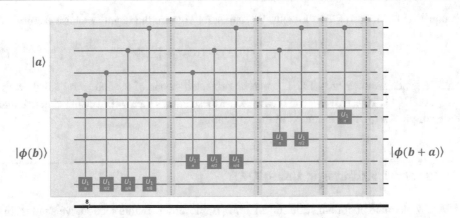

Fig. 7.23 The adder circuit sandwiched between QFT and inverse QFT layers

This circuit adds a systematic set of controlled U1 rotation gates to the quantum register b, which is now in a Fourier basis, due to the action of the earlier QFT stage. The controlled U1 gates add a phase factor to the register b.

From Eq. (5.49), and Sect. 5.4.6.6, the action of the cU1 gate on a system $|a\rangle^{\otimes N}|b\rangle^{\otimes N}$ can be written as:

$$cU1(\pi)|a\rangle^{\otimes N}|b\rangle^{\otimes N} = e^{\frac{i2\pi a \cdot b}{N}}|a\rangle^{\otimes N}|b\rangle^{\otimes N} \tag{7.115}$$

For now, let us drop the superscript notation and rewind a bit to apply the QFT on the register $|b\rangle$. Register $|b\rangle$ is now in the Fourier basis $|\phi(b)\rangle$

$$|a\rangle|b\rangle \xrightarrow{QFT(b)} \frac{1}{\sqrt{N}} \sum_{k=0}^{N-1} e^{\frac{i2\pi b \cdot k}{N}} |a\rangle|k\rangle, \tag{7.116}$$

where $N = 2^n$ and $n =$ number of qubits in the registers a and b.

The adder circuit in Fig. 7.23 is constructed in such a way that the rotational angles introduce a phase shift, which is equivalent to a *modulo N* addition in the Fourier basis. Assuming that we cascade this circuit with the 4-qubit QFT, the system state becomes:

$$\underrightarrow{CU1(\lambda, a, b)} \frac{1}{\sqrt{N}} \sum_{k=0}^{N-1} e^{\frac{i2\pi b \cdot k}{N}} \cdot e^{\frac{i2\pi a \cdot k}{N}} |a\rangle|k\rangle$$

$$\rightarrow \frac{1}{\sqrt{N}} \sum_{k=0}^{N-1} e^{\frac{i2\pi(a+b)k}{N}} |a\rangle|k\rangle \tag{7.117}$$

By applying an inverse QFT, we can transform the system into the computational basis.

$$\underrightarrow{IQFT} \frac{1}{N} \sum_{k,l=0}^{N-1} e^{\frac{i2\pi(a+b)k}{N}} \cdot e^{\frac{-i2\pi k \cdot l}{N}} |a\rangle|l\rangle$$

$$\rightarrow |a\rangle|(a+b) \bmod N\rangle \tag{7.118}$$

Thus, register b is transformed into $(a + b) \bmod N$.

Listing 7.18 contains the code that performs the addition using QFT. Use the import sections from previous exercises, and build a wrapper code that calls the method "add_qft." This code can be tried on the simulator as well as on a backend processor.

Code example for QDK is included in the Appendix.

```
# Sets a quantum register with an input value
# qc - quantum cirtuit
# input - the number to be stored.
# q - the qubit register
# n - width of the qubit register.

def set_reg(qc, input, q, n):
    for i in range (0, n):
        if ((1 << i) & input ):
            qc.x(q[i])

# the adder circuit
# qc - quantun circuit
# a - source register
# b - target register
# n - number of qubits in the registers

def add_cct(qc,a, b, n):
    while (n):
        for i in range(n, 0, -1):
            qc.cu1(math.pi/2**(n - i), a[i-1], b[n-1])
        n -= 1
        qc.barrier()

# The inverse QFT circuit
# qc - quantum circuit
# b  - the target register
# n - number of qubits in the register

def iqft_cct(qc, b, n):
    for i in range (n):
        for j in range (1, i+1):
            # for inverse transform, we have to use negative angles
            qc.cu1(  -math.pi / 2** ( i -j + 1 ), b[j - 1], b[i])
        # the H transform should be done after the rotations
        qc.h(b[i])
    qc.barrier()

# The forward QFT circuit
# qc - quantum circuit
# b  - the target register
# n - number of qubits in the register

def qft_cct(qc, b, n):
    while (n):
        qc.h(b[n-1])
        for i in range (n-1, 0, -1):
            qc.cu1(math.pi / 2** (n - i ), b[i - 1], b[n-1])
```

Listing 7.18 The QFT adder circuit block.

```
        n -= 1
        qc.barrier()

# Simple adder using QFT
# qc - quantum cirtuit
# input1 & input 2 - the numbers to add.
# must be less than n bits in size.
# a - qubit register bank a of size n qubits
# b - qubit register bank b of size n qubits
#      the sum ( a + b ) is stored in register a
# n - number of qubits in the registers & the inputs.

def add_qft(qc, input1, input2, a, b, n):
    set_reg(qc, input1, a, n)
    set_reg(qc, input2, b, n)

    qft_cct(qc, b, n)
    add_cct(qc, a, b, n)
    iqft_cct(qc, b, n)
```

Listing 7.18 (continued)

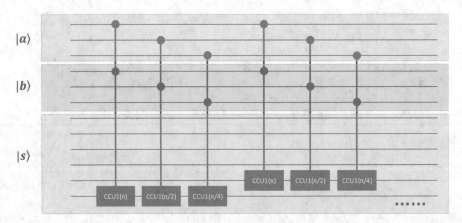

Fig. 7.24 Part of a controlled adder circuit. Note that this circuit is sandwiched between a QFT and inverse QFT parts

7.6.2 Controlled QFT Adder

Before developing the multiplication algorithm, we must construct the controlled QFT adder. Consider the adder circuit shown in Fig. 7.24. This circuit is implemented using a doubly controlled U1 gate (ccu1). Note that the development environments do not readily support the ccu1 gates. However, we can construct a ccu1 unitary ourselves.

This circuit has a property of providing the following output $|a + b + s\rangle$. If we start $|s\rangle$ in a state of $|0\rangle$, it produces an output of $|a + b\rangle$. Similarly, we can produce controlled versions of QFT and IQFT operations as well. The source code for implementing the doubly controlled U1 gate is provided in Listing 7.19.

```
# Implements a doubly controlled U1 gate.
# qc - quantum circuit
# theta - angle of rotation.
# q0 - Control qubit #1
# q1 - Control qubit #2
# q2 - Target qubit

def ccu1(qc, theta, q0, q1, q2):
    qc.cu1(theta/2, q1, q2)
    qc.cx(q0, q1)
    qc.cu1(-theta/2, q1, q2)
    qc.cx(q0, q1)
    qc.cu1(theta/2, q0, q2)


# The controlled adder circuit
# qc - quantun circuit
# a - source register
# b - target register
# offset - the starting qubit in register b to work from
# control - the control qubit
# n - number of qubits in the registers

def c_add_cct(qc, a, b, offset, control, n):
    while (n):
        for i in range(n, 0, -1):
            ccu1(qc,  math.pi/2**(n-i),   control,  a[i-1],\
            b[n-1+offset])
        qc.barrier()
        n -= 1

# The controlled inverse QFT circuit
# qc - quantum circuit
# b  - the target register
# offset - the starting qubit in register b to work from
# control - the control qubit.
# n - number of qubits in the register

def c_iqft_cct (qc, b, offset, control, n):
    for i in range (n):
        for j in range (1, i+1):
            # for inverse transform, we have to use negative angles
            ccu1(qc,-math.pi/2**(i-j+1),control,b[j-1+offset],\
                b[i+offset])
        # the H transform should be done after the rotations
        qc.ch(control, b[i+offset])
    qc.barrier()

# The forward QFT circuit
```

Listing 7.19 Controlled QFT adder modules.

```
# qc - quantum circuit
# b  - the target register
# offset - the starting qubit in register b to work from
# control - the control qubit.
# n - number of qubits in the register

def c_qft_cct(qc, b, offset, control, n):
    while (n):
        qc.ch(control, b[n + offset - 1])
        for i in range (n-1, 0, -1):
            ccu1(qc, math.pi / 2** (n - i), control, b[i - 1 +
            offset],\
                b[n + offset - 1])
        n -= 1
    qc.barrier()


# The controlled adder circuit block
# qc - quantum circuit
# a - the source register
# b  - the target register
# offset - the starting qubit in register b to work from
# control - the control qubit.
# n - number of qubits in the register

def c_add_qft(qc, a, b, offset, control, n):
    c_qft_cct(qc, b, offset, control, n)
    c_add_cct(qc, a, b, offset, control, n)
    c_iqft_cct(qc, b, offset, control, n)
```

Listing 7.19 (continued)

We can use the doubly controlled U1 gate to build a controlled adder circuit as a reusable block. The function *c_add_qft* in the Listing 7.19 explains the sequence of instructions needed to build this block. We begin with a controlled QFT and then apply the controlled addition, followed by the controlled inverse QFT.

Type the following code into a new cell in the Jupyter notebook. The code developed earlier can be reused to import the required components and set up IBM Q. The method "c_add_qft" is called to perform the controlled addition. Developing a unit testing code to verify whether the controlled addition works is a good exercise.

Code example for QDK is included in the Appendix.

7.6.3   QFT-Based Multiplier

We all know that multiplication is nothing but repeated addition. By stacking up a series of controlled QFT adder blocks, we can implement a multiplication module. Consider the block diagram illustrated in Fig. 7.25.

The above quantum circuit multiplies two n-bit numbers a and b by performing n successive controlled QFT additions. The result is stored in the register s with size $2n$.

The first controlled QFT adder block, labeled as $2^0\sum$, has the least significant qubit of register a_n as the control qubit. Register s is initially prepared in the state $|0\rangle$. The adder block first performs a QFT on register s, producing the state $|\phi(0)\rangle$. After this, the $2^0\sum$ block applies a series of conditional phase rotation gates to evolve the state into $|\phi(0 + b)\rangle$. This block is controlled by the least significant qubit of register a. So the state can be written as $|\phi(0 + a_n 2^0 b)\rangle$. Finally, the block applies an inverse QFT, moving the system state to $|0 + a_n 2^0 b\rangle$.

The qubit a_{n-1} controls the second controlled QFT adder $2^1\sum$. The scale factor for the phase addition at this level is 2^1. We can calculate the output state of this stage to be $|0 + a_n 2^0 b + a_{n-1} 2^1 b\rangle$.

We continue to apply the rest of the controlled QFT adder blocks in the circuit. When all the blocks are added, the output state becomes:

$$= |0 + a_n 2^0 b + a_{n-1} 2^1 b + \ldots + a_1 2^{n-1} b\rangle$$

$$= |0 + a \cdot b\rangle \tag{7.119}$$

$$= |a \cdot b\rangle$$

The code in Listing 7.20 can be used to test the Quantum Multiplier. Use the import modules from previous exercises. Use the controlled adder function "c_add_qft," and the set register function "set_reg" from previous exercises.

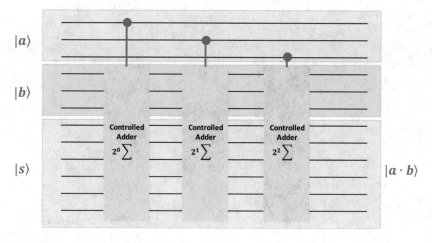

Fig. 7.25 Multiplication using controlled addition

```
# The quantum multiplier circuit
# calculates s = a * b
# input1 - multiplicand
# input2 - multiplied
# s - the qubit register holding product a * b
#       s has two times the length of a and b
# a - the qubit register for multiplicand
# b - the qubit register for multiplier
# n - length of the registers

def mult_cct(qc, input1, input2, a, b, s, n):
    set_reg(qc, input1, a, n)
    set_reg(qc, input2, b, n)

    for i in range (0, n):
        c_add_qft(qc, b, s, i,  a[i], n)

# Function to test the multiplier circuit
# input1 - multiplicand
# input2 - multiplied
# n - length of the registers

def test_mult( input1, input2, n):
    shots = 1024

    a = QuantumRegister(n , 'a')
    b = QuantumRegister(n , 'b')
    s = QuantumRegister(n*2 , 's')
    c = ClassicalRegister(n*2 , 'c')
    qc = QuantumCircuit(a, b, s, c)

    mult_cct(qc, input1, input2, a, b, s, n)

    for i in range (n*2):
        qc.measure(s[i],c[i])

    backend = Aer.get_backend("qasm_simulator")
    job = execute(qc, backend=backend, shots=shots)
    counts = job.result().get_counts()
    #print (counts)
    if (1 < len(counts)):
        print("More than one result!")
    else:
        scounts = str(counts)
        myproduct = \
    int("0b"+scounts[scounts.index("{")+2:scounts.index(":")-1], 2)
        #print(myproduct)
```

Listing 7.20 The quantum multiplier and code to test its function.

```
        return myproduct
    #
    # Test the multiplier and confirm it works...
    #
    for i in range (7):
        print("testing in the range of: ", i)
        for j in range (7):
            res = test_mult(i,j, 4)
            if ( res != (i*j)):
                print ("error, got", res, "expected:", i*j)
    print("done")
```

Listing 7.20 (continued)

Exercise 7.4

1. *Implement a quantum subtraction circuit.*
2. *Implement a generic quantum adder/subtractor that can handle overflows/underflows and negative integers.*
3. *Implement a quantum divider circuit.*

◀

Code example for QDK is included in the Appendix.

7.7 Modular Exponentiation

 Note: Reading Sect. 6.8 on RSA security may be helpful to read the following section in this chapter.

In the previous chapter, we learned about modular exponentiation when we discussed the RSA security. Recall from section 6.8.1 that modular exponentiation calculates the remainder, when an integer a raised to the xth power (that is, a^x) is divided by a positive integer N, called the modulus. Using symbols, we can write this as $c = a^x \bmod N$. Table 7.5 illustrates this operation by assuming $a = 2$ and $N = 15$.

We can calculate this using the direct method—calculate a^x first, and then perform *mod N*, using a calculator.

If we can state x in its binary form ($x_{n-1}x_{n-2}x_{n-3}\ldots x_0$), we can expand the modular exponentiation as follows:

$$a^x \bmod N = \left(a^{2^0} \bmod N\right)^{x_0} \cdot \left(a^{2^1} \bmod N\right)^{x_1} \cdots$$
$$\left(a^{2^{n-1}} \bmod N\right)^{x_{n-1}} \bmod N \tag{7.120}$$

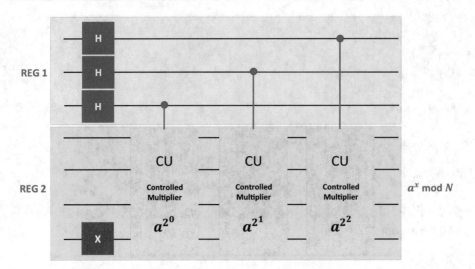

Fig. 7.26 Block diagram for modular exponentiation

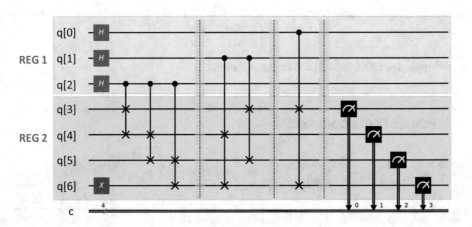

Fig. 7.27 Example circuit for modular exponentiation 2^x mod 15

This can be constructed as a circuit using controlled multiplier blocks, as shown in Fig. 7.26.

In this circuit, each of the controlled multiplier blocks performs a multiplication of the input REG 2, by a constant function a^{2^i} mod N. Each of these controlled multiplier blocks is controlled by the respective qubit x_i in REG 1. If we assume $|x\rangle$ as the control qubits (REG 1) and $|y\rangle$ as the target qubits (REG 2), this operation produces:

$$CU_{a^{2^i}}|x\rangle|y\rangle = |x\rangle \left(a^{2^i}\right)^x y \bmod N \tag{7.121}$$

Let us calculate the inner workings of this circuit with an example that performs 2^x mod 15. Consider the circuit shown in Fig. 7.27. In this circuit, REG 1 is formed by qubits q[0] to q[2]. REG 2 is formed by qubits q[3] to q[6].

Table 7.5 Modular arithmetic

$a^x \bmod N$	c	$a^x \bmod N$	c
$2^0 \bmod 15$	1	$2^2 \bmod 15$	4
$2^1 \bmod 15$	2	$2^3 \bmod 15$	8

We can write the state of the system after application of the Walsh–Hadamard transform and the X gate as follows:

$$|\psi\rangle = \frac{1}{\sqrt{8}} \begin{pmatrix} |000\rangle|0001\rangle + |001\rangle|0001\rangle + |010\rangle|0001\rangle \\ +|011\rangle|0001\rangle + |100\rangle|0001\rangle + |101\rangle|0001\rangle + |110\rangle|0001\rangle \\ +|111\rangle|0001\rangle \end{pmatrix} \tag{7.122}$$

The application of the first block of multiplier converts this state as follows:

$$= \frac{1}{\sqrt{8}} \begin{pmatrix} |000\rangle|0001\rangle + |001\rangle|0010\rangle + |010\rangle|0001\rangle + \\ |011\rangle|0010\rangle + |100\rangle|0001\rangle + |101\rangle|0010\rangle + \\ |110\rangle|0100\rangle + |111\rangle|0010\rangle \end{pmatrix} \tag{7.123}$$

After the application of the second block:

$$= \frac{1}{\sqrt{8}} \begin{pmatrix} |000\rangle|0001\rangle + |001\rangle|0010\rangle + |010\rangle|0100\rangle + \\ |011\rangle|1000\rangle + |100\rangle|0001\rangle + |101\rangle|0010\rangle + \\ |110\rangle|0100\rangle + |111\rangle|1000\rangle \end{pmatrix} \tag{7.124}$$

Application of the final block produces the following state:

$$= \frac{1}{\sqrt{8}} \begin{pmatrix} |000\rangle|1000\rangle + |001\rangle|0010\rangle + |010\rangle|0100\rangle + \\ |011\rangle|0001\rangle + |100\rangle|1000\rangle + |101\rangle|0010\rangle + \\ |110\rangle|0100\rangle + |111\rangle|0001\rangle \end{pmatrix} \tag{7.125}$$

At this stage, if we measure REG 2 (that is, the qubits q[3] to q[6] in the diagram), we can find that the system is projected to the following states: 0001, 0010, 0100, 1000 with equal probability. By comparing these values with Table 7.5, we can conclude that this circuit has performed modular exponentiation.

Exercise 7.5

1. *Implement this circuit using Qiskit and QDK.*
2. *Design a circuit for* 4^x *mod 15.*

7.8 Grover's Search Algorithm

Grover's algorithm [10] uses amplitude amplification, an interesting technique we have not seen so far. This algorithm was introduced by **Lov Grover** [9] in 1996. Notably, this algorithm exhibits quadratic speedup.

Imagine that we have access to a phone directory. We are given a phone number and asked to find to whom the phone number is associated. Phone directories are usually not arranged by the phone number. We have to go through each entry in the phone directory to determine whom the phone number belongs. At an average, it may take $\frac{N}{2}$ attempts to find the right answer, which is somewhat difficult if the phone book is large.

Grover's algorithm solves the problem of unstructured search. The search is called unstructured because we do not have any guarantee that the database is sorted in some fashion. The Grover's algorithm takes a number corresponding to an entry in the database as its input and performs a test to see if this is the special value ω being searched for. This search is done in about $\frac{\pi}{4}\sqrt{N}$ steps, a quadratic speedup compared to the classical algorithm described above. The Grover's search is defined as follows.

Given a set of N elements forming a set $X = \{x_1, x_2, x_3, \ldots, x_N\}$, and given a Boolean function $f : X \rightarrow \{0, 1\}$, the goal is to find an element ω in X such that $f(\omega) = 1$.

Before solving the problem, let us make some assumptions.

- The element ω, which satisfies the equation $f(\omega) = 1$, is unique. We shall relax this condition at the end of the discussion.
- The size of the database is a power of 2, that is, $N = 2^n$. Some padding may be required to achieve this.
- The data are labeled as n-bit Boolean strings $\{0, 1\}^n$.
- The Boolean function f maps $\{0, 1\}^n$ to $\{0, 1\}$.

Let us assume that we define an oracle function $\widehat{O}$ which produces a transformation function: $O \, |x\rangle \rightarrow |f(x)\rangle$. Unfortunately, this is not unitary, as it takes an n-bit string and produces a single bit as an output. The solution is to use an oracle black-box function, as shown in Fig. 7.10, which is quite familiar to us from our past exercises. Let us now define the oracle function $\widehat{O}$, acting on the n-qubit input register $|x\rangle$.

$$\widehat{O} \, |x\rangle = (-1)^{f(x)} |x\rangle = \begin{cases} |x\rangle & \text{if } f(x) = 0 \\ -|x\rangle & \text{if } f(x) = 1 \end{cases} \tag{7.126}$$

Consider the block diagram shown in Fig. 7.28. In this diagram, we start the system with all qubits initialized to state $|0\rangle$. We then apply a Walsh–Hadamard transform to put the qubits in an equal superposition state. The state of the system can be written as follows:

$$|\psi\rangle = \frac{1}{\sqrt{N}} \sum_{x \in \{0, 1\}^n} |x\rangle \tag{7.127}$$

All the qubits have the same amplitude $\frac{1}{\sqrt{N}}$ at this stage. We can plot this in a bar graph, as shown in Fig. 7.29.

The next step is to apply the oracle $\widehat{O}$. The oracle function negates the amplitude of the state ω. The rest of the states remain unaffected by the oracle. Equation (7.128) shows the new system state. We can graph the amplitudes of all possible states, and this is shown in Fig. 7.30.

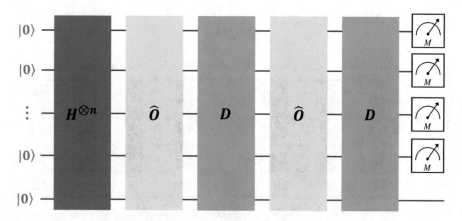

Fig. 7.28 Grover's search algorithm, block diagram

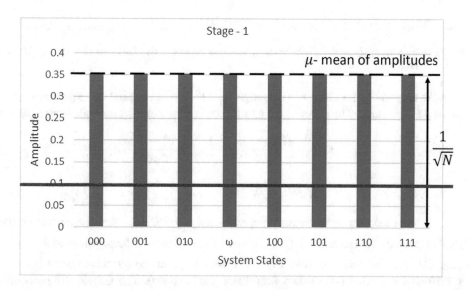

Fig. 7.29 Bar graph showing the amplitudes of the possible systems states at Stage-1, using a 3-qubit system for illustration

$$|\psi\rangle = -\frac{1}{\sqrt{N}} |\omega\rangle + \frac{1}{\sqrt{N}} \sum_{\substack{x \in \{0,1\}^n \\ x \neq \omega}} |x\rangle \qquad (7.128)$$

Now that the amplitude of the state $|\omega\rangle$ has changed, let us calculate the mean of the amplitudes to ascertain we are on the right path. The mean of the amplitudes μ is given by the following standard equation for calculating mean from a set of values

$$\mu = \frac{1}{N} \sum_{x \in \{0,1\}^n} a_x, \qquad (7.129)$$

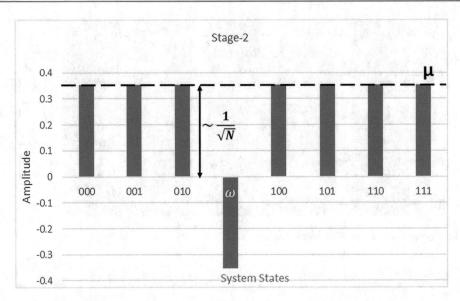

Fig. 7.30 Bar graph after applying the oracle function. Note that the amplitude of ω is negated. The amplitude of rest of the states remains almost the same

where a_x is the amplitude of a given x in the range $\{0, 1\}^n$. We can expand this equation to calculate the value of μ at this stage. The mean of the amplitudes is now given by the following equation:

$$\mu = \frac{1}{N}\left(\frac{(2^n - 1)}{\sqrt{N}} - \frac{1}{\sqrt{N}}\right) = \frac{2^n - 2}{N\sqrt{N}}$$
$$\approx \frac{N}{N\sqrt{N}} = \frac{1}{\sqrt{N}}$$

(7.130)

Hence, the amplitude for most terms is approximately $\frac{1}{\sqrt{N}}$. We see that the amplitude has not significantly changed, and the term $-\frac{1}{\sqrt{N}}$ is relatively insignificant for large values of n.

We can now proceed with the next stage. The next stage in the quantum circuit is called the **Grover diffusion operator** (also called **amplitude purification**). The Grover diffusion operator performs the following mapping:

$$\sum_{x\in\{0,1\}^n} a_x|x\rangle \mapsto \sum_{x\in\{0,1\}^n} (2\mu - a_x)|x\rangle$$

(7.131)

We can plot the bar graph of the system states to get an understanding of this operator. Figure 7.31 illustrates the system state at this stage.

From the graph, we find that the system states almost remain the same, except for the state ω. The amplitude of ω gets amplified by a factor of about 3.

The combination of these two steps – the oracle function and the Grover Diffusion Operator – is called **Grover Iteration**.

To amplify ω further, let us apply the Grover Iteration one more time. Figure 7.32 illustrates the state of the system after the second iteration.

By studying the graph, we can conclude that the amplitude of ω gets magnified linearly $\sim \frac{t}{\sqrt{N}}$ with each iteration (being the number of solutions.) By applying this cycle for $O(\sqrt{N})$ times, the amplitude of ω should exceed a threshold value. Beyond that point, the large negative amplitude of ω (when the

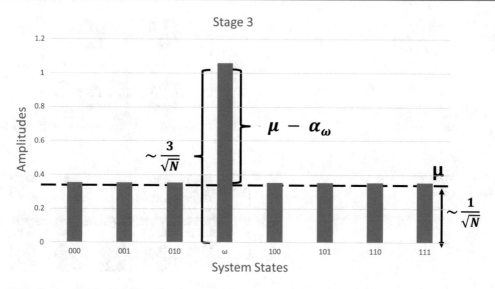

Fig. 7.31 Bar graph at Stage-3. Note that the amplitude of state ω is amplified greatly, while the amplitude of the rest of the states remains almost the same

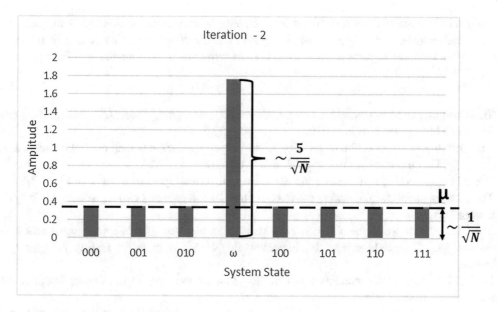

Fig. 7.32 System state after second iteration

oracle $\widehat{O}$ flips it) can reduce the overall mean value causing a reduction in the amplitude when the Grover Diffusion Operator is applied.

Note that Grover's algorithm magnifies the amplitude and not the probability amplitude. With $O(\sqrt{N})$ iterations, the Grover algorithm solves the problem, whereas a classical solution requires $O(N)$ steps. This quadratic speedup is because quantum parallelism computes $f(x)$ for all the $N = 2^n$ states in parallel. If in case there are multiple solutions k to the problem, we would need $O\left(\sqrt{N/k}\right)$ iterations.

Let us now implement a 3-qubit version of Grover's algorithm based on the oracle function and the Grover diffusion operator defined by C. Figgatt [8] et al.

Fig. 7.33 A 3-qubit
Toffoli gate implementing
Grover's oracle for a value
of $\omega = 010$

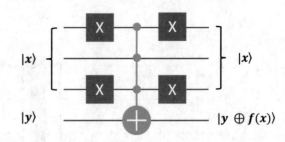

Fig. 7.34 Grover's
diffusion operator

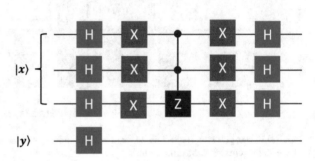

The goal of the oracle function is to implement a unitary transformation function $\widehat{O}$ that acts on an n-qubit input register and a 1-qubit output register. The function applies an X gate to the output register when the input register is equal to ω. We can write this transformation as follows:

$$\widehat{O}|x\rangle^{\otimes n}|y\rangle = |x\rangle^{\otimes n}|y \oplus f(x)\rangle \tag{7.132}$$

This function can be implemented as a multi-control Toffoli gate. Figure 7.33 illustrates a 3-qubit oracle with a hidden value of $\omega = 010$.

The Toffoli gate switches the target (that is, the output register), when all control registers are in state $|1\rangle$. For the circuit shown in Fig. 7.33, this is when the $|x\rangle = 010$.

The quantum circuit for implementing the Grover Diffusion Operator is given in Fig. 7.34.

The code fragments implementing the three control Toffoli gate and Grover's oracle are provided in Listing 7.21.

The CCZ gate shown in the circuit is not a standard implementation. We must implement it as a compound gate. The code snippet implementing this circuit is given in Listing 7.22 for quick reference.

Listing 7.23 puts all components together. Note that we must repeat the Grover Iteration three times to get the desired output.

We find that after the third iteration ω is amplified to the desired level. It is a good exercise to change the number of iterations and observe this growth. The histogram at the end of the cycle is shown in Fig. 7.35.

To complete our discussion on this section, we can derive the underlying math. We start the system with $|x\rangle^{\otimes 3}|y\rangle$ in state $|0\rangle$.

$$|\psi\rangle = |000\rangle|0\rangle \tag{7.133}$$

The next step is to apply a NOT gate to $|y\rangle$ register.

$$|\psi\rangle = |000\rangle|1\rangle \tag{7.134}$$

We then apply a Walsh–Hadamard transform to put the system in an equal superposition state.

```
# The three control Toffoli gate
# qc - the quantum register
# control1, control2, control3 - The control registers
# anc - a temporary work register
# target - the target register, where the transform is applied

def cccx(qc,control1, control2, control3, anc, target):
    qc.ccx(control1,control2,anc)
    qc.ccx(control3,anc,target)
    qc.ccx(control1,control2,anc)
    qc.ccx(control3,anc,target)

# The Grover's Oracle
# qc - the quantum register
# x1, x2, x3 - The input register x
# anc - a temporary work register
# target - the target register, where the transform is applied

def grover_oracle( qc, x1, x2, x3, anc, target):
    qc.x(x3)
    qc.x(x1)
    cccx(qc, x1, x2, x3, anc, target)
    qc.x(x1)
    qc.x(x3)
```

Listing 7.21 Code implementing the Grover's oracle and a 3 control Toffoli gate

$$= \frac{1}{\sqrt{8}} \left(|000\rangle + |001\rangle + |010\rangle + |011\rangle + |100\rangle + \right.$$
$$\left. |101\rangle + |110\rangle + |111\rangle \right) \otimes \frac{1}{\sqrt{2}}(|0\rangle - |1\rangle) \tag{7.135}$$

At this stage, we proceed with the application of the first Grover Iteration. The oracle function flips the state of ω. The resultant state can be derived by applying the gates, step by step.

$$|\psi\rangle = \frac{1}{\sqrt{8}} \left(|000\rangle + |001\rangle - |010\rangle + |011\rangle + |100\rangle + |101\rangle + |110\rangle + |111\rangle \right)$$
$$\otimes \frac{1}{\sqrt{2}} (|0\rangle - |1\rangle) \tag{7.136}$$

The next step is to apply the Grover Diffusion operator, which amplifies each state by a factor $(2\mu - a_i)$. The action of the H gate on register $|y\rangle$ changes it to state $|1\rangle$. Since $\mu = \frac{1}{\sqrt{8}}$, we find that the amplitudes of all states remain the same. However, the state of ω becomes $\frac{3}{\sqrt{8}}$. The new state can be written as:

$$|\psi\rangle = \left[\frac{1}{\sqrt{8}} \left(|000\rangle + |001\rangle + |011\rangle + |100\rangle + |101\rangle + |110\rangle + |111\rangle \right) + \frac{3}{\sqrt{8}} |101\rangle \right] \otimes |1\rangle \tag{7.137}$$

For the rest of our discussions, let us assume $\mu = \frac{1}{\sqrt{8}}$. For large values of n, the changes to μ is insignificant. At the end of the second iteration, the system changes to the following state.

```
# The CCZ gate
# qc - the quantum register
# control1, control2 - The control registers
# target - the target register, where the Z transform is applied

def ccz(qc, control1, control2, target):
    qc.h(target)
    qc.ccx(control1, control2, target)
    qc.h(target)

# The Grover's Diffusion Operator
# qc - the quantum register
# x1, x2, x3 - The input register x
# target - A temporary register

def grover_diffusion_operator(qc, x1, x2, x3, target):
    qc.h(x1)
    qc.h(x2)
    qc.h(x3)
    qc.h(target) # Bring this back to state 1 for next stages
    qc.x(x1)
    qc.x(x2)
    qc.x(x3)
    ccz(qc, x1, x2, x3 )
    qc.x(x1)
    qc.x(x2)
    qc.x(x3)
    qc.h(x1)
    qc.h(x2)
    qc.h(x3)
```

Listing 7.22 Code implementing the Grover Diffusion Operator

$$|\psi\rangle = \left[\frac{1}{\sqrt{8}}\left(|000\rangle + |001\rangle + |011\rangle + |100\rangle + |101\rangle + |110\rangle + |111\rangle\right) + \frac{5}{\sqrt{8}}|101\rangle\right] \otimes |1\rangle$$

$$(7.138)$$

Finally, at the end of the third state, the system changes to:

$$|\psi = \left[\frac{1}{\sqrt{8}}\left(|000\rangle + |001\rangle + |011\rangle + |100\rangle + |101\rangle + |110\rangle + |111\rangle\right) + \frac{7}{\sqrt{8}}|101\rangle\right] \otimes |1\rangle \quad (7.139)$$

From Eq. (7.139) we find that the amplitude of ω is magnified to $\frac{7}{\sqrt{8}}$. This proves Grover's algorithm.

```
#
# Grover's algorithm
#
shots = 1024
q = QuantumRegister(3 , 'q')
t = QuantumRegister(2 , 't')
c = ClassicalRegister(3 , 'c')
qc = QuantumCircuit(q,t, c)


qc.h(q[0])
qc.h(q[1])
qc.h(q[2])
qc.h(t[0])


grover_oracle (qc, q[0], q[1], q[2], t[1], t[0])
grover_diffusion_operator (qc, q[0], q[1], q[2],t[0])


grover_oracle (qc, q[0], q[1], q[2], t[1], t[0])
grover_diffusion_operator (qc, q[0], q[1], q[2],t[0])


grover_oracle (qc, q[0], q[1], q[2], t[1], t[0])
grover_diffusion_operator (qc, q[0], q[1], q[2],t[0])


qc.measure(q[0],c[0])
qc.measure(q[1],c[1])
qc.measure(q[2],c[2])


backend = Aer.get_backend("qasm_simulator")
job = execute(qc, backend=backend, shots=shots)
counts = job.result().get_counts()
plot_histogram(counts)
```

Listing 7.23 Grover's algorithm

Exercise 7.6

1. *Derive the complete math and workout changes to μ.*
2. *What happens when you perform a fourth iteration?*
3. *Develop a generic code that works for any value of n.*
4. *In what other systems could you implement Grover's algorithm?*

◄

Code example for QDK is included in the Appendix.

Fig. 7.35 The histogram
of the Grover's algorithm,
showing the magnified
amplitude of state $\omega = 010$

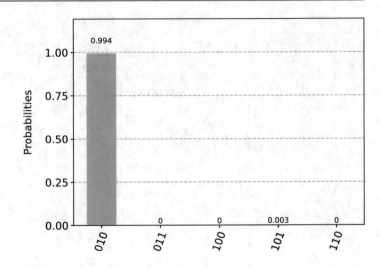

7.9   Shor's Algorithm

 Note: Reading Sect. 6.8 on RSA security may be helpful to read the following section in this chapter.

In the previous chapter, we learned about the factorization of integers and RSA security. We learned how the period r of a function $f(x) \equiv a^x \bmod N$ could be used to factorize an integer N into two prime numbers p and q, such that $N = p \times q$. Recall that a is a number that is coprime with N, and that $a^r \bmod N = 1$.

In 1994, the American mathematician **Peter Shor** proposed a novel quantum algorithm for factorizing large integers [11], which runs on a polynomial-time. Shor's algorithm is not just theoretical work. There is some experimental proof that Shor's algorithm does work. In 2001, a group of scientists at IBM factorized the integer 15, using Nuclear Magnetic Resonance (NMR). In 2012, the integer 21 was factorized, as of 2020, this is the most significant number to have been factorized using Shor's algorithm [4]. When large-scale quantum computers with error-corrected qubits are available, we may be able to factorize large integers [5]. We shall factorize 15 later in this section.

Shor solved the problem of factorizing integers by shifting the problem domain to finding the period of the modular exponentiation function. Shor's problem is stated as follows.

Given an integer N, and two prime numbers p and q such that $N = p \times q$. Let $a \in \mathbb{Z}_N$. Let a modular exponentiation function $f_a : \mathbb{Z}_+ \to \mathbb{Z}_N$ defined as $f_a(x) = a^x \bmod N$ satisfy the condition, $f(x) = f(x_0) \Leftrightarrow x = x_0 + kr$, where r is the period of f_a, and k is an integer. Shor's problem is determining the period r by querying f_a.

The block diagram of Shor's algorithm is given in Fig. 7.36.

The block diagram pretty much looks like Simon's algorithm, and it uses steps familiar to us now—modular exponentiation and inverse QFT.

An interesting feature of Shor's algorithm is the usage of inverse QFT to find the period r of the equation $f(x) \equiv a^x \bmod N$. The period r can be used to find the factors of N.

The algorithm uses two banks of qubit registers. The upper bank is called the source register, and the lower bank is called the target register. The oracle function performs modular exponentiation on

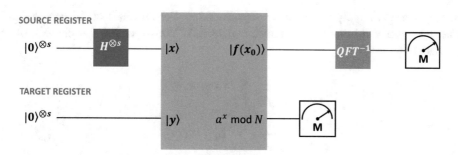

Fig. 7.36 Block diagram of Shor's algorithm

the source register and stores the values in the target register. This process is done in one step using quantum parallelism.

The number of qubits s in the source, and target registers must be in the following range:

$$N^2 \leq 2^s \leq 2N^2 \tag{7.140}$$

From this range, $\frac{2^s}{r} > N$ and it guarantees that when estimating the period r, N terms are contributing to the probability amplitude, even if r is close to $\frac{N}{2}$. This assumption creates a sufficiently large state-space in the source register, which can contain even the longest periods. By having a large value of s, we can explore a dense solution space, and the period r can be estimated as accurately as possible. The modular exponentiation circuits require additional qubits to store intermediate information. Hence, we have selected that the target register is also made of s qubits.

Before proceeding with the algorithm, we must perform some preprocessing tasks. First, we need to check whether N is a prime number. There are some classical methods to do the primality check, and it can be done efficiently on a classical computer. The second task is to identify a and confirm that N and a are coprime. The third task is to determine the number of qubits required. After completing these steps, the circuit can be constructed for execution.

Having constructed a quantum circuit, as shown in Fig. 7.36, let us examine the inner working of this algorithm. As usual, we start the system with both the registers initialized to state $|0\rangle$. The following equation describes the system state at this stage.

$$|\psi\rangle = |0\rangle^{\otimes s} |0\rangle^{\otimes s} \tag{7.141}$$

We then apply a Walsh–Hadamard transformation to the source register, transforming the source register into a single large superposition state comprising of all possible states of $|x\rangle$ from $|000\ldots0\rangle$ to $|111\ldots1\rangle$, including all the states in between.

$$
\begin{aligned}
&= H^{\otimes s} |0\rangle^{\otimes s} |0\rangle^{\otimes s} \\
&= \frac{1}{\sqrt{2^s}} (|0\rangle + |1\rangle)^{\otimes s} |0\rangle^{\otimes s} \\
&= \frac{1}{\sqrt{2^s}} (|000\ldots0\rangle + |000\ldots1\rangle + \cdots + |111\ldots1\rangle) |0\rangle^{\otimes s} \\
&= \frac{1}{\sqrt{2^s}} \sum_{x=0}^{2^s-1} |x\rangle |0\rangle^{\otimes s}
\end{aligned}
\tag{7.142}
$$

The next step is to apply the modular exponentiation. For all states of the source register, we calculate $y \oplus f(x)$ and store in the target register $|y\rangle$. We explained the concept of modular exponentiation in Sect. 7.7.

$$= \frac{1}{\sqrt{2^s}} \sum_{x=0}^{2^s-1} |x\rangle f(x) |0\rangle^{\otimes s}$$

$$= \frac{1}{\sqrt{2^s}} \sum_{x=0}^{2^s-1} |x\rangle |y \oplus f(x)\rangle \qquad (7.143)$$

$$= \frac{1}{\sqrt{2^s}} \sum_{x=0}^{2^s-1} |x\rangle \, |f(x)\rangle$$

After performing the modular exponentiation, we measure the target register. Let us assume that the target register projects to some value $f(x_0)$. Measuring the target register has a side effect of collapsing the superposition state of the source register to the states for which $f(x_0) = a^{x_0} \bmod N$. Since $f(x)$ is periodic in r, the possible values are of the form $(x_0 + kr)$. Note that x_0 is the least x such that $f(x)$ is what was measured in the target register. This ensures k is strictly nonnegative. The new state of the system is as follows:

$$= \frac{1}{\sqrt{\frac{2^s}{r}}} \sum_{k=0}^{\frac{2^s}{r}-1} |x_0 + kr\rangle |f(x_0)\rangle \qquad (7.144)$$

Since the period of $f(x)$ is r, the measurement reduces the range of the source register to $\frac{2^s}{r}$. There are 2^s states. Hence, the maximum number of periods we can have is $\frac{2^s}{r}$, which is the range the system collapses to when we measure the target register. At this stage, if we measure the source register, the probability of getting a particular state is shown in Fig. 7.37. These states are the ones for which $f(x) = f(x_0)$ when we measured the target register. Hence, when we measure the source register, we may get one of the values of $(x_0 + kr)$ for some value of k. Since we do not know the value of x_0, it is impossible to determine the value of r at this stage. Hence, we apply an inverse QFT to identify the period. We can drop the terms pertaining to the target register, as it is measured, and we do not need it anymore.

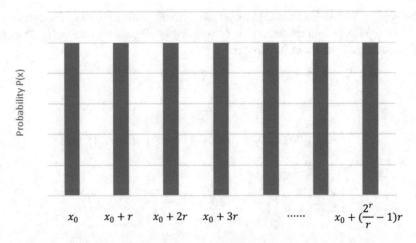

Fig. 7.37 Probability of measuring a certain state at the source register

As discussed, the next stage is to apply an inverse QFT to the source register. From Eq. (7.59),

$$|x\rangle = \frac{1}{\sqrt{N}} \sum_{y=0}^{(N-1)} e^{\frac{-2\pi i\, xy}{N}} |y\rangle, x = 0 : N - 1 \tag{7.145}$$

Applying the inverse QFT in Eq. (7.144) and dropping the term pertaining to the target register, we get:

$$|\psi\rangle = \frac{1}{\sqrt{2^s}} \frac{1}{\sqrt{Q}} \sum_{k=0}^{(Q-1)} \sum_{y=0}^{(2^s-1)} e^{\frac{-2\pi i\, (x_0+kr)\, y}{2^s}} |y\rangle, \tag{7.146}$$

where $Q = \frac{2^s}{r}$. The exponentials can be expanded now.

$$|\psi\rangle = \frac{1}{\sqrt{2^s}} \frac{1}{\sqrt{Q}} \sum_{k=0}^{(Q-1)} \sum_{y=0}^{(2^s-1)} \left(e^{\frac{-2\pi i x_0 y}{2^s}} \cdot e^{\frac{-2\pi i k r y}{2^s}} \right) |y\rangle \tag{7.147}$$

Interested readers can derive the next step in detail and obtain Eq. (7.148). In this equation, we have changed the order of the summation

$$|\psi\rangle = \frac{1}{\sqrt{2^s Q}} \sum_{y=0}^{(2^s-1)} e^{\frac{-2\pi i x_0 y}{2^s}} \sum_{k=0}^{(Q-1)} \left(e^{\frac{-2\pi i r y}{2^s}} \right)^k |y\rangle \tag{7.148}$$

The summation on the right-hand side is a geometric series. The sum of this term is non-zero, iff $e^{\frac{-2\pi i r y}{2^s}} = 1$. This is possible iff $\frac{ry}{2^s}$ is an integer. This implies that the values of y, which are multiples of $\frac{2^s}{r}$, can have non-zero amplitudes. In other words when $\frac{2^s}{r}$ is an integer, due to constructive quantum interference, there is a peak in the value of y.

We can now calculate the probability of measuring a certain value y.

$$P(y) = \frac{1}{2^s Q} \left| \sum_{k=0}^{Q-1} \left(e^{\frac{-2\pi i\, kr\, y}{2^s}} \right) \right|^2 \tag{7.149}$$

From our discussions above, when $\frac{ry}{2^s}$ is an integer, the exponentials add up nicely, and the phase is $=1$.

$$P(y) = \frac{r}{2^s 2^s} \left[\frac{2^s}{r} \frac{2^s}{r} \right] = \frac{1}{r}. \tag{7.150}$$

Let us now return to Eq. (7.148). The summation on the right-hand side reduces to a set of r values due to the quantum interference and the equation contracts.

Assume that we measure the source register, and it projects y into the classical register c. The classical register then reads a value, which is a multiple of $\frac{2^s}{r}$, that is, $c = z \frac{2^s}{r}$, where z is a random variable in the range $0 \ldots r - 1$. With a little bit of derivation, the new system state can be written as the following equation:

$$|\psi\rangle = \frac{1}{\sqrt{r}} \sum_{z=0}^{(r-1)} e^{\frac{-2\pi i x_0 z}{r}} \left| z \frac{2^s}{r} \right\rangle \tag{7.151}$$

The value of c is given by:

$$c = z\frac{2^s}{r} \tag{7.152}$$

Since both c and 2^s are known, the ratio of z over r can be determined.

$$\frac{c}{2^s} = \frac{z}{r} \tag{7.153}$$

From this, we can obtain r using continued fractions. If z and r have no common divisors, that is, $\gcd(z, r) = 1$, then,

$$\frac{c}{2^s} = a_0 + \cfrac{1}{a_1 + \cfrac{1}{a_2 + \cfrac{1}{a_3 + \cfrac{1}{\ldots}}}} \tag{7.154}$$

The approximation to the ratio $\frac{z}{r}$ can be calculated using the continued fraction expansion relations:

$$
\begin{aligned}
z_0 &= a_0 \\
r_0 &= 1 \\
z_1 &= a_0 a_1 + 1 \\
r_1 &= a_1 \\
z_n &= a_n z_{n-1} + z_{n-2} \\
r_n &= a_n r_{n-1} + r_{n-2}
\end{aligned} \tag{7.155}
$$

For each value of z_i and r_i, we check whether $\gcd(z_i, r_i) = 1$. After verifying this, we calculate the function $f(x) = a^x \bmod N$ by substituting x with r_i. If $f(r_i) = 1$, it means the period of the function $f(x)$ is r_i. If not, we go back and repeat the algorithm with another value of a or repeat the experiment.

Shor's algorithm runs partly on a classical computer and partly on a quantum computer. Primality check of N, choosing a value for a, and determining whether N and a are coprime are done as a preprocessing step on a classical computer. Continued fractional expansion and verification of r are done as postprocessing steps. Once r is known, one of the factors of N can be calculated from $\gcd\left(a^{r/2} \pm 1, N\right)$.

The fastest known classical algorithm that can factor large integers is the **General Number Field Sieve algorithm** [12], which works in a sub-exponential time of $O\left(e^{1.9(\log N)^{\frac{1}{3}}(\log\log N)^{\frac{2}{3}}}\right)$. According to a study done in 2010 [13], factoring a 768-bit RSA modulus using the number sieve can take about 1,500 years on a system with 2.2 GHz Opteron AMD processor with 2GB of RAM. Shor's algorithm solves this in polynomial time of $O((\log N)^2(\log\log N)(\log\log\log N))$, and it is an example of **Bounded-error Quantum Polynomial (BQP)** problem [15–18].

7.9.1   Factoring Number 15

In this section, we try to factorize 15 using Shor's algorithm. Here are the preprocessing steps.

- Select $N = 15$, 15 is a prime number.
- Let us select $a = 13$. $\gcd(13, 15) = 1$.

- Select the number of qubits. $2^4 > N$. So let's select two sets of 4-qubit registers. Ideally, we should go for 8-qubit registers, but since the range is small, 4-qubit registers should be sufficient.

After the preprocessing steps are completed, the quantum circuit is constructed, and we start the system by initializing both the registers to $|0\rangle$.

$$|\psi\rangle = |0000\rangle|0000\rangle \qquad (7.156)$$

Apply Walsh–Hadamard transform to the source register:

$$|\psi\rangle = \frac{1}{\sqrt{16}} \sum_{k=0}^{15} |k\rangle|0000\rangle \qquad (7.157)$$

Compute $13^k \bmod 15$ and apply it to the target register.

k	0	1	2	3	4	5	6	7	8	9	10	11	12	13	14	15
$f(k)$	1	13	4	7	1	13	4	7	1	13	4	7	1	13	4	7

From this table, we can easily find that the period is 4. But, let us continue with the algorithm and verify this at the end,

$$
\begin{aligned}
|\psi\rangle &= \frac{1}{\sqrt{16}} \sum_{k=0}^{15} |k\rangle|f(k)\rangle \\
&= \frac{1}{\sqrt{16}} \left(|0\rangle|f(0)\rangle + |1\rangle|f(1)\rangle + \ldots \right) \\
&= \frac{1}{\sqrt{16}} \left(
\begin{array}{l}
|0\rangle|1\rangle + |1\rangle|13\rangle + |2\rangle|4\rangle + |3\rangle|7\rangle + |4\rangle|1\rangle + \\
|5\rangle|13\rangle + |6\rangle|4\rangle + |7\rangle|7\rangle + |8\rangle|1\rangle + |9\rangle|13\rangle + \\
|10\rangle|4\rangle + |11\rangle|7\rangle + |12\rangle|1\rangle + |13\rangle|13\rangle + \\
|13\rangle|4\rangle + |15\rangle|7\rangle
\end{array}
\right)
\end{aligned}
\qquad (7.158)
$$

At this stage, we measure the target register. The target register can measure any of the 4 possible states – $|1\rangle$, $|13\rangle$, $|4\rangle$, $|7\rangle$ which are in equal probability.

Assume that we got a "7." This collapses the superposition state of the source register into the states for which $f(x) = 7$. The target register is measured. We don't need it anymore. Hence it is dropped from the equation. The new system state is as follows, after normalization.

$$|\psi\rangle = \sqrt{\frac{4}{16}} (|3\rangle + |7\rangle + |11\rangle + |15\rangle) \qquad (7.159)$$

The next step is to apply the QFT.

$$|\psi\rangle = \sqrt{\frac{4}{16}} \frac{1}{\sqrt{16}} \sum_{y=0}^{16} \left[e^{\frac{-2\pi i\, y\cdot 3}{16}} + e^{\frac{-2\pi i\, y\cdot 7}{16}} + e^{\frac{-2\pi i\, y\cdot 11}{16}} + e^{\frac{-2\pi i\, y\cdot 15}{16}} \right] |y\rangle \qquad (7.160)$$

The probability of measuring y is defined as follows:

$$P(y) = \left| \frac{1}{8} \left(e^{\frac{-2\pi i\, y\cdot 3}{16}} + e^{\frac{-2\pi i y\cdot 7}{16}} + e^{\frac{-2\pi i y\cdot 11}{16}} + e^{\frac{-2\pi i\, y\cdot 15}{16}} \right) \right|^2 \qquad (7.161)$$

When we solve this, we get an equal probability $\frac{1}{4}$ for the 4 states $|0\rangle$, $|4\rangle$, $|8\rangle$, $|12\rangle$. The rest of the states evaluate to zero probability. Interested readers can compute this in WolframAlpha at the link below.

https://www.wolframalpha.com/

Let's assume that we measured 12 with the source register. This means $12r = 16k$, $k = 3$. Therefore $r = 4$.

7.9.2   Quantum Circuit to Factorize 15

In this section, we construct a quantum circuit and demonstrate that Shor's algorithm works by factorizing 15. This circuit is based on the work done by Lieven M.K et al. in 2001 using Nuclear Magnetic Resonance [14]. Reading this early pioneering research paper can be an enchanting experience!

The circuit shown in Figure 7.38 finds the period of the function $f(x) = 7^x \bmod 15$. This circuit is an optimized version that uses 3 qubits for the source register. Since the period of this function is relatively smaller, the choice of 3 qubits holds good.

We start this circuit by preparing the source register in an equal superposition state. Modular exponentiation is done for all the states of the source register and stored in the target register. After this stage, we perform the inverse QFT and measure the source register. The X gate in qubit q[6] in the diagram provides for the phase kickback.

Let us now verify the modular exponentiation circuit in this implementation. The system state after the initialization stage is:

$$|\psi\rangle = \frac{1}{\sqrt{8}} \begin{pmatrix} |000\rangle|0001\rangle + |001\rangle|0001\rangle + \\ |010\rangle|0001\rangle + |011\rangle|0001\rangle + |100\rangle|0001\rangle + \\ |101\rangle|0001\rangle + |110\rangle|0001\rangle + |111\rangle|0001\rangle \end{pmatrix} \tag{7.162}$$

After preparing the register to the initial state, we apply a series of CNOT and CCNOT gates to perform the modular exponentiation. Here are the steps that perform this task.

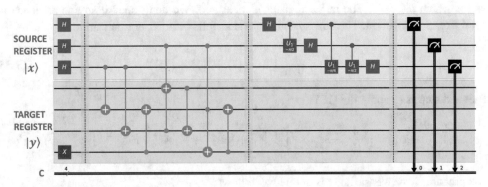

Fig. 7.38 Quantum circuit that factorizes 15. Based on the research by Lieven M.K et al. in 2001 using Nuclear Magnetic Resonance

STEP 1: The first CNOT gate uses q[2] as the control qubit and q[4] as the target qubit

$$= \frac{1}{\sqrt{8}} \begin{pmatrix} |000\rangle|0001\rangle + |001\rangle|0101\rangle + |010\rangle|0001\rangle+ \\ |011\rangle|0101\rangle + |100\rangle|0001\rangle+ \\ |101\rangle|0101\rangle + |110\rangle|0001\rangle + |111\rangle|0101\rangle \end{pmatrix} \quad (7.163)$$

STEP 2: The second CNOT gate uses q[2] as the control qubit and q[5] as the target qubit.

$$= \frac{1}{\sqrt{8}} \begin{pmatrix} |000\rangle|0001\rangle + |001\rangle|0111\rangle + |010\rangle|0001\rangle+ \\ |011\rangle|0111\rangle + |100\rangle|0001\rangle+ \\ |101\rangle|0111\rangle + |110\rangle|0001\rangle + |111\rangle|0111\rangle \end{pmatrix} \quad (7.164)$$

STEP 3: The third CNOT gate with q[6] as the control and q[4] as the target.

$$= \frac{1}{\sqrt{8}} \begin{pmatrix} |000\rangle|0101\rangle + |001\rangle|0011\rangle+ \\ |010\rangle|0101\rangle + |011\rangle|0011\rangle+ \\ |100\rangle|0101\rangle + |101\rangle|0011\rangle+ \\ |110\rangle|0101\rangle + |111\rangle|0011\rangle \end{pmatrix} \quad (7.165)$$

STEP 4: CCNOT gate with q[1] and q[5] as the control gates and q[3] as the target qubit.

$$= \frac{1}{\sqrt{8}} \begin{pmatrix} |000\rangle|0101\rangle + |001\rangle|0011\rangle + |010\rangle|0101\rangle+ \\ |011\rangle|1011\rangle + |100\rangle|0101\rangle+ \\ |101\rangle|0011\rangle + |110\rangle|0101\rangle + |111\rangle|1011\rangle \end{pmatrix} \quad (7.166)$$

STEP 5: The fourth CNOT gate with q[3] as the control and q[5] as the target.

$$= \frac{1}{\sqrt{8}} \begin{pmatrix} |000\rangle|0101\rangle + |001\rangle|0011\rangle + |010\rangle|0101\rangle+ \\ |011\rangle|1001\rangle + |100\rangle|0101\rangle+ \\ |101\rangle|0011\rangle + |110\rangle|0101\rangle + |111\rangle|1001\rangle \end{pmatrix} \quad (7.167)$$

STEP 6: CCNOT gate with q[1] and q[4] as the control and q[6] as the target.

$$= \frac{1}{\sqrt{8}} \begin{pmatrix} |000\rangle|0101\rangle + |001\rangle|0011\rangle + |010\rangle|0100\rangle+ \\ |011\rangle|1001\rangle + |100\rangle|0101\rangle+ \\ |101\rangle|0011\rangle + |110\rangle|0100\rangle + |111\rangle|1001\rangle \end{pmatrix} \quad (7.168)$$

STEP 7: The fifth CNOT gate with q[6] as the control and q[4] as the target.

$$= \frac{1}{\sqrt{8}} \begin{pmatrix} |000\rangle|0001\rangle + |001\rangle|0111\rangle + |010\rangle|0100\rangle+ \\ |011\rangle|1101\rangle|100\rangle|0001\rangle+ \\ |101\rangle|0111\rangle + |110\rangle|0100\rangle + |111\rangle|1101\rangle \end{pmatrix} \quad (7.169)$$

After performing the modular exponentiation, the target register is in equal probability with the states: $|1\rangle$, $|7\rangle$, $|4\rangle$, $|13\rangle$. This result very well aligns with the modular exponentiation shown in

Fig. 7.39 Factorization of
15. Plot of the possible
states of the source register
at the end

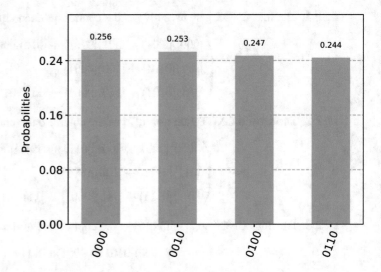

Eq. (7.170). We readily see that the period r is 4. Nevertheless, let us continue testing this circuit and validate this later.

$$
\begin{aligned}
7^0 \bmod 15 &= 1 \\
7^1 \bmod 15 &= 7 \\
7^2 \bmod 15 &= 4 \\
7^3 \bmod 15 &= 13 \\
7^4 \bmod 15 &= 1
\end{aligned}
\tag{7.170}
$$

The final step of Shor's algorithm is to perform the inverse QFT on the source register and measure it. When we do this experiment with 1,024 shots, we get a distributed measurement of states $|0\rangle$, $|2\rangle$, $|4\rangle$, $|6\rangle$ with almost equal probability. The outcome of this experiment is shown in the histogram Fig. 7.39.

By analyzing the distribution pattern, we can figure out that the periodicity in the amplitude of $|y\rangle$ is 2. Therefore, we can calculate r as $r = \frac{2^3}{2} = 4$, which we saw when we performed the modular exponentiation in Eq. (7.170). Now since r is known, we can calculate the factors of N using the brute force method.

$$
\begin{aligned}
\gcd\left(a^{r/2} + 1, 15\right) &= \gcd\left(7^2 + 1, 15\right) = 5 \\
\gcd\left(a^{r/2} - 1, 15\right) &= \gcd\left(7^2 - 1, 15\right) = 3
\end{aligned}
\tag{7.171}
$$

We conclude that the factors of 15 are 5 and 3. Shor's algorithm indeed works!

Listing 7.24 provides the source code to construct Shor's algorithm factorizing 15 using Qiskit. A version that runs on Q# or other systems can be easily constructed based on the logic developed.

Code example for QDK is included in the Appendix.

```
# The inverse QFT circuit
# qc - quantum circuit
# b  - the target register
# n - number of qubits in the register

def iqft_cct (qc, b, n):
    for i in range (n):
        for j in range (1, i+1):
            # for inverse transform, we have to use negative angles
            qc.cu1(  -math.pi / 2** ( i -j + 1 ), b[j - 1], b[i])
        # the H transform should be done after the rotations
        qc.h(b[i])
    qc.barrier()

# Shor's algorithm to factorize 15 using 7^x mod 15.
numberofqubits = 7
shots = 1024
q = QuantumRegister(numberofqubits , 'q')
c = ClassicalRegister(4 , 'c')
qc = QuantumCircuit(q, c)

# Initialize source and target registers
qc.h(0)
qc.h(1)
qc.h(2)
qc.x(6)
qc.barrier()

# Modular exponentiation 7^x mod 15
qc.cx(q[2],q[4] )
qc.cx(q[2],q[5] )
qc.cx(q[6],q[4] )
qc.ccx(q[1],q[5],q[3] )
qc.cx(q[3],q[5] )
qc.ccx(q[1],q[4],q[6] )
qc.cx(q[6],q[4] ) #
qc.barrier()

# IQFT. Refer to implementation from earlier examples
iqft_cct (qc, q, 3)

# Measure
qc.measure(q[0], c[0])
qc.measure(q[1], c[1])
qc.measure(q[2], c[2])
```

Listing 7.24 Code factorizing 15, using Shor's algorithm.

7.10   A Quantum Algorithm for K-Means

The k-means algorithm is perhaps the widely used unsupervised machine learning algorithm. Given input vectors $v_i = \mathbb{R}^d$; $i \in [N]$, the algorithm partitions the inputs into k clusters according to a similarity measure. The similarity measure (aka, the distance function) refers to the Euclidean distance between the data elements. Data elements are assigned to the nearest cluster according to the distance function used. The algorithm produces k cluster centers, called mean or centroid of the data elements in the cluster.

The general workflow for the classical k-means algorithm is as follows:

- Given an input dataset V of vectors $v_i = \mathbb{R}^d$; $i \in [N]$
- At each step t, let the k clusters are the sets C_j^t; $j \in [k]$. Let the centroids be denoted by the vector c_j^t.
- At each iteration, the data points v_i are assigned to C_j^t such that:

$$C_1^t \cup C_2^t \cup \cdots \cup C_k^t = V \text{ and } C_i^t \cap C_j^t = 0 \text{ for } i \neq j \qquad (7.172)$$

If we assume that the Euclidean distance between the vectors v_i and $c_j^t = d\left(v_i, c_j^t\right)$, then the label $l(v_i)^t$ corresponding to the closest centroid assigned to vectors v_i at step t is as follows:

$$l(v_i)^t = \text{argmin}_{j \in [k]} \left(d\left(v_i, c_j^t\right)\right) \qquad (7.173)$$

The centroids are updated so that the new centroid is the average of all points assigned to that cluster:

$$c_j^{t+1} = \frac{1}{\left|c_j^t\right|} \sum_{i \in c_j^t} v_i \qquad (7.174)$$

The algorithm converges once we achieve a small threshold ϵ such that

$$\frac{1}{k} \sum_{j=1}^{k} d\left(c_j^t, c_j^{t-1}\right) \leq \epsilon \qquad (7.175)$$

The loss function we are trying to minimize in this algorithm is the Residual Sum of Squares, and it is defined as follows:

$$\text{RSS} := \sum_{j \in [k]} \sum_{i \in C_j} d\left(c_j, v_i\right)^2 \qquad (7.176)$$

K-means is an example of an NP-hard problem. Each iteration runs with a complexity of $O(kNd)$. The worst-case complexity could be super-polynomial. The number of iterations is $2^{\omega^{\sqrt{N}}}$. Figure 7.40 illustrates this workflow.

The quantum version of the k-means algorithm [19] follows a similar workflow. It uses a quantum distance estimator and runs partially on a classical computer. Figure 7.41 illustrates the workflow for the quantum version of the k-means.

Fig. 7.40 The classical
k-means algorithm

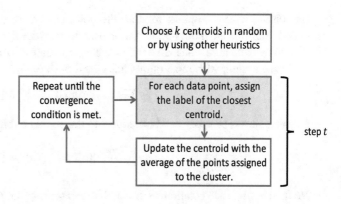

Fig. 7.41 The quantum
k-means algorithm

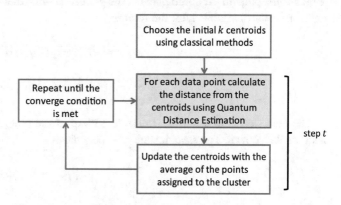

Fig. 7.42 The quantum
distance estimator

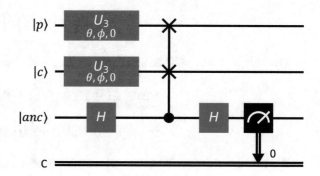

7.10.1 Quantum Distance Estimation

Recall the following equation that calculates the Euclidean distance between two points p and q.

$$\begin{aligned}
\| q - p \| &= \sqrt{\| p \|^2 + \| q \|^2 - 2p \cdot q} \\
&= \sqrt{2 - 2 \langle p, q \rangle} \\
&= d(p, q)
\end{aligned} \tag{7.177}$$

The goal of the quantum distance estimation circuit is to arrive at a state similar to the above equation. Consider the quantum circuit shown in Figure 7.42.

This circuit uses two qubits to store the point and the centroid, for which the distance is to be estimated. There is an ancilla qubit that is used to perform the **swap test**. The swap test determines how much two quantum states differ. Let us now calculate the inner workings of this circuit. As usual, we begin the system with the 3 qubits initialized to state $|0\rangle$.

$$|\psi\rangle = |p\rangle|c\rangle|anc\rangle = |0\rangle|0\rangle|0\rangle \tag{7.178}$$

The next step is to apply an H transform to the ancilla qubit. The new system state is:

$$|\psi\rangle \stackrel{H[\text{anc}]}{\to} |0\rangle|0\rangle \frac{1}{\sqrt{2}}(|0\rangle + |1\rangle) \tag{7.179}$$

We then apply a U3 rotation to the qubits representing the point and the centroid. The rotational angles correspond to the coordinates of the points. Assume that $|i\rangle$ and $|j\rangle$ are the resultant state of the point and centroid qubits after the rotations.

$$|\psi\rangle \stackrel{U3}{\to} |i\rangle|j\rangle \frac{1}{\sqrt{2}}(|0\rangle + |1\rangle)$$

$$\to \frac{1}{\sqrt{2}}(|0\rangle|i\rangle|j\rangle + |1\rangle|i\rangle|j\rangle) \tag{7.180}$$

The next step is to perform a controlled swap of the point and centroid qubits using the ancilla qubit as the control. The system state now becomes:

$$\stackrel{\text{swap}}{\to} \frac{1}{\sqrt{2}}(|0\rangle|i\rangle|j\rangle + |1\rangle|j\rangle|i\rangle) \tag{7.181}$$

For the sake of simplicity, let us write this equation as follows:

$$\to \frac{1}{\sqrt{2}}(|0\rangle|ij\rangle + |1\rangle|ji\rangle) \tag{7.182}$$

The next step is to perform an H gate on the ancilla qubit.

$$\xrightarrow{H[\text{anc}]} \frac{1}{2}(|0\rangle|ij\rangle + |1\rangle|ij\rangle + |0\rangle|ji\rangle - |1\rangle|ji\rangle)$$

$$\to \frac{1}{2}(|0\rangle(|ij\rangle + |ji\rangle)) + \frac{1}{2}(|1\rangle(|ij\rangle - |ji\rangle)) \tag{7.183}$$

If we measure the ancilla qubit, the probability of measuring a "0" in that register is given by:

$$|\langle 0|\psi\rangle|^2 = \frac{1}{4} \||ij\rangle + |ji\rangle|^2$$

$$= \frac{1}{4}[(\langle ij| + \langle ji|)(|ij\rangle + |ji\rangle)]$$

$$= \frac{1}{4}[2 + \langle ij|ji\rangle + \langle ji|ij\rangle] \tag{7.184}$$

$$= \frac{1}{2} + \frac{1}{2}|\langle i|j\rangle|^2$$

In the above equation, if $\langle i|j\rangle = 1$, then the probability of measuring a "0" in the ancilla qubit is "1." From Eq. (2.60), this is possible only if $i = j$. Therefore, if we measure a "0" in the ancilla qubit, it means the qubits $|p\rangle$ and $|c\rangle$ are close to each other. Note that the probability of measuring a "0" in the ancilla qubit is $\frac{1}{2}$, if the qubits $|p\rangle$ and $|c\rangle$ are in orthogonal states.

Let us calculate the probability of measuring a "1" in the ancilla qubit.

$$
\begin{aligned}
|\langle 1|\psi\rangle|^2 &= \frac{1}{4} \||ij\rangle - |ji\rangle\|^2 \\
&= \frac{1}{4}\left[(\langle ij| - \langle ji|)\,(|ij\rangle - |ji\rangle)\right] \\
&= \frac{1}{4}\left[2 - \langle ij|ji\rangle - \langle ji|ij\rangle\right] \\
&= \frac{1}{4}\left[2 - 2|\langle i|j\rangle|^2\right]
\end{aligned}
\tag{7.185}
$$

This equation is similar to Eq. (7.177), which describes the Euclidean distance between two points.

By comparing Eqs. (7.177) and (7.185), we can say that the probability of measuring a "1" in the ancilla qubit is proportional to the square of the distance between the qubits $|p\rangle$ and $|c\rangle$.

By repeating the experiment several times, we can approximate that the more the number of 1's we get, the farther the qubits are.

The code snippet Listing 7.25 implements the quantum distance estimator circuit.

Listing 7.26 contains the source code for the core algorithm that implements the k-means.

The code fragment in Listing 7.27 generates a random data set and invokes the *QuantumKMeans* method.

Executing this code generates the scatter plot, shown in Fig. 7.43a.

We can verify this with the Scikit-learn [20]. Type the code provided in Listing 7.28 in a new cell in the Jupyter Notebook and execute it. Note that we are using the data generated previously.

Upon executing the code in Listing 7.28, we get a scatter plot, as shown in Fig. 7.43b.

Note that both the scatter plots are similar. A good exercise is to verify the deviation between the quantum algorithm and the scikit-learn code programmatically and experiment with the threshold value, and the rotational angles.

This algorithm was first published by **S Lloyd** as an IEEE paper in 1982, "Least Squares Quantization in PCM."

The quantum version runs in time $O\left(\frac{kN \log d}{\epsilon}\right)$.

```python
# Implements the quantum distance estimation function
# pointA, pointB -> The point and the centroid
# processor - which processor to use
# shots - number of times the experiment is to be repeated
# jobMonitor - set to True when testing on a backend

def QuantumFindDistance( pointA, pointB, processor, shots,
jobMonitor):

    # Calculate the rotational angles
    phiPointA = (pointA[0]-1) * pi / 2
    thetaPointA = (pointA[1]-1) * pi / 2

    phiPointB = (pointB[0]-1) * pi / 2
    thetaPointB = (pointB[1]-1) * pi /2

    # Create a 3 qubit Quantum Register
    # qreg[0] - for encoding the point
    # qreg[1] - for encoding the centroid
    # qreg[2] - for the ancilla qubit
    qreg = QuantumRegister(3, 'qreg')

    # Create one bit of a classical register to do the measurement
    creg = ClassicalRegister(1, 'creg')

    # construct the Quantum Circuit now
    qc = QuantumCircuit(qreg, creg, name='qc')

    # Step 1: Put the ancella qubit in equal superposition state.
    qc.h(qreg[2])

    # Step 2: Encode the point and centroid as U3 rotations
    qc.u3(thetaPointA, phiPointA, 0, qreg[0])
    qc.u3(thetaPointB, phiPointB, 0, qreg[1])

    # Step 3: Apply Fredkin gate
    qc.cswap(qreg[2], qreg[0], qreg[1])

    # Step 4: Apply H gate again to the ancilla qubit
    qc.h(qreg[2])

    # Step 5: Measure the ancilla qubit now.
    qc.measure(qreg[2], creg[0])

    job = execute(qc, backend = processor, shots = shots)

    # need this, if running on a backend
    if ( True == jobMonitor ):
        job_monitor(job)

    return job.result().get_counts(qc)
```

Listing 7.25

```python
# Implements the procedure that determines the distance between
# the point and the centroids
# point - the point for which we need to measure
# centroids - current centroids.
# processor - where to execute the code
# shots - number of experiments, usually 1024
# jobMonitor - set this to True when running on a backend
# returns: Array of probability of measuring a 1.

def QuantumDistanceEstimator(point, centroids, processor, shots,
jobMonotor):

    k = len(centroids)

    results = []

    for i in range(k):
        counts = QuantumFindDistance( point, centroids[i], \
                                      processor,           shots,
jobMonitor)

        # If we projected a '1' add the measurement
        # to the list, or just 0
        if ( True == ('1' in counts )):
            results.append(counts['1'])
        else:
            results.append(0)

    return results

# Implements the Quantum K-means algorithm.
# processor - where to execute the code
# shots - number of experiments, usually 1024
# jobMonitor - set this to True when running on a backend
# Expects - the data to be prepared

def QuantumKMeans(processor, shots = 1024, jobMonitor = False):

    # Generate some random centers
    np.random.seed(0)
    centroids = np.random.normal(scale=0.6, size=[k, 2])
    centroids_prev = deepcopy(centroids)

    # storage for recording which cluster the point is closest to
    clusters = np.zeros(n)

    # setup a threshold value for the algorithm
    error = np.inf
```

Listing 7.26

```
        threshold = 0.04
        iteration = 0

        # Run until threshold is reached
        while error > threshold:

            distancefromcenters = []

            for x in X:

distancefromcenters.append(QuantumDistanceEstimator(x, \
                        centroids, processor, shots, jobMonitor
))

            # Find out which cluster the points are closest to
            clusters = np.argmin(distancefromcenters, axis = 1)
            centroids_prev = deepcopy(centroids)

            # Calculate mean distance for every cluster
            # and update the centroids
            for i in range(k):
                if np.sum(clusters == i) != 0:
                    # find the new center based on points now
                    # belonging to the cluster
                    centroids[i] = np.mean(X[clusters == i], axis=0)
                else:
                    # assign a new center, since it is not in the range
                    centroids[i] = np.random.normal(scale=0.6,  \
                                size=centroids[i].shape)

            # find the norm
            error = np.linalg.norm(centroids - centroids_prev)
            iteration += 1

            print("Current iteration: ", iteration, " Norm: ", error)

        # Plot the graph
        for i in range(n):
            plt.scatter(X[i,   0],   X[i,1],   s   =   10,   color   =
colors[int(y[i])])

        plt.scatter(centroids[:,0], centroids[:,1], marker='*', c='g',
                    s=150)
        plt.show()
```

Listing 7.26 (continued)

```
import numpy as np
from numpy import pi
from copy import deepcopy
from matplotlib import pyplot as plt
from sklearn.datasets import make_blobs
from sklearn.cluster import KMeans
from sklearn.preprocessing import MinMaxScaler

# Assume the number of clusters and sample size
k = 3
n = 100

# Generate some random data
X, y = make_blobs(n_samples=100, centers=k, cluster_std=0.60,
                  random_state=0)

# bring this to a range.
scaler = MinMaxScaler(feature_range=(-0.8, 1))
scaler.fit(X)
X = scaler.transform(X)

# define some colors
colors=['red', 'green', 'blue', 'yellow', 'black']

backend = Aer.get_backend('qasm_simulator')
QuantumKMeans(backend)
```

Listing 7.27 Code to test the quantum k-means algorithm

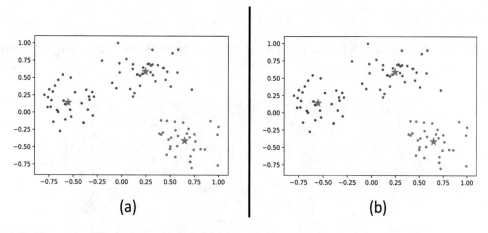

Fig. 7.43 The output of K-means algorithms for 3 clusters of 100 samples. (**a**) The quantum version. (**b**) The Scikit-learn version

```
from sklearn.cluster import KMeans
kmeans = KMeans(n_clusters = k)
# Fitting with inputs
kmeans = kmeans.fit(X)
# Predicting the clusters
labels = kmeans.predict(X)
# Getting the cluster centers
C = kmeans.cluster_centers_

for i in range(n):
    plt.scatter(X[i,    0],    X[i,    1],    s    =    10,    color    =
colors[int(y[i])])
    plt.scatter(C[:, 0], C[:, 1], marker='*', c='g', s=150)
```

Listing 7.28 K-means, using the scikit-learn

7.11 Quantum Phase Estimation (QPE)

Given an eigenvector $|\psi\rangle$ of a unitary operator U, such that $U|\psi\rangle = e^{2\pi i\theta}|\psi\rangle$, $0 \leq \theta \leq 1$, the goal of the quantum phase estimation algorithm is to find the eigenvalue $e^{2\pi i\theta}$ of $|\psi\rangle$. This operation is equivalent to finding the phase θ to a certain accuracy. The phase estimation algorithm is used in Shor's algorithm and in solving system of linear equation, etc. The block diagram shown in Fig. 7.44 illustrates the quantum phase estimation procedure. The quantum phase estimation procedure is similar to the method we used to perform arithmetic in the Fourier basis. Each of the successive blocks of the controlled U gates kickbacks a phase $\frac{\theta}{2^n}$ into the upper block, called the counting register. The counting register sums up to a value proportional to the phase factor $e^{2\pi i\theta}$. An inverse QFT is applied to the counting register to convert this to the computational basis, which we can measure and post-process (Fig. 7.44).

The system is started with the registers in the default state.

$$|\psi\rangle \rightarrow |0\rangle^{\otimes n}|\psi\rangle \tag{7.186}$$

A Walsh–Hadamard transform is applied to the counting register to put that in an equal superposition state.

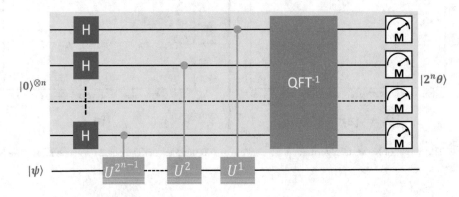

Fig. 7.44 Quantum Phase Estimation

$$\xrightarrow{\ H(\text{counting register})\ } \frac{1}{\sqrt{2^n}}(|0\rangle + |1\rangle)^{\otimes n}|\psi\rangle \tag{7.187}$$

Since U is an unitary operator, $U^{2^n}|\psi\rangle = U^{2^{n-1}}e^{2\pi i\theta}|\psi\rangle = U^{2^{n-2}}e^{2\pi i\theta 2^1}e^{2\pi i\theta 2^0}|\psi\rangle$. Applying this logic into the previous equation,

$$\rightarrow \frac{1}{\sqrt{2^n}}\left[\begin{array}{l}(|0\rangle + |1\rangle)e^{2\pi i\theta 2^{n-1}}|\psi\rangle \otimes (|0\rangle + |1\rangle)e^{2\pi i\theta 2^{n-2}}|\psi\rangle \cdots \\ \cdots \otimes (|0\rangle + |1\rangle)e^{2\pi i\theta 2^1}|\psi\rangle \otimes (|0\rangle + |1\rangle)e^{2\pi i\theta 2^0}|\psi\rangle\end{array}\right] \tag{7.188}$$

which can be rewritten as:

$$\rightarrow \frac{1}{\sqrt{2^n}}\left[\left(|0\rangle + e^{2\pi i\theta 2^{n-1}}|1\rangle\right)|\psi\rangle \otimes \cdots \otimes \left(|0\rangle + e^{2\pi i\theta 2^1}|1\rangle\right)|\psi\rangle \otimes \left(|0\rangle + e^{2\pi i\theta 2^0}|1\rangle\right)|\psi\rangle\right] \tag{7.189}$$

And then as a summation:

$$\rightarrow \frac{1}{\sqrt{2^n}}\sum_{k=0}^{2^n-1}e^{2\pi i\theta k}|k\rangle \otimes |\psi\rangle, \tag{7.190}$$

where k is a binary expansion. Applying an inverse QFT to the counting registers, we get

$$\rightarrow \frac{1}{\sqrt{2^n}}\sum_{k=0}^{2^n-1}e^{2\pi i\theta k}\frac{1}{\sqrt{2^n}}\sum_{y=0}^{2^n-1}e^{\frac{-2\pi iky}{2^n}}|y\rangle \otimes |\psi\rangle$$

$$\rightarrow \frac{1}{2^n}\sum_{y=0}^{2^n-1}\sum_{k=0}^{2^n-1}e^{2\pi i\theta k}\,e^{\frac{-2\pi iky}{2^n}}|y\rangle \otimes |\psi\rangle \tag{7.191}$$

$$\rightarrow \frac{1}{2^n}\sum_{y=0}^{2^n-1}\sum_{k=0}^{2^n-1}e^{\frac{-2\pi ik}{2^n}(y-2^n\theta)}|y\rangle \otimes |\psi\rangle$$

A close examination of the above equation reveals that when $y = 2^n\theta$, the exponential term evaluates to 1 and a measurement results in:

$$\rightarrow |2^n\theta\rangle \otimes \ |\psi\rangle \tag{7.192}$$

From this value, θ can be estimated. Let us now calculate the accuracy to which the phase estimation can be done. Since the measurement is done in the computational basis, we can add a correction parameter $\Delta\theta$, which rounds off $2^n\theta$ to get a $\{0,1\}$. Equation (7.191) can be written as:

$$|\psi\rangle \rightarrow \frac{1}{2^n}\sum_{y=0}^{2^n-1}\sum_{k=0}^{2^n-1}e^{\frac{-2\pi ik}{2^n}(y-2^n\theta)}\,e^{2\pi ik\Delta\theta}|y\rangle \otimes |\psi\rangle \tag{7.193}$$

The probability of measuring $2^n\theta$ is given by the following equation:

$$P(2^n\theta) = \left|\left\langle 2^n\theta\left|\frac{1}{2^n}\sum_{y=0}^{2^n-1}\sum_{k=0}^{2^n-1}e^{\frac{-2\pi ik}{2^n}(y-2^n\theta)}\,e^{2\pi ik\Delta\theta}\right|y\right\rangle\right|^2$$

$$= \frac{1}{2^{2n}}\left|\sum_{k=0}^{2^n-1}e^{2\pi ik\Delta\theta}\right|^2 \tag{7.194}$$

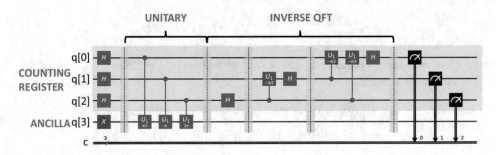

Fig. 7.45 Quantum circuit for performing Phase Estimation. Shown here is the circuit for the S gate which performs $\pi/2$ rotation

When $\Delta\theta = 0$, we measure this with the highest probability. When $\Delta\theta \neq 0$, the above equation reduces to:

$$P(2^n\theta) = \frac{1}{2^{2n}} \left| \frac{1 - e^{2\pi i 2^n \Delta\theta}}{1 - e^{2\pi i \Delta\theta}} \right|^2$$

$$= \frac{1}{2^{2n}} \left| \frac{\sin \pi 2^n \Delta\theta}{\sin \pi \Delta\theta} \right|^2$$

$$\approx \frac{1}{2^{2n}} \frac{|2 \times 2^n \Delta\theta|^2}{|\pi \Delta\theta|^2} \tag{7.195}$$

$$= \frac{4}{\pi^2} = 40.5\%$$

Thus, the phase estimation can be done to an accuracy of about 40.5% (Fig. 7.45).

We can try this algorithm with the S gate from Sect. 5.4.1.6, which performs a $\frac{\pi}{2}$ rotation about the z-axis. From Eq. (5.27), the S gate adds the phase $e^{i\frac{\pi}{2}}$ when the qubit is at state $|1\rangle$, hence we must apply an X gate to the ancilla qubit at the initialization state. Construct the quantum circuit for phase estimation, using Listing 7.29.

Executing the code generates the histogram shown in Figure 7.46.

From the histogram, we find that we measured 2 with a 100% probability. The circuit was constructed with three qubits. Therefore, $n = 3$. So $2^3\theta = 2$. Hence, $\theta = \frac{1}{4}$. For the S gate, the phase is $e^{2\pi i\theta} = e^{i\frac{\pi}{2}}$. $\theta = \frac{1}{4}$ in this equation. This matches with our observation from the experimental results. Hence we can say that the QPE algorithm works. Note that the algorithm will work only if the ancilla qubit, that is, $|\psi\rangle$ is an eigenstate of the rotational operator.

Code example for QDK is included in the Appendix.

7.12 HHL Algorithm for Solving Linear Equations

The HHL algorithm is due to **Aram Harrow**, **Avinatan Hassidim**, and **Seth Lloyd**. The goal of the problem is given a matrix A and a vector $\vec{b}$, to find a vector $\vec{x}$ such that $A\vec{x} = \vec{b}$. The HHL algorithm uses the QPE to perform matrix inversion. The algorithm needs three sets of qubit registers. A single qubit ancilla, an n qubit register to store the eigenvalues of A and a quantum memory register of size $O(\log(N))$ to store $\vec{b}$ and then evolve $\vec{x}$.

```
# Inverse QFT code
def qft_inv (qc, n_qubits):
    for i in range (n_qubits - 1, -1, -1):
        for qubit in range (i, n_qubits-1):
            # for inverse transform, we have to use negative angles
            qc.cu1( - math.pi / 2** ( qubit + 1 - i ), qubit + 1, i)
        # the H transform should be done after the rotations
        qc.h(i)
        qc.barrier()

# Quantum Phase Estimation
# S gate, which performs pi/2 rotation about the z axis

countingRegister = 3
q = QuantumRegister(countingRegister + 1 , 'q')
c = ClassicalRegister(countingRegister , 'c')
qc = QuantumCircuit(q, c)

# Prepare the initial state
qc.h(q[0])
qc.h(q[1])
qc.h(q[2])
qc.x(q[3])
qc.barrier()

# Perform the unitaries
rotation = math.pi/2
for i in range (countingRegister):
    qc.cu1(2**i * rotation, q[i],   q[3])
qc.barrier()

# Inverse QFT

qft_inv(qc, countingRegister)

# measure the counting register
qc.measure(q[0], c[0])
qc.measure(q[1], c[1])
qc.measure(q[2], c[2])

#fig = qc.draw('mpl')
#fig.savefig('qpe.svg', bbox_inches='tight', dpi=300)

backend = Aer.get_backend("qasm_simulator")
job = execute(qc, backend=backend, shots=1024)
counts = job.result().get_counts()
fig = plot_histogram(counts)
fig.savefig('qpe_histogram.svg', bbox_inches='tight', dpi=300)
```

Listing 7.29 Quantum phase estimation

Fig. 7.46 Histogram for
the QPE algorithm for $\pi/2$
rotation

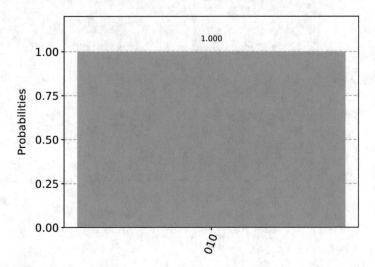

Consider the following example:

$$A = \begin{bmatrix} 2 & 1 \\ 3 & -1 \end{bmatrix} \quad \vec{b} = \begin{bmatrix} x \\ y \end{bmatrix} \quad \vec{x} = \begin{bmatrix} 15 \\ 5 \end{bmatrix} \tag{7.196}$$

This system of matrices can be written as a set of linear equations

$$\begin{aligned} 2x + y &= 15 \\ 3x - y &= 5 \end{aligned} \tag{7.197}$$

The solution to this equation is straightforward, $x = 4$ and $y = 7$. A system of linear expressions can be generalized as a set of equations with N variables and M constraints. In the above example, both N and M are two. We shall consider the system of equations where $N = M$, the matrix A is invertible, and the equations have a unique solution. Let us also assume that A is a Hermitian matrix and s-sparse.[2] We shall also assume that the matrix is well conditioned. We assume that the singular values[3] of the matrix are between $\frac{1}{k}$ and 1, where k is the condition number of the matrix.

Supposing, if A is not Hermitian, we can define $A^{'}$ which is Hermitian as follows. However, we shall restrict our discussions to Hermitian matrices in this section.

$$A' = \begin{bmatrix} 0 & A \\ A^{\dagger} & 0 \end{bmatrix}, \quad b' = \begin{bmatrix} 0 \\ b \end{bmatrix}, \quad x' = \begin{bmatrix} x \\ 0 \end{bmatrix} \tag{7.198}$$

$$A\vec{x} = \vec{b} \text{ implies that } \vec{x} = A^{-1}\vec{b} \tag{7.199}$$

Assume that the vectors $\vec{b}$ and $\vec{x}$ are encoded as normalized quantum states as follows.

---

[2] A sparse matrix is a matrix with most elements zero. An s-sparse matrix has at most s nonzero entries per row. For a given row i and   column j, the entries A_{ij}can be computed in time $O(s)$.

[3] Singular values of an $M \times N$ matrix A is defined as the square roots of the eigenvalues of the symmetric $N \times N$ matrix $A^T A$ arranged in decreasing order.

$$|b\rangle = \frac{\sum_{j=0}^{N-1} b_j |j\rangle}{\left\| \sum_{j=0}^{N-1} b_j |j\rangle \right\|} \quad \text{and} \quad |x\rangle = \frac{\sum_{j=0}^{N-1} x_j |j\rangle}{\left\| \sum_{j=0}^{N-1} x_j |j\rangle \right\|} \tag{7.200}$$

The matrix A can be written as its spectral decomposition.

$$A = \sum_{j=0}^{N-1} \lambda_j |u_j\rangle \langle u_j|, \tag{7.201}$$

where λ_j is the jth eigenvalue of the corresponding eigenvector $|u_j\rangle$. We can define the inverse matrix of A as follows:

$$A^{-1} = \sum_{j=0}^{N-1} \lambda_j^{-1} |u_j\rangle \langle u_j| \tag{7.202}$$

With these definitions, in the eigenbasis of A, we can write the following equations:

$$|b\rangle = \sum_{j=0}^{N-1} b_j |u_j\rangle \tag{7.203}$$

$$|x\rangle = A^{-1}|b\rangle = \sum_{j=0}^{N-1} \lambda_j^{-1} b_j |u_j\rangle \tag{7.204}$$

Figure 7.47 illustrates the HHL algorithm. The algorithm requires a procedure for state preparation and encoding the coefficients into $|b\rangle$. For brevity, this step is not detailed in this section. We assume that there is a unitary B that prepares the state $|b\rangle$.

Given $|b\rangle$, our goal is to find $|x\rangle = A^{-1}|b\rangle$. Therefore, we need to find an efficient method for determining A^{-1}. From Eq. (7.202), we can infer that if we can find the mapping $|u_j\rangle \rightarrow \frac{1}{\lambda_j}|u_j\rangle \forall |u_j\rangle$, we can say that we have found A^{-1}.

The first block of registers – the input register – nb contains the vector solution. The second block of registers – the clock register – ne is used to store the binary representation of the eigenvalues of A. The ancilla qubit is used to perform the eigenvalue inversion. Let us assume that $N = 2^{nb}$.

The system starts with all qubits initialized to the default state. To keep the math less cluttered, we will bring the ancilla qubit into the equation a little later.

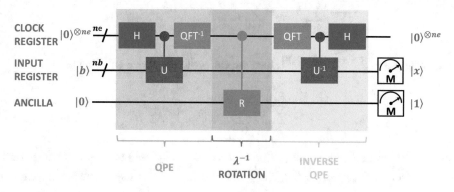

Fig. 7.47 The block diagram of the HHL algorithm

$$|\psi\rangle = |0\rangle_{ne} \otimes |0\rangle_{nb} \tag{7.205}$$

The data set $|b\rangle$ is loaded into the register nb.

$$\rightarrow |0\rangle_{ne} \otimes |b\rangle_{nb} \tag{7.206}$$

From Eq. (7.203), we can substitute for $|b\rangle_{nb}$.

$$= \sum_{j=0}^{N-1} |0\rangle_{ne} \otimes b_j |u_j\rangle_{nb}$$
$$= \sum_{j=0}^{N-1} b_j |0\rangle_{ne} |u_j\rangle_{nb} \tag{7.207}$$

In this equation $|u_j\rangle_{nb}$ represents the jth eigenvector of the matrix A. To this state, we apply the QPE algorithm, with the unitary $U = e^{iAt} = \sum_{j=0}^{N-1} e^{i\lambda_j t} |u_j\rangle\langle u_j|$. The QPE algorithm first applies the Walsh–Hadamard transform on the clock register.

$$\xrightarrow{H^{\otimes ne}} \sum_{j=0}^{N-1} b_j \frac{1}{\sqrt{2^{ne}}} \sum_{l=0}^{2^{ne}-1} (|0\rangle + |1\rangle)^{\otimes ne} |u_j\rangle_{nb} \tag{7.208}$$

Applying the controlled U gate $\sum_{j=0}^{N-1} e^{i\lambda_j t} |u_j\rangle\langle u_j|$, we get:

$$\rightarrow \sum_{j=0}^{N-1} b_j \frac{1}{\sqrt{2^{ne}}} \sum_{l=0}^{2^{ne}-1} \left(|0\rangle + e^{i\lambda_j t}|1\rangle\right)^{\otimes ne} |u_j\rangle_{nb}$$
$$\rightarrow \sum_{j=0}^{N-1} b_j \frac{1}{\sqrt{2^{ne}}} \sum_{l=0}^{2^{ne}-1} e^{i\lambda_j t} |l\rangle |u_j\rangle_{nb} \tag{7.209}$$

The process of applying e^{iAt}, given matrix A, is known as **Hamiltonian simulation**. This method applies e^{iAt} to the input $|b\rangle$ over a period t and decomposes $|b\rangle$ into the eigenbasis of A to find the corresponding values of λ_j using phase estimation.

The next step is to apply the inverse QFT.

$$\xrightarrow{\text{QFT}^{-1}} \sum_{j=0}^{N-1} b_j \frac{1}{2^{ne}} \sum_{k=0}^{2^{ne}-1} \sum_{l=0}^{2^{ne}-1} e^{i\lambda_j lt} e^{\frac{-2\pi ilk}{2^{ne}}} |k\rangle_{ne} |u_j\rangle_{nb} \tag{7.210}$$

If we assume $\alpha_{k|l} = \frac{1}{2^{ne}} \sum_{l=0}^{2^{ne}-1} e^{il\left[\lambda_j t - \frac{2\pi k}{2^{ne}}\right]}$, then we can write the above equation as:

$$\rightarrow \sum_{j=0}^{N-1} b_j \sum_{k=0}^{2^{ne}-1} \alpha_{k|l} |k\rangle_{ne} |u_j\rangle_{nb} \tag{7.211}$$

In the scenario where the phase estimation is perfect $\lambda_j t = \frac{2\pi k}{2^{ne}}$, and $\alpha_{k\,|\,l} = 1$. If we set set $k = \frac{k}{2^{ne}}$, then $\lambda_j = \frac{2\pi k}{t}$. Hence, $|k\rangle_{ne} |u_j\rangle_{nb} = |\lambda_j\rangle_{ne} |u_j\rangle_{nb}$. The ket $|k\rangle_{ne}$ at this stage contains the amplitudes of the basis states after the Fourier transform. λ_j represents the jth eigenvalue of the matrix A. With this, we can rewrite the above equation.

$$\rightarrow \sum_{j=0}^{N-1} b_j \left|\lambda_j\right\rangle_{ne} \left|u_j\right\rangle_{nb}, \tag{7.212}$$

where $\left|\lambda_j\right\rangle_{ne}$ contains the eigenvalue estimate. At this stage, we can introduce the ancilla qubit into the equation. Performing a conditional rotation with the $\left|\lambda_j\right\rangle_{ne}$ qubit as the control qubit, we get the following system state

$$\left|\psi\right\rangle \rightarrow \sum_{j=0}^{N-1} b_j \left|\lambda_j\right\rangle_{ne} \left|u_j\right\rangle_{nb} \left(\sqrt{1 - \frac{C^2}{\lambda_j^2}} \left|0\right\rangle + \frac{C}{\lambda_j}\left|1\right\rangle\right), \tag{7.213}$$

where the normalization constant $C = O\left(\frac{1}{k}\right)$. The rotation finds the inverse of the eigenvalues λ_j.

The next step is to uncompute $\left|\lambda_j\right\rangle_{ne}$ by performing an inverse QPE operation. The resultant system state is given below:

$$\left|\psi\right\rangle \rightarrow \sum_{j=0}^{N-1} b_j \left|0\right\rangle_{ne} \left|u_j\right\rangle_{nb} \left(\sqrt{1 - \frac{C^2}{\lambda_j^2}} \left|0\right\rangle + \frac{C}{\lambda_j}\left|1\right\rangle\right) \tag{7.214}$$

Now, if we measure a $\left|1\right\rangle$ at the ancilla qubit, the resultant state of the system is given by:

$$\left|x\right\rangle \approx \left(\frac{1}{\sqrt{\sum_{j=0}^{N-1} \frac{c^2 |b_j|^2}{|\lambda_j|^2}}}\right) \sum_{j=0}^{N-1} \frac{c\, b_j}{\lambda_j} \left|0\right\rangle_{ne} \left|u_j\right\rangle_{nb} \tag{7.215}$$

Up to a normalization factor, this equation corresponds to Eq. (7.204). The state of the nb register is $\sum_{j=0}^{N-1} b_j \lambda_j^{-1} \left|u_j\right\rangle_{nb}$ up to a normalization factor which is equal to the solution $\left|x\right\rangle$ we are looking for.

Finally, the qubits nb are measured. The expectation value of this measurement M, $\langle x|M|x\rangle$ corresponds to the $\vec{x}$ we wanted to compute.

Several improvements have been suggested to the HHL algorithm, which includes amplitude amplification [23–26]. A detailed error analysis is provided in the original research publication [22].

Let us study a pedagogical example of the HHL algorithm using a 2×2 matrix. The example we are studying is a simplified version, which performs a direct phase estimation using the knowledge of the eigenstates of the matrix A.

Consider the following system of linear equations:

$$A \cdot \vec{x} = \vec{b}: \qquad \begin{bmatrix} \frac{3}{4} & \frac{1}{4} \\ \frac{1}{4} & \frac{3}{4} \end{bmatrix} \cdot \begin{vmatrix} x_1 \\ x_2 \end{vmatrix} = \begin{vmatrix} 2 \\ 0 \end{vmatrix} \tag{7.216}$$

By finding the inverse of matrix A, we can rewrite the above equation as follows to find the classical solution:

$$\vec{x} = A^{-1}\vec{b} : \qquad \begin{bmatrix} \dfrac{3}{2} & -1 \\ -1 & \dfrac{3}{2} \end{bmatrix} \cdot \begin{vmatrix} 2 \\ 0 \end{vmatrix} = \begin{vmatrix} 3 \\ -1 \end{vmatrix} \tag{7.217}$$

We can create a normalized ket $\frac{|b\rangle}{\||b\rangle\|}$ from the classical vector $\vec{b}$ and solve the above equation in quantum mechanics.

$$|x\rangle = \frac{A^{-1}|b\rangle}{\|A^{-1}|b\rangle\|} = \frac{1}{\sqrt{10}} \begin{vmatrix} 3 \\ -1 \end{vmatrix} \tag{7.218}$$

From the above equation, it can be found that, up to a factor, the classical solution and the quantum solution are the same. In the following part of this section, we shall apply the HHL algorithm to the linear system of equations defined in (7.126). We shall also construct a quantum circuit to compare experiment with theory.

Let us begin with the definition of A and $|b\rangle$.

$$A = \begin{bmatrix} \dfrac{3}{4} & \dfrac{1}{4} \\ \dfrac{1}{4} & \dfrac{3}{4} \end{bmatrix} \text{ and } |b\rangle = \begin{vmatrix} 1 \\ 0 \end{vmatrix} \text{ (after normalization)} \tag{7.219}$$

With some math work or using an online tool, the eigenvalues of the matrix A can be calculated as $\lambda_1 = 1$ and $\lambda_2 = \frac{1}{2}$. Let us assume $t = 2 \cdot 2\pi$ and $C = 2$. With these assumptions, we can calculate $\frac{\lambda_1 t}{2\pi} = 2$ and $\frac{\lambda_2 t}{2\pi} = 1$. These two values correspond to the states $|01\rangle$ and $|10\rangle$ of the ne register. The eigenvectors of the A matrix are $|u_1\rangle = \frac{1}{\sqrt{2}} \begin{bmatrix} 1 \\ 1 \end{bmatrix}$ and $|u_2\rangle = \frac{1}{\sqrt{2}} \begin{bmatrix} 1 \\ -1 \end{bmatrix}$ in the matrix form. We can define the ket $|b\rangle_{nb} = \sum_{j=0}^{1} |u_i\rangle_{nb}$. Figure 7.48 illustrates the quantum circuit we construct to implement the simplified version of the HLL algorithm.

The state of the system after the QPE stage can be written as:

$$|\psi\rangle = \frac{1}{\sqrt{2}} |01\rangle_{ne} |u_1\rangle_{nb} + \frac{1}{\sqrt{2}} |10\rangle_{ne} |u_2\rangle_{nb} \tag{7.220}$$

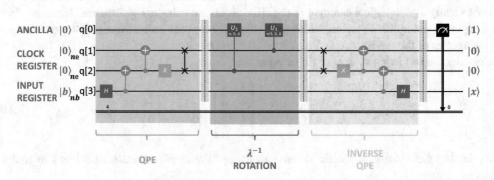

Fig. 7.48 A simplified quantum circuit implementing the HLL algorithm for a 2 x 2 matrix

The next step is to perform the rotations of the ancilla, the system state after the rotations can be written as:

$$|\psi\rangle = \frac{1}{\sqrt{2}}|01\rangle_{ne}|u_1\rangle_{nb}\left(\sqrt{1 - \frac{2^2}{1^2}}|0\rangle + \frac{2}{1}|1\rangle\right)$$
$$+ \frac{1}{\sqrt{2}}|10\rangle_{ne}|u_2\rangle_{nb}\left(\sqrt{1 - \frac{2^2}{\left(\frac{1}{2}\right)^2}}|0\rangle + \frac{2}{\left(\frac{1}{2}\right)}|1\rangle\right)$$

(7.221)

This equation reduces to the following:

$$|\psi\rangle = \frac{1}{\sqrt{2}}|01\rangle_{ne}|u_1\rangle_{nb}\left(\sqrt{-3}|0\rangle + 2|1\rangle\right)$$
$$+ \frac{1}{\sqrt{2}}|10\rangle_{ne}|u_2\rangle_{nb}\left(\sqrt{-16}|0\rangle + 4|1\rangle\right)$$

(7.222)

After performing the rotations, we put the system through an inverse QPE. The resultant state can be written as:

$$|\psi\rangle = \frac{1}{\sqrt{2}}|00\rangle_{ne}|u_1\rangle_{nb}\left(\sqrt{-3}|0\rangle + 2|1\rangle\right)$$
$$+ \frac{1}{\sqrt{2}}|00\rangle_{ne}|u_2\rangle_{nb}\left(\sqrt{-16}|0\rangle + 4|1\rangle\right)$$

(7.223)

Now, if we measure a $|1\rangle$ in the ancilla qubit, the system collapses to:

$$\rightarrow \frac{\frac{1}{\sqrt{2}}|00\rangle_{ne}|u_1\rangle_{nb}2|1\rangle + \frac{1}{\sqrt{2}}|00\rangle_{ne}|u_2\rangle_{nb}4|1\rangle}{\sqrt{\left(\left(\frac{1}{\sqrt{2}}\right)^2 \times 2^2\right) + \left(\left(\frac{1}{\sqrt{2}}\right)^2 \times 4^2\right)}}$$

(7.224)

$$= \frac{\frac{2}{\sqrt{2}}|u_1\rangle_{nb} + \frac{4}{\sqrt{2}}|u_1\rangle_{nb}}{\sqrt{10}}$$

Substituting the values of $|u_1\rangle_{nb}$ and $|u_2\rangle_{nb}$ in the matrix form,

$$= \frac{1}{\sqrt{10}}\left(\frac{2}{\sqrt{2}}\cdot\frac{1}{\sqrt{2}}\begin{bmatrix}1\\1\end{bmatrix} + \frac{4}{\sqrt{2}}\cdot\frac{1}{\sqrt{2}}\begin{bmatrix}1\\-1\end{bmatrix}\right)$$
$$= \frac{1}{\sqrt{10}}\left(\begin{bmatrix}1\\1\end{bmatrix} + 2\begin{bmatrix}1\\-1\end{bmatrix}\right)$$

(7.225)

$$= \frac{1}{\sqrt{10}}\begin{bmatrix}3\\-1\end{bmatrix}$$

This result is the same as the Eq. (7.218). Listing 7.30 contains the source code that implements the quantum circuit in qiskit.

```python
# HHL Algorithm for the 2 x 2 matrix
# [ 3/4   1/4]
# [1/4    3/4]

q = QuantumRegister(4 , 'q')
c = ClassicalRegister(4 , 'c')
qc = QuantumCircuit(q, c)

# QPE
qc.h(q[3])
qc.cx(q[3], q[2])
qc.cx(q[2], q[1])
qc.x(q[2])
qc.swap(q[2], q[1])
qc.barrier()

# Lambda inverse
qc.cu3(math.pi, 0,0,q[2], q[0])
qc.cu3(math.pi/3, 0,0,q[1], q[0])
qc.barrier()

# Inverse QPE
qc.swap(q[2], q[1])
qc.x(q[2])
qc.cx(q[2], q[1])
qc.cx(q[3], q[2])
qc.h(q[3])
qc.barrier()

# measure the ancilla qubit
qc.measure(q[0], c[0])

#qc.draw('mpl')

backend = Aer.get_backend('statevector_simulator')
job = execute(qc, backend=backend, shots=1024).result()
statevector = job.get_statevector()

# extract the statevector of the nb qubit
sv = np.zeros((2,), dtype=statevector.dtype)
sv[0] = statevector[1]
sv[1] = statevector[9]
sv.imag = 0.0

# Find the solution
result = math.sqrt(10) * sv[:2]
print("|x> using HLL algorithm: {}".format(result))
```

Listing 7.30 Source code for the HLL algorithm shown in Fig. 7.48.

Upon executing the circuit, we get the following result. Note that the circuit may need to be tried a few times to get the correct result.

```
|x> using HLL algorithm: [ 3.+0.j -1.+0.j]
```

The result matches the theoretical value. Hence, we can say that the HHL algorithm works. The algorithm works well for the matrices with a relatively less condition number. However, the authors of the original research paper have given some methods to handle ill-conditioned matrices. The performance of the run time is $k^2 \log (N)/\epsilon$, where ϵ is the additive system error. In cases where k and $1/\epsilon$ are polylog(N), the algorithm exhibits exponential speedup [22]. The algorithm requires O(log N) qubits if direct phase estimation and Hamiltonian evolution are used.

A number of applications have been reported for the HHL algorithm, which is an essential milestone in the discovery of quantum algorithms. Extensions in solving nonlinear differential equations have also been reported [28].

Code example for QDK is included in the Appendix.

7.13 Quantum Complexity Theory

The extended Church-Turing thesis states, *"Any physically-realistic computing device can be simulated by a deterministic (or maybe probabilistic) Turing machine, with at most polynomial overhead in time and memory."* This thesis forms the basis for all complexity theories.

The Quantum Complexity Theory focuses on the study of quantum complexity classes and the degree of hardness of computational problems between the classical and quantum implementations. Let us begin our discussions in this section by recalling the classical complexity classes (Fig. 7.49).

Fig. 7.49 The overlap of the complexity classes

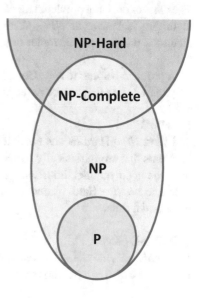

Polynomial Complexity (P)

An algorithm is said to be in polynomial-time if it solves a problem in kn^c time, where k and c are some constants, and n is the size of an instance—a measure of the number of bits required to specify the instance. All classes of problems for which deterministic algorithms that run in polynomial time belong to the P - class.

Nondeterministic Polynomial Time (NP)

A nondeterministic algorithm produces different outputs for the same input each time it is run. The class of NP problems are decision problems where a nondeterministic algorithm can solve the problem in polynomial time. Usually, there are two parts to this—the nondeterministic algorithm provides a solution, and then the solution is verified using a deterministic algorithm.

Given a string x and a decision problem L, the prover sends a string $\pi \in \{0, 1\}^n$ to a deterministic polynomial algorithm called a verifier V. $V(x, \pi)$ runs in polynomial time and returns "1" if x is a YES-instance and "0" if x is a NO-instance.

The complexity class NP is defined as follows:

$$NP = \bigcup_{k \in N} \text{NTIME}(n^k), \tag{7.226}$$

where $\text{NTIME}(n^k)$ is the set of decision problems that can be solved by a nondeterministic Turning machine in time $O(n^k)$.

Note that the class of P problems also belongs to the class of NP. However, this is an unsolved problem in computer science, whether $P = NP$.

NP-Complete

The class of decision problems for which the solutions can be verified for correctness by a polynomial algorithm is known as NP - Complete. The traveling salesman problem, bin-packing problem, knapsack problem are examples of problems belonging to this class.

A problem belongs to NP - Complete, if it belongs to NP, and there exists a problem in class NP in polynomial time that can be reduced to it.

NP-Hard

The term NP - Hard means nondeterministic polynomial acceptable problems. These are decision problems that are intrinsically harder than those that can be solved using a nondeterministic Turing machine in polynomial time. Some search and optimization problems belong to this class. A problem is said to be NP - Hard, if there exists a problem in class NP - complete that can be reduced to it in polynomial time.

EXPTIME (EXP)

EXP represents the class of decision problems solvable by a deterministic turning machine in $O(2^{p(n)})$ time, where $p(n)$ is a polynomial function of n.

Probabilistic Polynomial Time (PP)
The class of decision problems that can be solved by a probabilistic Turing machine in polynomial time, with an error probability of $\frac{1}{2}$ for all instances.

Bounded-Error, Probabilistic Polynomial time (BPP)
BPP represents the class of problems that can be solved by a probabilistic Turing machine in polynomial time, with an error probability $\geq \frac{2}{3}$ for all instances. For problems not in BPP, the Turing machine may solve it with a probability $\leq \frac{1}{3}$

PSPACE
PSPACE is the set of all decision problems that can be solved by a Turing machine in polynomial complexity.

Note that $P \subseteq NP \subseteq \text{PSPACE} \subseteq \text{EXPTIME}$.

Merlin Arthur (MA)
In an NP problem, if the verifier V is a BPP algorithm, we get a new complexity class called **Merlin Arthur**. A decision problem L is in MA iff there exists a probabilistic polynomial time verifier V (called Arthur) such that when x is a YES-instance of L, there is π (which we get from Merlin, the prover) for which the probability $Pr(V(x,\pi) = 1) \geq \frac{3}{4}$ and when x is a NO-instance of L, for every π, the probability $Pr(V(x,\pi) = 1) \leq \frac{1}{4}$.

Note that $NP \subseteq MA$.
With these definitions, we can learn about the quantum complexity classes.

Bounded Error, Quantum Polynomial Time (BQP)
Bounder error, Quantum Polynomial Time is the class of problems (Fig. 7.50) that can be solved by a polynomial-time quantum Turing machine with error probability $\leq 1/3$ using polynomial-size quantum circuits built with universal quantum gates.

Fig. 7.50 The BQP complexity class

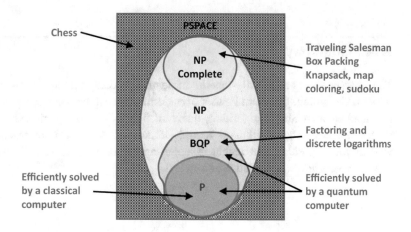

The BQP class is somewhat equivalent to the BPP class, as the problems of this class can be solved by probabilistic Turing machines.

The following properties are applicable to the BQP class.

- $P \subseteq$ BQP. This draws from the fact that we can simulate all classical circuits in quantum.
- BPP $\subseteq$ BQP, but BPP $\neq$ BQP and BQ$P \nsubseteq$ BPP The first part is true because the quantum systems are probabilistic. Hence, quantum computers can solve any classical probabilistic problem. The second and third parts are because we still cannot prove that the quantum computer is better than the classical computer.
- BQP $\subseteq$ PSPACE. Quantum computers can efficiently solve all problems solved by deterministic classical computers. This set does not include problems with polynomial complexity that cannot be solved by classical computers.

Quantum Merlin Arthur (QMA)

If the verifier V is in QBP, then the class of problems is called QMA. The QMA class is defined as follows.

QMA is the set of decision problems that can be solved with a quantum algorithm V such that for all possible inputs x, if x is a YES-instance, then there is $|\phi\rangle$ for which $V(x, |\phi\rangle) = 1$ with probability $\geq \frac{3}{4}$, and if x is a NO-instance, for every $|\phi\rangle$, the probability that $V(x, |\phi\rangle) = 1$ is $\leq \frac{1}{4}$.

Note that $MA \subseteq$ QMA, $NP \subseteq$ QMA, and QMA $\subseteq PP$.

Finally, if the prover can send only classical bits, and if the verifier is a quantum algorithm, we have another complexity class called QCMA, Quantum Classical Merlin Arthur. QCMA $\subseteq$ QMA.

The abbreviation for QIP is **Quantum Interactive Polynomial Time**. It is the quantum equivalent of the classical complexity class IP. IP is an interactive proof system with a polynomial verifier and a computationally unbound prover, in which the prover and the verifier can interact polynomially for many rounds. It has been reported that QIP(1) = QMA and QIP(2) = QMA.

The relation between various complexity classes is summarized as follows:

$$P \subseteq NP \subseteq MA \subseteq \text{QCMA} \subseteq \text{QMA} \subseteq PP \subseteq \text{PSPACE}$$

7.14   Summary

This chapter was another milestone in our learning—quantum algorithms. We started by building a ripple adder circuit. Quantum Fourier Transform was a significant concept we learned in this chapter and applied it in Shor's algorithm, quantum phase estimation, and HHL algorithm. We learned several oracle-based algorithms—Deutsch–Jozsa, Bernstein–Vazirani, and Simon's. Grover's search was another algorithm that taught us about amplitude magnification. Quantum k-means was yet another algorithm, which used the swap test.

We completed this chapter by learning about the quantum complexity theory.

The following chapter introduces another quantum computing model—adiabatic quantum computing, which is quite a deviation from the gate model of quantum computation.

Practice Problems

1. Design a quantum circuit to measure the decoherence of qubits. (Hints: Flip the qubit using an X gate, apply a certain number of Identity gates and measure.) Perform this with a number of backends available.
2. Develop a modular exponentiation circuit to factorize numbers 21 and 35.
3. Develop a quantum comparator circuit which can compare two n bit unsigned integers and determine which number is larger. Extend this logic for signed integers.
4. Design and develop a quantum divider circuit. Develop unit testing code to verify that the circuit works.
5. Satisfiability problems can test whether a given Boolean expression can evaluate true for a given set of input values. For example, the problem $f(x, y, z) = (x \wedge \neg y \vee z) \wedge (\neg x \wedge y \vee \neg z) \wedge (x \wedge y \wedge z) \wedge (\neg x \vee y \neg z)$ is a 3-SAT problem with four clauses (the brackets) and three literals (variables inside the brackets). The symbol $\neg$ denotes a NOT operation. $\wedge$ is an AND operation and $\vee$ is an OR operation. Evaluate for what values of x, y, z, the function $f(x, y, z)$ evaluates TRUE. Use Grover's algorithm to solve this problem.
6. In literature [21], it is known that Grover's Algorithm can be used in minimizing **Quadratic Unconstrained Binary Optimization** Problems (**QUBO.**) Use this method to optimize $2x + 4y + x^2 + y^2$.
7. An oracle function implements a secret key $k \in \{0, 1\}^n$ which opens a secret locker in the national bank. Implement a quantum circuit that can reveal the secret key in one step.
8. Develop a working code to implement the **Group Leader Optimization Algorithm** [27, 28], to decompose unitary matrices into quantum circuits. Verify the code with the unitary matrix of the Toffoli gate and compare it with the literature.
9. Improve the HHL example discussed in Sect. 7.12 using amplitude amplification. Explore implementing the method using Hamiltonian simulation.
10. Implement the HHL procedure outlined in [23] to solve a linear system that can be described using a 4 × 4 matrix.
11. Measure the state vector of the qubits, at each step of the Grover's iteration and plot a graph. For measuring the state vector, you may use the code example from the HHL algorithm.

References

1. Addition on a Quantum Computer, Thomas G. Draper, June 15, 2000, https://arxiv.org/pdf/quant-ph/0008033.pdf
2. For evaluating Boolean expressions, the following website comes handy: https://www.dcode.fr/boolean-expressions-calculator

3. On the power of quantum computation, Daniel R. Simon, October 1997, https://www.researchgate.net/publication/2822536_On_the_Power_of_Quantum_Computation

4. Experimental demonstration of Shor's algorithm with quantum entanglement, B. P. Lanyon et al, 16-OCT-2007, https://arxiv.org/pdf/0705.1398.pdf

5. Realization of a scalable Shor algorithm, T. Monz et al, https://arxiv.org/pdf/1507.08852.pdf

6. Fast quantum modular exponentiation architecture for shor's factoring algorithm, Archimedes Pavlidis et al, 4-Nov-2013, https://arxiv.org/pdf/1207.0511.pdf

7. Quantum arithmetic with the Quantum Fourier Transform, Lidia et al, 2-May-2017, https://arxiv.org/pdf/1411.5949.pdf

8. Complete 3-Qubit Grover Search on a Programmable Quantum Computer, C. Figgatt et al, 30-March-2017, https://arxiv.org/pdf/1703.10535.pdf

9. A fast quantum mechanical algorithm for database search, Lov K. Grover, 19-Nov-1996, https://arxiv.org/pdf/quant-ph/9605043.pdf

10. Grover search algorithm, Eva Borbely, https://arxiv.org/ftp/arxiv/papers/0705/0705.4171.pdf

11. Simplified Factoring Algorithms for Validating Small-Scale Quantum Information Processing Technologies, Omar Gamel et al, 14-Nov-2013, https://arxiv.org/pdf/1310.6446v2.pdf

12. https://mathworld.wolfram.com/NumberFieldSieve.html

13. Factorization of a 768-bit RSA modulus, Thorsten Kleinjung et al, 18-February-2010, https://eprint.iacr.org/2010/006.pdf

14. Experimental realization of Shor's quantum factoring algorithm using nuclear magnetic resonance, Lieven M.K et al, 30-Dec-2001, https://arxiv.org/pdf/quant-ph/0112176.pdf

15. Quantum factorization of 56153 with only 4 qubits, Nikesh S. Dattani, Nathaniel Bryans, 27-Nov-2014, https://arxiv.org/pdf/1411.6758.pdf

16. Quantum Annealing for Prime Factorization, Shuxian Jiang et al, 11-June-2018, https://arxiv.org/pdf/1804.02733.pdf

17. Pretending to factor large numbers on a quantum computer, John A. Smolin et al, 29-June-2013, https://arxiv.org/pdf/1301.7007.pdf

18. Experimental realisation of Shor's quantum factoring algorithm using qubit recycling, Enrique Mart'ın-L'opez et al, 24-Oct-2012, https://arxiv.org/pdf/1111.4147.pdf.

19. Further reading and reference on the quantum k-means algorithm: (a) q-means: A quantum algorithm for unsupervised machine learning, *Iordanis Kerenidis, Jonas Landman, Alessandro Luongo and Anupam Prakash,* December 12, 2018, https://arxiv.org/pdf/1812.03584.pdf. (b) Quantum algorithms for supervised and unsupervised machine learning, *Seth Lloyd, Masoud Mohseni, Patrick Rebentrost, Nov*ember 4, 2013, https://arxiv.org/pdf/1307.0411.pdf. (c) Quantum Algorithms for Nearest-Neighbor Methods for Supervised and Unsupervised Learning, *Nathan Wiebey, Ashish Kapoor, and Krysta M. Svorey,* July 18, 2014, https://arxiv.org/pdf/1401.2142.pdf. (d) Quantum machine learning: distance estimation for k-means clustering, Sashwat Anagolum, March 26, 2019, https://towardsdatascience.com/quantum-machine-learning-distance-estimation-for-k-means-clustering-26bccfbfcc76. (e) K-means clusterization algorithm with Quantum Circuit - Part 2, https://developer.ibm.com/recipes/tutorials/kmeans-clusterization-algorithm-with-quantum-circuit-part-2/

20. Scikit-learning, https://scikit-learn.org/stable/index.html

21. Grover Adaptive Search for Constrained Polynomial Binary Optimization, Austin Gilliam et al, 10-Aug-2020, https://arxiv.org/pdf/1912.04088.pdf.

22. Quantum algorithm for linear systems of equations, Aram W. Harrow, Avinatan Hassidim, and Seth Lloyd, 30-Sep-2009, https://arxiv.org/pdf/0811.3171.pdf

23. Quantum Circuit Design for Solving Linear Systems of Equations, Yudong Cao et al, 16-Apr-2012, https://arxiv.org/pdf/1110.2232v2.pdf

24. Quantum Circuit Design for Solving Linear Systems of Equations, Yudong Cao et al, 10-Aug-2013, https://arxiv.org/pdf/1110.2232v3.pdf

25. Demonstration of a Quantum Circuit Design Methodology for Multiple Regression, Sanchayan Dutta et al, 17-Nov-2019, https://arxiv.org/pdf/1811.01726.pdf

26. Solving systems of linear algebraic equations via unitary transformations on quantum processor of IBM Quantum Experience, S.I. Doronin et al, 1-Jan-2020, https://arxiv.org/pdf/1905.07138.pdf

27. Group Leaders Optimization Algorithm, Anmer Daskin, Sabre Kais, 26-Jan-2011, https://arxiv.org/pdf/1004.2242. pdf
28. Further reading on HHL algorithm and derivate work. (a) Decomposition of Unitary Matrices for Finding Quantum Circuits: Application to Molecular Hamiltonians, Anmer Daskin, Sabre Kais, 23-Feb-2013, https://arxiv.org/pdf/ 1009.5625.pdf. (b) Concrete resource analysis of the quantum linear system algorithm used to compute the electromagnetic scattering cross section of a 2D target, Artur Scherer et al, 27-July-2016, https://arxiv.org/pdf/ 1505.06552.pdf. (c) Quantum algorithms and the finite element method , Ashley Montanaro and Sam Pallister, 25-Feb-2016, https://arxiv.org/pdf/1512.05903.pdf. (d) A quantum algorithm to solve nonlinear differential equations, Sarah, K Leyton and Tobias J. Osborne, 23-Dec-2008, 337

Adiabatic Optimization and Quantum Annealing

8

Note: Reading Sect. 3.6 on quantum tunneling may be helpful to read this chapter.

Adiabatic quantum computing (AQC) is an alternate quantum computing method based on the quantum adiabatic theorem [2]. A related concept is quantum annealing. In an AQC based computation [1, 3], an initial Hamiltonian in its ground state is evolved slowly to a final Hamiltonian whose ground state encodes the solution to the problem. AQC is used in solving optimization problems [4, 5] and developing heuristic algorithms.

The quantum adiabatic theorem was first introduced by the German physicist **Max Born** and the Soviet physicist **Vladimir Fock** in 1928. The adiabatic theorem is stated formally as follows:

The instantaneous state of a quantum system that begins in the nondegenerate ground state of a time-dependent Hamiltonian will remain in its ground state, provided the Hamiltonian changes sufficiently slowly. However, the wavefunction may pick up a phase.

Imagine that an equestrian is riding a horse with a glass full of wine in his/her hand. If the horse animates sufficiently slowly, the equestrian can continue to hold the glass without the wine spilling!

▶ **Learning Objectives**
- Quantum Adiabatic Theorem
- Proof of the Quantum Adiabatic Theorem
- Adiabatic Optimization
- Quantum Annealing

© The Author(s), under exclusive license to Springer Nature Switzerland AG 2021   365
V. Kasirajan, *Fundamentals of Quantum Computing*, https://doi.org/10.1007/978-3-030-63689-0_8

8.1   Adiabatic Evolution

In quantum mechanics, we are generally interested in the time evolution of the quantum states. Following the Hamiltonian is a way to generate the time evolution of the quantum states. If $|\psi(t)\rangle$ is the state of the system at time t, then the general form of the Schrödinger equation is given as follows:

$$i\frac{\partial}{\partial t}|\psi(t)\rangle = H|\psi(t)\rangle \tag{8.1}$$

The solution to this equation at time $t = 0$ is straightforward, especially if the Hamiltonian is time-independent. The eigenstate of the Hamiltonian simply acquires a phase.

$$|\psi(t)\rangle = e^{\frac{-iHt}{\hbar}}|\psi(0)\rangle \tag{8.2}$$

If the Hamiltonian is dependent on time, then the eigenfunction and the eigenvalues are dependent on time. Assume a spin qubit in its initial state $|\psi(0)\rangle = |0\rangle$, pointing to the $+z$ direction (see Fig. 3.11). Suppose, if we apply an X rotational field that slowly rotates the qubit to point to the $-z$ direction. The spin starts to precess about the direction of the rotational field, slowly moving towards the $-z$-axis. If the time T is made sufficiently large, the rotational field changes its direction at an infinitesimally slower rate. We can say that the spin simply tracks the field and remains in its ground state $H(t)$. If we assume $|0\rangle \rightarrow e^{i\alpha}|1\rangle$ and $|1\rangle \rightarrow e^{i\beta}|0\rangle$, the instantaneous Hamiltonian for the X rotation is given by the following equation [13]:

$$H(t) = \frac{\pi\hbar}{T}\left(N - \frac{1}{2}\right)[\cos(\gamma)\,\sigma_x + \sin(\gamma)\,\sigma_z], \tag{8.3}$$

where N is an arbitrary constant and $\gamma = (\alpha + \beta)/2$.

If the time T is arbitrarily large, then $H(t)$ varies arbitrarily slowly as a function of t. The evolution of the initial quantum state $|\psi(0)\rangle$ can be written as follows:

$$i\frac{\partial}{\partial t}|\psi(t)\rangle = H(t)|\psi(t)\rangle \tag{8.4}$$

Let us assume that the nondegenerate ground state $\psi(0)$ is the eigenstate of $H(0)$. Let $s \in [0, 1]$. Let $H(s)$ varies smoothly as a function of s, such that $s = t/T$. If the ground state of $H(s)$ is nondegenerate for all $s \in [0, 1]$, then according to the adiabatic theorem, in the limit $T \rightarrow \infty$, the final state $|\psi(T)\rangle$ will be in the ground state of $H(1)$.

8.2   Proof of the Adiabatic Theorem

Note: Readers can skip the equations and read the text without compromising the understanding.

Since Bohr and Fock, several authors have provided proof for the adiabatic theorem. We follow a pedagogical method for its simplicity. Let us start with the instantaneous nondegenerate eigenstates:

$$H(t)|\psi_n(t)\rangle = E_n(t)|\psi_n(t)\rangle, \tag{8.5}$$

where $E_n(t)$ is the nth eigenstate of the Hamiltonian $H(t)$.

Continuing our discussions from the previous section, since the Hamiltonian changes with time, the corresponding eigenfunctions and eigenvalues are time-dependent. Nevertheless, at any given instant, they form a complete orthonormal set.

$$\langle \psi_n(t) | \psi_m(t) \rangle = \delta_{nm} \tag{8.6}$$

The solution to Eq. (8.4) can be written as a linear combination of an orthonormal set.

$$|\psi(t)\rangle = \sum_n c_n(t) |\psi_n(t)\rangle, \tag{8.7}$$

where $c_n(t)$ is a function of time that needs to be determined. Substituting Eq. (8.7) in Eq. (8.4), we get:

$$i\hbar \sum_n \left(\dot{c}_n |\psi_n(t)\rangle + c_n |\dot{\psi}_n(t)\rangle \right) = \sum_n c_n E_n(t) |\psi_n(t)\rangle \tag{8.8}$$

The dot accent (˙) denotes the time derivative. By taking an inner product with ψ_m and using the orthonormality,

$$i\hbar \left(\dot{c}_m + \sum_n c_n \langle \psi_m | \dot{\psi}_n \rangle \right) = c_m E_k \tag{8.9}$$

Note that we are dropping '(t)' for now to make the equations simple. We shall reintroduce this once the equation simplifies.

$$i\hbar \dot{c}_m = E_m c_m - i\hbar \sum_n \langle \psi_m | \dot{\psi}_n \rangle c_n \tag{8.10}$$

Upon rewriting,

$$i\hbar \dot{c}_m = \left(E_m - i\hbar \langle \psi_m | \dot{\psi}_m \rangle \right) c_m - i\hbar \sum_{n \neq m} \langle \psi_m | \dot{\psi}_n \rangle c_n \tag{8.11}$$

The time derivative of Eq. (8.5) is:

$$\dot{H}\psi_n + H\dot{\psi}_n = \dot{E}_n \psi_n + E_n \dot{\psi}_n \tag{8.12}$$

Taking an inner product with ψ_m,

$$\langle \psi_m | \dot{H} | \psi_n \rangle + \langle \psi_m | H | \dot{\psi}_n \rangle = \dot{E}_n \delta_{mn} + E_n \langle \psi_m | \dot{\psi}_n \rangle \tag{8.13}$$

For $n \neq m$, $\langle \psi_m | H | \dot{\psi}_n \rangle = E_m \langle \psi_m | \dot{\psi}_n \rangle$. Hence, we can rewrite the above equation.

$$\langle \psi_m | \dot{H} | \psi_n \rangle = (E_n - E_m) \langle \psi_m | \dot{\psi}_n \rangle \tag{8.14}$$

Or,

$$\langle \psi_m | \dot{\psi}_n \rangle = \frac{\langle \psi_m | \dot{H} | \psi_n \rangle}{(E_n - E_m)} \tag{8.15}$$

Substituting this in Eq. (8.11), we get:

$$i\hbar \dot{c}_m = \left(E_m - i\hbar \langle \psi_m | \dot{\psi}_m \rangle \right) c_m - i\hbar \sum_{n \neq m} \frac{\langle \psi_m | \dot{H} | \psi_n \rangle}{(E_n - E_m)} c_n \tag{8.16}$$

Applying adiabatic approximation, if $\dot{H}$ is infinitesimally small, the second term vanishes, and $|c_m| = 1$. Hence,

$$i\hbar\dot{c}_m = (E_m - i\hbar \langle\psi_m|\dot{\psi}_m\rangle) \tag{8.17}$$

The solution to this equation is:

$$c_m(t) = c_m(0) \exp\left(\frac{1}{i\hbar}\int_0^t (E_m(t') - i\hbar \langle\psi_m|\dot{\psi}_m\rangle)\, dt'\right)$$

$$= c_m(0)\exp\left(\frac{1}{i\hbar}\int_0^t E_m(t')dt'\right) \exp\left(i\int_0^t i\langle\psi_m|\dot{\psi}_m\rangle\, dt'\right) \tag{8.18}$$

$$= c_m(0)\, e^{i\theta_m(t)}\, e^{i\gamma_m(t)},$$

where the dynamic phase $\theta_m(t)$, and geometric phase $\gamma_m(t)$ are defined as:

$$\theta_m(t) \equiv -\frac{1}{\hbar}\int_0^t E_m(t')dt'$$

$$\gamma_m(t) \equiv \int_0^t i\langle\psi_m|\dot{\psi}_m\rangle dt' \tag{8.19}$$

If the quantum system started in the nth eigenstate, $c_n(0) = 1$, and $c_m(0) = 0$ for $m \neq n$, the Eq. (8.7) becomes,

$$|\psi_n(t)\rangle = c_n(t)\, e^{i\theta_n(t)}\, e^{i\gamma_n(t)}|\psi_n(t)\rangle + \sum_{m\neq n}c_m(t)\, e^{i\theta_m(t)}\, e^{i\gamma_m(t)}|\psi_m(t)\rangle \tag{8.20}$$

$$|\psi_n(t)\rangle = e^{i\theta_n(t)}\, e^{i\gamma_n(t)}|\psi_n(t)\rangle \tag{8.21}$$

This means the system remains in the nth eigenstate up to a global phase. Hence, the proof. QED.

8.2.1 Berry Phase

Note that in Eq. (8.19), the geometric and dynamic phases are real quantities, as long as the Hamiltonian is real. Let us assume that the Hamiltonian changes with time, due to a generalized parameter λ that changes with time. The geometric phase can then be written as follows.

$$\gamma_m = \int_{t1}^{t2} i \left\langle\psi_m\left|\frac{\partial}{\partial\lambda}\right|\psi_m\right\rangle d\lambda = \int_{t1}^{t2} f(\lambda)d\lambda \tag{8.22}$$

During the system evolution, if λ changed as a set of values $\lambda = (\lambda_1, \lambda_2, \ldots)$, then the above equation becomes:

$$\gamma_m = \int_{t1}^{t2} i \langle\psi_m|\nabla_\lambda|\psi_m\rangle d\lambda = \int_{t1}^{t2} f(\lambda_1)d\lambda_1 + f(\lambda_2)d\lambda_2 + \ldots \tag{8.23}$$

Suppose if the system returned to the original state it started with, the total geometric phase acquired must be zero. From the above equation, it is evident that the geometric phase is not necessarily zero, unless $\nabla_\lambda \times f = 0$.

The geometric phase of an adiabatic system in a closed loop is known as **Berry phase**. The Berry Phase can be written as:

$$\gamma_m(C) = \oint_C i \langle \psi_m | \nabla_\lambda | \psi_m \rangle d\lambda, \tag{8.24}$$

where λ is an adiabatic process following a cyclic path C. It is the phase acquired by slow moving systems.

Berry phase has applications in condensed matter physics and quantum computing.

8.3 Adiabatic Optimization

The objective of an optimization problem is to find the best solution from a set of feasible solutions. In a finite-dimensional optimization problem, the choices are from a set of real number variables, called **decision variables**. The decision variables are labelled $x_1, x_2, \ldots, x_n$. Each possible choice is a point $x = (x_1, x_2, \ldots, x_n)$ in the solution space $\mathbb{R}^n$. The **feasible set** is a subset $C \subset \mathbb{R}^n$ containing the choices $x = (x_1, x_2, \ldots, x_n)$. The **objective function** $f(x)$ is to be minimized or maximized over C, that is, for all $x \in C, f(x_0) \leq f(x)$ for a minimizing function and $f(x_0) \geq f(x)$ for a maximizing function. In other words, the objective function determines how good a choice is. Optimization problems may have **constraints**, which are conditions that specify the feasible set. For example, the range is a constraint.

The quantum adiabatic theorem is used in solving optimization problems. The solutions to the minimizing problem are encoded in the ground state of a Hamiltonian. The goal of the algorithm is to adiabatically evolve into this ground state from a known ground state. According to the adiabatic theorem, if the evolution is sufficiently slow, the probability of finding the solution close to the ground state is very high.

Consider the following Hamiltonian with eigenvalues $h(z)$:

$$H_P = \sum_{s \in \{0,1\}^n} h(z)|z\rangle \langle z| \tag{8.25}$$

We call this the problem Hamiltonian. It is defined in the computational basis as a diagonal matrix. The ground state of this Hamiltonian contains the states z, such that $h(z)$ is minimized.

We define a beginning Hamiltonian H_B, whose ground state can be prepared relatively easily. Let $H_D(t)$ be a smooth varying time-dependent driver Hamiltonian such that $H_D(0) = H_B$, and $H_D(T) = H_P$, where T is a sufficiently large evolution time. The goal is to evolve the system adiabatically from H_B to H_P. The gradual change of the Hamiltonian can be described by a smooth monotonic function $f(x)$ (called **schedule or annealing schedule**) such that $f(0) = 0$ and $f(1) = 1$. The Hamiltonian at the time t is given by:

$$H_D(t) = \left(1 - f\left(\frac{t}{T}\right)\right) H_B + f\left(\frac{t}{T}\right) H_P \tag{8.26}$$

Note that $H_D(0) = H_B$ and $H_D(1) = H_P$. Hence, $f(x)$ must vary from 0 to 1 as time t changes from 0 to T. The choice of the initial Hamiltonian and the final Hamiltonian is given in the following equation.

$$H_B = -\sum_{i=1}^{n} X_i$$

$$H_P = H_{Ising} = -\sum_{i}^{n} h_i X_i - \sum_{i,j}^{n} J_{ij} Z_i Z_j, \tag{8.27}$$

where X and Z are Pauli operators acting on the ith qubit. h_i is the external transverse magnetic field. J_{ij} is spin-spin interaction energy between nearest qubits i and j, respectively. The Hamiltonian H_P represents an Ising model in a transverse field. The first term represents the interaction between the applied field and the qubits. The second term represents the interaction energy needed to bring about an ordered state.

In the computational basis, the quantum state of n qubits prepared in the ground state is described as follows. We should be familiar with this expression by now.

$$|\psi(0)\rangle = \frac{1}{\sqrt{2^n}} \sum_{x}^{2^n-1} |x\rangle^{\otimes n} \tag{8.28}$$

This system of qubits evolves as per the Schrödinger Eq. (8.4) from time $t = 0$ to $t = T$. The time T is chosen in such a way that the changes in $\psi(t)$ are slow relative to the inverse of the minimum energy gap of $H(t)$ with the instantaneous eigenspectrum defined by Eq. (8.5). The minimum energy gap Δ_{min} is defined as the energy difference between the instantaneous ground state and the first excited states. If the time $T \gg \Delta_{min}^{-a}$ for $a = 2, 3$, then the qubits remain in the ground state of the instantaneous Hamiltonian. At the end of the evolution $H(T) = H_P$.

After the preparation of the final state $\psi(T)$, the qubits are measured in the computational basis. Note that H_P is diagonal in the computational basis. Hence, the measurement results correspond to the ground state, which is the solution to the problem.

A numerical simulation example given in the reference [6] is a good starting point.

8.4   Quantum Annealing

Quantum annealing [8–10] is closely related to a method used in solving combinatorial optimization problems in discrete search spaces where there are many local minima. The search attempts to find a fairly good solution to the problem, but not necessarily the most optimal one. The traveling salesman (TSP) is an example of such a problem. The annealing process draws from naturally occurring phenomena, such as the formation of diamonds over millennia, and freezing water. When water is frozen slowly, the resulting ice crystals are almost perfect than those formed by rapid cooling. The reason is that those ice crystals cooled slowly are in a lower energy meta-state, a natural process of a system minimizing its free energy. Diamonds are formed when carbon-rich rocks deep beneath the earth's surface are put through heat and pressure for thousands of years. This is a slow geological process that produces a highly ordered crystalline structure of the diamond. Anything short term can only produce coal! Annealing is a widely used manufacturing process, for example, in the steel forging industry. Annealing improves the machinability of steel by softening it for cold working. Steel is first heated to a specific temperature and held at that temperature for a while. This recrystallizes steel, which helps in the formation of a new grain-free structure and makes it defect free. After this stage, steel is cooled slowly in a controlled environment. Cooling steel in a controlled environment results in a refined microstructure ensuring softness and ductility.

Fig. 8.1 Thermal jump
and quantum tunneling

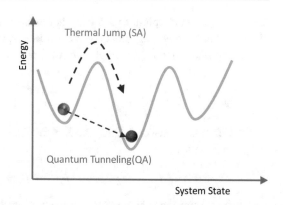

The classical simulated annealing (SA) is an equivalent process and uses the classical Monte Carlo sampling (of thermodynamic systems) based on the **Metropolis-Hastings Algorithm**. Simulated annealing works by increasing the energy of the system and allowing it to cool down gradually. The thermal fluctuations lead the system to transit between states over intermediate energy barriers to search for the lowest energy state possible. The system anneals reaching a solution state, through free energy minimization:

$$F[p(x)] = E[p(x)] - TS[p(x)], \tag{8.29}$$

where the thermodynamic potentials F- free energy, E-inner energy, S-entropy are functions of the probability distribution of the system states x. When the temperature is high, the free energy F is minimized for maximum entropy S, a maximally mixed state $p(x)$ of a wide range of possible minima. As the system cools down, E modifies the entropy such that the initial $p(x)$ settles down in a fairly good solution to the optimization problem.

However, this method doesn't find the local minima in scenarios where there may be a high energy barrier preceding the lowest energy state.

Quantum annealing is similar to simulated annealing (SA) but uses **quantum tunneling.** Quantum tunneling (Sect. 3.6) is a consequence of quantum mechanics, where a particle can tunnel through a potential barrier and appear on the other side with the same energy.

Figure 8.1 illustrates the simulated annealing and the quantum annealing. As illustrated in the diagram, SA requires the system to overcome energy barriers. QA uses quantum tunneling.

Quantum annealing proceeds by replacing the current state (a candidate solution) with a randomly chosen neighbor state for which the target function computes the least energy. The **quantum field strength** controls the exploration of the neighborhood, also called the **quantum tunneling width**. At first, the tunneling radius is so vast that the neighborhood contains the entire search space. As the system evolves, the neighborhood gets gradually shrunk until the current state is not too variant from a randomly chosen neighbor state. Quantum annealing converges at this stage.

Assume that the solution to the problem is encoded in the ground state of an Ising Hamiltonian in a transverse field described by the following equation. The quantum Ising model [11] describes a lattice structure whose nearest-neighbor interactions are due to the alignment or antialignment of the spins along the z-axis and the external magnetic field along the x-axis (the transverse field.)

$$H_{Ising} = -\sum_{i}^{n} J_i Z_i - \sum_{i,j}^{n} J_{ij} Z_i Z_j - \sum_{i,j,k}^{n} J_{ijk} Z_i Z_j Z_k - \dots \tag{8.30}$$

In this equation, Z is the Pauli operator σ_z. The subscripts i, j, k refer to the lattice sites. J represents the coupling between adjacent qubits. n is the number of spins. The eigenvalues of the Pauli operators are ± 1, which correspond to the Ising spin states. An optimization problem is coded into the Ising Hamiltonian by mapping binary variables into spin states. To induce quantum annealing, a time-dependent transverse field is applied to the system.

$$H_{TF}(t) \equiv -\Gamma(t) \sum_{i=1}^{n} X_i, \tag{8.31}$$

where $\Gamma(t)$ is the transverse field, and X is the Pauli operator σ_x. The total Hamiltonian of the system can be written as:

$$H(t) = H_{Ising} + H_{TF}(t) \tag{8.32}$$

The transverse field causes quantum fluctuations between two spin states with eigenvalues $Z_i = +1$ and $Z_i = -1$, which induces the quantum search of the phase space.

When the algorithm starts, $\Gamma(t)$ is chosen with a very high value, such that in Eq. (8.32) the second term is the most significant. This state corresponds to the high-temperature state of SA. A high-temperature value is chosen because a high-temperature state in SA is a mixture of all possible states that are almost equally probable. As the computation proceeds, the transverse field $\Gamma(t)$ is adiabatically reduced to '0', at this point H_{Ising} is the most significant parameter in the equation.

The system state follows the Schrödinger Eq. (8.4), evolving from the trivial initial ground state described by the transverse field Hamiltonian Eq. (8.31) to the non-trivial Ising Hamiltonian described by Eq. (8.30), which encodes the solution to the problem.

Note that some quantum systems may still have entropy even at absolute zero because of the zero-point energy. Due to this reason, the adiabatic evolution must be slow enough. However, too slow annealing comes at the cost of computing, so in practice, a balance between the quality of annealing and the computing time should be sought. If the annealing is rapid, it is possible that some interstate energy barriers may cause quantum jumps leading to the algorithm getting stuck at a local minimum. So, it is important to decrease $\Gamma(t)$ in such a way that the system state is close to the instantaneous ground state of the total Hamiltonian Eq. (8.32).

8.5   Summary

In this chapter, we learned about adiabatic quantum computing, an alternate form of quantum computing, and quantum annealing a closely related method. The advantage of quantum annealing over simulated annealing is evident from numerous studies [7], and there is proof that the probability of the system being in the lowest state is high for quantum annealing. This does not mean that all NP-Hard problems can be efficiently performed using quantum annealing, and there is no evidence of quantum supremacy in this space yet. However, the use cases for QA are in plenty. A number of combinatorial optimization problems can be readily solved using QA.

D-Wave	Google
MDR	MIT Lincoln Lab
NEC	Northrop Grumman
Qilimanjaro	

There is much activity in this base, both in the industry and in academia. The above table illustrates the list of organizations involved in developing adiabatic/annealing methods [12].

Practice Problems

1. The Hamiltonian of a qubit in initial state $|0\rangle$ is given by $H = \begin{bmatrix} 0 & a \\ a & 0 \end{bmatrix}$. What is the probability of the qubit to be in state $|1\rangle$ at time $t > 0$.

2. The Hamiltonian of a hypothetical qubit is defined by the matrix $H = \begin{bmatrix} \hbar\omega & 0 \\ 0 & -\hbar\omega \end{bmatrix}$, where ω is the qubit frequency. The qubit is prepared in the equal superposition state initially, that is, $|\psi(0)\rangle = \frac{1}{\sqrt{2}}(|0\rangle + |1\rangle)$. Evolve the system from time $t = 0$ to $t = 2\tau$. At the end of the evolution, the qubit is measured using a basis $\{|+\rangle, |-\rangle\}$ defined by $|+\rangle = \frac{1}{\sqrt{2}}(|0\rangle + |1\rangle)$ and $|-\rangle = \frac{1}{\sqrt{2}} \times (|0\rangle - |1\rangle)$. Find the probability of measuring the qubit in one of the states. (Hints: use: $|\psi(t)\rangle = e^{\frac{-iHt}{\hbar}}|\psi(0)\rangle$ and $P(a) = |\langle a|\psi\rangle|^2$.)

3. Can a pure state (adiabatically) evolve into a mixed state? Explain your answer.

4. Adiabatic invariants are the properties of a physical system, that remain constant when a change occurs. Prove that $\frac{E}{\omega}$ of a simple harmonic oscillator is adiabatically invariant.

5. What happens to von Neumann's entropy under an adiabatic evolution? Explain your answer.

References

1. Adiabatic Quantum Computing, Tameem Albash et al, 2-Feb-2018, https://arxiv.org/pdf/1611.04471.pdf
2. Geometry of the adiabatic theorem, Augusto C'esar Lobo et al, 8-Jun-2012, https://arxiv.org/pdf/1112.4442.pdf
3. Notes on Adiabatic Quantum Computers, Boaz Tamir, Eliahu Cohen, 7-Dec-2016, https://arxiv.org/pdf/1512.07617.pdf
4. Designing Adiabatic Quantum Optimization: A Case Study for the Traveling Salesman Problem, Bettina Heim et al, 21-Feb-2017, https://arxiv.org/pdf/1702.06248.pdf
5. Experimental implementation of an adiabatic quantum optimization algorithm, Matthias Steffen et al, 14-Feb-2003, https://arxiv.org/pdf/quant-ph/0302057.pdf
6. Adiabatic Quantum Optimization for Associative Memory Recall, Seddiqi and Humble, 12-Dec-2014, https://arxiv.org/pdf/1407.1904.pdf
7. Study of Optimization Problems by Quantum Annealing, Tadashi Kadowaki, 5-May-2002, https://arxiv.org/pdf/quant-ph/0205020.pdf
8. Solving Optimization Problems Using Quantum Annealing, Motasem Suleiman and Nora Hahn, December 20, 2018, http://www.henryyuen.net/fall2018/projects/annealing.pdf
9. Mathematical Foundation of Quantum Annealing, Satoshi Morita and Hidetoshi Nishimori, 11-Jun-2008, https://arxiv.org/pdf/0806.1859.pdf
10. An introduction to quantum annealing, Diego de Falco et al, 5-Jul-2011, https://arxiv.org/pdf/1107.0794.pdf
11. Transverse Ising Model, Glass and Quantum Annealing, Bikas K. Chakrabarti and Arnab Das, 15-May-2006, https://arxiv.org/pdf/cond-mat/0312611.pdf
12. https://quantumcomputingreport.com/qubit-technology/
13. Hamiltonians for Quantum Computing, Vladimir Privman et al, 13-May-1997, https://arxiv.org/pdf/quant-ph/9705026.pdf

Quantum Error Correction

9

> "I'm sure I 'm not Ada," she said, "for her hair goes in such long ringlets, and mine doesn't go in ringlets at all; and I 'm sure I can 't be Mabel, for I know all sorts of things, and she, oh! she knows such a very little! Besides, she's she, and I'm I, and—oh dear, how puzzling it all is! I 'll try if I know all the things I used to know. Let me see: four times five is twelve, and four times six is thirteen, and four times seven is—oh dear! I shall never get to twenty at that rate! However, the Multiplication Table don't signify: let's try Geography. London is the capital of Paris, and Paris is the capital of Rome, and Rome—no, that's all wrong, I 'm certain! I must have been changed for Mabel! I'll try and say 'How doth the little—'" and she crossed her hands on her lap, as if she were saying lessons, and began to repeat it, but her voice sounded hoarse and strange, and the words did not come the same as they used to do.
>
> — Lewis Carroll, Alice's Adventures in Wonderland

In previous chapters, we learned that the qubits undergo a process called "decoherence." By interacting with the environment, the qubits systematically lose information. Some of the codes that we developed run successfully on the simulators, but they do not run quite well on the backends, especially the algorithms with an extended circuit depth. In this chapter, we explore the practicality of quantum computation. The primary learning in this chapter is about the kinds of quantum errors that can typically occur and the measures to correct them. Our exploration starts with a review of the classical error correction methods and proceeds with correcting bit and phase errors. After building a basic theoretical framework, this chapter introduces the concept of stabilizer formalism on which the Gottesman–Knill theorem is based. The stabilizer formalism is quite a deviation from the amplitude formalism we have been using so far. The usefulness of the stabilizer formalism is seen when we apply Shor's nine-qubit code and Steane's seven-qubit code in correcting quantum errors. The last few sections of this chapter focus on surface codes and protected qubits, which protect the qubits at the physical level. Before concluding this chapter, we explore a scheme that helps in solving measurement errors.

▶ **Learning Objectives**
- Communication on a classical channel, classical error correction, and Shannon's entropy
- Quantum error correction
- Pauli group and correcting bit and phase errors
- Stabilizer formalism
- Shor's nine-qubit error correction code, CSS code, and Steane's seven-qubit error correction code

© The Author(s), under exclusive license to Springer Nature Switzerland AG 2021
V. Kasirajan, *Fundamentals of Quantum Computing*, https://doi.org/10.1007/978-3-030-63689-0_9

- The path towards fault-tolerant quantum computation
- Toric code and Surface code
- Protected qubits

9.1 Classical Error Correction

Ever since **John Bardeen**, **Walter Brattain**, and **William Shockley** invented the first transistor in 1947, electronics has far advanced and greatly influenced our lives. No other technology has revolutionized the way we live, transact, socialize, and feel secure. The first transistor was about the size of our palm. Since then, the circuit sizes have shrunken to the range of 7–10 nm. Communication has evolved from two-way radios and fixed telephones to the intranet that connects everyone in the world. Today, we can transfer terabits of data in seconds.[1] As more things get packed at the circuit level, and the more we exploit communication channels, the more prone to errors our transactions are. The classical errors are due to electromagnetic interference, design problems, signal attenuation over distance, and inherent material deficiencies. Classical circuits employ several schemes to overcome these errors, and we shall see a few of them briefly in this section.

9.1.1 Shannon's Entropy

In 1927, **Ralph Hartley** at Bell Labs proposed that the number of digits required to represent a message can be expressed in terms of logarithms. By 1940, Shannon recognized the relationship between the thermodynamic entropy and informational entropy. He called the negative logarithm of the probability of an event as the entropy.

Let p be the probability of an event. We wish to use a message to communicate the occurrence of this event. Let the measure $\frac{1}{p}$ represents the "surprise" brought by the communication of this event. If the event p is composed of two independent events q and r, such that $p = qr$, then the following inequality is true:

$$\frac{1}{p} \neq \frac{1}{q} + \frac{1}{r} \tag{9.1}$$

If we measure the surprise by the logarithm of $\frac{1}{p}$, then the following equality is true:

$$\log \frac{1}{p} = \log \frac{1}{q} + \log \frac{1}{r} \tag{9.2}$$

We can extend this to a probability distribution of various events $\sum_i p_i = 1$. The uncertainty in this distribution is equal to the average surprise $\sum_i p_i \log \frac{1}{p}$.

---

[1] May 22, 2020, researchers have reported data rate of 44.2 Terra bits per second (Tbps) from a single light source. Ref: https://www.sciencedaily.com/releases/2020/05/200522095504.htm

In **Information Theory**, **entropy** (or **Shannon's Entropy**) is the measure of uncertainty of the outcomes of a random variable. If we are given a random variable X, with a possible outcome x_i, each with a probability $p_X(x_i)$, the entropy $H(X)$ is defined as follows:

$$H(X) = -\sum_i p_X(x_i) \, \log \, p_X(x_i) \tag{9.3}$$

9.1.2 Communication on a Classical Channel

On a classical channel, the original message goes through an information encoding process. It is encoded at the source and decoded at the destination. The information encoding process helps in performing critical tasks such as error correction, compression, and security. Information coding happens at two levels, source coding and channel coding, as explained in the sections below.

Source Coding
A source may be producing a continuous stream of messages composed of the source alphabet. The source encoder splits the messages into blocks of fixed size. The source decoder assembles the blocks in the right order and delivers the output message in the output alphabet.

Channel Coding
This process is responsible for transmitting the message generated by the source through the channel. The channel encoder accepts input messages of fixed length and maps the source alphabet into a channel alphabet. It then adds a set of redundancy symbols (usually error correction codes, timing signatures, packet numbers, etc.) and sends the message through the channel. The channel decoder at the receiving end first determines if the message has any errors, and performs the error correction. It then removes the redundancy symbols and maps the channel alphabet into the source alphabet. Finally, it sends the message to the source decoder. This workflow is illustrated in Fig. 9.1.

Shannon experimented with entropy to parametrize communication. He used entropy to explain the limit to which information can be losslessly compressed on a perfect lossless channel. This formed his famous **source coding theorem**. According to this theorem, if $l_X(n)$ be the number of bits required to encode n symbols, then $l_X(n)$ is in the following range:

$$H(X) \leq l_X(n) = H(X) + \frac{1}{n} \tag{9.4}$$

He also proposed the **noisy-channel coding theorem** (or **Shannon's theorem** or **Shannon's limit**), which proved that on a noisy channel, it is possible to transmit digital data error-free up to a computable maximum rate.

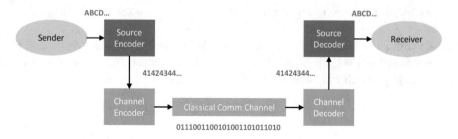

Fig. 9.1 Communication on a classical channel

Let us assume the following:

- Let X be the input alphabet. The input symbol $x \in X$ is modeled by the random variable X, with the probability density function $p_X(x)$.
- Let Y be the output alphabet. The input symbol $y \in Y$ is modeled by the random variable Y, with the probability density function $p_Y(y)$.
- Let the probability of observing the output y for the input x be defined as $p(y|x)$.

If we assume that the channel is memoryless (that is, output depends upon input), then the channel capacity can be defined as $C = \max_{p_X(x)} I(X;Y)$, where $I(X; Y)$, the **mutual information** is the mutual dependency between the two random variables.

The channel coding theorem guarantees that on a noisy channel with capacity C, with $0 < \varepsilon < 1$, there exists a coding scheme to transmit information with a rate close to C, with a probability of error less than ε.

The capacity of a noisy channel is represented as $C = B \times \log\left(1 + \frac{\text{signal}}{\text{noise}}\right)$, and measured in decibels (dB.) Here, B is the bandwidth (maximum transmission rate), *signal* is the average power of the signal, and *noise* is the average power of the noise.

The *fidelity* of the channel is defined as $F(p_X(x), p_Y(x)) = \sum_x \sqrt{p_X(x)p_Y(x)}$.

9.1.3   Classical Error Correction Codes

Parity

Parity is an example of a **systematic code**, where the information symbols are embedded into the codeword. Implementing a parity check is easy. Say, for example, on an 8-bit system, we add an additional bit to ensure that the total number of 1s is even. For example, the alphabet "A" is represented by the 8-bits "01000001" in the ASCII encoding system. The number of 1s in the 8-bits is 2. Hence, we add a 9th bit, which is zero. The resultant codeword is "010000010." At the receiving end, this can be checked easily.

Unfortunately, parity does not correct bit errors, and it cannot detect if more than one bit has errors.

Block Code

Block codes use a Hamming distance to identify errors. A block code C of length n over the alphabet A with q symbols is a set of M codewords (or n-tuples). Each of these codewords uses symbols from A. **Hamming distance** is defined as the number of positions by which two codewords differ. For example, if $x = (0, 1, 0, 1)$ and $y = (1, 1, 0, 0)$, the Hamming distance $d(x, y) = 2$. **Hamming weight** is defined as the distance from a position 0. In our example, $wt(x) \equiv d(x, 0) = 2$. Note that $d(x, y) = wt(x + y)$.

A block code with a block length of n bits, k information symbols (message length), and Hamming distance d is denoted as $[n, k, d]$. Here, $M = q^k$. The rate of a block code R is defined as the ratio between the message length and block length, that is, $R = k/n$.

A small subset of codewords with **Hamming codes** is inserted into the transmission. We can detect errors by checking whether the Hamming codes are received or not. The error correction codewords are chosen in such a way that the hamming distance is large enough, and it is not possible for a valid codeword to transform into another valid codeword. The error correction code is used to fix the corrupted codewords.

Linear Code

A linear code encodes k bits of information in an n bit code space. The code space is defined by an $n \times k$ generator matrix G whose elements are $\{0, 1\}$. The matrix G maps messages into an encoded form. The k bit message x is encoded as Gx. For example, $G(0) = (0,0,0)$ and $G(0,1) = (0,0,0,1,1,1)$. The generator matrix for the repetition code described in Sect. 9.2.1 can be written as follows:

$$G = \begin{bmatrix} 1 \\ 1 \\ 1 \end{bmatrix} \tag{9.5}$$

The repetition code is an example for a [3,1] code, and it requires 3 bits (n) to encode 1 bit (k) of information. Linear codes are also used to create **parity check matrices**. An $[n, k]$ code has n-element vectors x in $\{0, 1\}$. It is defined as follows:

$$Hx = 0, \tag{9.6}$$

where H is an $(n - k) \times n$ matrix called the parity check matrix. To generate H, we need to choose $(n - k)$ linearly independent vectors orthogonal to G and set them as the rows of H. For the repetition code, we choose two linearly independent vectors [1] (0,1,0) and (0,1,1). The party check matrix is:

$$H = \begin{bmatrix} 0 & 1 & 0 \\ 0 & 1 & 1 \end{bmatrix} \tag{9.7}$$

We can easily verify that $Hx = 0$ for the codewords $x = (0, 0, 0)$ and $x = (1, 1, 1)$.

Error Syndrome

Given an $[n, k]$ linear code C with the parity check matrix H, the error syndrome S of the n-tuple v is defined as:

$$S(v) = Hv^T \text{ where } v^T \text{ is the transpose of } v \tag{9.8}$$

The syndrome is used to determine if there is an error, and if so, the syndrome identifies the bit in error.

Generator Matrix

The generator matrix of a $[n, k]$ linear code C is a $k \times n$ matrix. Its elements are the coefficients of a generator polynomial $g(x)$. It is explained in the below equation:

$$G = \begin{bmatrix} g(x) \\ x \cdot g(x) \\ x^2 \cdot g(x) \\ \vdots \\ x^{k-1} \cdot g(x) \end{bmatrix}$$

$$= \begin{bmatrix} g_0 & g_1 & \cdots & g_{n-k-1} & g_{n-k} & 0 & \cdots & 0 \\ 0 & g_0 & g_1 & \cdots & g_{n-k-1} & g_{n-k} & \cdots & 0 \\ 0 & 0 & g_0 & g_1 & \cdots & g_{n-k-1} & \cdots & 0 \\ \vdots & \vdots & \vdots & \cdots & \vdots & \vdots & \cdots & \vdots \\ 0 & 0 & 0 & \cdots & g_0 & \cdots & g_{n-k-1} & g_{n-k} \end{bmatrix}$$

(9.9)

Example 9.1

Given $n = 7$ and $g(x) = 1 + x + x^2 + x^3$, calculate the generator matrix. Encode the message $y = (1101)$ using the generator matrix.

Solution:

Here, $n - k = 3$. Since $n = 7$, we can say that $k = 4$. From Eq. (9.9), we can write the following matrix:

$$G = \begin{bmatrix} 1 & 1 & 1 & 1 & 0 & 0 & 0 \\ 0 & 1 & 1 & 1 & 1 & 0 & 0 \\ 0 & 0 & 1 & 1 & 1 & 1 & 0 \\ 0 & 0 & 0 & 1 & 1 & 1 & 1 \end{bmatrix}$$

(9.10)

To encode the message, we just multiply it with the generator matrix.

$$[1 \ 1 \ 0 \ 1] \begin{bmatrix} 1 & 1 & 1 & 1 & 0 & 0 & 0 \\ 0 & 1 & 1 & 1 & 1 & 0 & 0 \\ 0 & 0 & 1 & 1 & 1 & 1 & 0 \\ 0 & 0 & 0 & 1 & 1 & 1 & 1 \end{bmatrix} = [1 \ 0 \ 0 \ 1 \ 0 \ 1 \ 1]$$

(9.11)

Note that this is done in modulo 2.

Dual Code

Given an $[n, k]$ linear code C, the orthogonal complement or the dual of C, denoted as $C^{\perp}$ consists of vectors (or n-tuples) orthogonal to every vector in C. Note that, the generator matrix of $C^{\perp}$ is the parity check matrix of C.

Hamming Code

Hamming codes are linear codes that can detect 2-bit errors and correct 1-bit errors. Assume an integer $r \geq 2$. There exists a code with block length $n = 2^r - 1$ and message length $k = 2^r - r - 1$. The parity check matrix $[2^r - 1, 2^r - r - 1]$ of a Hamming code is constructed with columns of length r, which are non-zero. Any two columns are linearly independent, therefore $d = 3$. The number of parity check symbols are $m = n - k$. Such Hamming code $[n, n - r, d]$ are also known as **Binary Hamming**

Code. The following is a Hamming code [7, 4, 3] of order $r = 3$. This Hamming code can check single bit errors.

$$H = \begin{bmatrix} 0 & 0 & 0 & 1 & 1 & 1 & 1 \\ 0 & 1 & 1 & 0 & 0 & 1 & 1 \\ 1 & 0 & 1 & 0 & 1 & 0 & 1 \end{bmatrix} \tag{9.12}$$

Supposing if we received a vector $x = [1\ 1\ 1\ 0\ 1\ 1\ 0]$, the error syndrome $S(x) = Hx^T$ can be calculated as follows. Note that the arithmetic is to be done in modulo 2.

$$S(x) = \begin{bmatrix} 0 & 0 & 0 & 1 & 1 & 1 & 1 \\ 0 & 1 & 1 & 0 & 0 & 1 & 1 \\ 1 & 0 & 1 & 0 & 1 & 0 & 1 \end{bmatrix} \begin{bmatrix} 1 \\ 1 \\ 1 \\ 0 \\ 1 \\ 1 \\ 0 \end{bmatrix} = \begin{bmatrix} 0 \\ 1 \\ 1 \end{bmatrix} \tag{9.13}$$

This matches with column 3 of the parity check matrix, indicating that the third bit is in error. Since this is a classical bit, we can conclude that the third bit is toggled, and correct the received vector as $x = [1\ 1\ 0\ 0\ 1\ 1\ 0]$.

Hamming Sphere
The n-tuples have 2^n elements. We can partition them into 2^k disjoint sets called Hamming spheres, each with a radius t. Thus, we can encode 2^k messages. The center of the spheres forms codewords of a $(2t + 1)$ error-correcting code. Erroneous codewords will be inside or on the surface of a Hamming sphere. Therefore, the codewords inside a Hamming sphere of radius t can come from a maximum of t errors. This helps in identification of code words that can be corrected with error correcting codes.

Cyclic Codes
Cyclic codes use circular shifts of the codeword. Each shifted codeword generates another codeword of the alphabet. Cyclic codes are efficient ways of checking and correcting errors. Cyclic codes can correct single or double errors. They can also correct burst errors.

Reed Solomon Codes
Reed Solomon codes are widely used in storage media, QR codes, and wireless networks. Reed Solomon codes work on a block of data. It adds t check symbols to the data and detects t erroneous symbols or corrects $t/2$ symbols with errors. Reed Solomon codes can also be used to correct burst errors.

9.2 Quantum Error Codes

In the previous section, we learned that errors do happen in classical systems and that they are efficiently handled. The quantum error correction [8] draws from classical error correction. However, we have fundamental challenges. Because of the way the quantum systems work, the qubits eventually interact with the environment, resulting in decoherence and decay of the encoded quantum information. Quantum gates are unitary transformations that cannot be implemented perfectly at the physical systems. All these effects accumulate, causing erroneous computation. So, we must combat against decoherence and gate errors.

In the following sections of this chapter, we discuss how quantum information can be efficiently encoded in a noisy environment, error corrected, and retrieved.

9.2.1   Quantum Error Correction

Assume a quantum system A $|\varphi_A\rangle$ in the presence of noise. Let us define another quantum system B $|\varphi_B\rangle$, prepared in the state $|0\rangle$. Let there be an encoding E, which maps the joint state of A and B into a new state:

$$|\varphi_A\rangle \otimes |\varphi_B\rangle \rightarrow |\varphi_{AB}^E\rangle = E(|\varphi_A\rangle \otimes |\varphi_B\rangle) \tag{9.14}$$

Assume that this system goes through a noisy process:

$$|\varphi_{AB}^E\rangle \rightarrow \sum_s e_s |\varphi_{AB}^E\rangle \tag{9.15}$$

A decoding operator $E^\dagger$ will be able to recover $|\varphi_A\rangle$ and reconstruct the state A, iff:

$$E^\dagger |\varphi_{AB}^E\rangle \rightarrow |\varphi_A\rangle \otimes |\varphi_B\rangle \tag{9.16}$$

The goal of quantum error correction is to identify an encoding E, such that the set of correctable errors $S = \{e_s\}$ are the most likely errors that can happen. In this definition, we assume that there are no errors caused during the operation of E or $E^\dagger$. This is possible, only if the quantum circuits performing the operations E or $E^\dagger$ are ideal.

Let us now include another quantum system, ancilla $|\varphi_C\rangle$, prepared in the state $|0_C\rangle$. If we can identify a recovery operator R, which transfers the noise to the ancilla without decoding the state of A and B.

$$R\, e_s\left(|\varphi_{AB}^E\rangle \otimes |0_C\rangle\right) \rightarrow |\varphi_{AB}^E\rangle \otimes |S_C\rangle \qquad \forall e_s \in S \tag{9.17}$$

We shall apply this formalism in subsequent sections. Let us now look at the error correction process.

Figure 9.2 illustrates the quantum error correction process. The quantum error correction code encodes k-qubits in state $\rho_i^{(k)}$ into n-qubits of state $\rho^{(n)}(0)$. These n-qubits then go through a noise process. At time t, their state is $\rho^{(n)}(t)$. The n-qubits are then decoded into k-qubits $\rho_o^{(k)}$.

Fig. 9.2 Quantum error correction process. (Reference: [2])

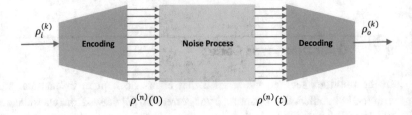

As an example, the quantum state $|\psi\rangle = a|0\rangle + b|1\rangle$ can be encoded as $|\psi\rangle = a|000\rangle + b|111\rangle$. The set of 3-qubits $|000\rangle$ and $|111\rangle$ are called codewords. This type of encoding is known as "**repetition code**," as we encode the message by repeating it many times. The decoder can use a simple error correction scheme such as the "majority rule." The decoder can correct $|100\rangle$ as $|000\rangle$, and $|101\rangle$ as $|111\rangle$.

9.2.2 Quantum Code

A quantum error occurs due to the interaction of the qubits with the environment. Assume an error operator E, which transforms an initial state $|\psi\rangle$ to $|\varphi\rangle$ through the mapping:

$$|\varphi\rangle = E|\psi\rangle \tag{9.18}$$

If we can correct some arbitrary errors E_a and E_b, then we should be able to correct linear combinations of them $E = aE_a + bE_b$. Recall from our earlier discussions that we can distinguish only orthogonal states, that is, $\langle\psi_i| \psi_j\rangle = 0$; $i \neq j$.

Applying this logic to errors E_a and E_b, we should be able to distinguish the states resulting from the application of these error operators, that is, $E_a|\psi_i\rangle$ and $E_b|\psi_j\rangle$. This requires that these two error states must be orthogonal, that is,

$$\langle\psi_i|E_a^\dagger E_b|\psi_j\rangle = 0 \tag{9.19}$$

If we measure the states $|\psi_i\rangle$ or $|\psi_j\rangle$, the system collapses. Hence, we gather information about $|\psi_i\rangle$ by measuring $\langle\psi_i| E_a^\dagger E_b| \psi_i\rangle$. This is possible if $\langle\psi_i| E_a^\dagger E_b| \psi_i\rangle$ is a constant for all basis states, that is, for all $i \in 2^k$ basis vectors of the quantum code $[n, k, d]$. This can be summarized as the necessary condition for a quantum code.

$$\langle\psi_i|E_a^\dagger E_b|\psi_j\rangle = C_{i,j}\delta_{i,j}, \quad \text{where } C \text{ is a constant of } i \text{ and } j. \tag{9.20}$$

Assume that we have a quantum error correction code with $k = 1$, which maps one logical qubit into n physical qubits in Hilbert space $\mathcal{H}_{2^n}$ (we discuss this in Sect. 9.3 on the stabilizer formalism, see Eq. (9.40)):

$$|0\rangle_L = \sum_i a_i|i\rangle \text{ and } |1\rangle_L = \sum_i b_i|i\rangle; L \text{ stands for logical qubits} \tag{9.21}$$

These two logical qubits are entangled in Hilbert space $\mathcal{H}_{2^n}$. The quantum error correction code must coherently map the 2D space spanned by $|0\rangle_L$ and $|1\rangle_L$ into the 2D Hilbert spaces required to correct the three types of errors (bit-flip, phase-flip, and a combination of both, which we discuss in the next section) for each of the n physical qubits. The Hilbert space $\mathcal{H}_{2^n}$ must be sufficiently large to contain two sets of $(3n + 1)$ mutually orthogonal 2D subspaces for the syndromes (one for the subspace without errors, and 3 2D subspaces for bit-flips, phase-flips, and bit and phase-flip errors) for the n-qubits. The two sets required are for each of the logical qubits. This requirement defines the **quantum Hamming bound** as the inequality: $2(3n + 1) \leq 2^n$.

From this inequality, we can say that five qubits are required at a minimum to encode the logical qubits $|0\rangle_L$ and $|1\rangle_L$ to recover them from one of the error types.

9.2.3   Pauli Group

Let us assume we have a qubit in state $|\psi\rangle = a|0\rangle + b|1\rangle = \begin{bmatrix} a \\ b \end{bmatrix}$. Due to the interaction with the environment, the qubit is transformed into a final state $|\psi'\rangle = a'|0\rangle + b'|1\rangle = \begin{bmatrix} a' \\ b' \end{bmatrix}$. We can use the Pauli matrices to model the effects of the environment on the qubit.

No Errors:
We can use σ_I to describe the scenario when there is no error on the qubit.

$$\sigma_I|\psi\rangle = \begin{bmatrix} 1 & 0 \\ 0 & 1 \end{bmatrix} \begin{bmatrix} a \\ b \end{bmatrix} = \begin{bmatrix} a \\ b \end{bmatrix} = a|0\rangle + b|1\rangle \tag{9.22}$$

Amplitude or Bit-Flip Errors:
Bit-flip errors cause the qubit state to switch between $|0\rangle$ and $|1\rangle$. Recall from Eq. (5.5) that this is a Pauli-X rotation. The effect of this error on the qubit is as follows:

$$\sigma_x|\psi\rangle = \begin{bmatrix} 0 & 1 \\ 1 & 0 \end{bmatrix} \begin{bmatrix} a \\ b \end{bmatrix} = \begin{bmatrix} b \\ a \end{bmatrix} = b|0\rangle + a|1\rangle \tag{9.23}$$

Phase Error or Phase-Flip:
The phase-flip error causes a Pauli-Z rotation on the qubit. This causes the qubit to change from $|0\rangle + |1\rangle$ to $|0\rangle - |1\rangle$ and vice versa.

$$\sigma_z|\psi\rangle = \begin{bmatrix} 1 & 0 \\ 0 & -1 \end{bmatrix} \begin{bmatrix} a \\ b \end{bmatrix} = \begin{bmatrix} a \\ -b \end{bmatrix} = a|0\rangle - b|1\rangle \tag{9.24}$$

Bit and Phase-Flip Error:
This type of error causes a bit flip and a phase-flip to occur. This error is a rotation by Pauli-Y.

$$\sigma_y|\psi\rangle = i \begin{bmatrix} 0 & -1 \\ 1 & 0 \end{bmatrix} \begin{bmatrix} a \\ b \end{bmatrix} = i \begin{bmatrix} -b \\ a \end{bmatrix} = i(a|1\rangle - b|0\rangle) \tag{9.25}$$

The Pauli matrices have interesting properties. Two Pauli matrices are equivalent by a factor of ± 1 or $\pm i$ or by a complex number of unit magnitude. The set of Pauli matrices along with their multiplicative factors ± 1 or $\pm i$ is called the 1-qubit Pauli Group $\mathcal{G}_1$.

$$\begin{aligned} \mathcal{G}_1 &= \{\pm\sigma_I, \pm i\sigma_I, \pm\sigma_x \pm i\sigma_x, \pm\sigma_y, \pm i\sigma_y, \pm\sigma_z, \pm i\sigma_z\} \\ &= \langle X, Y, iI \rangle, \end{aligned} \tag{9.26}$$

where $\langle X, Y, iI \rangle$ is the generator of $\mathcal{G}_1$.

A set of n elements $\langle g_1, g_2, \ldots, g_n \rangle$ is called the **generator** of a group G, if the element g_i can be written as a product of elements from the set $\forall g_i \in G$. We leave this as an exercise to our readers to derive the generator of $\mathcal{G}_1$.

Exercise 9.1

Prove that the generator of G_1 is $\langle X, Y, i\mathbf{I} \rangle$.

◀

9.2.4 Correcting Bit-Flips

Figure 9.3 illustrates the general steps followed in correcting single bit-flip errors.
The parity check matrix for this repetitive code is:

$$H = i \begin{bmatrix} 0 & 1 & 1 \\ 1 & 0 & 1 \end{bmatrix} \tag{9.27}$$

Step 1: Encoding the k-Qubits

Consider the circuit in Fig. 9.4 which encodes the state $|\psi\rangle = a|0\rangle + b|1\rangle$ as $|\psi\rangle = a|000\rangle + b|111\rangle$.
The inner working of this circuit is straight forward. The initial state of the system is as follows:

$$\begin{aligned}
|\psi\rangle &= (a|0\rangle + b|1\rangle) \otimes |0\rangle \otimes |0\rangle \\
&= a|0\rangle|0\rangle|0\rangle + b|1\rangle|0\rangle|0\rangle = a|000\rangle + b|100\rangle
\end{aligned} \tag{9.28}$$

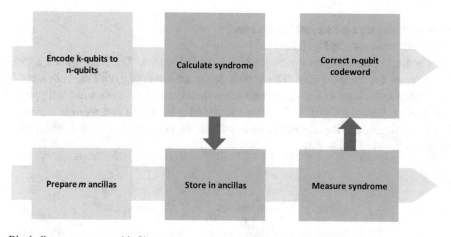

Fig. 9.3 Block diagram to correct bit-flip errors

Fig. 9.4 The 3-qubit encoder for the bit-flip error correction code

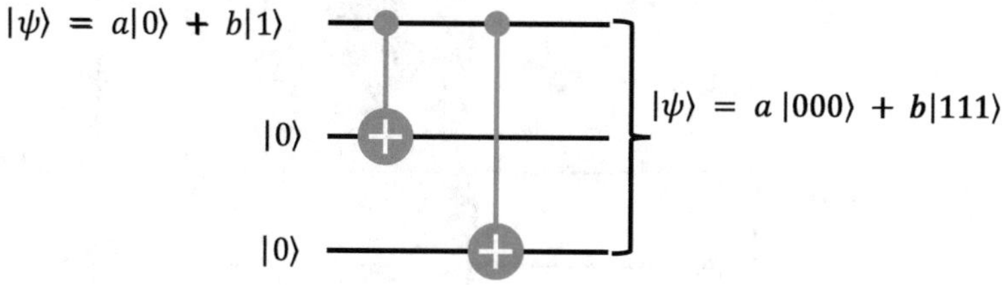

$|\psi\rangle = a|0\rangle + b|1\rangle$

$|0\rangle$

$|0\rangle$

$|\psi\rangle = a|000\rangle + b|111\rangle$

To this state, we apply two CNOT gates. Both the CNOT gates have $q[0]$ as the control qubit and $q[1]$, $q[2]$ as the target qubits, respectively. Application of the CNOT gates changes the system state to the following:

$$\text{CNOT}_{1,2}\text{CNOT}_{1,3}|\psi\rangle \rightarrow a|000\rangle + b|111\rangle \tag{9.29}$$

Step 2: Calculation of the Syndrome

If we assume that these three qubits go through a noise process, there are four possible scenarios. The best scenario is when the qubits are not affected. The other possibilities are when one of the qubits is bit flipped. We assume that at most, one qubit is bit flipped. The four possible scenarios are summarized below.

$$
\begin{aligned}
|\psi\rangle &= a|000\rangle + b|111\rangle; \quad \text{No Error} \\
|\psi_1\rangle &= a|100\rangle + b|011\rangle; \quad \text{The first qubit is flipped} \\
|\psi_2\rangle &= a|010\rangle + b|101\rangle; \quad \text{The second qubit is flipped} \\
|\psi_3\rangle &= a|001\rangle + b|110\rangle; \quad \text{The third qubit is flipped}
\end{aligned}
\tag{9.30}
$$

These four possible scenarios are called **syndromes**. Unfortunately, we cannot detect which qubit is flipped without doing a measurement. Performing a measurement will collapse the system state, and that is not what we want. Hence, we must seek alternatives. One way to solve the problem is to store the syndromes in ancilla qubits.

Step 3: Store the Syndrome in Ancilla Qubits

Consider the circuit shown in Fig. 9.5.

In this circuit, we transfer the syndrome to the ancilla qubits. The three qubits containing the codeword serve as the control qubits. The ancilla qubits are entangled with the codeword. The least significant ancilla qubit is set to 1, if the second or the third qubit of the code word is flipped. The most significant ancilla qubit is set to 1, if the first or the third qubit of the codeword is flipped.

With this arrangement, the correspondence of the ancilla qubits with the codewords can be summarized, as listed in Table 9.1.

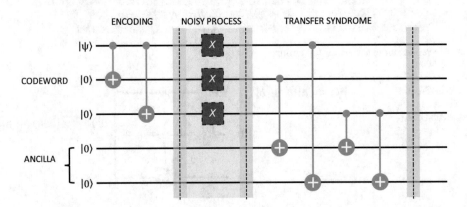

Fig. 9.5 Circuit showing the ancilla qubits transferring the syndrome. The noisy process may flip one of the bits of the codeword

Table 9.1 Codewords

Codeword	Ancilla
$|\psi\rangle$	$|00\rangle$
$|\psi_1\rangle$	$|01\rangle$
$|\psi_2\rangle$	$|10\rangle$
$|\psi_3\rangle$	$|11\rangle$

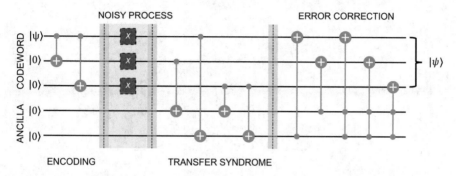

Fig. 9.6 Quantum error correction code for single bit-flips

Step 4: Correct the Codeword

Once we know how the syndrome is encoded in the ancilla qubits, we can easily correct the codeword using the logic shown in the quantum circuit in Fig. 9.6.

The recovery (error correction) logic is simple. We apply the CNOT gate to the respective codewords to correct bit-flips for the first and second qubits. If both the ancilla bits are set, then we need to apply CCNOT gates to all the qubits in the codeword.

Listing 9.1 contains the Qiskit code that constructs this circuit.

9.2.5 Correcting Phase-Flips

Recall from Eq. (9.24) that a phase-flip error is a σ_z rotation. The classical bits do not have any phase. Hence, they do not undergo phase errors. One way to correct a phase-flip is to convert the basis into the polar basis $\{|+\rangle, |-\rangle\}$, using H gates. After the basis is converted, we can use the same logic developed in correcting the bit-flips to correct the phase-flips. Recall from our earlier discussions on H gates. From Eqs. (5.11) and (5.12), Eqs. (5.18) and (5.19), the action of the H gate on pure states can be summarized as follows:

$$H|1\rangle = |+\rangle \quad H|0\rangle = |-\rangle \quad H|+\rangle = |1\rangle \quad H|-\rangle = |0\rangle \tag{9.31}$$

We can apply this logic to the encoder circuit in Fig. 9.4. Consider the circuit shown in Fig. 9.7. The four possible syndromes from this circuit are summarized in the following equation:

$$
\begin{aligned}
|\psi\rangle &= a|+++\rangle + b|---\rangle; \text{No Error} \\
|\psi_1\rangle &= a|-++\rangle + b|+--\rangle; \text{The first qubit is flipped} \\
|\psi_2\rangle &= a|+-+\rangle + b|-+-\rangle; \text{The second qubit is flipped} \\
|\psi_3\rangle &= a|++-\rangle + b|--+\rangle; \text{The third qubit is flipped}
\end{aligned} \tag{9.32}
$$

```
# Source code to illustrate bit-flip error correction
q = QuantumRegister(3, 'q')        #codeword
a = QuantumRegister(2, 'a')        #ancilla
c = ClassicalRegister(3, 'c')      #classical register for measuring
codeword
qc = QuantumCircuit(q,a,c)

# use the following statements to setup the initial condition
#qc.x(q[0])
#qc.h(q[0])

#Encode the codeword
qc.cx(q[0], q[1])
qc.cx(q[0], q[2])


qc.barrier()

# use one of the following to simulate noisy process
#qc.x(q[0])
#qc.x(q[1])
#qc.x(q[2])


qc.barrier()

# Transfer syndrome to ancilla
qc.cx(q[1], a[0])
qc.cx(q[2], a[0])

qc.cx(q[0], a[1])
qc.cx(q[2], a[1])


qc.barrier()

# Perform recovery
qc.cx(a[1], q[0])
qc.cx(a[0],q[1])
qc.ccx(a[0], a[1], q[0])
qc.ccx(a[0], a[1], q[1])
qc.ccx(a[0], a[1], q[2])

# measure the codeword
qc.measure(q[0],c[0])
qc.measure(q[1],c[1])
qc.measure(q[2],c[2])
```

Listing 9.1   Bit-flip error correction code

Recovery is made by applying Hadamard conjugation. From Eq. (5.86), we know the following identities. The error correction circuit applies this logic to return the qubits to the original basis.

$$HXH = Z \quad HZH = X \tag{9.33}$$

The quantum circuit shown in Fig. 9.8 illustrates the phase-flip error correction code.
Source code for implementing the phase-flip correction circuit is provided in Listing 9.2.

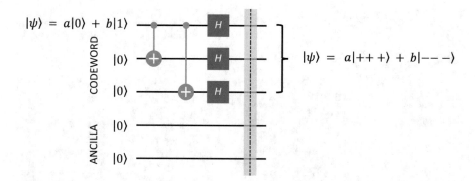

Fig. 9.7 The encoder circuit for phase-flip error correction

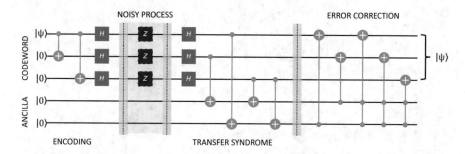

Fig. 9.8 Phase-flip correction circuit

9.3   Stabilizer Formalism

The stabilizer formalism was introduced by **Daniel Gottesman**, and helps us to define quantum codes as quantum operators that correct all possible errors occurring on the codewords. Assume that we have n-qubit registers in state $|\psi_i\rangle$ representing m codewords of a quantum code Q. Let there be M_i operators, which can detect errors occurring on the codewords. Assume that the operators can find whether any of the codewords contained errors. If the errors can be defined by Pauli matrices, as seen in the previous section, we can define a stabilizer S as a subgroup of G_n, that is, $S \subset G_n$. The generators of the subgroup can be defined as:

$$M = \{M_1, M_2, \ldots\} \tag{9.34}$$

The eigenvalues of the generators are $+1$, if they are codewords of Q and -1, if they are affected by errors, that is, $M_i|\psi_i\rangle = |\psi_i\rangle$ for the codewords Q. If there were some errors such that $|\varphi_i\rangle = E|\psi_i\rangle$, then we can say $M_i|\varphi_i\rangle = -|\varphi_i\rangle$. In this definition, S is the stabilizer of Q.

Let us explain this with the help of an EPR state.

$$|\phi^+\rangle = \frac{1}{\sqrt{2}}(|00\rangle + |11\rangle) \tag{9.35}$$

Let us assume couple of transforms $M_x = \sigma_x \otimes \sigma_x$, and $M_z = \sigma_z \otimes \sigma_z$. It is easy to prove that:

$$M_x |\phi^+\rangle = |\phi^+\rangle \text{ and } M_z |\phi^+\rangle = |\phi^+\rangle \tag{9.36}$$

```
# Code for Phase-flip correction

q = QuantumRegister(3, 'q')
a = QuantumRegister(2, 'a')
c = ClassicalRegister(3, 'c')
qc = QuantumCircuit(q,a,c)

# use one of the following statements to setup the initial condition
#qc.x(q[0])
#qc.h(q[0])

#Encode the codeword
qc.cx(q[0], q[1])
qc.cx(q[0], q[2])
qc.h(q[0])
qc.h(q[1])
qc.h(q[2])

qc.barrier()
# use one of the following to simulate noisy process
#qc.z(q[0])
#qc.z(q[1])
#qc.z(q[2])

qc.barrier()

qc.h(q[0])
qc.h(q[1])
qc.h(q[2])

# Transfer syndrome to ancilla
qc.cx(q[1], a[0])
qc.cx(q[2], a[0])
qc.cx(q[0], a[1])
qc.cx(q[2], a[1])

qc.barrier()

# Apply error correction
qc.cx(a[1], q[0])
qc.cx(a[0],q[1])
qc.ccx(a[0], a[1], q[0])
qc.ccx(a[0], a[1], q[1])
qc.ccx(a[0], a[1], q[2])

# measure the codeword
qc.measure(q[0],c[0])
qc.measure(q[1],c[1])
qc.measure(q[2],c[2])

%matplotlib inline
qc.draw(output="mpl")
```

Listing 9.2 Qiskit code for phase-flip correction

This means $|\phi^+\rangle$ is an eigenvector of M_x and M_z. The corresponding eigenvalue is +1, and these two generators commute.

The errors $E = \{E_1, E_2, \ldots, E_e\}$ occurring on the m codewords is also a subgroup of G_n, that is, $E \subset G_n$. Each operator E_i in this set is a tensor product of the Pauli matrices, with its weight equal to the number of errors in the codeword. The error operators anticommute with the generators G_n. If we compute the eigenvectors of the generators, those with an eigenvalue of -1 have errors.

In a $[n, k, d]$ code C, the cardinality of M is $n - k$ and the cardinality of the stabilizer S is 2^{n-k}.

The weight of $M \in S$ is the number of elements of M different from I. Here, I refers to σ_I. If any of the elements (other than I) of the stabilizer S has a weight less than d, then the stabilizer code is said to be degenerate.

The GHZ and Bell states are good examples of stabilizer states. The GHZ state described by Eq. (5.158) can be described by the following linearly independent generators of the $|GHZ\rangle_3$ stabilizer group:

$$
\begin{aligned}
M_1 &= \sigma_x \otimes \sigma_x \otimes \sigma_x \equiv XXX \\
M_2 &= \sigma_z \otimes \sigma_z \otimes \sigma_I \equiv ZZI \\
M_3 &= \sigma_I \otimes \sigma_z \otimes \sigma_z \equiv IZZ
\end{aligned}
\tag{9.37}
$$

The four Bell states are stabilized by the following operators:

$$
\begin{aligned}
M_1 &= (-1)^a XX \\
M_2 &= (-1)^b ZZ, \text{where } [a, b] = \{0, 1\}
\end{aligned}
\tag{9.38}
$$

The Bell states can be written in terms of combinations of eigenstates of ± 1 of these two operators.

$$
|\Phi^+\rangle = \begin{pmatrix} M_1 = XX \\ M_2 = ZZ \end{pmatrix} \quad |\Phi^-\rangle = \begin{pmatrix} M_1 = -XX \\ M_2 = ZZ \end{pmatrix}
$$
$$
|\Psi^+\rangle = \begin{pmatrix} M_1 = XX \\ M_2 = -ZZ \end{pmatrix} \quad |\Psi^-\rangle = \begin{pmatrix} M_1 = -XX \\ M_2 = -ZZ \end{pmatrix}
\tag{9.39}
$$

The stabilizer formalism helps design correction circuits and identifies the logical operations that can be applied directly to encoded data. In multi-qubit systems, the stabilizer states are defined as a subspace within the larger Hilbert space of a multi-qubit system. For example, 2-qubit systems require four dimensions in the Hilbert space. If two ZZ operators can represent them, then we need only two orthogonal states to represent the system:

$$
|0\rangle_L \equiv \frac{1}{\sqrt{2}}(|00\rangle + |11\rangle) \quad |1\rangle_L \equiv \frac{1}{\sqrt{2}}(|00\rangle - |11\rangle)
\tag{9.40}
$$

This reduces to a logical qubit. Effectively, multi-qubit systems can be represented as single qubit systems for convenience sake, by reducing the dimensionality of the associated Hilbert space.

The stabilizer formalism can also be used to describe quantum gates such as the H gate, phase gates, and the CNOT gate. We shall look here an example interpretation from Nielsen and Chuang [1].

Let U be a unitary operator of the quantum code Q that is stabilized by S. Assume that $|\psi_i\rangle$ be a state of Q, and let g_i be an element of S. We can then write:

$$U|\psi_i\rangle = Ug_i|\psi_i\rangle = Ug_iU^{\dagger}U|\psi_i\rangle \qquad (9.41)$$

We can therefore say $Ug_iU^{\dagger}$ is the stabilizer of $U|\psi_i\rangle$. We can as well expand this and say that $USU^{\dagger}$ is the stabilizer of UQ. As an application of this concept, let us look at the Hadamard conjugation of Z, that is, $HZH^{\dagger} = X$. We can interpret the conjugation as follows: the Hadamard gate applied to the state stabilized by Z (say for example $|0\rangle$) is the state stabilized by X (in this case it is $|+\rangle$.) This interpretation can be expanded to n-qubits. If we have stabilizers $\langle Z_1, Z_2, \ldots, Z_n \rangle$ for the state $|0\rangle^{\otimes n}$, the Hadamard gates transform them to a state with stabilizers $\langle X_1, X_2, \ldots, X_n \rangle$. Note that, this is a compact notation when compared with the equal superposition state described by 2^n states.

9.3.1 Gottesman–Knill Theorem

From the above argument, if we must simulate a quantum computation on a classical computer, it is sufficient if we update the generators of the stabilizers as the system evolves unitarily. State preparation, Clifford group, controlled operations, and measurements can be performed in $O(n^2)$ steps. If we have m operations on n qubits, this is done in $O(m \times n^2)$ steps. These are some tasks classical computers can easily perform. Therefore, according to Gottesman–Knill theorem [4–7], some basic tasks like quantum entanglement, superdense coding, quantum teleportation, can be efficiently performed on a classical computer. Formally, the theorem states that:

A classical computer can efficiently simulate a quantum circuit based on the following elements:

- *Preparation and measurement of qubits in the computational basis*
- *Quantum gates from the Clifford group*

Gottesman–Knill theorem shows us that simply having a system in an entangled state is not enough to give quantum speedup. However, not all quantum computations can be efficiently described by the stabilizer formalism. The system state properties that are not described by polynomial sized stabilizers seem to have exponential speedup [2]. Besides, it just seems to be the formalism that is chosen—amplitude formalism or stabilizer formalism.

The classical computer is a classical physical system whose parts interact with one another in a locally causal, spatiotemporally continuous way. Thus, a description of a classical computer simulation of a quantum system, even when it involves communication, is a local hidden variables description of the problem [3].

This is a subject for study. Nevertheless, there is much more to quantum computing than the power of entanglement [1].

9.3.2 Shor's 9-Qubit Error Correction Code

In the preceding sections, we saw single-qubit error correction codes for bit and phase-flips. Peter Shor devised a 9-qubit error correction code that can correct bit-flip, phase-flip, or a combination of both on a single qubit. The algorithm is quite simple. We first encode the codewords for phase-flip recovery and then encode each qubit for bit-flip recovery. Consider the quantum circuit shown in Fig. 9.10. This circuit is known as "**concatenation**" as it encodes using a hierarchy of levels. The phase-flip circuit encodes the qubits as: $|0\rangle \rightarrow |+++\rangle$ and $|1\rangle \rightarrow |---\rangle$. The second part of the circuit is the bit-flip circuit, which encodes the states as follows: $|+\rangle \rightarrow \frac{1}{\sqrt{2}}(|000\rangle + |111\rangle)$ and $|-\rangle \rightarrow \frac{1}{\sqrt{2}}(|000\rangle - |111\rangle)$.

The resultant encoding of the 9-qubits are the following codewords:

$$
\begin{aligned}
|0\rangle \rightarrow |0_L\rangle &\rightarrow \frac{(|000\rangle + |111\rangle) \otimes (|000\rangle + |111\rangle) \otimes (|000\rangle + |111\rangle)}{\sqrt{8}} \\
|1\rangle \rightarrow |1_L\rangle &\rightarrow \frac{(|000\rangle - |111\rangle) \otimes (|000\rangle - |111\rangle) \otimes (|000\rangle - |111\rangle)}{\sqrt{8}}
\end{aligned}
\tag{9.42}
$$

Shor's 9-qubit code can fix one bit-flip in any of the three circuit blocks and a single bit phase error in any of the nine qubits. If there is a combination of a bit-flip and a phase-flip occurring on a qubit, the bit-flip correction circuit fixes the bit-flip error while the phase-flip corrects the phase-flip errors. Figure 9.9 illustrates the Shor's 9-qubit encoder.

Source code for the circuit in Fig. 9.10 is provided in Listing 9.3. Readers can introduce single-qubit errors and verify the circuit works. Readers can also set the initial system state to basis state or pure state, and try various combinations.

The parity check matrix for Shor's code is:

$$
H = \begin{bmatrix}
0 & 0 & 0 & 0 & 0 & 0 & 0 & 1 & 1 \\
0 & 0 & 0 & 1 & 1 & 1 & 1 & 0 & 0 \\
0 & 1 & 1 & 0 & 0 & 1 & 1 & 0 & 0 \\
1 & 0 & 1 & 0 & 1 & 0 & 1 & 0 & 1
\end{bmatrix}
\tag{9.43}
$$

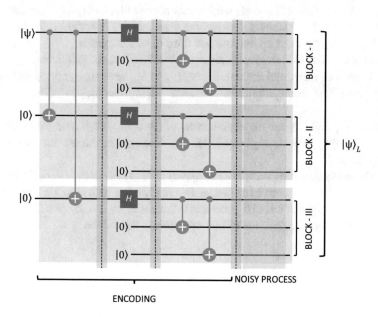

Fig. 9.9 Encoder for Shor's 9-qubit code

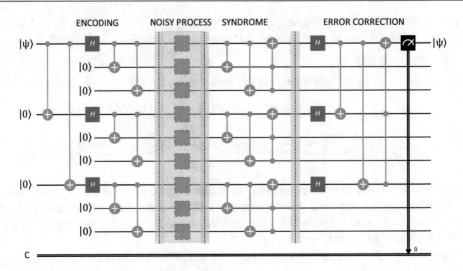

Fig. 9.10 A simplified version of the 9-qubit Shor's code

The logical operators X and Z for Shor's encoded qubits are defined as follows:

$$\overline{X} \equiv Z_1 Z_2 Z_3 Z_4 Z_5 Z_6 Z_7 Z_8 Z_9$$
$$\overline{Z} \equiv X_1 X_2 X_3 X_4 X_5 X_6 X_7 X_8 X_9$$

$$(9.44)$$

We can now create the stabilizers for the nine qubits.

Let us define eight operators $\{M_1, M_2, \ldots, M_8\}$ that are the generators of the group G that stabilize the nine qubits. Table 9.2 summarizes the action of the eight operators on the nine qubits.

The table arranges the operators in a particular order. The stabilizers M_1 and M_2 work on the first block of qubits. The stabilizers M_3 and M_4 work on the second block of qubits. The stabilizers M_5 and M_6 work on the third block of qubits. Likewise, the stabilizer M_7 operates on the first two blocks of qubits. The stabilizer M_8 operates on the first and third blocks of qubits.

We have two classes of stabilizers in this arrangement: The first six stabilizers $M_1 . . M_6$ have a pair of σ_z operators in each of them, and the stabilizers M_7 and M_8 have six σ_x operators in each of them. We can easily prove that these stabilizers commute among themselves within in the respective classes.

The codewords $|0\rangle_L$ and $|1\rangle_L$ are the eigenvectors of the operators M_i and the eigenvalues are +1, that is, $M_i |0\rangle_L = (+1)|0\rangle_L$ and $M_i |1\rangle_L = (+1)|1\rangle_L$. As discussed in the previous section, this can be treated as a logical qubit.

From the table, the operator $M_1 = Z_0 \otimes Z_1 \otimes I_2 \otimes I_3 \otimes I_4 \otimes I_5 \otimes I_6 \otimes I_7 \otimes I_8$. This operator works on the first block of qubits. The remaining blocks are not changed, as it is an identity matrix operating on them. Note that the weight of this first generator is 2. Hence, we can say that Shor's 9-qubit code is degenerate. Let us examine the function of this operator on the qubits.

```python
# Shor's 9-qubit code
# full circuit

q = QuantumRegister(9, 'q')
c = ClassicalRegister(1, 'a')
qc = QuantumCircuit(q,c)

# use one of the following statements to setup the initial condition
#qc.x(q[0])
#qc.h(q[0])

#Encode the codeword
qc.cx(q[0], q[3])
qc.cx(q[0], q[6])

qc.h(q[0])
qc.h(q[3])
qc.h(q[6])

qc.cx(q[0], q[1])
qc.cx(q[0], q[2])
qc.cx(q[3], q[4])
qc.cx(q[3], q[5])
qc.cx(q[6], q[7])
qc.cx(q[6], q[8])

# introduce noise here
qc.barrier()

#qc.x(q[1])
#qc.z(q[0])

qc.barrier()

# Transfer syndrome to ancilla
qc.cx(q[0], q[1])
qc.cx(q[0], q[2])
qc.ccx(q[2], q[1], q[0])

qc.cx(q[3], q[4])
qc.cx(q[3], q[5])
qc.ccx(q[5], q[4], q[3])

qc.cx(q[6], q[7])
qc.cx(q[6], q[8])
qc.ccx(q[8], q[7], q[6])

qc.barrier()

# Error recovery
qc.h(q[0])
qc.h(q[3])
qc.h(q[6])

qc.cx(q[0], q[3])
qc.cx(q[0], q[6])
qc.ccx(q[6], q[3], q[0])

qc.measure(q[0], c[0])

%matplotlib inline
qc.draw(output="mpl")
```

Listing 9.3   9-qubit Shor's quantum error correction code

Table 9.2 The 8-stabilizers of the 9-qubits in Shor's code

Qubits→ ↓Operators	0	1	2	3	4	5	6	7	8
M_1	Z	Z	I	I	I	I	I	I	I
M_2	Z	I	Z	I	I	I	I	I	I
M_3	I	I	I	Z	Z	I	I	I	I
M_4	I	I	I	Z	I	Z	I	I	I
M_5	I	I	I	I	I	I	Z	Z	I
M_6	I	I	I	I	I	I	Z	I	Z
M_7	X	X	X	X	X	X	I	I	I
M_8	X	X	X	I	I	I	X	X	X

$$M_1|0_L\rangle \rightarrow \frac{1}{\sqrt{8}} \left[(Z|0\rangle|0\rangle Z|0\rangle + Z|1\rangle|1\rangle Z|1\rangle) \otimes I^{\otimes 3}(|000\rangle + |111\rangle) \otimes I^{\otimes 3}(|000\rangle + |111\rangle) \right]$$

$$\rightarrow \frac{1}{\sqrt{8}} \left[((|0\rangle|0\rangle|0\rangle + (-|1\rangle|1\rangle - |1\rangle))) \otimes (|000\rangle + |111\rangle) \otimes (|000\rangle + |111\rangle) \right]$$

$$\rightarrow \frac{1}{\sqrt{8}} \left[((|0\rangle|0\rangle|0\rangle + |1\rangle|1\rangle|1\rangle) \otimes (|000\rangle + |111\rangle) \otimes (|000\rangle + |111\rangle) \right]$$

$$\rightarrow |0_L\rangle$$

$$M_1|1_L\rangle \rightarrow \frac{1}{\sqrt{8}} \left[(Z|0\rangle|0\rangle Z|0\rangle - Z|1\rangle|1\rangle Z|1\rangle) \otimes I^{\otimes 3}(|000\rangle - |111\rangle) \otimes I^{\otimes 3}(|000\rangle - |111\rangle) \right] \tag{9.45}$$

$$\rightarrow \frac{1}{\sqrt{8}} \left[((|0\rangle|0\rangle|0\rangle - (-|1\rangle|1\rangle - |1\rangle))) \otimes (|000\rangle - |111\rangle) \otimes (|000\rangle - |111\rangle) \right]$$

$$\rightarrow \frac{1}{\sqrt{8}} \left[((|0\rangle|0\rangle|0\rangle - |1\rangle|1\rangle|1\rangle) \otimes (|000\rangle - |111\rangle) \otimes (|000\rangle - |111\rangle) \right]$$

$$\rightarrow |1_L\rangle$$

We can similarly prove for the other operators. Hence, the eigenvalue of these operators for uncorrupted states is +1.

Let us look at the action of an error operator on the first qubit of Block-I.

$$E_x^{(0)} = X_0 \otimes I_1 \otimes I_2 \otimes I_3 \otimes I_4 \otimes I_5 \otimes I_6 \otimes I_7 \otimes I_8 \tag{9.46}$$

This operator performs a bit-flip operation on the first qubit. We can easily prove that this operator anticommutes with the stabilizer M_1.

$$\begin{aligned} M_1 E_x^{(0)} &= (Z_0 \otimes Z_1 \otimes I_2 \otimes I_3 \otimes I_4 \otimes I_5 \otimes I_6 \otimes I_7 \otimes I_8) \\ &\quad (X_0 \otimes I_1 \otimes I_2 \otimes I_3 \otimes I_4 \otimes I_5 \otimes I_6 \otimes I_7 \otimes I_8) \\ &= (Z_0 \otimes X_0) \otimes I_1 \otimes I_2 \otimes I_3 \otimes I_4 \otimes I_5 \otimes I_6 \otimes I_7 \otimes I_8 \\ &= -(X_0 \otimes Z_0) \otimes I_1 \otimes I_2 \otimes I_3 \otimes I_4 \otimes I_5 \otimes I_6 \otimes I_7 \otimes I_8 \\ &= -E_x^{(0)} M_1 \end{aligned} \tag{9.47}$$

The anticommuting property means $E_x^{(0)}$ is an eigenvector of the stabilizer M_1 with an eigenvalue -1. Hence, if the eigenvalue of the stabilizer is -1, it indicates the presence of error E in the codeword. Note that M_1 can detect the presence of a bit-flip error in the second qubit. However, M_1 cannot detect the bit-flip errors in other qubits (because it commutes with the error operator of those qubits). We can similarly prove for a phase-flip, with operator M_7, for example, with the first two blocks of qubits.

To conclude our discussions on this topic, let us recap how this works. Assume the following state:

$$|0\rangle_L \rightarrow \frac{1}{\sqrt{8}}[((|000\rangle + |111\rangle)) \otimes (|000\rangle + |111\rangle) \otimes (|000\rangle + |111\rangle)] \quad (9.48)$$

Assume that an electromagnetic interference sets the second qubit with a bit-flip and phase-flip error. The new state of the system is as follows:

$$\rightarrow \frac{1}{\sqrt{8}}[((|010\rangle - |101\rangle)) \otimes (|000\rangle + |111\rangle) \otimes (|000\rangle + |111\rangle)] \quad (9.49)$$

We apply the bit-flip stabilizers $M_1..M_6$ first. From Table 9.2, we can say that stabilizer M_1 returns -1, while all other stabilizers return $+1$. We can conclude that the second qubit has a bit-flip error and an $I_0 \otimes X_1 \otimes I_2 \otimes I_3 \otimes I_4 \otimes I_5 \otimes I_6 \otimes I_7 \otimes I_8$ transform can be applied to correct this error. After this correction, the state of the system is as follows:

$$\rightarrow \frac{1}{\sqrt{8}}[((|000\rangle - |111\rangle)) \otimes (|000\rangle + |111\rangle) \otimes (|000\rangle + |111\rangle)] \quad (9.50)$$

At this stage, the phase-flip stabilizers M_7 and M_8 are applied. From Table 9.2, we can also say that both the stabilizers return -1, which means, the first block has a phase error. The phase error can be corrected by applying a Z gate to the qubits in the first block.

Figure 9.11 shows the implementation of the stabilizer circuit, and it is based on Table 9.2.

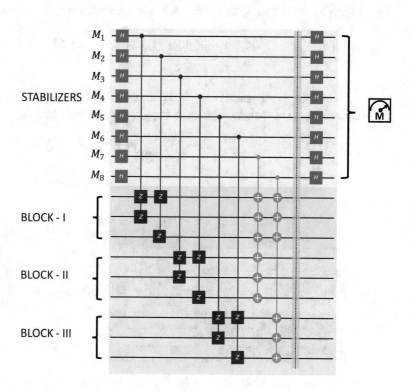

Fig. 9.11 The stabilizer circuit for Shor's 9-qubit code

9.3.3 CSS Codes

Robert Calderbank, **Peter Shor**, and **Andrew Steane** invented the CSS codes. The CSS codes are a type of stabilizer codes. In this section, we shall provide the definition [9] of the CSS codes. In the next section, we shall study Steane's 7-bit code as a case of CSS codes.

Let C be a linear $[n, k, d]$ code of a field $\mathcal{F}_2^n$ (a binary field of n dimensions.) This code can correct $t = \frac{d-1}{2}$ errors. The generator matrix G is a $k \times n$ matrix. The rows of this matrix are independent n-tuples spanning the code C. The parity check matrix H is a $(n - k) \times n$ matrix, such that $Hc^T = 0; \forall c \in C$. The code C is spanned by k independent n-tuples, that is, the cardinality of C is k. Code C has 2^k codewords.

Let $C^\perp$ be the dual code of C, that is, the n-tuples of $C^\perp$ are orthogonal to the codewords of C. $C^\perp$ is spanned by $(n - k)$ independent n-tuples, that is, the cardinality of $C^\perp$ is $(n - k.)$ Code $C^\perp$ has $2^{n - k}$ codewords.

The parity check matrix H of C is the generator of $C^\perp$.

The generator G of C is the parity check matrix of $C^\perp$.

If $C^\perp \subset C$, then C is **weakly self-dual**.

The quantum code Q for the weakly self-dual classical code C is defined by creating superpositions of the codewords of C.

We first construct a c-basis, as follows:

$$|c_w\rangle = 2^{\frac{-k}{2}} \sum_{v \in C} (-1)^{v+w} |v\rangle; \quad w \in \mathcal{F}_2^n \ , \tag{9.51}$$

where the " $+$ " sign is a modulo 2 addition of v and w.

We then construct an s-basis. The codewords of this basis are the set of $|c_w\rangle \forall w \in C^\perp$. We can get this by applying an H transform to the n qubits of the codewords (Fig. 9.12).

$$|0\rangle \rightarrow \frac{1}{\sqrt{2}} (\,|0\rangle + |1\rangle), \ \ |1\rangle \rightarrow \frac{1}{\sqrt{2}} (\,|0\rangle - |1\rangle) \tag{9.52}$$

Phase errors $|0\rangle \rightarrow |0\rangle$ and $|1\rangle \rightarrow -|1\rangle$ in the c-basis become bit errors in the s-basis. After applying the change of basis to each qubit in $|c_w\rangle$, the new state is:

$$|s_w\rangle = 2^{\frac{k-n}{2}} \sum_{u \in C^\perp} |u + w\rangle; \quad w \in \mathcal{F}_2^n \tag{9.53}$$

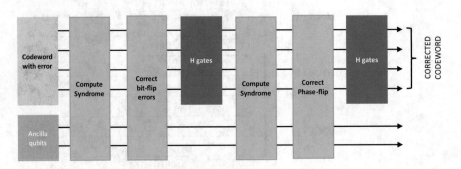

Fig. 9.12 Diagram illustrating the workflow for CSS Code. (From Dan C. Marinescu and Gabriela Marinescu)

The cardinality of the quantum code Q is $k - (n - k) = 2k - n$. So, the quantum code Q has $2^{2k - n}$ codewords.

To summarize, in this scheme, we first correct bit-flip errors in the c-basis. We then change the system to s-basis. The change in basis causes the phase-flip errors to become bit-flip errors. They can now be corrected as bit-flip errors. We can correct at most t bit-flip and phase-flip errors with the CSS code.

9.3.4 Steane's 7-Qubit Error Correction Code

Steane's 7-qubit error correction code is based on a [7,4,3] Hamming code C. Its generator matrix is given by:

$$G_c = \begin{bmatrix} 1 & 0 & 0 & 0 & 0 & 1 & 1 \\ 0 & 1 & 0 & 0 & 1 & 0 & 1 \\ 0 & 0 & 1 & 0 & 1 & 1 & 0 \\ 0 & 0 & 0 & 1 & 1 & 1 & 1 \end{bmatrix} \tag{9.54}$$

Note that the generator matrix is composed of a 4×4 identity matrix, and the matrix

$$A = \begin{bmatrix} 0 & 1 & 1 \\ 1 & 0 & 1 \\ 1 & 1 & 0 \\ 1 & 1 & 1 \end{bmatrix}.$$

In this error correction code, there are $2^k = 2^4 = 16$ codewords. The codewords are generated by multiplying the 4-tuples with the generator matrix (Table 9.3).

The following equation shows the parity check matrix:

$$H = \begin{bmatrix} A^T I_{n-k} \end{bmatrix} = \begin{bmatrix} A^T I_3 \end{bmatrix} = \begin{bmatrix} 0 & 1 & 1 & 1 & 1 & 0 & 0 \\ 1 & 0 & 1 & 1 & 0 & 1 & 0 \\ 1 & 1 & 0 & 1 & 0 & 0 & 1 \end{bmatrix} \tag{9.55}$$

The codewords of $C^\perp$ can be derived from the H matrix. Note that the H matrix is the generator of $C^\perp$, and we get the codewords by multiplying the n-tuples of $C^\perp$ with the H matrix. We advise our readers to verify the tables by calculating the codewords (Table 9.4).

Note that $C^\perp \subset C$. We can also establish that $C^\perp = (C^\perp)^\perp = C$. So, both C and $C^\perp$ are $d = 3$ codes. C is a [7, 4, 3] code, and $C^\perp$ is a [7, 3, 3] code. Therefore, CSS$(C, C^\perp)$ is a [7, 1, 3] quantum code.

The logical zero ($|0\rangle_L$) qubit in Steane's code is defined as the superposition of all even codewords of the Hamming code. This qubit is described below:

Table 9.3 Codewords of the code C.

Tuple	Codeword	Tuple	Codeword	Tuple	Codeword	Tuple	Codeword
0000	0000000	0100	0100101	1000	1000011	1100	1100110
0001	0001111	0101	0101010	1001	1001100	1101	1101001
0010	0010110	0110	0110011	1010	1010101	1110	1110000
0011	0011001	0111	0111100	1011	1011010	1111	1111111

Table 9.4 Codewords of $\mathcal{C}^{\perp}$.

Tuple	Codeword	Tuple	Codeword
000	0000**000**	100	0111**100**
001	1101**001**	101	1010**101**
010	1011**010**	110	1100**110**
011	0110**011**	111	0001**111**

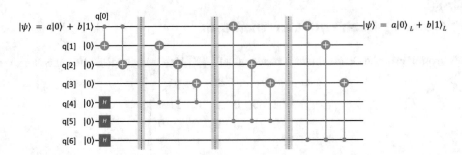

Fig. 9.13 Steane's 7-qubit encoder circuit. Note that the ordering of the qubits. The topmost qubit is the first qubit q[0], and the bottom-most qubit is the seventh qubit q[6]

$$|0\rangle_L \rightarrow \frac{1}{\sqrt{8}}\Big[|0000000\rangle+|0001111\rangle+|0110011\rangle+|0111100\rangle+|1010101\rangle+|1011010\rangle+$$
$$|1100110\rangle+|1101001\rangle\Big] \tag{9.56}$$

Similarly, the logical one ($|1\rangle_L$) qubit is defined by the superposition of all odd codewords of the Hamming code.

$$|1\rangle_L \rightarrow \frac{1}{\sqrt{8}}\Big[|0010110\rangle+|0011001\rangle+|0100101\rangle+|0101010\rangle+|1000011\rangle+|1001100\rangle+$$
$$|1110000\rangle+|1111111\rangle\Big] \tag{9.57}$$

Figure 9.13 illustrates the circuit that encodes the Steane's 7-qubit code. Let us look at the inner workings of this circuit. We use two CNOT gates to encode the initial state into the superposition state of the Hamming codewords. We used this technique in our earlier circuits as well. Then, we use three H gates to prepare the last three qubits in an equal superposition state. The system state at this stage can be written as:

$$|\psi\rangle_L \rightarrow \Big(a|000\rangle+b|111\rangle\Big)\otimes|0\rangle \otimes \frac{1}{\sqrt{8}}\Big[\big(|0\rangle+|1\rangle\big) \otimes \big(|0\rangle+|1\rangle\big) \otimes \big(|0\rangle+|1\rangle\big)\Big] \tag{9.58}$$

This equation can be expanded:

$$\rightarrow \frac{1}{\sqrt{8}}a(|0000000\rangle + |0000001\rangle + |0000010\rangle + |0000011\rangle + |0000100\rangle + |0000101\rangle$$
$$+|0000110\rangle + |0000111\rangle))+$$
$$\frac{1}{\sqrt{8}}b(|1110000\rangle + |1110001\rangle + |1110010\rangle + |1110011\rangle + |1110100\rangle + |1110101\rangle \tag{9.59}$$
$$+|1110110\rangle + |1110111\rangle))$$

The first block of CNOTs gates is applied with q[4] as the control qubit and q[1], q[2], q[3] as the target qubits. This operation transforms the equation as:

$$
\stackrel{CNOT(4\to1,2,3)}{\longrightarrow} \frac{1}{\sqrt{8}} a(|0000000\rangle + |0000001\rangle + |0000010\rangle + |0000011\rangle + |0111100\rangle
$$
$$
+|0111101\rangle + |0111110\rangle + |0111111\rangle)+
$$
$$
\frac{1}{\sqrt{8}} b(|1110000\rangle + |1110001\rangle + |1110010\rangle + |1110011\rangle + |1001100\rangle + |1001101\rangle
$$
$$
+|1001110\rangle + |1001111\rangle)) \tag{9.60}
$$

The second block of CNOTs is applied with q[5] as the control qubit and q[0], q[2], q[3] as the target qubits. The new system state is:

$$
\stackrel{CNOT(5\to0,2,3)}{\longrightarrow} \frac{1}{\sqrt{8}} a(|0000000\rangle + |0000001\rangle + |1011010\rangle + |1011011\rangle + |0111100\rangle
$$
$$
+|0111101\rangle + |1100110\rangle + |1100111\rangle)+
$$
$$
\frac{1}{\sqrt{8}} b(|1110000\rangle + |1110001\rangle + |0101010\rangle + |0101011\rangle + |1001100\rangle + |1001101\rangle
$$
$$
+|0010110\rangle + |0010111\rangle)) \tag{9.61}
$$

The third block of CNOTs gates is applied with q[6] as the control qubit and q[0], q[1], q[3] as the target qubits:

$$
\stackrel{CNOT(6\to0,1,3)}{\longrightarrow} \frac{1}{\sqrt{8}} a(|0000000\rangle + |1101001\rangle + |1011010\rangle + |0110011\rangle + |0111100\rangle
$$
$$
+|1010101\rangle + |1100110\rangle + |0100111\rangle)+
$$
$$
\frac{1}{\sqrt{8}} b(|1110000\rangle + |0011001\rangle + |0101010\rangle + |1000011\rangle + |1001100\rangle + |0100101\rangle
$$
$$
+|0010110\rangle + |1111111\rangle)) \tag{9.62}
$$

By comparing with Eqs. (9.56) and (9.57):

$$
|\psi\rangle_L \to \frac{1}{\sqrt{8}} \left(a|0\rangle_L + b|1\rangle_L \right) \tag{9.63}
$$

This equation establishes that the circuit works as intended. We can now apply the stabilizer formalism to Steane's code. The generators of the stabilizer code are defined in Table 9.5.

Table 9.5 Stabilizer group for Steane's code

Qubits→ ↓Operators	0	1	2	3	4	5	6
M_1	I	I	I	X	X	X	X
M_2	I	X	X	I	I	X	X
M_3	X	I	X	I	X	I	X
M_4	I	I	I	Z	Z	Z	Z
M_5	I	Z	Z	I	I	Z	Z
M_6	Z	I	Z	I	Z	I	Z

Table 9.6 Summary of Steane's code

Error	Generators	Syndrome	Error	Generators	Syndrome	Error	Generators	Syndrome
X_1	M_6	000001	Z_1	M_3	001000	Y_1	M_3, M_6	001001
X_2	M_5	000010	Z_2	M_2	010000	Y_2	M_2, M_5	010010
X_3	M_5, M_6	000011	Z_3	M_2, M_3	011000	Y_3	$M_2, M_3,$ M_5, M_6	011011
X_4	M_4	000100	Z_4	M_1	100000	Y_4	M_1, M_4	100100
X_5	M_4, M_6	000101	Z_5	M_1, M_3	101000	Y_5	$M_1, M_3,$ M_4, M_6	101101
X_6	M_4, M_5	000110	Z_6	M_1, M_2	110000	Y_6	$M_1, M_2,$ M_4, M_5	110110
X_7	M_4, M_5, M_6	000111	Z_7	M_1, M_2, M_3	111000	Y_7	$M_1, M_2, M_3,$ M_4, M_5, M_6	111111

The stabilizers of the Steane's code are a balanced group. The stabilizers have a uniform weight of 4 and distance 3. The single-qubit errors that can happen in this code are summarized in Table 9.6.

The Pauli operators Z and X for the encoded states in Steane's code are as follows, and they are written as the operators of the unencoded qubits.

$$\bar{Z} \equiv Z_1 Z_2 Z_3 Z_4 Z_5 Z_6 Z_7$$
$$\bar{X} \equiv X_1 X_2 X_3 X_4 X_5 X_6 X_7$$
(9.64)

Exercise 9.2

Develop a complete quantum circuit that implements Steane's code.

◄

9.4   The Path Forward: Fault-Tolerant Quantum Computing

It is needless to say that the classical computing industry has grown exponentially because of the fault tolerance of the transmission and computing systems. Errors do occur in these systems routinely, but they are detected and corrected efficiently. Communication protocols and hardware designs have implemented elements to effectively perform error detection and recovery schemes with very high reliability that under ordinary usage, we do not encounter them at all.

Therefore, quantum error correction is critical for mass adoption of quantum computing and quantum communication. As we have seen in the past chapters, quantum errors can happen at various levels. Any circuit we build is no exception to these errors!

- Transmission of quantum information over quantum communication channels
- Decoherence of the qubits over time
- Gate errors caused by noisy unitary transformations on the qubits
- Measurement errors
- State preparation errors

While we can learn a lot from the classical error correction schemes, we cannot apply them directly to quantum error correction, because of how quantum systems work.

- Classical error correction schemes have some levels of redundancy. We cannot have redundancy in quantum systems, because qubits in arbitrary unknown states cannot be cloned.
- We can distinguish with certainty only orthogonal quantum states. If the errors push the system into nonorthogonal quantum states, we cannot distinguish the states. Hence, errors cannot be corrected. This problem is not there in classical systems as the state of the bits can be distinguished correctly.
- In a quantum system, measurement causes the system to collapse. So without collapsing the system, we cannot detect errors. In classical bits, information can be read from the bit any number of times. Hence, it is easy to detect and correct errors.
- Qubits can undergo bit-flip, phase-flip, or a combination of both. The error states of quantum systems are complex when compared to classical bits. Classical bits can only undergo bit-flip.
- In the case of multi-qubit entanglement, errors can happen with some of the entangled qubits alone. This poses additional challenges.
- Quantum errors can keep adding up from erroneous qubits and erroneous gates.

Quantum error correction comes handy. QEC schemes we have come across can be used to correct errors as they dynamically evolve at every level in the system. The **quantum accuracy threshold theorem** states that arbitrarily long computation can be carried out reliably, provided all sources of noise acting a quantum computer are weaker than the accuracy threshold p_c. A lower bound on this threshold is $\epsilon > 1.9 \times 10^{-4}$.

The QEC schemes use syndrome circuits to get as much information as possible about the quantum error. Once the syndrome is determined, it is used to relate the error to a superposition of the Pauli operators or the identity matrix so that the erroneous qubits can be corrected by applying Pauli rotations. However, the issue with QEC is that the error correction circuits itself can introduce errors into the encoded qubits. So, the error correction circuits must be designed specifically that they do not introduce many errors into the encoded data [15].

9.4.1 Performing a Fault-Tolerant Procedure

In a fault-tolerant procedure, we replace each qubit in the original circuit with an encoded block of qubits, which are error corrected. The gate operations in the original circuit need to be replaced with encoded gates. The encoded gates must be properly designed so that they do not propagate errors. Even, if they propagate, it should be to a small subset of qubits in the encoded blocks in such a way that the errors can be isolated and removed. Let us examine this possibility in some detail by creating encoded gates of the Clifford group.

From Eq. (5.86), we can say that the Z and X gates can be interchanged under H gate conjugation. By referring to Eq. (9.64), and by drawing a parallel, we can create an encoded Hadamard gate for the Steane's 7-qubit code as $\overline{H} \equiv H_1 H_2 H_3 H_4 H_5 H_6 H_7$. This encoded gate can be implemented as a parallel circuit $\overline{H} = H^{\otimes 7}$ (Fig. 9.14).

Let us assume that while performing this operation, we over- or under-rotated a qubit (which is an example of a gate error) or an electromagnetic interference happened on a qubit. Let us assume that the error can be modeled as a phase error. We can write this incidence as $HZ \equiv XH$. This is equivalent to applying an H gate first and then an X gate. In other words, we can say that following a perfect H gate, a noise occurred on a qubit. This modeling is an example of a fault-tolerant process. The reason being, the error that happened on the faulty qubit did not propagate to other qubits. It just resulted in one error in the block of qubits. Besides, this single-qubit error can easily be corrected.

Fig. 9.14 An encoded
H gate operating on a
logical qubit

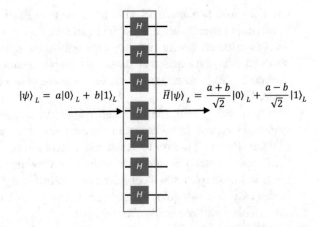

$$|\psi\rangle_L = a|0\rangle_L + b|1\rangle_L \qquad \overline{H}|\psi\rangle_L = \frac{a+b}{\sqrt{2}}|0\rangle_L + \frac{a-b}{\sqrt{2}}|1\rangle_L$$

Fig. 9.15 An encoded
CNOT gate

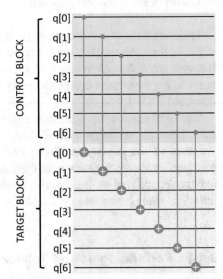

The process of implementing an encoded gate using bitwise operations is called the **transversality** of an encoded gate. From Eq. (9.64), encoded Pauli operators $\overline{X}$ and $\overline{Z}$ are implemented as transverse operations. If we must make the encoded gates universal, we must implement the rest of the Clifford group. This compliance requires us to develop encoded S, T, and CNOT gates. From Sect. 5.4.8.2, $SXS = Y$, $SYX = -X$, $SZS = Z$. We can implement an encoded S gate $\overline{S} \equiv S_1 S_2 S_3 S_4 S_5 S_6 S_7$, which is transversal. The conjugation of Y yields $-X$. The negative phase in the conjugation of Y can be corrected with an iZ gate. Therefore, we can confirm that $ZS^{\otimes 7}$ implements an encoded S gate. Implementing an encoded CNOT gate is straight forward. Before describing that, let us briefly look at how a CNOT gate can propagate errors.

The CNOT gate has a control qubit and a target qubit. The state of the target qubit is altered depending upon the state of the control qubit. The error propagates in the *forward* direction to the target qubit, if the control qubit is in error. Now consider the circuit shown in Fig. 5.30, and the Eq. (5.101). In this circuit, we showed how the roles of the target and control qubits could be reversed. The H gates rotate the basis in such a way that a bit-flip error is transformed into a phase-flip error. Hence, if a phase error happens in the target qubit, the error can propagate *backwards* to the control qubit!

Returning to our discussions, consider the circuit shown in Fig. 9.15, which implements the encoded CNOT gate.

Fig. 9.16 Fault-tolerant encoded T-gate

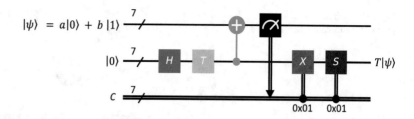

Assuming Steane's 7-qubit QEC code, the encoded CNOT gate uses seven controlled-NOT operations working on two blocks. Assume that one of the CNOT gates fail. Even if we implement the circuit in parallel, the CNOT gates couple only two qubits. In our example, the faulty CNOT gate couples one qubit from the control block into another qubit in the target block. This coupling results in one error in both the blocks, irrespective of the direction of propagation of the error. Again, this is a single-qubit error that can be efficiently corrected using the QEC schemes.

Recall the T gate from our discussions in Sect. 5.4.2.4. The T gate performs a $\frac{\pi}{4}$ rotation about the z-axis. Our next goal is to implement an encoded T gate to make our discussions complete on the universality of encoded gates. Once universality is achieved, fault-tolerant quantum circuits can be constructed with a combination of the encoded gates.

Figure 9.16 explains how we may construct a fault-tolerant T gate.

For the sake of simplicity, the diagram shows only one block of the T gate. It should be constructed in a block of seven for the Steane's code. Let us derive the inner workings of this circuit for one block. The derivation can be extended to the rest of the blocks. We start the second qubit in a state of $|0\rangle$ and apply an H gate, which is followed by a T gate. The first qubit is in a state of $a|0\rangle + b|1\rangle$. The resultant state of the system can be written as follows:

$$(a|0\rangle + b|1\rangle) \otimes \frac{1}{\sqrt{2}} \left(|0\rangle + e^{\frac{i\pi}{4}}|1\rangle \right)$$
$$\rightarrow \frac{1}{\sqrt{2}} \left[a|0\rangle|0\rangle + a|0\rangle e^{\frac{i\pi}{4}}|1\rangle + b|1\rangle|0\rangle + b|1\rangle e^{\frac{i\pi}{4}}|1\rangle \right] \tag{9.65}$$

After applying the CNOT, the state is transformed as follows:

$$\rightarrow \frac{1}{\sqrt{2}} \left[a|0\rangle|0\rangle + e^{\frac{i\pi}{4}}a|1\rangle|1\rangle + b|1\rangle|0\rangle + e^{\frac{i\pi}{4}} b|0\rangle|1\rangle \right]$$
$$\rightarrow \frac{1}{\sqrt{2}} \left[|0\rangle \left(a|0\rangle + e^{\frac{i\pi}{4}}b|1\rangle \right) + |1\rangle \left(b|0\rangle + e^{\frac{i\pi}{4}}a|1\rangle \right) \right] \tag{9.66}$$

Now, there are two possibilities. If we measure a $|0\rangle$ with qubit q[0], then the second qubit is in a state $a|0\rangle + e^{\frac{i\pi}{4}}b|1\rangle$, which is $T|\psi\rangle$. If we measure a $|1\rangle$, then the second qubit is in a state $b|0\rangle + e^{\frac{i\pi}{4}}a|1\rangle$, to this state, we apply an XS gate. If we ignore a global phase that builds up, this operation is equivalent to a $T|\psi\rangle$ gate (this can be verified easily.) There are some challenges with this circuit, however. The second qubit must be prepared in a perfect $|0\rangle$ state, and the measurement needs to be perfect. Errors happening in the first qubit do not impact the measurement. So, we apply the XS gate depending upon the measurement. A single-qubit error occurring on the second qubit propagates to one qubit in each of the two blocks, causing a single error in the output of the encoded gate.

Periodic error correction still needs to be performed on the encoded states to prevent errors from building up. From the implementation of the T gate, it is evident that similar procedures for state preparation and measurement are needed as well!

Fig. 9.17 Circuit for
projective measurement

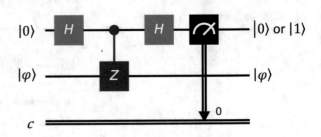

9.4.2   Performing Error Corrected Measurements

Consider the circuit in Fig. 9.17, which can perform a projective measurement into the eigenvalues
(± 1) of a Pauli operator that squares to the identity matrix.

The system shown in the diagram starts with the first qubit q[0] in state $|0\rangle$ and the second qubit
q[1] in state $|\varphi\rangle = a|0\rangle + b|1\rangle$. The collective state is given by:

$$|\psi\rangle = |0\rangle \otimes (a|0\rangle + b|1\rangle) \tag{9.67}$$

We apply an H gate to the first qubit:

$$\rightarrow \frac{1}{\sqrt{2}} (|0\rangle + |1\rangle) \otimes (a|0\rangle + b|1\rangle)$$

$$\rightarrow \frac{1}{\sqrt{2}} [|0\rangle a|0\rangle + |0\rangle b|1\rangle + |1\rangle a|0\rangle + |1\rangle b|1\rangle] \tag{9.68}$$

To this state, we apply a CZ gate. The effect of this gate is as follows:

$$\rightarrow \frac{1}{\sqrt{2}} [|0\rangle a|0\rangle + |0\rangle b|1\rangle + |1\rangle a|0\rangle - |1\rangle b|1\rangle] \tag{9.69}$$

The next step is to apply an H gate to qubit #1.

$$\rightarrow \frac{1}{2} [(|0\rangle + |1\rangle)a|0\rangle + (|0\rangle + |1\rangle)b|1\rangle + (|0\rangle - |1\rangle)a|0\rangle - (|0\rangle - |1\rangle)b|1\rangle] \tag{9.70}$$

This reduces to:

$$\rightarrow |0\rangle \otimes (a|0\rangle + b|1\rangle) \tag{9.71}$$

Note that this circuit does not alter the state of the second qubit, but performs a projective
measurement of the input register onto the Pauli operator's eigenvalues. The circuit measures 0 for
the eigenvalue +1 and 1 for the eigenvalue -1. Readers can try this logic with other Pauli operators
that square to identity and different input states. Let us try to apply this logic to measure a transverse Z
gate on Steane's code. The circuit for this measurement is shown in Fig. 9.18.

This circuit does not work as expected. Any error occurring on the ancilla qubit will propagate to
all of the encoded qubits, which is not in agreement with our criteria for fault-tolerance. We expect at
most single-qubit errors in a code block. So, this circuit is not fault-tolerant.

The solution to this problem is to create n number of cat states ($n = 7$ for Steane's code)
$\frac{1}{\sqrt{2}}(0^{\otimes n} + 1^{\otimes n})$, and then kick back the phase of the Pauli measurement operator into the measurement
circuit. Recall our discussion of the cat state in Sect. 3.4.1.1. When we make the final measurement,
the system collapses into one of the cat states: $\frac{1}{\sqrt{2}}(0^{\otimes n} + 1^{\otimes n})$ or $\frac{1}{\sqrt{2}}(0^{\otimes n} - 1^{\otimes n})$. We can achieve the

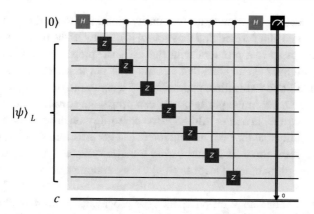

Fig. 9.18 Quantum circuit for measuring a transverse Z gate

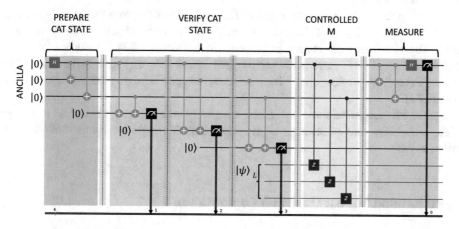

Fig. 9.19 Fault-tolerant measurement

cat states with the help of ancilla qubits. Figure 9.19 illustrates this concept in detail. For the sake of simplicity, the circuit is shown with three encoded data qubits. Recall that we need a minimum of five qubits. Ideally, Steane's 7-qubit code is used.

In this circuit, the top three qubits are the ancilla qubits. We prepare these qubits in the cat state by preparing the first qubit in an equal superposition state with the help of an H gate. Then, two successive CNOT gates put the ancilla qubits into the cat state.

$$\rightarrow |000\rangle \xrightarrow{H^{(0)}} \frac{1}{\sqrt{2}}(|000\rangle + |100\rangle) \xrightarrow{\text{CNOT}^{(0\rightarrow 1,2)}} \frac{1}{\sqrt{2}}(|000\rangle + |111\rangle) \tag{9.72}$$

The next stage is to verify the cat state. We can verify the cat state with the help of three parity check qubits. We do this with the help of two CNOT gates with two ancilla qubits serving as the control qubit. Let us illustrate how this works by evaluating the first parity check operation.

$$\frac{1}{\sqrt{2}}(|000\rangle|0\rangle + |111\rangle|0\rangle) \xrightarrow{\text{CNOT}^{(0\rightarrow 3)}} \frac{1}{\sqrt{2}}(|000\rangle|0\rangle + |111\rangle|1\rangle)$$
$$\xrightarrow{\text{CNOT}^{(2\rightarrow 3)}} \frac{1}{\sqrt{2}}(|000\rangle|0\rangle + |111\rangle|0\rangle) \tag{9.73}$$

From the above equation, if we measure the parity qubit, it should return a state $|0\rangle$. All parity check qubits are measured and verified that they return $|0\rangle$. If any of the parity qubits return a state $|1\rangle$, we can conclude that there is a problem with the cat state preparation. We can then discard the preparation and start all over again. Note that this circuit is not implemented in a fault-tolerant way, the CNOT gates can fail due to bit-flips, phase-flips, or a combination of both.

Any phase error occurring in the parity qubits can propagate forward into the ancilla qubits and cause phase errors in two ancilla qubits. Although the phase-flips do not propagate to the encoded data qubits, there is a possibility that the final measurement can get affected. We can solve this problem by repeating the measurement three times and taking a majority vote. This additional step gives us a possibility to reduce the probability of error to $O(p)^2$, if we assume that the circuit can fail at most two times.

The bit-flip or the combination of bit-phase flips can propagate to the encoded data. However, after the cat state verification, there can be at most only one of these errors. Hence, the overall implementation is fault-tolerant.

After the verification of parity, we perform the controlled execution of the M gates (Z in the circuit.) The cat state ensures that the errors occurring in any of the M gates do not propagate to the other gates. Therefore, there can at most be one error in the encoded data. Finally, we "un-do" the cat state with the series of CNOT gates in the ancilla qubit, followed by an H gate in the first ancilla qubit. We measure a 0 or 1, depending upon the eigenvalue, the encoded data produces with the measurement operator M.

The experiment is repeated three times, and a majority vote is taken. This gives us an option to reduce the probability rate of error to $O(p)^2$. Note that this logic works for fault-tolerant state preparation as well.

9.4.3 Performing Robust Quantum Computation

From the discussions so far, it is well understood that to perform reliable quantum computation, we need to reduce the error rate to the threshold level. Implementing the QEC schemes discussed so far is one way to increase fault tolerance. However, QEC schemes come at the cost of increased physical qubits. The more the physical qubits are, the more error-prone the circuit is. Assume that the probability that an individual qubit (or a quantum gate/preparatory circuit/measurement circuit) may fail is p, then the probability of an encoded operation failing is cp^2, where c is a constant. If $p \le \frac{1}{c}$, then the probability gets lower than p. We use concatenation to reduce this further. Concatenation, in this context, works by taking the original circuit and concatenate with itself. By doing so, we make the circuit at the second level operated by the fault-tolerant operation of the first level of encoding. Concatenation causes the probability of failing to reduce from cp^2 to $c(cp^2)^2$. If we concatenate l times, then the number of qubits in the circuit is n^l and the error rate is $\epsilon = \frac{1}{c}cp^{2^l}$. When the number of qubits increases exponentially in l, the error rate decreases doubly exponentially in l. If there are d operations in a fault-tolerant procedure, then the size of the circuit is given by the following equation.

$$d = O\left(\text{poly}\left(\log \frac{1}{c\epsilon}\right)\right) \tag{9.74}$$

Note that each gate has to be implemented with an accuracy of $\frac{\epsilon}{n^l}$. If the probability of error is less than $\frac{1}{c}$, then we can perform robust quantum computation. The increase in the circuit size is polylogarithmic in the accuracy with which we want to perform the computation.

While the error rate goes down rapidly when compared with the number of qubits, this is still not realistic [10]. We need physical qubits with error rates significantly less than the threshold value. In the next section, we discuss another novel and emerging error correction scheme—surface codes.

9.5 Surface Codes

Surface codes are stabilizer circuits known for the tolerance to local errors (deformations). Alexei Kitaev first proposed a surface model to describe topological order using qubits placed on the surface of a toroidal geometry. The toric code is pedagogical, and included in this chapter for its sheer elegance in describing topological protection against local deformations. The planar code evolved from the works of **Michael H. Freedman**, **David A. Meyer** [12], and **Sergey B. Bravyi** and **Alexei Kitaev** [13].

> 📖 **Note:** Reading Chap. 4 may be helpful to read the following section in this chapter.

9.5.1 Toric Code

Toric code is a toy model proposed by Alexei Kitaev. We can obtain a toric code of genus 1 (the genus is the number of holes in the torus) by taking a square lattice with periodic boundary conditions, and joining the top and bottom edges, left and right edges. This process embeds the n edge graph Γ of the square lattice into the torus $\mathbb{T}^2$ (Fig. 9.20).

A qubit is placed at each edge of the graph, the states of which form the codewords. The total number of qubits in the lattice is $2n^2$.

For each vertex v of the graph Γ, let us define an operator A_v (v denotes vertex) as a tensor product of Pauli X matrices and Identity matrices, such that A_v performs a Pauli X gate on each of the four qubits corresponding to the four edges adjacent to v. Otherwise, A_v acts as an identity gate. Similarly, we define one more operator B_p (p denotes plaquette) such that for each of the four qubits enclosing

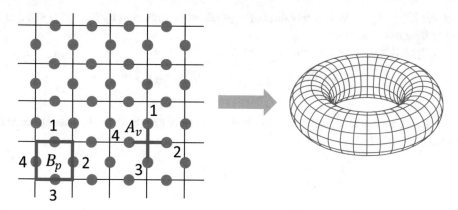

Fig. 9.20 Toric code

the plaquette, B_p performs a Pauli Z gate. Otherwise, B_p acts as an identity gate. Thus, the stabilizers are defined as the loops around each of the vertices or plaquettes of the lattice:

$$A_v = X_1 X_2 X_3 X_4 \text{ and } B_p = Z_1 Z_2 Z_3 Z_4 \tag{9.75}$$

For two vertices v, v', and for two plaquettes p, p', we can say that $[A_v, A_{v'}] = [B_p, B_{p'} = 0], \forall v, v', p, p'$, since the Pauli operators commute with each other and with the identity. If v and p are not adjacent, then A_v and B_p naturally commute. If they are adjacent, they share two qubits, and hence they still commute, that is, $[A_v, B_p] = 0, \forall v, p$. Hence, we can create a subgroup of commuting generators.

$$S = \langle A_{v1}, A_{v2}, A_{v3}, \ldots, B_{p1}, B_{p2}, B_{p3}, \ldots \rangle \tag{9.76}$$

We have n edges in the lattice, so there are n qubits. However, it is a characteristic of the torus that we have $\frac{n}{2}$ plaquettes and $\frac{n}{2}$ vertices. So, there is actually $\frac{n}{2} A_v$ and $\frac{n}{2} B_p$ operators. Therefore, S is a generator of $\frac{n}{2} A_v$ and $\frac{n}{2} B_p$ operators of size n. However, they are not entirely independent generators. The eigenvalues of A_v and B_p are ± 1. So, we have,

$$\Pi_v A_v = \Pi_p B_p = +1 \tag{9.77}$$

Note that, each qubit is included in two vertices and two plaquettes. Due to this property, one of A_v and one of B_p must be linearly dependent on all others, and we lose two degrees of freedom from the n generators.

The size of the code space is $2^{n-k} = 2^{n-(n-2)} = 2^2 = 4$. So, we can say that the toric code is a $[n^2, 2, n]$ stabilizer code. Hence, we can have two data qubits in a system of n physical qubits. The Hamiltonian of the toric code is derived from the interaction terms of the vertices and plaquettes,

$$H = -\sum_v A_v - \sum_p B_p \tag{9.78}$$

The ground state $|\xi\rangle$ is a common eigenstate $+1$ of A_v and B_p $\forall v, p$. In the computational Z-basis, the ground state can be written as [11]:

$$|\xi\rangle = \Pi_v \frac{1}{\sqrt{2}} (I + A_v) |000\ldots 0\rangle, \tag{9.79}$$

where the ket $|000\ldots 0\rangle$ is the superposition of a ground state stabilized by B_p, and $(I + A_v)$ projects that state to the ground state of A_v.

In the computational basis, the ground state constraint of B_p is:

$$\varphi_p \equiv \sum_{i \in C_p} |z_i\rangle = |z_1\rangle + |z_2\rangle + |z_3\rangle + |z_4\rangle = 0 , \tag{9.80}$$

where C_p is the boundary of a plaquette p. Similarly, we can define the ground state constraint for an adjacent plaquette p'.

$$\varphi_{p'} \equiv \sum_{i \in C_{p'}} |z_i\rangle = |z_2\rangle + |z_5\rangle + |z_6\rangle + |z_7\rangle = 0 \tag{9.81}$$

If we merge the plaquettes p and p',

$$\varphi_{pm} \equiv \sum_{i \in C_{pm}} |z_i\rangle = |z_1\rangle + |z_3\rangle + |z_4\rangle + |z_5\rangle + |z_6\rangle + |z_7\rangle = 0 \tag{9.82}$$

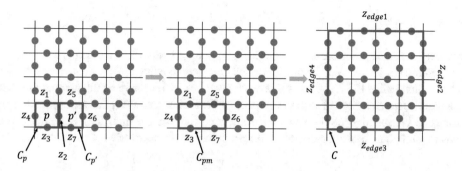

Fig. 9.21 Merging adjacent plaquettes and expanding into the entire lattice

Note that we have removed $|z_2\rangle$. We can expand this procedure to encapsulate the entire lattice (Fig. 9.21).

$$\varphi \equiv \sum_{i \in C} |z_i\rangle = |z_{\text{edge}1}\rangle + |z_{\text{edge}2}\rangle + |z_{\text{edge}3}\rangle + |z_{\text{edge}4}\rangle = 0 \tag{9.83}$$

When we introduced the lattice Γ, we said that it is bound by periodic boundary conditions. The boundary conditions lead to the fact that $|z_{\text{edge}1}\rangle = |z_{\text{edge}3}\rangle$ and $|z_{\text{edge}2}\rangle = |z_{\text{edge}4}\rangle$.

Therefore, the ground state can be characterized by two independent quantities describing the four degenerate states.

$$\xi_1 = \prod_{i \in \text{edge}1} Z_i = \pm 1 \text{ and } \xi_2 = \prod_{i \in \text{edge}2} Z_i = \pm 1 \tag{9.84}$$

The ground state $|\xi\rangle$ corresponds to the vacuum (that is, absence of anyons) state **1**, and the lattice is a source of $e(X)$, $m(Z)$ anyons, and ϵ fermions. Let us assume that we apply a Z gate to a qubit on an edge. This creates a pair of quasiparticle excitations in two adjacent vertices. If we measure the A_v operator on the two adjacent vertices, we get a -1 eigenstate. The quasiparticles created in this process are called e anyons. Similarly, we can create pairs of m anyons in the plaquette B_p using an X gate. Composite ϵ anyons can be created by the action of Z and X gates. If we apply the same gate two times, the overlapping anyon fusion results in the eigenstate of $+1$, which is the vacuum. The fusion rules are:

$e \times e = m \times m = \epsilon \times \epsilon = \mathbf{1}$, when created on the same vertex or plaquette, they annihilate.

$e \times m = \epsilon$, $m \times \epsilon = e$, $\epsilon \times e = m$, if created in the same vertex or plaquette.

$$\tag{9.85}$$

We can detect the presence of the e, m, ϵ anyons by measuring the eigenvalues of the operators A_v and B_p.

By applying Z rotations to consecutive qubits, we can move the anyons. This creates an open path, with the e anyons at the boundaries. If the path is closed, the anyons annihilate. Let us create a loop operator Z_C that performs this task.

$$Z_C = \prod_{i \in C} Z_i, \text{ where C is the path} \tag{9.86}$$

Note that the loop operator Z_C measures the fusion outcome of m anyons. A loop around a plaquette is the operator B_p, and larger loops can be split into plaquettes with overlapping edges. Hence, we can generalize Z_C as follows:

$$Z_C = \prod_{p \in C} B_p \tag{9.87}$$

We can try the same process using X rotations, but it does not work as expected, as the X operator does not commute with some of the B_p operators. One way to solve this problem is to introduce the concept of a dual lattice Γ'. The dual lattice has its vertices in the center of the plaquettes of the first lattice, and its plaquettes aligned with the vertices of the first lattice. With this arrangement, applying X rotations to consecutive qubits can create a closed-loop in the dual lattice, which causes the m anyons to annihilate. The loop operator corresponding to the successive X rotations is:

$$X_{C'} = \prod_{i \in C'} X_i \tag{9.88}$$

Here again, we can say that the loop operator X_C measures the fusion outcome of e anyons. In other words, the loop operators measure the parity of e or m anyons inside the loop. Following our previous argument, we can generalize X_C as:

$$X_{C'} = \prod_{v \in C'} A_v \tag{9.89}$$

Note that Z_C and X_C are **contractible operators** in the sense that they don't loop around the torus in either of the directions.

If in case, the loop encircles the torus horizontally, then it is equivalent to the **noncontractible operator** ξ_1. Similarly, if the loop encircles the torus vertically, it is equivalent to the non-contractible operator ξ_2. This is evident from the edges these two operators work on from Eq. (9.84).

For simplicity, let us label the horizontal and vertical trajectories as 1 and 2, respectively. We can also name the horizontal Z-loop operator as Z_1 and the vertical Z-loop operator as Z_2, with eigenvalues ± 1, as this naming convention is easy to use. Since $[Z_1, Z_2] = 0$, the four combinations of the eigenvalues are representative of the possible two-qubit manifold in the torus.

Following the above argument, let us create a horizontal X-loop operator X_1, and a vertical X-loop operator X_2, with eigenvalues ± 1.

In a scenario where we have Z_1 and X_1loops in the torus, $[X_1, Z_1] = 0$. The reason is we can displace one of the loops so that they do not have to share a common edge. The same logic holds good for Z_2 and X_2 loops, and $[X_2, Z_2] = 0$.

However, if we have a combination of Z_1 and X_2, or Z_2 and X_1, the two loops must share one edge. If we can label the common edge as 0, then the anticommutators are:

$$\{X_1, Z_2\} = \{\sigma_0^x, \sigma_0^y\} = 0 \text{ and } \{X_2, Z_1\} = \{\sigma_0^x, \sigma_0^y\} = 0 \tag{9.90}$$

From the above logic, we can deduce the correspondence with the logical Pauli operators on the torus,

$$\begin{aligned} X_1 \to \sigma_2^x \quad X_2 \to \sigma_1^x \\ Z_1 \to \sigma_1^z \quad Z_2 \to \sigma_2^z \end{aligned} \tag{9.91}$$

Note that the horizontal and vertical trajectories 1 and 2 are noncontractible. Since the anyons annihilate at the end of these loops, there are no resultant excitations. Therefore, they should represent ground states. The four ground states can be defined as follows:

$$\begin{aligned} |\psi_1\rangle = |\xi\rangle \quad\quad |\psi_2\rangle = Z_1 |\xi\rangle \\ |\psi_3\rangle = Z_2 |\xi\rangle \quad\quad |\psi_4\rangle = Z_1 Z_2 |\xi\rangle \end{aligned} \tag{9.92}$$

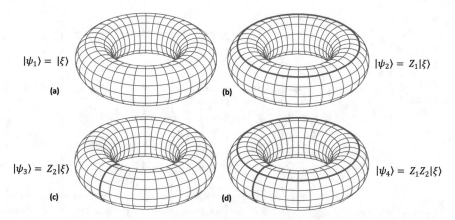

Fig. 9.22 The fourfold degeneracy of the ground state of the toric code. (a) ground state. Note that the noncontractible loops in (b) and (c) can be displaced. The loops in (d) share one edge in common

Since $|\psi_1\rangle$ starts at the ground state $|\xi\rangle$, the Z_1 and Z_2 operators act on $|\xi\rangle$ and create e anyons that move along the paths 1 and 2, respectively, to get annihilated at the end of the loops. The four states are linearly independent and correspond to the winding directions of the torus. From this, we can create linearly dependent states that use the loop operators corresponding to the e or m anyons. Hence, the torus represents a 4D Hilbert space, encoding two qubits. Figure 9.22 illustrates the fourfold degeneracy of the ground states, where information can be encoded.

Using the ground state degeneracy, the four different states of the two logical qubits can be defined as follows:

$$|00\rangle = |\xi\rangle \qquad |01\rangle = Z_1\ |\xi\rangle$$
$$|10\rangle = Z_2\ |\xi\rangle \qquad |11\rangle = Z_1 Z_2\ |\xi\rangle \tag{9.93}$$

Note that the loops 1 and 2 can be displaced, as long as they go around the torus.

Since we have established the ground state degeneracy, one important property we must explore is how the ground state degeneracy is topologically protected.

Consider a *local* operator that operates on a random sequence $O \equiv X_i Z_j Z_k \ldots$, on nearby sites $i, j,$ $k \ldots$ Let us define some noncontractible loops C_1, C_2, C'_1, C'_2 with corresponding operators $Z_1, Z_2, X_1,$ X_2. If we choose the operators in such a way that they do not share a common edge with O, then we can say that $[O, X_1] = [O, X_2] = [O, Z_1] = [O, Z_2] = 0$.

From Eq. (9.91), we know that the operators X_1, X_2, Z_1, Z_2 correspond to Pauli rotations σ_x and σ_y. This means O commutes with the Pauli operators. From Example 2.8, we know that this is possible only if O is a multiple of I.

From Eq. (9.79), we know that the ground state $|\xi\rangle$ is a superposition of all possible elementary loops such as C_1, C_2, C'_1, C'_2. Hence, the application of the contracting local operator O on the ground state gives rise to a multiple of the same state. We can ignore the multiple, as the loops can be displaced. Due to this reason, the excited state is an equal superposition state of all possible paths that connect the anyons. Hence, the ground state degeneracy is topologically protected (Fig. 9.23).

Errors on the toric code are caused by stray operators, which can cause anyonic excitations. One bit-flip error creates a pair of anyons, causing a path along two consecutive edges of the dual lattice—the endpoints of the path project to eigenvalue -1. The eigenvalue of a normal plaquette is $+1$, so a change in this is an error. Hence, by measuring the stabilizers, we can easily find the error states and their locations. Once the location is identified, we can use Z or X gates to move the anyons along the

Fig. 9.23 In this figure, two *m* anyons at the endpoints of a straight line is created by X operators. Their state is invariant, if the shape of the loop is deformed, as long as the endpoints remain fixed [11]

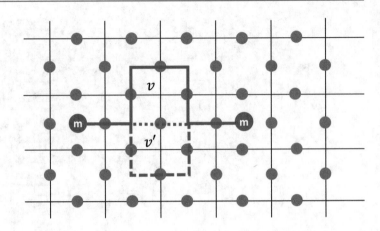

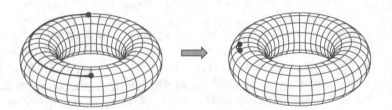

Fig. 9.24 Error correction on a toric code

shortest path and annihilate them, thereby returning the system to the ground state. The error correction works if the error path is $< \frac{n}{2}$, where n is the linear length of the torus. If the path is $> \frac{n}{2}$, the correction path may be equivalent to a noncontractible path of a logical gate. However, with large torus sizes, we can prevent the error paths from going around the torus [11] (Fig. 9.24).

9.5.1.1 Summary

The logical gates encoded by the torus provide only phase information. Hence, the anyonic toric code is not suitable for developing universal quantum gates. Furthermore, irrespective of the torus' size, they can store only two qubits that are topologically protected. Due to this reason, the toric code has a very low code rate of $\frac{2}{n}$. For large values of n, this is not so useful. However, the toric code is a very good candidate for developing quantum memories. The toric code is still an experimental model. There is a possibility of implementing toric code using arrays of Josephson junctions.

9.5.2   Planar Code (or Surface Code)

Consider a 2D lattice [10] as shown in Fig. 9.25.

Physical qubits form the vortices of the 2D lattice. Unshaded circles are data qubits. The dark shaded circles are measurement qubits that stabilize the data qubits. The measurement qubits marked Z are the Z-syndrome qubits, and the qubits marked X are the X-syndrome qubits. Each data qubit is coupled with two Z-measurement qubits, and two X-measurement qubits. In a similar arrangement, each measurement qubit is coupled with four data qubits. Each measurement qubit functions as the corresponding stabilizer of the neighboring qubits, measuring them into the eigenstates of the measurement operator products. If we can label the data qubits surrounding a measurement qubit as the set $\{a, b, c, d\}$, then,

Fig. 9.25 2D surface code

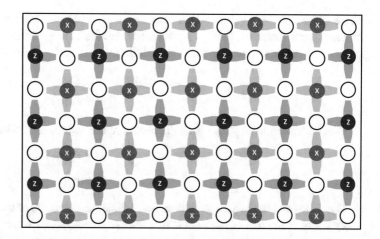

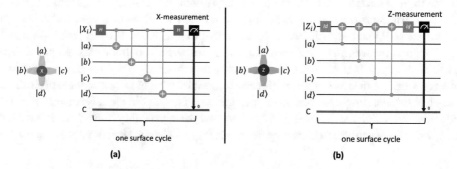

Fig. 9.26 Surface code cycles. (**a**) X-measurement and (**b**) Z-measurement

$$Z = Z_a Z_b Z_c Z_d \quad \text{and} \quad X = X_a X_b X_c X_d \tag{9.94}$$

Due to this arrangement, any error occurring on a data qubit reflects on two adjacent X-measurement qubits and two adjacent Z-measurement qubits. Note that the lattice terminates with X boundaries at the top and bottom and Z boundaries on the left and right. The X boundaries are known as smooth boundaries, and the Z boundaries are known as rough boundaries.

The measurement qubits operate on the data qubits in a specific order. The X-measurement starts at state $|0\rangle$, and performs an H gate on the measurement qubit. Then, four CNOT operations are performed on the neighboring data qubits with the measurement qubit as the control qubit. After performing the CNOT gates, another H gate is applied to the measurement gate, before it can be measured. The Z-measurement starts first by initializing the measurement qubit to state $|0\rangle$, and then performing an I gate. The I gate provides for the necessary waiting time for the H operation to complete on the X-measurement. After the I gate, four CNOT gate operations are performed with the neighboring data qubits serving as the control qubits. Finally, an I gate is applied to the measurement qubit, before it can be measured. This sequence is shown in the quantum circuits featured in Fig. 9.26.

The measurement operation is performed for all measurement qubits in the lattice. This task projects the data qubits into a quiescent state, which is not a fully entangled state of the lattice. While some of the local qubits may be highly entangled, a small degree of entanglement may be present with distant qubits.

If the measurement is repeated the quiescent state should remain the same, unless perturbed by a bit-flip, phase-flip, or a combination of both occurring on one or more qubits. Let us assume that a phase-flip occurs on a qubit a. This error causes the quiescent state to project to $|\psi\rangle \rightarrow Z_a|\psi\rangle$. If we repeat the measurement, there is a possibility that the lattice returns to the original state $|\psi\rangle$ erasing the error. Let us examine the second possibility with X and Z stabilizers.

$$
\begin{aligned}
X(Z_a|\psi\rangle) &= Z_a(X|\psi\rangle) \\
&= -Z_a|\psi\rangle
\end{aligned}
\tag{9.95}
$$

This equation means $Z_a|\psi\rangle$ is an eigenstate of the stabilizer X, with an opposite sign.

$$
\begin{aligned}
Z(Z_a|\psi\rangle) &= Z_a(Z|\psi\rangle) \\
&= Z_a|\psi\rangle
\end{aligned}
\tag{9.96}
$$

Thus, the error in the data qubit does not change the outcomes of the Z-measurement. These results apply for both the sets of adjacent Z or X-measurement qubits. Since the phase error reflects on the two adjacent X-measurement qubits, we should be able to identify the qubit a. We can fix this by applying Z_a to the affected qubit. However, the fix is done in the classical governing software by flipping the sign of the corresponding X-measurement qubits in subsequent operations. Doing this in software avoids additional error injection by an error-prone Z_a gate. Bit-flip errors are handled by flipping the sign of the neighboring Z-measurement in software. If both bit-flip and phase-flip errors occur simultaneously, it is reflected in both the two neighboring Z- and X-measurement qubits.

If an error occurs in the measurement process, it should vanish in the subsequent cycle. There is a possibility that the measurement error can repeat for the same qubit in the subsequent cycle, but with a lower probability. For the following measurements, the probability should be further reduced. Therefore, we need several surface code cycles to identify single and sequential measurement errors.

9.5.2.1 Logical Qubit

Consider the lattice shown in Fig. 9.27 with two logical measurement operators X_L and Z_L [10].

The surface code shown in the figure has 41 data qubits and 40 measurement qubits. In the computational basis, there are 2×41 degrees of freedom for the data qubits and 2×40 constraints

Fig. 9.27 Logical operators of the surface code

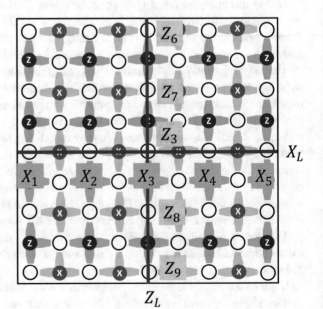

from the stabilizer measurements. The stabilizers form a linearly independent set. Hence, the 2×40 constraints are linearly independent. The remaining two unconstrained degrees of freedom can be used to construct a logical qubit using the logical operators X_L and Z_L shown in the diagram.

Consider an operation X_1 on the leftside of the boundary. This change is detected by the Z-measurement qubit on the right-hand side of this data qubit that is bit-flipped. If we perform an X_2 operation on the qubit to the right, we can say that the first Z stabilizer commutes with the two X operations. Now, the Z-measurement to the right of this finds a change, and we add an X_3 operation to this. We can repeat this cycle until we reach the right side boundary with the X_5 operation. With this, we have created a concurrent operator $X_L = X_1 X_2 X_3 X_4 X_5$, which connects the two X-boundaries. We find that every other Z-measurement in the lattice interacts with the data qubits along X_L, to which it commutes. Hence, X_L forms one of the degrees of freedom. It can be shown that any other operator X_L' which we may create, similarly, can be written as X_L multiplied by the product of X stabilizers used. Hence, there can be only one linearly independent X_L operator of the lattice. Similarly, we can create another operator $Z_L = Z_6 Z_7 Z_3 Z_8 Z_9$, as shown in the diagram.

The two logical operators X_L and Z_L anticommutes (we leave it to our readers to work this out) just like the physical qubit gates X and Z do. We can show that $X_L^2 = Z_L^2 = I$, and we can create the sequence $Y_L = Z_L X_L$. Therefore, we can say that the lattice behaves like a logical qubit.

In our example, the number of data qubits was higher than the number of stabilizer qubits. Hence, the quiescent state $|\psi\rangle$, as we have seen, is not fully constrained. We can, therefore, describe the quiescent state as $|\psi\rangle = |Q\rangle|q_L\rangle$. Here, $|Q\rangle$ represents a constrained state in the Hilbert space of N dimensions, determined by the N stabilizer measurement outcomes. The remaining degrees of freedom are with $|q_L\rangle$. The stabilizers have no effect on $|q_L\rangle$, and the operators X_L and Z_L have no effect on $|Q\rangle$. The eigenstates of Z_L are $|q_L\rangle = |0_L\rangle$, and $|q_L\rangle = |1_L\rangle$. The eigenstates of X_L are $|q_L\rangle = |\pm_L\rangle$. Therefore, we can use a computational basis or polar basis for the logical qubit $|q_L\rangle$.

Similarly, we can create more than one logical qubit in the lattice by creating additional degrees of freedom. Each logical qubit added to the system increases the dimensionality of the Hilbert space by two. In a system of n qubits, the dimensionality of the Hilbert space is 2^n. For the lattice shown in Fig. 9.27, $|q_L\rangle$ is in a 2D Hilbert space defining a single qubit.

Additional degrees of freedom are created by turning off an intermittent Z-measure qubit (called **Z-cut-hole**), which creates two additional degrees of freedom. Logical qubits created using this method are called **smooth defect** or **dual defect** qubits in literature. Similarly, we can create an **X-cut-hole** by turning off an intermittent X-measure qubit. Logical qubits created using this method are called **rough defect** or **primal defect** qubits. It is also possible to create double **Z-cut** or double **X-cut** logical qubits [10].

9.5.2.2 Error Handling

Errors can happen at any time during the computation. We saw bit-flip, phase-flip, and the combination of bit-flip and phase-flip occurring on physical qubits. Errors can also happen during qubit initialization and measurement. Besides, gate operations are also prone to errors. Not just one-time errors, we can also expect error chains—errors occurring on a series of qubits. If the errors can be identified, the surface code is a good candidate to address the errors. Once identified, the classical governing software can track and apply fixes to the measurement outcomes.

Let p denotes the number of errors per surface code cycle, and the logical error rate P_L denotes the number of X_L errors occurring per surface code cycle. The relation between these two parameters depends on the distance parameter d. The distance parameter is the number of physical qubits required to define X_L or Z_L. Note that $d = 5$ in our example. When p is less than a threshold value, that is, when $p < P_{th}$, P_L decreases exponentially with d. When $p > P_{th}$, P_L increases with d. For error rates $p < P_{th}$, numerical simulations indicate that $P_L \sim p^{d_e}$ [10]. The error dimension d_e for odd d is given by $d_e = (d + 1)/2$, and for even d, it is $d_e = d/2$. With this, the logical error rate is approximated as [10]:

$$P_L \cong 0.03 \left(\frac{p}{P_{th}}\right)^{d_e} \tag{9.97}$$

The total number of data and measurement qubits is given by

$$n_q = (2d - 1)^2 \tag{9.98}$$

We find that n_q increases rapidly as p approaches P_{th}. If we set a target for the gate fidelity to be above 99.9% $\left(\text{that is, } p \lesssim 10^{-3}\right)$, the number of physical qubits required to form a logical qubit is in the range 10^3 to 10^4. This brings the logical error rates below to the range 10^{-14} to 10^{-15}. When the error rate is brought down to this level, complex algorithms such as Shor's can be run with a good success.

9.5.3   Summary

The tolerance of the surface code to errors for each step of the operation is about 1%. Steane and Bacon-Shor codes in 2D lattices with nearest-neighbor coupling configurations require a per-step threshold of 2×10^{-5}. This requirement is higher than the surface codes by order of three magnitudes [10]. The less stringent requirement for error tolerance and the ability to do a 2D layout with nearest-neighbor coupling makes the planar code (or the surface code, interchangeably) the most realistic of the surface code family.

However, this comes at the cost of the number of physical qubits required. We need a minimum of 13 physical qubits to implement one logical qubit. For a reasonable fault-tolerant logical qubit on a surface code, we need about 14,500 physical qubits for an error rate, which is 1/10th of the threshold rate (0.75% per operation.)

Quantum computers with error-corrected qubits that can perform complex tasks such as the integer factorization are not too far from now! In the following section, we examine some of the emerging qubit designs with inbuilt fault-tolerance. Hope is not yet crushed!

9.6   Protected Qubits

Note: Reading Chap. 4 may be helpful to read the following section in this chapter.

We learned that the surface codes implement quantum error correction using stabilizer circuits, confining the system state into a subspace protected from local noises. Such schemes are stepping stones toward fault-tolerant quantum computing. However, from the preceding arguments, we need physical qubits with error rates less than a threshold value to implement fault-tolerant quantum computers. Therefore, it is essential that fault tolerance should stem from physical qubits. In this section, we shall explore a few selected protected qubit designs with inbuilt fault-tolerance, which have been reported by researchers in recent times. A study of these designs should give an idea, where qubit technologies are heading to and re-emphasize the promise of quantum supremacy [14].

9.6.1 0–π Qubit

The $0 - \pi$ qubit is an intelligent design that uses a two-lead superconducting device, whose energy is minimized when the superconducting phase between the leads is either 0 or π (and hence the name.) Due to the design, the $0 - \pi$ qubits have a higher barrier that prevents bit-flips. The degeneracy of the basis states suppresses the phase-errors. The potential to minimize the effects of local noises provides the necessary protection to the encoded quantum information. The $0 - \pi$ qubit design is due to **Peter Brooks, Alexei Kitaev**, and **John Preskill**. This section is a condensed version of the original research paper. We first establish that the energy stored in a system of two identical Josephson circuits connected by a capacitor is a function of the phases and then apply the concept to a qubit design by rewiring the circuit. We finally demonstrate how a protected phase gate can be constructed by coupling the qubit with an oscillator.

9.6.1.1 Qubit Physics

Consider the superconducting circuit [16] shown in Fig. 9.28, formed by two identical stages having Josephson junctions with Josephson energy J and intrinsic capacitance C. The two stages are connected by high capacitance $C_1 \gg C$. Besides, each stage in the circuit has a super inductance L in series. The values of L and C are chosen such that $\sqrt{L/C}$ is much larger than $\frac{4\pi}{c}$, the impedance of free space, c being the speed of light. φ_1 and φ_2 are the superconducting phases on both sides of the capacitor. θ_1, θ_2, θ_3, and θ_4 are the phases at the respective leads, as shown in the diagram.

The Hamiltonian of this circuit can be written as:

$$H = \frac{Q^2}{2C} + \frac{\varphi^2}{2L}, \tag{9.99}$$

where $Q = \frac{q}{2e}$, $2e$ is the Cooper pair charge, and $\varphi = \frac{2e}{\hbar} \Phi$ is the superconducting phase. The phase $\varphi = 2\pi$ corresponds to $\frac{h}{2e}$ of the magnetic flux.

The ground state of the Hamiltonian has the energy $E_0 = \frac{1}{2}\sqrt{LC}$. Its wavefunction can be described in terms of the Gaussian function:

$$|\psi(\varphi)|^2 \propto e^{\frac{-\varphi^2}{2\langle\varphi^2\rangle}}, \tag{9.100}$$

where $\langle\varphi^2\rangle = \frac{1}{2}\sqrt{\frac{L}{C}}$. We can now define:

Fig. 9.28 Two-stage superconducting circuit. (Reference: [16])

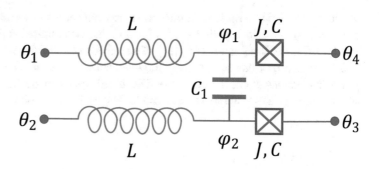

$$\langle \cos \varphi \rangle = e^{\frac{-\langle \varphi^2 \rangle}{2}} = \exp\left(\frac{-1}{4}\sqrt{\frac{L}{C}}\right). \tag{9.101}$$

If $\sqrt{\frac{L}{C}} \gg 1$, the phase φ has large fluctuations in the ground state, and the ground state of the wavefunction is broad. The above function averages out, except for a small exponential correction factor.

Let us examine the sum and difference of the superconducting phases at the leads now.

The phase $\varphi_+ = (\varphi_1 + \varphi_2)/2$ is not dependent on C_1, and it is called the "light" variable with large fluctuations. The phase $\varphi_- = (\varphi_1 - \varphi_2)$ is dependent on C_1, and it is a localized "heavy" variable. If the phase-slip[2] between the inductors is suppressed, $\varphi_\pm$ are real variables.

From the circuit, the effective capacitance controlling the phase φ_+ is $2C$, and the effective inductance is $L/2$. Therefore,

$$\langle \varphi_+^2 \rangle = \frac{1}{4}\sqrt{\frac{L}{C}} \text{ and} \tag{9.102}$$

$$\langle \cos \varphi_+ \rangle = \exp\left(\frac{-1}{8}\sqrt{\frac{L}{C}}\right). \tag{9.103}$$

If $\sqrt{\frac{L}{C}} \gg 1$, this is an ignorable value. Hence, we need to rely on the dynamics of the heavy variable φ_-, which can be written as a function of the external phases of the leads:

$$\begin{aligned} \varphi_- &= \varphi_1 - \varphi_2 = (\theta_4 - \theta_1) - (\theta_3 - \theta_2) \\ &= (\theta_4 + \theta_2) - (\theta_1 - \theta_3) \end{aligned} \tag{9.104}$$

The energy stored in the circuit is:

$$E = f(\theta_4 + \theta_2 - \theta_1 - \theta_3) + O(\langle \cos \varphi_+ \rangle), \tag{9.105}$$

where $f(\theta)$ is a periodic function with period 2π.

To construct a qubit, we rewire the circuit as shown in Fig. 9.29, connecting θ_4 and θ_2, θ_1 and θ_3. A large capacitance C_1 is added between θ_1 and θ_2 to prevent quantum tunneling.

The energy of the resultant circuit is:

$$E = f(2(\theta_2 - \theta_1)) + O(\langle \cos \varphi_+ \rangle), \tag{9.106}$$

where the correction term is insignificant. Hence, the energy of the circuit is a function of the phase difference $\theta_2 - \theta_1$ with a period π. This function has two minima at $\theta = 0$ and $\theta = \pi$, which can be considered as the basis states $\{|0\rangle, |1\rangle\}$ of an encoded qubit (Fig. 9.30).

Kitaev theorized that by replacing the inductance with a long chain of Josephson junctions, the phase fluctuations along the chain are distributed across many junctions. Hence, the information about the basis states is not locally accessible. The chain of Josephson junctions is shown in Fig. 9.31.

---

[2] Quantum phase-slip is a phenomenon that occurs in superconducting nanowires. It is a process in which the superconducting phase difference between two superconducting regions changes by 2π in a short time.

Fig. 9.29 The 0–π qubit.
(Reference: [16])

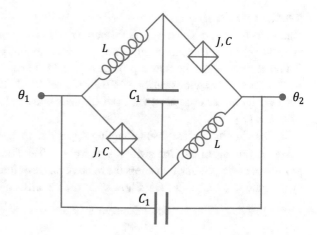

Fig. 9.30 The energy
diagram of the 0–π qubit
illustrating the basis states.
(Reference: [16])

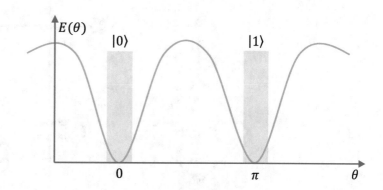

Fig. 9.31 Chain of
Josephson Junctions

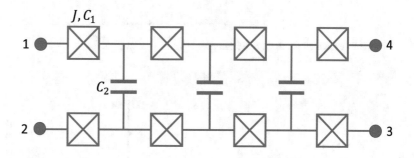

9.6.1.2 Measurement

Z-measurement can be performed by connecting the two leads of the qubit with a Josephson junction
and applying a $\frac{1}{4}$ of a flux quantum through the loop. The current through the circuit is proportional to
$\sin\left(\theta_2 - \theta_1 - \frac{\pi}{2}\right)$. By measuring the direction of the current through the circuit, we can determine
whether the qubit is in state $|0\rangle$ or $|1\rangle$.

For X-measurement, the suggestion is to break the circuit at the connection between θ_1 and θ_3 and
measure the charge $\theta_1 - \theta_3$. The charge is an even or odd multiple of $\frac{1}{2}$, depending upon the
eigenvalue of the measurement.

9.6.1.3 Protected Phase Gate

The protected phase gate design is to couple the 0–π qubit with an oscillator, which acts as a reservoir that absorbs the entropy induced by noise. The oscillator, however, needs to drain, if it is too excited. In this section, we show how a protected phase gate can be implemented using this scheme. Phase gates with exponential precision can be achieved by coupling the qubits with a super inductive LC circuit (a harmonic oscillator) via a switch made of a tunable loop containing two identical Josephson junctions (Fig. 9.32).

In this circuit, the Josephson coupling $J(t)$ is due to the magnetic flux $\frac{\eta}{2\pi}\Phi_0$, where $\Phi_0 = \frac{h}{2e}$ is the magnetic flux quantum. The switch is on for $\eta = 0$ and off for $\eta = \pi$. We shall later explain that the process of cycling the switch alters the relative phase of the basis states. The relative phase introduced by the gate locks at $\frac{\pi}{2}$. The Hamiltonian of the circuit is given by:

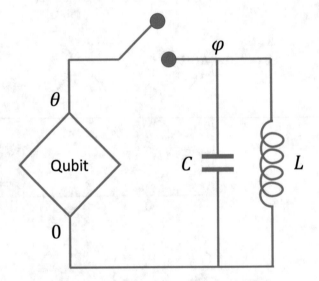

Fig. 9.32 Protected phase gate. (Reference: [16])

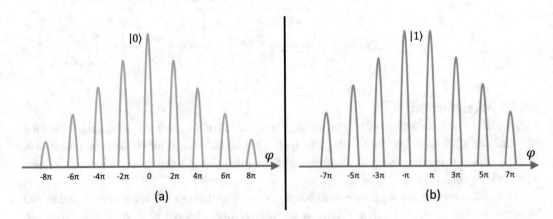

Fig. 9.33. Diagram illustrating the grid state of the oscillator, when the Josephson coupling $J(t)$ is turned on: (**a**) when the qubit is in state $|0\rangle$ and (**b**) when the qubit is in state $|1\rangle$

$$H(t) = \frac{Q^2}{2C} + \frac{\varphi^2}{2L} - J(t)\cos(\varphi - \theta), \tag{9.107}$$

where $J(t)$ is the time-dependent Josephson coupling, θ is the phase difference across the qubit, and φ is the phase difference across the inductor. If the qubit is in state $|0\rangle$, the phase θ is zero. If the qubit is in state $|1\rangle$, the phase θ is π. Hence, the above equation reduces to:

$$H(t) = \frac{Q^2}{2C} + \frac{\varphi^2}{2L} \mp J(t)\cos(\varphi), \tag{9.108}$$

When the Josephson coupling $J(t)$ is turned on, the initial state of the oscillator becomes a grid state—a superposition of narrowly peaked functions governed by a broad envelope function, as shown in Fig. 9.33.

If the qubit is in state $|0\rangle$, the narrow peaks on the grid state occur when φ is an even multiple of π. We can denote this grid state as $|0_C\rangle$ of the code.

If the qubit is in state $|1\rangle$, the narrow peaks on the grid state occur when φ is an odd multiple of π. We can denote this grid state as $|1_C\rangle$ of the code.

If the initial state of the qubit is $a|0\rangle + b|1\rangle$, then the combined state of the qubit and the oscillator evolves as follows when $J(t)$ is turned on:

$$(a|0\rangle + b|1\rangle)|\psi^{IN}\rangle = a|0\rangle \otimes |0_C\rangle + b|1\rangle \otimes |1_C\rangle, \tag{9.109}$$

$J(t)$ starts at zero, ramps up, and maintains a steady-state J_0 for a period $t = \frac{L\tilde{t}}{\pi}$, where $\tilde{t}$ is a rescaled time variable. $J(t)$ then ramps down. When $J(t)$ is at the steady-state, each of the narrow peaking function is stabilized by the cosine potential. However, the state is modified by the Gaussian function $e^{-i\tilde{t}\frac{\varphi^2}{2\pi}}$. As t increases, the oscillator states $|0_C\rangle$ and $|1_C\rangle$ evolve, but when $\tilde{t} = 1$, the oscillator states return to the initial states, excepting for a geometric phase. For the oscillator state $|0_C\rangle$, peaks in φ occur at $\varphi = 2\pi n$. This change is trivial. For the oscillator state $|1_C\rangle$, peaks in φ occur at $\varphi = 2\pi(n + \frac{1}{2})$, which modifies the phase by $-i$. The new state is:

$$\rightarrow a|0\rangle \otimes |0_C\rangle - i\, b|1\rangle \otimes |1_C\rangle, \tag{9.110}$$

When $J(t)$ ramps down, the coupling turns off, and the oscillator states $|0_C\rangle$ and $|1_C\rangle$ have evolved to $\left|\psi_0^{fin}\right\rangle$ and $\left|\psi_1^{fin}\right\rangle$, respectively. The joint state of the qubit and the oscillator can be written as:

$$\rightarrow |0\rangle \otimes \left|\psi_0^{fin}\right\rangle - i\, b|1\rangle \otimes \left|\psi_1^{fin}\right\rangle \tag{9.111}$$

Thus, a phase gate $\exp\left(i\frac{\pi}{4}Z\right)$ has been applied to the qubit. Under appropriate conditions, we can insist that $\langle\psi_1^{fin}|\psi_0^{fin}\rangle \approx 1$, which means that the phase gate is perfect.

Gate errors may occur, if the coupling between the qubit and the oscillator remains on for a too shorter or longer time, (that is, $\tilde{t} = 1 + \epsilon$). If $|\epsilon| < 2\pi\left(\frac{L}{C}\right)^{\frac{-3}{4}}$, the gate error is $\exp\left(-\frac{1}{4}\sqrt{\frac{L}{C}}\right) + O(1)$ and the timing errors do not affect the performance of the gate [16].

Due to the degradation of the oscillator, errors may happen due to slight over or under rotation of the qubit. However, this may not change the sensitivity of the final state of the oscillator to the qubit and thus may not reduce the fidelity of the gate.

Both these two types of errors can be characterized, and the qubit calibrated for near- perfect performance.

We have essentially embedded a qubit with 2D Hilbert space in a harmonic oscillator with an infinite-dimensional Hilbert space. The position φ and momentum operator Q of the oscillator satisfies $[\varphi, Q] = i$. The code space of the setup can be described in terms of the commuting generators of the stabilizer group:

$$M_Z = e^{2i\varphi}$$
$$M_X = e^{-2\pi i Q} \tag{9.112}$$

The corresponding Pauli operators are:

$$Z = e^{i\varphi}$$
$$X = e^{-\pi i Q} \tag{9.113}$$

It can be verified that the Pauli operators commute with the generators and anticommute among themselves. The eigenstate $Z = 1$ represents $|0_C\rangle$ and requires that φ be an integer multiple of 2π. $M_X = 1$ requires that the codeword is invariant under the displacement of φ by 2π. Hence, $|0_C\rangle$ can be described in the code space of the position eigenstates φ as integer multiples of 2π.

$$|0_C\rangle = \sum_{n=-\infty}^{\infty} |\varphi = 2\pi n\rangle \tag{9.114}$$

Similarly, we can define for the eigenstate $Z = -1$ as $|1_C\rangle = X|0_C\rangle$ as displacements of π in the code space.

$$|1_C\rangle = \sum_{n=-\infty}^{\infty} \left|\varphi = 2\pi\left(n + \frac{1}{2}\right)\right\rangle \tag{9.115}$$

The eigenstates $X = \pm 1$ correspond to $|\pm_C\rangle$ and are invariant under the displacement $Q = 2$. In the code space of the momentum eigenstates Q, they can be written as follows:

$$|+_C\rangle = \sum_{n=-\infty}^{\infty} |Q = 2n\rangle$$
$$|-_C\rangle = \sum_{n=-\infty}^{\infty} \left|Q = 2\left(n + \frac{1}{2}\right)\right\rangle \tag{9.116}$$

Note that any displacement of φ is detectable if it is $\pm\frac{\pi}{2}$, and can be corrected. Similarly, any displacement of Q can be detected and corrected, as long as it is $\pm\frac{1}{2}$. These are somewhat large values. Error detection is performed by measuring M_Z and checking φ mod π. Similarly, we can measure M_X and check for Q mod 1. Errors can be corrected by shifting φ or Q to the nearest value. The eigenvalue of the measurement operator is then determined from the shifted value (Fig. 9.34).

9.6.2 Fluxon-Parity Protected Superconducting Qubit

Recently, symmetry-protected superconducting qubit that offers simultaneous exponential suppression of energy decay from charge and flux noise, and dephasing from flux noise is reported [17]. The Bifluxon protected qubit design is due to the scientists **Konstantin Kalashnikov, Wen Ting Hsieh, Wenyuan Zhang, Wen-Sen Lu, Plamen Kamenov, Agustin Di Paolo, Alexandre Blais, Michael E. Gershenson**, and **Matthew Bell**. This introductory section is an extract of the original research

Fig. 9.34 Codewords. (a) Codewords pertaining to the eigenvalues $Z = \pm 1$. (b) Codewords pertaining to the eigenvalues $X = \pm 1$

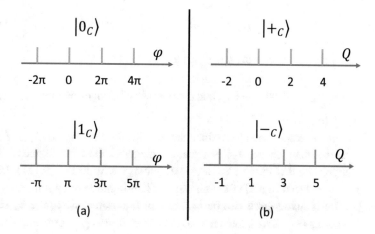

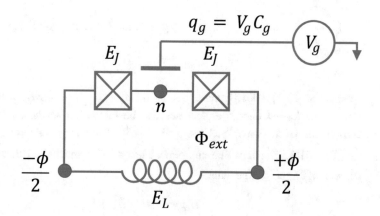

Fig. 9.35 The schematic of the bifluxon qubit. (Reference: [17])

paper. In this section, we establish that the two degrees of freedom of the qubit design can be used to protect against charge and flux decay and dephasing simultaneously.

9.6.2.1 Qubit Physics

Consider the circuit shown in Fig. 9.35 [17]. The circuit consists of a Cooper-pair box (CPB) shunted by a superinductor, thus forming a superconducting loop. Assume that the Josephson energy E_J and charging energy E_C are associated with the CPB. Let E_{CL} denote the charging energy of the superinductance and let $E_L = (\Phi_0/2\pi)^2/L$ be its inducive energy. Here, $\Phi_0 = h/2e$ is the magnetic flux quantum. The device can be controlled by two parameters: the offset charge on the superconducting island q_g or the external magnetic flux Φ_{ext}. The two control parameters of the circuit are described by a 2D Hamiltonian. This helps in simultaneous suppression of decay and dephasing via localization of the qubit's wavefunctions in disparate regions of the phase space.

The Hamiltonian of this circuit has three degrees of freedom, namely, the superconducting phase φ of the CPB island, the sum ϕ_+, and the difference ϕ_- of the phases at the leads of the superconducting loop. Like the $0 - \pi$ qubit, at high circuit frequencies ϕ_+ is not excited. Hence, the Hamiltonian can be written in terms of the remaining degrees of freedom. In charge basis, the Hamiltonian is [17]:

$$H = \sum_n \left[4E_C(n - n_g)^2 |n\rangle\langle n| \right] - E_J \cos\left(\frac{\phi}{2}\right)(\sigma_n^+ + \sigma_n^-)$$
$$-4E_{CL}\partial_\phi^2 + \frac{E_L}{2}(\phi - \varphi_{ext})^2,$$

(9.117)

where $\varphi_{ext} = \frac{2\pi\Phi_{ext}}{\Phi_0}$, $n_g = \frac{q_g}{2e}$, n is the number of Cooper pairs in the CPB island, $\sigma_n^+ = |n+1\rangle\langle n|$, and $\sigma_n^- = (\sigma_n^+)^\dagger$.

For the rest of the discussions, let us consider the regime $E_C \gg E_J$.

If the offset charge n_g is set near an integer N, the CPB degree of freedom is "frozen" close to the charge state that minimizes the kinetic-energy term in Eq. (9.117). In this case, the circuit Hamiltonian is reduced to a 1D fluxonium-like Hamiltonian with a normalized Josephson energy $E_J^2/4E_C$.

The bifluxon qubit can be operated in the protected regime by setting the offset charge as half integer $n_g \approx \frac{1}{2}$. If we consider a 2D subspace $|0\rangle$ and $|1\rangle$ corresponding to the computational basis for the CPB, the Hamiltonian in this protected 2D subspace is:

$$H_r = 4E_C\left(\frac{1}{2} - n_g\right)\sigma^z - E_J\cos\left(\frac{\phi}{2}\right)\sigma^x$$
$$-4E_{CL}\partial_\phi^2 + \frac{E_L}{2}(\phi - \varphi_{ext})^2,$$

(9.118)

where $\sigma^z = |1\rangle\langle 1| - |0\rangle\langle 0|$ and $\sigma^x = \sigma^+ + \sigma^-$. If $n_g = \frac{1}{2}$, the above equation is diagonal in the σ^x basis. Hence, the lowest energy eigenstates can be factorized as the tensor product of charge and flux components of the wavefunctions $|\psi^n\rangle \otimes |\psi^\phi\rangle$. The charge-like component results in states $|\pm^n\rangle = \frac{1}{\sqrt{2}}(|0\rangle + |1\rangle)$. The flux like component has a 1D Hamiltonian that has a potential energy that is dependent on the charge state.

$$H_\pm = -4E_{CL}\partial_\phi^2 + V_\pm$$
$$V_\pm = \mp E_J\cos\left(\frac{\phi}{2}\right) + \frac{E_L}{2}(\phi - \varphi_{ext})^2.$$

(9.119)

The local minima of this fluxonium[3] like potential are centered around $\phi_m = 2\pi m$, where m is an integer, which is even for V_+ and odd for V_-. The wavefunction of the fluxon[4] excitation can be written as follows, which is essentially a harmonic oscillator.

$$\psi_m(\phi) \cong \exp\left(-\frac{1}{4}\sqrt{\frac{E_J}{E_{CL}}}(\phi - \phi_m)^2\right)$$

(9.120)

The wavefunction localized at the mth minimum can be associated with a fluxon $|m\rangle$. In the fluxon representation, the eigenstates of Eq. (9.118) can be written as $|m\rangle = \{|2k\rangle\} \cup \{|2k+1\rangle\}$, where $|2k\rangle = |+^n, \psi_{2k}^\phi\rangle$ and $|2k+1\rangle = |-^n, \psi_{2k+1}^\phi\rangle$ are the even and odd number of fluxons that may be present in the loop.

---

[3] Fluxoniums are anharmonic artificial atoms which use an array of Josephson Junctions to shunt the CPB for protection against charge noise.

[4] Fluxons are the quantum of electromagnetic flux. Fluxons are formed as a small whisker of normal phase surrounded by superconducting phase in a type-ii superconductor when the magnetic field is between H_{C1} and H_{C2}. Supercurrents flow around the center of a fluxon.

The fluxon energy bands of neighboring fluxons intersect at half-integer intervals of $\frac{\varphi_{ext}}{2\pi}$. This essentially preserves the fluxon-parity. Therefore, when $n_g = \frac{1}{2}$, the design is a fluxonium qubit made up of a 4π-periodic Josephson element. Hence, the qubit is called "bifluxon."

9.6.2.2 Protection

The symmetry of the wavefunction forbids the single phase-slip processes (SPS) connecting $|m\rangle$ and $|m+1\rangle$. The double phase-slip (DPS) processes mix fluxon states with the same parity $|m\rangle$ and $|m+2\rangle$ opening energy gaps in the spectrum.

$$E_{sps} = \langle m|H_r|m+1\rangle \propto \langle +^n|-^n\rangle = 0$$
$$E_{dps} = \langle m|H_r|m+2\rangle = \hbar\sqrt{8E_JE_{CL}}\exp\left(-\pi^2\sqrt{2E_J/E_{CL}}\right) \tag{9.121}$$

The symmetry of states with distinct fluxon parity makes the qubit immune to energy decay due to both flux and charge noises. The researchers [17] observed a tenfold increase of the decay time, up to 100 μs by turning on the protection.

The weak sensitivity to charge noise can be suppressed by stronger localization of the single-well excitations by increasing the E_J/E_{CL} ratio. The charge–noise-induced dephasing time in the protected state was found in excess of 1 μs [17].

The flux dispersion of the qubit can be reduced by increasing the superinductance value. This enables wider delocalization of the qubit wavefunctions with different fluxon parities in disjoint subspaces. Hence, the bifluxon qubit protects against flux-noise dephasing.

By design, the bifluxon is a charge-sensitive device, a single qubit does not offer a protection against the charge–noise-induced dephasing. A small array of bifluxon qubits can provide a polynomial increase of the dephasing time and help to overcome this limitation.

9.6.3 Parity Protected Superconductor–Semiconductor Qubit

A parity-protected qubit design based on voltage-controlled semiconductor nanowire Josephson junctions was published recently. This qubit design is due to **T. W. Larsen, M. E. Gershenson, L. Casparis, A. Kringhoj, N. J. Pearson, R. P. G. McNeil, F. Kuemmeth, P. Krogstrup, K. D. Petersson**, and **C. M. Marcus** [18]. The researchers created a $\cos 2\varphi$ element using a pair of gate-tunable semiconductor JJs based on In As nanowires grown with epitaxial superconducting Al. The proposed qubit design uses two tunable high transmission Josephson junctions in a SQUID configuration to realize a parity-protected qubit.

9.6.3.1 Qubit Design

Consider the qubit circuit [18] shown in Fig. 9.36, the geometry of which resembles transmons. The design has a superconducting island with two Josephson junctions in a SQUID configuration. The charging energy E_C connects the SQUID to the ground. The electrostatic gate voltages V_k $\{k = 1, 2\}$ can tune the Josephson junction transmission by modulating the electron density in the junction region. The magnetic flux is controlled by changing the current through a nearby shorted transmission line. The microwave excitations are driven using an open transmission line. The JJs are modeled in the short-junction regime.[5] The Josephson coupling is mediated by a number ($i = 1, 2, \ldots$) of

---

[5] In the short-junction regime, the junction length $L \ll \xi$, the superconducting coherence length.

Andreev bound states (ABS),[6] each characterized by a transmission coefficient $T_i^{(k)}$. The energy-phase relation of each JJ can be calculated by summing over the i energies of the bound states.

$$U_k(\varphi_k) = -\Delta \sum_i \sqrt{1 - T_i^{(k)} \sin^2(\varphi_k/2)} \ , \tag{9.122}$$

where Δ is the superconducting gap, and φ_k is the superconducting phase difference across the kth JJ. The total Hamiltonian of the qubit is:

$$H = -4\,E_C \hat{n}^2 - U_1(\varphi) - U_2\left(\varphi - 2\pi\frac{\Phi}{\Phi_0}\right) , \tag{9.123}$$

where Φ is the applied magnetic flux through the SQUID, and Φ_0 is the superconducting flux quantum.

When the applied magnetic flux is one half of the flux quantum (that is, $\Phi = \frac{\Phi_0}{2}$), the terms $-U_1(\varphi) - U_2\left(\varphi - 2\pi\frac{\Phi}{\Phi_0}\right)$ are suppressed, leaving $\cos 2\varphi$ and its higher harmonics. This results in a qubit with a π periodic potential with coherent transport across the SQUID occurring in pairs of Cooper pairs ($4e$). Under this condition, the SQUID forms a symmetric double-well potential due to the higher harmonics in the energy-phase relation. When the asymmetry between the JJs is increased, the coupling between the potential wells is increased. This results in a potential like that of a flux qubit. When the asymmetry is strong, (that is, when $\frac{U_2(\varphi)}{U_1(\varphi)} \to 0$), the potential is a single well and resembles the transmon.

Fig. 9.36 Schematic of a parity protected qubit. (Reference: [18])

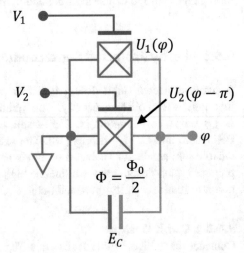

V_1

$U_1(\varphi)$

V_2

$U_2(\varphi - \pi)$

φ

$\Phi = \dfrac{\Phi_0}{2}$

E_C

---

[6] Andreev Bound States (ABS) are topologically trivial near-zero energy excitations. These states exhibit behavior similar to Majorana Bound States (MBS), and arise in superconductor–semiconductor nanowires in the presence of a smooth varying chemical potential.

9.6.3.2 Protection

The qubit can be tuned into the protected qubit regime by adjusting the gate voltages to balance the two junctions such that single-Cooper-pair transport across the SQUID is suppressed, forming a double-well potential with minima separated by $\varphi \sim \pi$. In this balanced configuration, energy states are strongly localized to each of the wells. This demonstrates the protection of coherence in the symmetric regime.

At $\Phi = \frac{\Phi_0}{2}$, the potential forms a double-well potential with minima at $\varphi \sim \frac{\pi}{2}$ with two nearly degenerate ground states in the charge basis. This clearly shows the separation in parity with either odd or even numbers of Cooper pairs, protecting the 4e-parity states of the qubit.

The researchers found a tenfold suppression of the relaxation in the protected regime [18].

9.6.4 Summary

This section came across some novel qubit designs that took existing designs and improved noise suppression by adding new circuit elements. Further research may be required to stabilize the designs. However, these emerging qubit designs have improved fault-tolerance at the physical design. Undoubtedly, they provide the path toward fault-tolerant quantum computing.

Practice Problems

1. Can a quantum system be simulated by a probabilistic classical computer? Explain your answer in terms of space and time complexity.
2. A hamming code encodes 1 bit of information into 2 bits. Generate the codeword.
3. Verify whether the code corresponding the following generator matrix is self-dual.

$$G = \begin{bmatrix} 1 & 0 & 0 & 1 & \omega & \omega \\ 0 & 1 & 0 & \omega & 1 & \omega \\ 0 & 0 & 1 & \omega & \omega & \omega \end{bmatrix}$$

4. Derive the minimum hamming distance for the code described by the following generator matrix.

$$G = \begin{bmatrix} 1 & 0 & 0 & 0 & 0 & 1 & 1 \\ 0 & 1 & 0 & 0 & 1 & 0 & 1 \\ 0 & 0 & 1 & 0 & 1 & 1 & 0 \\ 0 & 0 & 0 & 1 & 1 & 1 & 1 \end{bmatrix}$$

5. An electromagnetic interference accidentally applied an H gate to a qubit. Use the any of the error correction codes to correct the error.
6. Explain whether Shor's 9-qubit error correction code is degenerate or nondegenerate. Verify the same for Steane's code.
7. Develop a classical code that implements the surface cycle on a 16-qubit backend, which you can logically wire on a 2D lattice.
8. Establish that it does not matter whether a qubit error occurs before or after a CNOT gate.
9. Compare the effects of a qubit error followed by a quantum gate with that of the quantum gate followed by a qubit error.

10. Develop a concept how concatenated circuits and quantum repeaters can be used to transmit quantum information over large distances. What is the error probability anticipated in your design?
11. Which of the quantum gates would you implement in software? Why?
12. Create a logical qubit, using any of the methods described in this chapter using an available backend.
13. Implement a fully fault tolerant Toffoli gate using the methods described in this chapter.

References

1. Quantum Computation and Quantum Information: 10th Anniversary Edition, Nielsen, and I Chuang.
2. On the role of entanglement in quantum computational speed-up, Jozsa and Linden, 8-Mar-2002, https://arxiv.org/pdf/quant-ph/0201143.pdf
3. On the Significance of the Gottesman-Knill Theorem, Michael E. Cuffaro, http://philsci-archive.pitt.edu/12869/1/sufficiency_of_entanglement.pdf
4. A Class of Quantum Error-Correcting Codes Saturating the Quantum Hamming Bound, Daniel Gottesman, 24-Jul-1996. https://arxiv.org/pdf/quant-ph/9604038.pdf
5. Stabilizer Codes and Quantum Error Correction, Daniel Gottesman, 28-May-1997, https://arxiv.org/pdf/quant-ph/9705052.pdf
6. Fault-Tolerant Quantum Computation with Higher-Dimensional Systems, Daniel Gottesman, 2-Feb-1998, https://arxiv.org/pdf/quant-ph/9802007.pdf
7. A Theory of Fault-Tolerant Quantum Computation, Daniel Gottesman, 18-Feb-1997, https://arxiv.org/pdf/quant-ph/9702029.pdf
8. Quantum Error Correction for Beginners, Devitt et al., 21-Jun-2013, https://arxiv.org/pdf/0905.2794.pdf
9. From Classical to Quantum Information, Dan C. Marinescu and Gabriela M. Marinescu, March 30, 2010
10. Surface codes: Towards practical large-scale quantum computation, Austin G. Fowler et al., October 26, 2012, https://arxiv.org/ftp/arxiv/papers/1208/1208.0928.pdf
11. Introduction to Topological Quantum Computation, Jiannis K. Pachos
12. Projective Plane And Planar Quantum Codes, Michael H. Freedman, David A. Meyer, https://arxiv.org/pdf/quant-ph/9810055.pdf
13. Quantum codes on a lattice with boundary, Sergey B. Bravyi, Alexei Yu. Kitaev, https://arxiv.org/pdf/quant-ph/9811052.pdf
14. Superconducting Qubits: Current State of Play, Morten Kjaergaard et al., 21-April-2020, https://arxiv.org/pdf/1905.13641.pdf
15. Technical Roadmap for Fault-Tolerant Quantum Computing, Amir Fruchtman and Iris Choi, October 2016, https://nqit.ox.ac.uk/content/technical-roadmap-fault-tolerant-quantum-computing
16. Protected gates for superconducting qubits, Peter Brooks, Alexei Kitaev, John Preskill, 17-Feb-2013, https://arxiv.org/abs/1302.4122v1
17. Bifluxon: Fluxon-Parity-Protected Superconducting Qubit, Konstantin Kalashnikov et al., 9-Oct-2019, https://arxiv.org/pdf/1910.03769.pdf
18. A Parity-Protected Superconductor-Semiconductor Qubit, T. W. Larsen et al., 31-Jul-2020, https://arxiv.org/abs/2004.03975v2

Conclusion

10

"This here young lady," said the Gryphon, "she wants for to know your history, she do."
"I'll tell it her," said the Mock Turtle in a deep, hollow tone: "sit down, both of you, and don't speak a word till I've finished."
So they sat down, and nobody spoke for some minutes. Alice thought to herself, "I don't see how he can EVER finish, if he doesn't begin." But she waited patiently.

— Lewis Carroll, *Alice's Adventures in Wonderland*

From the definition of matter to creating surface codes and designing protected physical qubits, we have traveled quite so far. We started this journey by asking a question, and we should probably end this journey by answering some. Our journey will be unfinished if we do not find the answers to the most frequently asked questions! The top three questions that come to our mind are—How many qubits do we need? Can fault-tolerant quantum computers of commercial-grade be ever built? What does the future hold? We have witnessed many technological revolutions, and we know that classical computing did not reach its peak overnight. It took more than seven decades for classical computing to reach the current level. The first transistor was about our palm's size, and we did not build a supercomputer with that. The first commercial device that rolled out with the transistor was a hearing aid. With quantum computing, we are somewhat in an early transistor-like period. However, the great thing is that a handful of quantum algorithms is already established, and there are quantum devices currently available to experiment with. According to a 2019 interview statistics [1], about 110,000 users have run about 7 million experiments on the IBM Quantum Experience publishing about 145 research papers based on the research done on the devices. Given the complexity of the subject, this outreach is phenomenal. With such large-scale research and commercial interest, it is natural that this technology receives much endowment, accelerating further research and early commercialization.

In this conclusion, we relook at the promise of quantum supremacy and the answer to the most frequently asked questions.

10.1 How Many Qubits Do We Need?

Shor's factoring algorithm is one extreme application for quantum computing. While it may be fascinating to break cybersecurity, primetime quantum computing has several peaceful applications that can help humanity—for example, molecular synthesis. The original idea of Feynman was to use

V. Kasirajan, *Fundamentals of Quantum Computing*, https://doi.org/10.1007/978-3-030-63689-0_10

Table 10.1 Qubits needed for quantum applications

Application	Number of qubits required
Proof of concept	Available today
Secret sharing	A few long-lived qubits
Quantum co-processor	A few error-corrected qubits
Quantum chemistry	≈ 150 error-corrected qubits
Quantum supremacy	50 error-corrected qubits
Machine learning	A few hundred error-corrected qubits
A full fault-tolerant quantum computer	1,000,000 ++ error-corrected qubits

quantum computers to simulate quantum systems. Simulating the molecular structure of experimental drugs could be an extreme task for the supercomputers, as the variables grow exponentially with the number of electrons in each molecule. Quantum computers do not suffer from this fate because the internal state of a quantum system itself is exponentially large. Such applications have varying qubit requirements—from a few qubits to a few hundred qubits. Listed in Table 10.1 is a study [2] done by researchers at the University of Oxford. Table 10.1 provides the minimum number of qubits required for some of the application areas, according to the study.

From Table 10.1, we can guess that the existing technology has the potential to be used in many promising application areas. It is no wonder 110,000 researchers ran about 7 million experiments! We cannot dismiss the thought that some of those experiments may have a near-commercial potential.

Let us return to Shor's algorithm for a while and evaluate what it takes to implement it.

To factorize a 2000-bit number (600 decimal digits) using the Shor's algorithm, we need about 4000 qubits. The 4000 logical qubits require $14500 \times 4000 = 58$ million physical qubits. The heart of Shor's algorithm is the modular exponentiation. According to one estimate, 2.2×10^{12} states are required by this circuit. Implementing this in the surface code requires about 800,000 physical qubits. If we assume each state requires 100 μs to process, it takes about 26.7 hours to perform the modular exponentiation [3]. Note that these estimates do not consider reuse of qubits and potential optimization that could be done by transpilers.

Improving the error rate reduces the number of physical qubits required. For example, if we reduce the error rate by 1/10th, the number of physical qubits required is about 130 million, although the execution time remains the same. For a detailed calculation, please see Appendix M, **Austin G. Fowler** [3].

We should also consider how fast we can perform the surface code using classical systems. High-fidelity (99.7%) two-qubit gates with 60 ns gate timings have been reported recently [4]. If we assume a gate time in the range of 10–100 ns, a surface code cycle at about 200 ns is a good target for the current classical systems. Slower cycles will result in a higher qubit overhead. For example, if the cycle time is 2 ms, factoring a 600-digit prime number will take 30 years! [3].

Another challenge is implementing the high-speed interconnects to the qubits. The cabling requirements restrict the density of the qubits on an implement. In recent times, some progress has been made in this area. An ultra-low-power CMOS-based quantum control platform that takes digital commands as input and generates many parallel qubit control signals, removing the need for separate control lines for every qubit has been reported [5].

With superconducting or semiconductor-based qubits, with the technology available today, we should be able to fabricate a qubit in a cavity of 100 μm. If we can achieve this, implementing a $10^4 \times 10^4$ array of physical qubits needs a physical wafer of area 1 m^2 [3], which can easily fit in a dilution fridge! IBM Quantum Experience is edging forward with announcements of new generations of quantum systems of the future. IBM's future roadmap includes a promising growth of qubit count

Table 10.2 The number of qubits simulated by some of the TOP 500 classical systems today

System	Memory (PB)	Number of qubits
TACC Stampede	0.192	43
Titan	0.71	45
K Computer	1.4	46
APEX-2020	4–10	48–49

Acknowledgments: Mikhail Smelyanskiy, Nicolas P. D. Sawaya, and Alán Aspuru-Guzik

shortly: Eagle platform - 127 qubits by 2021, Osprey - 133 qubits by 2022, Condor - 1121 qubits by 2023, and 1K to 1M+ qubits moving forward through the year 2026 [14].

Therefore, we do have the technology that can scale-up soon.

10.2 Classical Simulation

We learned that quantum states could be efficiently simulated in the classical systems when we discussed the Gottesman-Knill theorem. The natural question that comes to our mind is that if classical computers can simulate quantum states, why all the trouble of creating error-prone qubits? Could we run (at least some of the) quantum algorithms on classical simulators? The answer to this question is evident from Table 10.2 [6]. Table 10.2 illustrates the number of qubits that some of the TOP 500 classical computers available today can simulate.

From Table 10.2, we can infer that the amount of available memory limits the number of qubits that can be simulated in a classical system, and it grows exponentially as more qubits are added. The above numbers were taken with double-precision arithmetic. If more precision is required, the number of qubits that can be simulated reduces considerably. On a quantum computer, the precision depends on the instrumentation used to setup the rotational pulses and the measurement circuits. Also, quantum computers can efficiently perform multi-qubit operations than simulators. Though we can simulate quantum systems classically, it is not possible to scale that beyond a point. This threshold is the operating ground where quantum supremacy sets in, where quantum computers outperform classical computers.

10.3 Backends Today

Some of the backends available at this time of writing in 2020 have higher degrees of reliability and repeatability [7]. Furthermore, several application areas promoting the usage of backends with real-time data have been identified and published by organizations [8]. Such use cases can be tried by anyone today on the available backends [9]. The tools and development frameworks are superior, easy to use, and keep improving [10]. IBM Quantum Experience has recently announced enhancements such as mid-circuit measurement, dynamic circuits that can implement the much-awaited "if" statement, and a new quantum runtime, reducing the reentry queue wait-time. In December 2020, an IBM engineer indicated that these enhancements helped cut down the total execution time of a Lithium-Hydride chemical simulation to 16 hours from the previous benchmark of 111 days [14]. All these possibilities are augmented by advanced technology research happening at universities worldwide. In 2019 alone, 5,558 articles were published in arXiv on quantum physics [11]. These advancements enable the NISQ processors available today to open the space for quantum computing truly.

10.4   Future State

While quantum computing will be making a significant impact in the society, harnessing its disruptive potential will take time until scalable systems are available. For the foreseeable future, small and medium enterprises may not be able to buy quantum computers and install them in their campuses, unless they have some mission-critical applications. Most organizations may be buying the computing time from a cloud provider such as Microsoft Azure Quantum [12] or the Amazon Braket [13].

The first quantum computing applications will probably be in the area of quantum communications and post-quantum cryptography. There are many areas classical computers are good at, such as most current enterprise applications, transactional databases, long-term storage, social media, and the internet. Quantum computers will not replace classical computers in these areas. Nevertheless, quantum computers have the potential to function as co-processors serving as the computing workhorse beneath several such applications.

An important application is to use quantum computing for quantum chemistry. Quantum chemistry has some critical use cases in molecular synthesis, which can benefit humanity by discovering new drugs, inventing new materials, and understanding chemical reactions that occur in nature. These are hard problems even for the supercomputers of date. The number of variables grows exponentially that classical computers cannot model even the simplest of the molecules.

While classical algorithms are improving, quantum computers can perform well in specific vital applications such as signal processing, combinatorial optimizations, financial modeling, and artificial intelligence. Research organizations are putting much work in developing solutions in these areas. Most software algorithms we use today were invented as classical computers became available and affordable. Similarly, many quantum algorithms are waiting to be discovered. The possibility always exists that some of our readers may be able to discover new quantum procedures that can cleverly solve classical problems, establishing quantum supremacy.

Such opportunities are available in plenty to the creative minds of our readers!

Now this is not the end. It is not even the beginning of the end. But it is, perhaps, the end of the beginning.

— *Sir Winston Churchill, "The end of the beginning", November 10, 1942.*

References

1. https://singularityhub.com/2019/02/26/quantum-computing-now-and-in-the-not-too-distant-future/
2. qHiPSTER: The Quantum High Performance Software Testing Environment, Mikhail Smelyanskiy et al, 12-May-2016, https://arxiv.org/pdf/1601.07195.pdf
3. Surface codes: Towards practical large-scale quantum computation, Austin G.Fowler et al, October 26, 2012, https://arxiv.org/ftp/arxiv/papers/1208/1208.0928.pdf
4. Superconducting Qubits: Current State of Play, Morten Kjaergaard et al, 21-April-2020, https://arxiv.org/pdf/1905.13641.pdf
5. A Cryogenic Interface for Controlling Many Qubits, S. J. Pauka et al, 3-Dec-2019, https://arxiv.org/pdf/1912.01299.pdf
6. Technical Roadmap for Fault-Tolerant Quantum Computing, Amir Fruchtman and Iris Choi, October 2016, https://nqit.ox.ac.uk/content/technical-roadmap-fault-tolerant-quantum-computing
7. For recent claims of high performing quantum computers, please refer to (a) https://quantumcomputingreport.com/qubit-quality/. (b) https://www.honeywell.com/en-us/newsroom/news/2020/06/the-worlds-highest-performing-quantum-computer-is-here. (c) https://newsroom.ibm.com/2020-08-20-IBM-Delivers-Its-Highest-Quantum-Volume-to-Date-Expanding-the-Computational-Power-of-its-IBM-Cloud-Accessible-Quantum-Computers
8. https://quantumcomputingreport.com/a-quantum-computing-application-roadmap-from-ibm/
9. https://quantumcomputingreport.com/qubit-implementation-dashboard/

10. Overview and Comparison of Gate Level Quantum Software Platforms, Ryan LaRose, 20-Mar-2019, https://arxiv.org/pdf/1807.02500.pdf
11. https://arxiv.org/year/quant-ph/19
12. https://azure.microsoft.com/en-us/services/quantum/
13. https://aws.amazon.com/braket/
14. https://quantumcomputingreport.com/ibm-introduces-new-capabilities-for-their-quantum-machines/

Appendix

Appendix

Braid words

By convention, the $(i)^{\text{th}}$ string passing below the $(i+1)^{\text{th}}$ string is labelled σ_i^{-1}. See Fig. 11.1 for an example.

Aharonov–Bohm effect

Aharonov–Bohm effect is often called the seventh wonder of quantum mechanics. Aharonov–Bohm effect is a quantum mechanical phenomenon, in which a particle experiences electromagnetic field, even when traveling in a space where electric and magnetic potentials are zero. This effect can be used to measure extremely small variations in magnetic flux.

Implementing U1, U2, and U3 rotations in QDK:

These three methods are not directly supported in the QDK. Use the following equations to implement them in QDK.

$$U_1(\lambda) = R_z(\lambda)$$

$$U_2(\phi, \lambda) = R_z(\phi)R_y\left(\frac{\pi}{2}\right)R_z(\lambda)$$

$$U_3(\theta, \phi, \lambda) = R_z(\phi)R_x\left(-\frac{\pi}{2}\right)R_z(\theta)R_x\left(\frac{\pi}{2}\right)R_z(\lambda)$$

Implementing controlled gates in QDK:

To implement controlled gates in QDK, use the "*Controlled*" functor. An example is given below. This example implements a controlled-U1 gate with qubit q[0] as the control qubit and q[1] as the target qubit. For more details, refer to the link: https://docs.microsoft.com/en-us/quantum/user-guide/using-qsharp/operations-functions

```
Controlled R ( [q[0]], (PauliZ, PI()/2.0, q[1]) );
```

The HHL algorithm (Listing 11.8) illustrates implementing U3 and CU3 gates using the QDK.

Fig. 11.1 Sample braid
with the strings passing
below each other

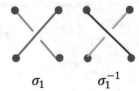

$\sigma_1 \qquad \sigma_1^{-1}$

Code example to perform QFT using QDK (Sect. 7.2.3) (Listing 11.1):

```
namespace Qft {
    open Microsoft.Quantum.Canon;
    open Microsoft.Quantum.Intrinsic;
    open Microsoft.Quantum.Arrays;
    open Microsoft.Quantum.Math;
    open Microsoft.Quantum.Convert;

    // Swap registers
    // q - array of qubits
    // n - total number of qubits in the array
    operation SwapReg(q:Qubit[], n:Int ):Unit {
        for (i in 0..(n/2)-1)
        {
            SWAP(q[i], q[n-i-1]);
        }
    }

    // Sets the qubits with a binary string
    // q - array of qubits
    // n - total number of qubits in the array
    // input - the data to be set
    operation SetReg(q:Qubit[], n:Int, input:Int ):Unit {
        for (i in 0..BitSizeL(IntAsBigInt(input)) )
        {
            if ( 0 != ( ( 1 <<< i) &&& input ) )
            {
                X(q[n-i-1]);
            }
        }
    }

    // Constructs the forward QFT circuit
    // q - array of qubits
    // m - total number of qubits in the array
    operation qft_cct (q:Qubit[], m:Int):Unit {
        mutable i = 0;
        mutable n = m;
```

Listing 11.1   QFT using QDK

```
        repeat {
            H(q[i]);
            set n = n - 1;
            set i = i + 1;

            for ( qubit in 0..n-1){
                Controlled R([q[qubit + i]], (PauliZ, PI() /
PowD(2.0,
                        IntAsDouble(qubit + 1 )), q[i-1]));
            }
        } until (n == 0);

}


// Performs the QFT
// q - array of qubits
// n - total number of qubits in the array
// input - Input binary number to be QFTed
operation qft (q:Qubit[], n:Int, input:Int):Unit {
    SetReg(q,n, input);
    qft_cct(q,n);
    SwapReg(q,n);
}


// Performs inverse of the QFT
// q - array of qubits
// m - total number of qubits in the array
operation qft_inv (q:Qubit[], n:Int):Unit {
    for ( i in RangeReverse(0..n-1)){
        for (qubit in i..n-2 ){
            Controlled R([q[qubit + 1]], (PauliZ, -PI() /
PowD(2.0,
                            IntAsDouble(qubit + 1 - i )), q[i]));
        }
        H(q[i]);
    }
}


// Code to test the QFT by performing IQFT and comparing
// input with output.
// input - input number for processing
operation testqft(input:Int):Int {
    mutable r = 0;
    mutable n = BitSizeL(IntAsBigInt(input));
```

Listing 11.1 (continued)

```
            // Assume a minmum number of qubits to use
            if (  n < 4)
            {
                set n = 4;
            }

            using ( q = Qubit[n] ){
                qft(q, n, input );
                SwapReg(q,4);
                qft_inv(q, n);
                SwapReg(q,4);
                for ( i in 0..n-1)
                {
                    if (One == M(q[i])) { set r = r ||| (1 <<< i); }
                }
                ResetAll(q);
            }
            Message($"Input: {input} Measured:  {r} ");
            return r;
        }
}
```

Listing 11.1 (continued)

Code example for Simon's problem in QDK (Sect. 7.5) (Listing 11.2):

```
namespace Simons {
    open Microsoft.Quantum.Canon;
    open Microsoft.Quantum.Intrinsic;
    open Microsoft.Quantum.Arrays;
    open Microsoft.Quantum.Math;
    open Microsoft.Quantum.Convert;

    // Simon's Algorithm for hidden string 3
    operation Simons (q:Qubit[], a:Int): Int {

        mutable n = Length(q)/2;
        mutable reg1 = 0;
        mutable reg2 = 0;

        // Put the first reg in equal superpoision state.
        for ( i in 0..n-1)
        {
            H(q[i]);
        }

        // Build the oracle function
        for (i in 0..n-1){
            if ( 0!= (( 1 <<< i) &&& a)){
                for ( j in 0..n-1){
                    CNOT(q[i], q[j+n]);
                }
            }
        }

        // Measure REG 2
        for ( i in n..Length(q)-1){
            if (One == M(q[i])) { set reg2 = reg2 ||| (1 <<< i); }
        }

        // Apply H transform to REG 1
        for ( i in 0..n-1)
        {
            H(q[i]);
        }

        // Measure REG 1
        for ( i in 0..n-1){
            if (One == M(q[i])) { set reg1 = reg1 ||| (1 <<< i); }
        }

        // Message($"Simons input:  {a}, Measured: {reg1} {reg2}");
```

Listing 11.2 Simon's problem

```
            ResetAll(q);

            return reg1;
      }

      // Code to test the Simon's algorithm.
      operation testSimons():Unit{
          mutable n = 6;  // Number of qubits
          mutable a = 3;  // Hidden string
          mutable res = new Int[8];
          mutable r = 0;
          mutable shots = 100;

          Message($"Test Simons");

          using (q=Qubit[n])
          {
              for ( i in 0..shots-1){
                  set r = Simons(q,a);
                  set res w/= r <- res[r] + 1;
              }
              // Print the number of hits for each output states
              Message($"Simons {res}");
          }
      }
}
```

Listing 11.2 (continued)

QDK Code example for the simple QFT Adder in Sect. 7.6.1 (Listing 11.3):

```
namespace QftAdder {
    open Microsoft.Quantum.Canon;
    open Microsoft.Quantum.Intrinsic;
    open Microsoft.Quantum.Arrays;
    open Microsoft.Quantum.Math;
    open Microsoft.Quantum.Convert;

    // Swap registers
    // q - array of qubits
    // n - total number of qubits in the array
    operation SwapReg(q:Qubit[], n:Int ):Unit {
        for (i in 0..(n/2)-1)
        {
            SWAP(q[i], q[n-i-1]);
        }
    }
```

Listing 11.3 Simple QFT Adder for the QDK

```
// Sets the qubits with a binary string
// q - array of qubits
// n - total number of qubits in the array
// input - the data to be set
operation set_reg(q:Qubit[], n:Int, input:Int ):Unit {
    for (i in 0..BitSizeL(IntAsBigInt(input)) )
    {
        if ( 0 != ( ( 1 <<< i) &&& input ) )
        {
            X(q[i]);
        }
    }

}


// Adder circuit
// adds a to b in the Fourier basis
operation add_cct (a:Qubit[], b:Qubit[], m:Int):Unit {
    mutable n = m;
    repeat {
        for (i in RangeReverse(1..n)){
            Controlled R([a[i-1]], (PauliZ, PI() / PowD(2.0, \
                         IntAsDouble(n-i )), b[n-1]));
        }
        set n -= 1;
    } until (n == 0);
}


// Inverse QFT
operation iqft_cct (t:Qubit[], n:Int):Unit {
    for ( i in 0..n-1 ){
        for ( j in 1..i) {
            Controlled R([t[j-1]], (PauliZ, -PI() / PowD(2.0, \
                IntAsDouble(i-j+1 )), t[i]));
        }
        H(t[i]);
    }
}


// Forward QFT
operation  qft_cct (t:Qubit[], m:Int):Unit {
    mutable n = m;
    repeat{
        H(t[n-1]);
```

Listing 11.3 (continued)

```
                for (i in RangeReverse(1..n-1)){
                    Controlled R([t[i-1]], (PauliZ, PI() / PowD(2.0, \
                            IntAsDouble(n-i )), t[n-1]));
            }
            set n -= 1;
        } until (n == 0);
    }


    // QFT based adder
    // b = a + b
    //
    operation add_qft( input1:Int, input2:Int, a:Qubit[], b:Qubit[],
n:Int): Int {
        mutable r = 0;

        set_reg(a, n, input1);
        set_reg(b, n, input2);

        qft_cct(b, n);
        add_cct(a, b, n);
        iqft_cct(b, n);

        for ( i in 0..n-1)
        {
            if (One == M(b[i])) { set r = r ||| (1 <<< i); }
        }

        return r;
    }


    // Testing code
    operation test_qftadder():Unit{
        mutable r = 0;
        mutable input1 = 7;
        mutable input2 = 8 ;

        using ( a = Qubit[4] ) {
            using( b = Qubit[4] ) {
                set r = add_qft(input1, input2, a, b, 4);
                Message($"Inputs: {input1} + {input2} = output:
{r}");

                ResetAll(a);
                ResetAll(b);
            }
        }
    }
}
```

Listing 11.3 (continued)

QDK Sample code for the QFT Multiplier Sect. 7.6.3 (Also contains the controlled QFT Adder in Sect. 7.6.2) (Listing 11.4):

```
namespace QftMultiplier {
    open Microsoft.Quantum.Canon;
    open Microsoft.Quantum.Intrinsic;
    open Microsoft.Quantum.Arrays;
    open Microsoft.Quantum.Math;
    open Microsoft.Quantum.Convert;


    // Swap registers
    // q - array of qubits
    // n - total number of qubits in the array
    operation SwapReg(q:Qubit[], n:Int ):Unit {
        for (i in 0..(n/2)-1)
        {
            SWAP(q[i], q[n-i-1]);
        }
    }


    // Sets the qubits with a binary string
    // q - array of qubits
    // n - total number of qubits in the array
    // input - the data to be set
    operation set_reg(q:Qubit[], n:Int, input:Int ):Unit {
        for (i in 0..BitSizeL(IntAsBigInt(input)) )
        {
            if ( 0 != ( ( 1 <<< i) &&& input ) )
            {
                X(q[i]);
            }
        }
    }


    //
    // Controlled Addition and Multiplication
    //


    // Doubly controlled U1 Gate
    // theta - angle of rotation
    // q0 and q1 - the two control qubits
    // q2 - target qubit
    operation ccu1(theta:Double, q0:Qubit, q1:Qubit, q2:Qubit):Unit
{
        Controlled R([q1], (PauliZ, theta/2.0, q2));
        CNOT(q0, q1);
        Controlled R([q1], (PauliZ, -theta/2.0, q2));
        CNOT(q0, q1);
        Controlled R([q0], (PauliZ, theta/2.0, q2));
```

Listing 11.4

```
    }

    // The controlled adder circuit
    // a - source register
    // b - target register
    // offset - the starting qubit in register b to work from
    // control - the control qubit
    // m - number of qubits in the registers
    operation c_add_cct(a:Qubit[],b:Qubit[], offset:Int,
                        control:Qubit, m:Int): Unit{
        mutable n = m;
        repeat {
            for ( i in RangeReverse(1..n)){
                ccu1( PI() / PowD(2.0, IntAsDouble(n-i)), control,
                    a[i-1], b[n-1+offset]);
            }
            set n -= 1;
        } until (n == 0);
    }


    // The controlled inverse QFT circuit
    // b  - the target register
    // offset - the starting qubit in register b to work from
    // control - the control qubit.
    // n - number of qubits in the register

    operation c_iqft_cct  (b:Qubit[],  offset:Int,  control:Qubit,
n:Int):Unit {
        for (i in 0..n-1){
            for (j in 1..i){
                // for inverse transform, we have to use negative
angles
                ccu1(-PI()/PowD(2.0, IntAsDouble(i-j+1)), control,
                    b[j-1+offset], b[i+offset]);
            }
            // the H transform should be done after the rotations
            Controlled H([control], b[i+offset]);
        }
    }

    // The forward QFT circuit
    // b  - the target register
    // offset - the starting qubit in register b to work from
    // control - the control qubit.
    // m - number of qubits in the register
```

Listing 11.4 (continued)

```
    operation   c_qft_cct(b:Qubit[],      offset:Int,      control:Qubit,
m:Int): Unit{
        mutable n = m;
        repeat {
            Controlled H([control], b[n + offset - 1]);

            for (i in RangeReverse (1..n-1)){
                ccu1(PI() / PowD(2.0, IntAsDouble(n - i)),
                    control, b[i - 1 + offset], b[n + offset - 1]);
            }
            set n -= 1;
        } until (n == 0);
    }


    // The controlled adder circuit block
    // a - the source register
    // b  - the target register
    // offset - the starting qubit in register b to work from
    // control - the control qubit.
    // n - number of qubits in the register

    operation c_add_qft(a:Qubit[], b:Qubit[], offset:Int,
                        control:Qubit, n:Int):Int {
        mutable r = 0;

        c_qft_cct(b, offset, control, n);
        c_add_cct(a, b, offset, control, n);
        c_iqft_cct(b, offset, control, n);

        for ( i in 0..n-1)
        {
            if (One == M(b[i])) { set r = r ||| (1 <<< i); }
        }

        return r;
    }


    // The quantum multiplier circuit
    // calculates s = a * b
    // input1 - multiplicand
    // input2 - multiplied
    // s - the qubit register holding product a * b
    // s has two times the length of a and b
    // a - the qubit register for multiplicand
    // b - the qubit register for multiplier
    // n - length of the registers
```

Listing 11.4 (continued)

```
operation mult_cct(input1:Int, input2:Int, a:Qubit[],
                    b:Qubit[], s:Qubit[], n:Int):Int{
    mutable r = 0;

    set_reg(a, n, input1);
    set_reg(b, n, input2);

    for (i in 0..n-1) {
        set r = c_add_qft(b, s, i,  a[i], n);
    }

    for ( i in 0..(n*2)-1)
    {
        if (One == M(s[i])) { set r = r ||| (1 <<< i); }
    }

    ResetAll(a);
    ResetAll(b);
    ResetAll(s);

    return r;
}

// Testing code
operation testControlledMultiplier( ): Unit {
    mutable n = 4;
    mutable shots = 100;
    mutable input1 = 5;
    mutable input2 = 5;
    mutable r = 0;

    using (a = Qubit[n] )  {
        using (b = Qubit[n] ) {
            using (s = Qubit[n*2] ) {
                set r = mult_cct(input1, input2, a, b, s, n);
            }
        }
    }
    Message($"{input1} * {input2} = {r}");
}
}
```

Listing 11.4 (continued)

QDK Code Example for Grover's search, from Sect. 7.8 (Listing 11.5):

```
namespace Grovers {
    open Microsoft.Quantum.Canon;
    open Microsoft.Quantum.Intrinsic;
    open Microsoft.Quantum.Arrays;
    open Microsoft.Quantum.Math;
    open Microsoft.Quantum.Convert;

    // The three control Toffoli gate
    // control1, control2, control3 - The control registers
    // anc - a temporary work register
    // target - the target register, where the transform is applied
    operation cccx(control1:Qubit, control2:Qubit,
                  control3:Qubit, anc:Qubit, target:Qubit): Unit{
        CCNOT(control1,control2,anc);
        CCNOT(control3,anc,target);
        CCNOT(control1,control2,anc);
        CCNOT(control3,anc,target);
    }


    // The CCZ gate
    // control1, control2 - The control registers
    // target - the target register, where the Z transform is
applied
    operation           ccz(control1:Qubit,           control2:Qubit,
target:Qubit):Unit{
        H(target);
        CCNOT(control1, control2, target);
        H(target);
    }


    // The Grover's Oracle
    // x1, x2, x3 - The input register x
    // anc - a temporary work register
    // target - the target register, where the transform is applied
    operation grover_oracle( x1:Qubit, x2:Qubit, x3:Qubit,
                            anc:Qubit, target:Qubit):Unit{
        X(x3);
        X(x1);
        cccx(x1, x2, x3, anc, target);
        X(x1);
        X(x3);
    }


    // The Grover's Diffusion Operator
    // x1, x2, x3 - The input register x
    // target - A temporary register
```

Listing 11.5 Implementing Grover's search in QDK

```
    operation grover_diffusion_operator(x1:Qubit, x2:Qubit,
                                    x3:Qubit, target:Qubit):Unit
{
        H(x1);
        H(x2);
        H(x3);
        H(target); // Bring this back to state 1 for next stages
        X(x1);
        X(x2);
        X(x3);
        ccz(x1, x2, x3 );
        X(x1);
        X(x2);
        X(x3);
        H(x1);
        H(x2);
        H(x3);
    }


    // Implements the Grovers algorithm
    operation Grovers():Unit {
        mutable r = 0;
        mutable shots = 100;
        mutable res = new Int[8];


        using(q = Qubit[3]) {
            using(t = Qubit[2]){

                for ( j in 0..shots-1) {
                    H(q[0]);
                    H(q[1]);
                    H(q[2]);
                    H(t[0]);

                    grover_oracle (q[0], q[1], q[2], t[1], t[0]);
                    grover_diffusion_operator        (q[0],        q[1],
q[2],t[0]);

                    grover_oracle (q[0], q[1], q[2], t[1], t[0]);
                    grover_diffusion_operator        (q[0],        q[1],
q[2],t[0]);

                    grover_oracle (q[0], q[1], q[2], t[1], t[0]);
                    grover_diffusion_operator        (q[0],        q[1],
q[2],t[0]);

                    for ( i in 0..2)
```

Listing 11.5 (continued)

```
                                  {
                                      if (One == M(q[i])) { set r = r ||| (1 <<<
i); }
                                  }
                                  set res w/= r <- res[r] + 1;

                                  ResetAll(q);
                                  ResetAll(t);
                          }
                      }
                  Message($"Output of Grovers is: {res}");
              }
          }
}
```

Listing 11.5 (continued)

QDK code for Shor's algorithm to factorize 15, using a fixed exponential modulation, Sect. 7.9.2 (Listing 11.6):

```
namespace Shors {

    open Microsoft.Quantum.Canon;
    open Microsoft.Quantum.Intrinsic;
    open Microsoft.Quantum.Arrays;
    open Microsoft.Quantum.Math;
    open Microsoft.Quantum.Convert;


    // Inverse QFT
    // t: qubit array for which IQFT is to be done.
    // n: width of the array
    operation iqft_cct (t:Qubit[], n:Int):Unit {
        for ( i in 0..n-1 ){
            for ( j in 1..i) {
                Controlled R([t[j-1]], (PauliZ, -PI() / PowD(2.0,
                              IntAsDouble(i-j+1 )), t[i]));
            }
            H(t[i]);
        }
    }


    // Implements the shortened form of Shor's factorization
    // Using a precoded modular exponentiation
    operation Shors15():Int{
        mutable r = 0;
```

Listing 11.6 Shor's algorithm for the QDK

```
using (q = Qubit[7]){

    // Initialize source and target registers
    H(q[0]);
    H(q[1]);
    H(q[2]);
    X(q[6]);

    //Modular exponentiation 7^x mod 15
    CNOT(q[2],q[4] );
    CNOT(q[2],q[5] );
    CNOT(q[6],q[4] ) ;
    CCNOT(q[1],q[5],q[3] );
    CNOT(q[3],q[5] );
    CCNOT(q[1],q[4],q[6] );
    CNOT(q[6],q[4] );

    // # IQFT. Refer to implementation from earlier examples
    iqft_cct (q, 3);

    // Measure REG1
    for ( i in 0..2)
    {
        if (One == M(q[i])) { set r = r ||| (1 <<< i); }
    }

    ResetAll(q);
}

return r;
}


// Test Shors algorithm
operation testShors():Unit{
    mutable shots = 100;
    mutable res = new Int[8];
    mutable r = 0;

    for (i in 0..shots-1){
        set r = Shors15();
        set res w/= r <- res[r] + 1;
    }

    Message($"Output of Shors is: {res}");
}

}
```

Listing 11.6 (continued)

QDK code for Quantum Phase Estimation example, Sect. 7.11 (Listing 11.7):

```
namespace Qpe {

    open Microsoft.Quantum.Canon;
    open Microsoft.Quantum.Intrinsic;
    open Microsoft.Quantum.Arrays;
    open Microsoft.Quantum.Math;
    open Microsoft.Quantum.Convert;

    // Inverse QFT (regular)
    // t: qubit array for which IQFT is to be done.
    // n: width of the array
    operation qft_inv (q:Qubit[], n:Int):Unit {
        for ( i in RangeReverse(0..n-1)){
            for (qubit in i..n-2 ){
                Controlled R([q[qubit + 1]], (PauliZ,
                        -PI() / PowD(2.0, IntAsDouble(qubit + 1 - i
)), q[i]));
            }
            H(q[i]);
        }
    }

    // The QPE Algorithm,
    // Condensed version
    operation QuantumPhaseEstimation():Int {
        mutable r = 0;
        mutable rotation = 0.0;

        using ( q = Qubit[4]) {

            // Prepare the initial state
            H(q[0]);
            H(q[1]);
            H(q[2]);
            X(q[3]);

            // Perform the unitaries
            for (i in 0..2) {
                set  rotation  =  PowD(2.0,  IntAsDouble(i))  *
(PI()/2.0);
                Controlled R([q[i]], (PauliZ, rotation , q[3] ));
            }

            // Inverse QFT
            qft_inv(q, 3);
```

Listing 11.7 Quantum Phase Estimation

```
                // Measure
                for ( i in 0..2)
                {
                    if (One == M(q[i])) { set r = r ||| (1 <<< i); }
                }

                ResetAll(q);
            }
        return r;
    }

    // Test QPE
    operation testQpe():Unit {
        mutable res = new Int[8];
        let shots = 100;
        mutable r = 0;

        for (i in 0..shots-1){
            set r = QuantumPhaseEstimation();
            set res w/= r <- res[r] + 1;
        }
        Message($"Output of QPE is: {res}");
    }
}
```

Listing 11.7 (continued)

QDK code for HHL Algorithm, described in Sect. 7.12 (Listing 11.8):

```
namespace Hhl {

    open Microsoft.Quantum.Canon;
    open Microsoft.Quantum.Intrinsic;
    open Microsoft.Quantum.Arrays;
    open Microsoft.Quantum.Math;
    open Microsoft.Quantum.Convert;
    open Microsoft.Quantum.Diagnostics;

    // Implements a U3 rotation
    operation    u3(theta:Double,      phi:Double,      lambda:Double,
target:Qubit): Unit {
        Rz(phi, target);
        Rx(-PI()/2.0, target);
        Rz(theta, target);
        Rx(PI()/2.0, target);
        Rz(lambda, target);
    }
```

Listing 11.8 HHL algorithm for the QDK

```
// Implements a controlled U3 rotation
operation cu3(theta:Double, phi:Double, lambda:Double,
             control:Qubit, target:Qubit): Unit {
    Rz((lambda + phi)/2.0, control);
    Rz((lambda - phi)/2.0, target);
    CNOT(control, target);
    u3(-theta/2.0, 0.0, -(phi +lambda)/2.0, target);
    CNOT(control, target);
    u3(theta/2.0, phi, 0.0, target);
}

// Implements the HHL algorithm
operation hhl():Unit {
    mutable r = 0;
    mutable b = Zero;

    using ( q = Qubit[4]){
        // QPE, shortened version
        H(q[3]);
        CNOT(q[3], q[2]);
        CNOT(q[2], q[1]);
        X(q[2]);
        SWAP(q[2], q[1]);

        // Lambda inverse
        cu3(PI(), 0.0, 0.0, q[2], q[0]);
        cu3(PI()/3.0, 0.0, 0.0, q[1], q[0]);

        // Shortened Inverse QPE
        SWAP(q[2], q[1]);
        X(q[2]);
        CNOT(q[2], q[1]);
        CNOT(q[3], q[2]);
        H(q[3]);

        // Measure the ancilla
        set b = M(q[0]);

        Message($"ancilla reads {b}");

        // Dumpregisters to examine the statevector
        DumpRegister((),q);

        ResetAll(q);
    }
}
}
```

Listing 11.8 (continued)

Executing the above code generates the following dump. Note that the dump could be slightly different when executed at different times.

```
# wave function for qubits with ids (least to most significant): 0;1;2;3
 |0):   0.000000 + 0.000000 i ==                          [0.000000]
 |1): -0.948683 + 0.000000 i == ******************* [0.900000] --- [ 3.14159 rad ]
 |2):   0.000000 + 0.000000 i ==                          [0.000000]
 |3):   0.000000 + 0.000000 i ==                          [0.000000]
 |4):   0.000000 + 0.000000 i ==                          [0.000000]
 |5):   0.000000 + 0.000000 i ==                          [0.000000]
 |6):   0.000000 + 0.000000 i ==                          [0.000000]
 |7):   0.000000 + 0.000000 i ==                          [0.000000]
 |8):   0.000000 + 0.000000 i ==                          [0.000000]
 |9):   0.316228 + 0.000000 i == **                       [0.100000] --- [ 0.00000 rad ]
|10):   0.000000 + 0.000000 i ==                          [0.000000]
|11):   0.000000 + 0.000000 i ==                          [0.000000]
|12):   0.000000 + 0.000000 i ==                          [0.000000]
|13):   0.000000 + 0.000000 i ==                          [0.000000]
|14):   0.000000 + 0.000000 i ==                          [0.000000]
|15):   0.000000 + 0.000000 i ==                          [0.000000]
```

Ignoring the imaginary values, we can extract the results $\begin{bmatrix} -0.948683 \\ 0.316228 \end{bmatrix}$. When this is multiplied by the normalizer $\sqrt{10}$, we get the results $\begin{bmatrix} -3 \\ 1 \end{bmatrix}$. Up to a phase, this is the result we expected.

Index

© The Author(s), under exclusive license to Springer Nature Switzerland AG 2021   457
V. Kasirajan, *Fundamentals of Quantum Computing*, https://doi.org/10.1007/978-3-030-63689-0

Printed in the United States
by Baker & Taylor Publisher Services